Nickel Catalysis in Organic Synthesis

Nickel Catalysis in Organic Synthesis

Methods and Reactions

Edited by
Sensuke Ogoshi

WILEY-VCH

Editor

Prof. Sensuke Ogoshi
Osaka University
Department of Applied Chemistry
1-1 Yamadaoka, Suita
565-0871 Osaka
Japan

Library of Congress Card No.:
applied for

British Library Cataloguing-in-Publication Data
A catalogue record for this book is available from the British Library.

Bibliographic information published by the Deutsche Nationalbibliothek

The Deutsche Nationalbibliothek lists this publication in the Deutsche Nationalbibliografie; detailed bibliographic data are available on the Internet at <http://dnb.d-nb.de>.

Print ISBN: 978-3-527-34407-9
ePDF ISBN: 978-3-527-81379-7
ePub ISBN: 978-3-527-81381-0
oBook ISBN: 978-3-527-81382-7

Typesetting SPi Global, Chennai, India
Printing and Binding Markono Print Media Pte Ltd, Singapore

Printed on acid-free paper

10 9 8 7 6 5 4 3 2 1

Contents

Preface

Nickel has been recognized as one of the most uncontrollable transition metals as a catalyst in homogeneous reactions and is sometimes called as a spirited horse. Therefore, chemists have developed new ligands, new reaction conditions, and new apparatus to control the reactivity of nickel catalysts, which allows us to apply nickel catalysts to a variety of bond forming reactions.

In fact, nickel-catalyzed reactions have been getting more popular and important for the past 10 years, since a variety of transformations of organic compounds via single-electron transfer processes as well as oxidative cyclization have been developed. These reactions are broadly applicable to the synthesis of complicated molecules and unprecedented bond formation.

The latest book focusing on nickel-catalyzed reactions had been published in 2005, "Modern Organonickel Chemistry," edited by Professor Yoshinao Tamaru.

In Chart 1, we can see that the number of publications titled palladium-catalyzed reactions is two times higher than that of nickel-catalyzed reactions in 2018. However, the number of the publications of palladium-catalyzed reactions has been decreasing in the recent three years. On the other hand, that of nickel-catalyzed reactions has been increasing. The increasing number of publications of nickel-catalyzed reactions could be due to the improvement of reactors as well as the improvement of the quality of commercially available nickel catalysts (complexes). For example, yellow shine crystal of $Ni(cod)_2$ is commercially available nowadays.

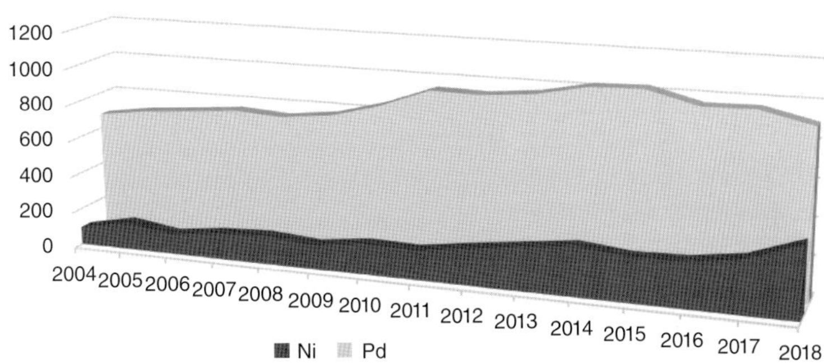

Chart 1.

Under these circumstances, I am convinced that the publication of a new book dealing with the recent progress in nickel catalysis would be very helpful to develop new reactions.

Finally, the authors hope this book will encourage graduate students who can control a spirited horse in the near future.

31 March 2019
Osaka

Sensuke Ogoshi

Part I

Reactions via Nickelacycles

1

Formation of Nickelacycles and Reaction with Carbon Monoxide

Sensuke Ogoshi

Osaka University, Department of Applied Chemistry, Suita, Osaka 565-0871, Japan

1.1 Introduction

Oxidative cyclization is one of the most efficient methods to form carbon–carbon bonds in one step. For example, the trimerization of alkynes to afford arenes catalyzed by a variety of transition metals has been well known as a part of Reppe chemistry, in which the oxidative cyclization of two alkynes with a transition metal to generate a metalacyclopentadiene has been believed to be a key step and might be the most common oxidative cyclization reaction. The reaction of metalacyclopentadiene with carbon monoxide is one of the logical extensions of trimerization of alkynes. The first Pauson–Kahnd-type reaction was reported in 1960 – the reaction of $Fe(CO)_5$ with acetylene to afford the corresponding (η^4-cyclopentadienone)iron complex (Figure 1.1) [1]. In general, the reaction of metal carbonyls with alkynes gives η^4-cyclopentadienone metal complexes. However, the coordination of η^4-cyclopentadienone is too strong to dissociate spontaneously, which makes the expansion of a stoichiometric reaction to a catalytic reaction difficult. On the other hand, both stoichiometric and catalytic Pauson–Kahnd reactions have been reported in 1973 at the same time (Figure 1.2) [2]. This reaction has been developed as a step economy method for the construction of cyclopentenones in one step, including optically active complicated compounds [3]. The minimum requirement for transition metals to promote the Pauson–Kahnd-type reactions is the formation of metalacycles and the following insertion of carbon monoxide and reductive elimination to give the corresponding carbonyl compounds. Thus, the synthesis of a variety of cyclic carbonyl compounds such as lactams and lactone by the catalytic hetero-Pauson–Kahnd reactions seems a logical extension.

1.2 Formation of Hetero-nickelacycles from Nickel(0)

As mentioned above, the first catalytic Pauson–Kahnd reaction had been reported at the same time as the first stoichiometric Pauson–Kahnd reaction. In contrast, to date, a very limited number of examples of the catalytic

Nickel Catalysis in Organic Synthesis: Methods and Reactions,
First Edition. Edited by Sensuke Ogoshi.

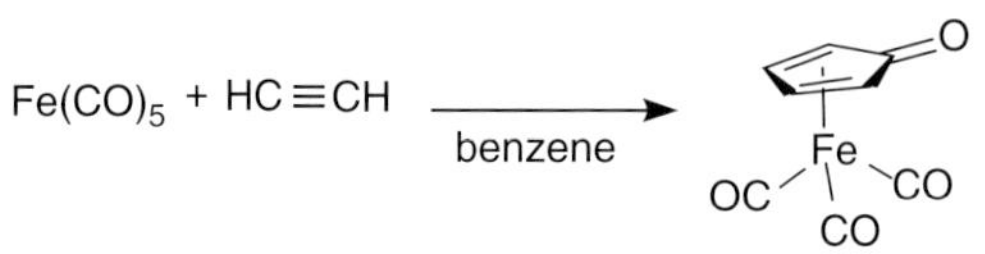

Figure 1.1 Formation of (η^4-cyclopentadienone)iron complex.

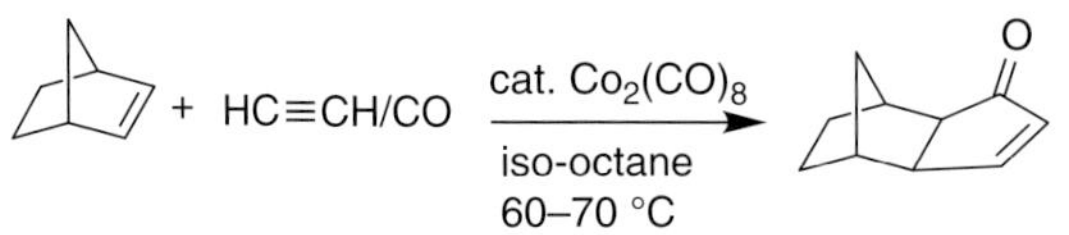

Figure 1.2 Cobalt-catalyzed Pauson–Kahnd reaction.

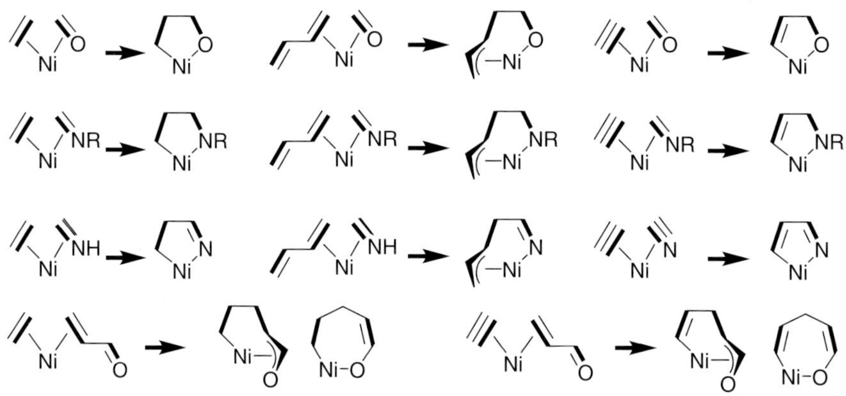

Figure 1.3 Formation of nickelacycles by oxidative cyclization.

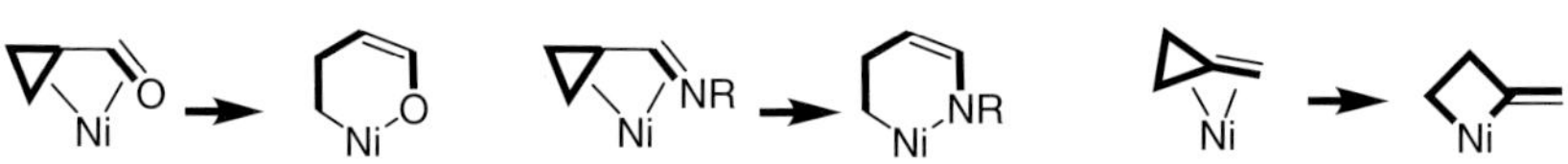

Figure 1.4 Formation of nickelacycles by oxidative addition.

hetero-Pauson–Kahnd reactions to afford lactones and lactams have been reported, due to the less transition metals that can form hetero-metalacycles by oxidative cyclization with carbonyl compounds or imines (Figure 1.3) [4–8]. In that sense, nickel is one of the most promising transition metals as a catalyst for the hetero-Pauson–Kahnd reaction. In fact, the formation of a variety of hetero-nickelacycles has been reported. In addition, oxidative addition of strained small ring compounds also affords nickelacycles (Figure 1.4) [9–11].

1.3 Stoichiometric Reaction of Hetero-nickelacycles with Carbon Monoxide

In general, nickel tetra-carbonyl complex, $Ni(CO)_4$, might be the most famous metal carbonyl complex owing to the very high toxicity and its usage for the purification of nickel metal. In addition, Tolman's electronic parameter (TEP) has been evaluated by using a variety of nickel carbonyl complexes, $Ni(CO)_3L$ or $Ni(CO)_2L_2$. These facts indicate that organonickel complexes can react with carbon monoxide to give $Ni(CO)_3L$ or $Ni(CO)_2L_2$ easily. In fact, the reaction of hetero-nickelacycles with carbon monoxide affords a variety of expected cyclic carbonyl compounds quantitatively (Figures 1.5–1.7) [12–15]. However, the oxidative cyclization is a reversible process; thus, there are some exceptions in that the coordination of carbon monoxide promotes carbon–carbon bond cleavage to regenerate the starting unsaturated compounds in the reaction with carbon monoxide. In fact, η^3-allyl-nickelacycle, which is prepared by the reaction of 2,3-dimethylbutadiene and benzaldehyde with $Ni(cod)_2$ and PCy_3, underwent a reductive ring-opening reaction to regenerate 2,3-dimethylbutadiene and benzaldehyde quantitatively, concomitant with the quantitative formation of $Ni(CO)_3(PCy_3)$ (Figure 1.8).

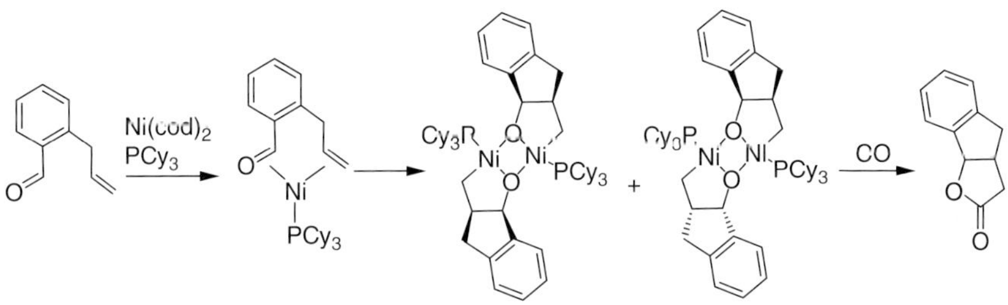

Figure 1.5 Formation of gamma lactone.

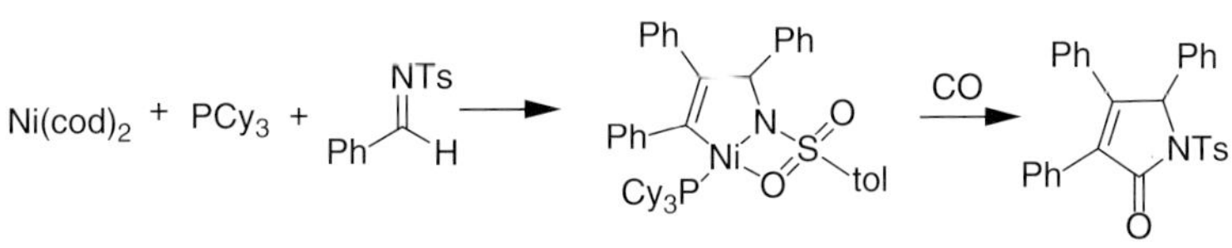

Figure 1.6 Formation of gamma lactam.

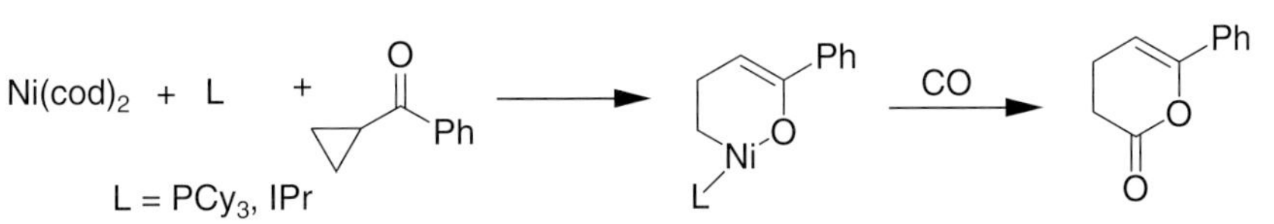

Figure 1.7 Formation of delta lactone.

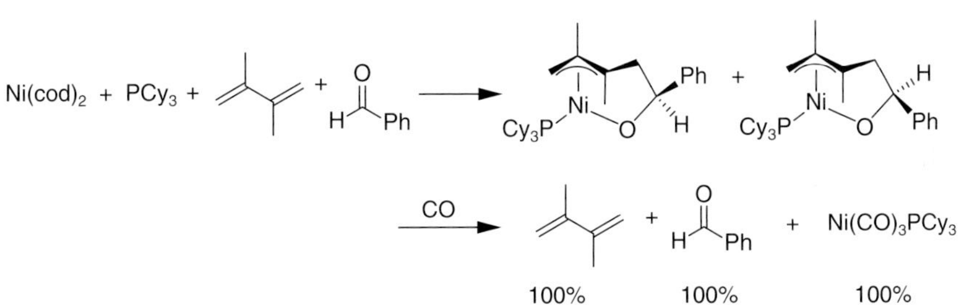

Figure 1.8 Carbon–carbon bond dissociation under carbon monoxide atmosphere.

These reports suggest that nickel is one of the most ideal transition metals to synthesize cyclic carbonyl compounds by using carbon monoxide, if the carbonylation step is irreversible. Nevertheless, neither nickel-catalyzed Pauson–Kahnd reaction nor hetero-Pauson–Kahnd reaction had been reported until 2014. It might be due to the formation of nickel-carbonyl complexes that are too stable to undergo oxidative cyclization to give hetero-nickelacycles. In fact, most of catalytic Pauson–Kahnd reactions have been carried out under relatively low carbon monoxide pressure and high reaction temperature due to the strong coordination of carbon monoxide to the metal center, which inhibits the coordination of either alkynes or alkenes. Thus, chemists have developed methods to control the concentration of carbon monoxide *in situ* or ex situ [16, 17].

The first nickel-catalyzed hetero-Pauson–Kahnd reaction has been achieved [18] by using *in situ* slow generation of carbon monoxide from phenyl formate and triethyl amine (Figure 1.9) [19, 20]. A variety of lactams have been prepared via aza-nickelacyles. The reaction intermediate complex was isolated and the molecular structure confirmed by X-ray crystallography. The reaction of (**X**) with carbon monoxide afforded the corresponding lactam quantitatively. This reaction can be applied to the reaction of norbornene and imine with carbon monoxide. The key to success is the control of the concentration level of carbon monoxide high enough to promote insertion of carbon monoxide and low enough to suppress the formation of $Ni(CO)_3(PCy_3)$, which is too stable to undergo oxidative cyclization. In fact, $Ni(CO)_3(PCy_3)$ is generated at the end of the reaction, which reacts with neither imines nor alkynes under mild conditions.

The most economical carbonylation reaction has to use carbon monoxide gas itself as the carbon monoxide source, in which the formation of nickel carbonyl complexes is inevitable. Thus, the reaction substrates have to be able to react with nickel carbonyl complexes to afford the corresponding carbonyl compound in high yield. Thus, bidentate substrates are appropriate for the carbonylation reaction, since the coordination of the bidentate substrate is an entropically favorable step. In fact, ene-imines can react with $Ni(CO)_3(PCy_3)$ to afford the desired lactam quantitatively (Figure 1.10). The reaction might proceed via an aza-nickelacycle, which has been confirmed by the stoichiometric reaction of isolated aza-nickelacycle with carbon monoxide.

The optimization of the reaction conditions has been conducted in an autoclave with regular equipment. However, only a catalytic amount of lactam X was obtained. To confirm the chemical species in the reaction mixture, the reaction was monitored by NMR. The reaction in the NMR tube afforded the

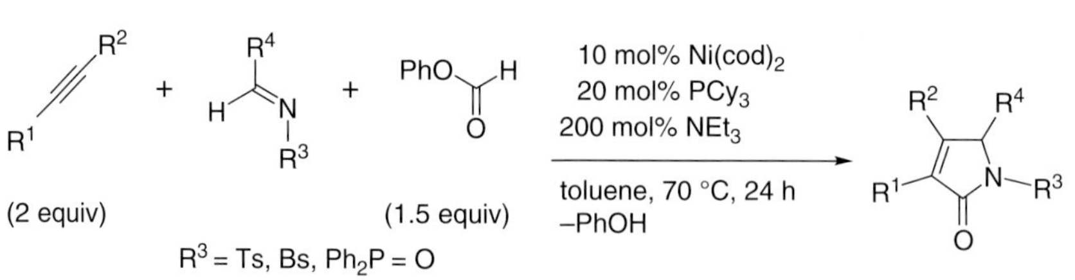

Figure 1.9 Nickel-catalyzed reaction of alkynes and imines with carbon monoxide.

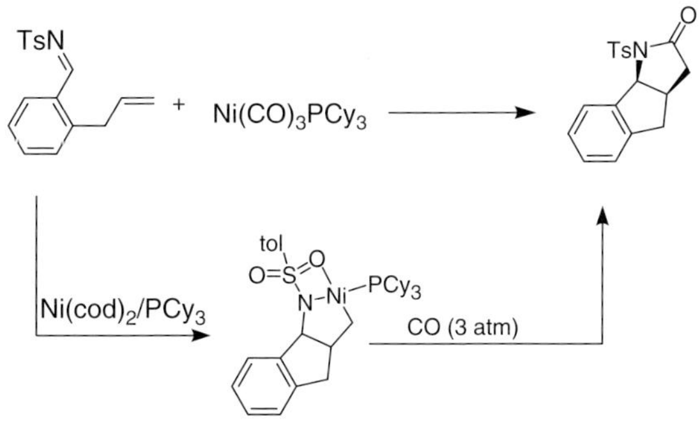

Figure 1.10 Stoichiometric reaction of en-imines with $Ni(CO)_3PCy_3$.

carbonylation product, which was generated quantitatively. This result indicates that stirring the reaction mixture suppresses the formation of nickelacycle. In fact, the reaction in an autoclave without stirring afforded the corresponding lactam quantitatively. The stoichiometric reaction of isolated nickelacycle with carbon monoxide proceeded smoothly to give the lactam quantitatively. A proposed reaction path is depicted in Figure 1.11. Not stirring the reaction mixture keeps the concentration of carbon monoxide in solution low enough to promote the coordination of the ene-imine to $Ni(CO)_3(PCy_3)$. With stirring, the coordination of the substrate to the nickel center is completely inhibited by rapid re-coordination of carbon monoxide to regenerate $Ni(CO)_3(PCy_3)$ under higher concentration of carbon monoxide. As a transition metal catalyzed the Pauson–Kahnd type reaction, the reaction temperature is exceptionally low. Thus, the reaction has been expanded to an asymmetric reaction.

The reaction product is the key synthetic intermediate [21] of strigolactam that is expected as a modified plant hormone. Thus, the nickel-catalyzed

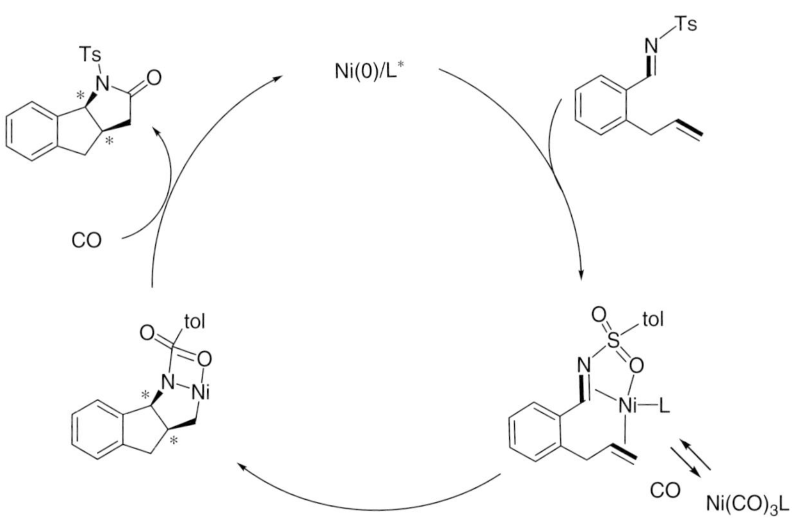

Figure 1.11 Nickel-catalyzed reaction of alkynes and imines with carbon monoxide.

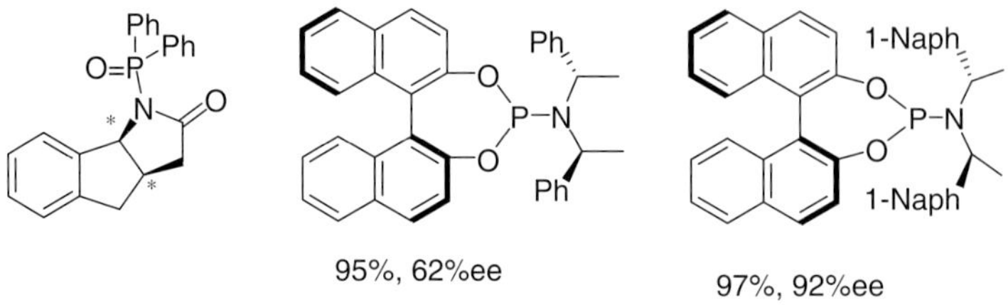

Figure 1.12 Application to asymmetric reaction.

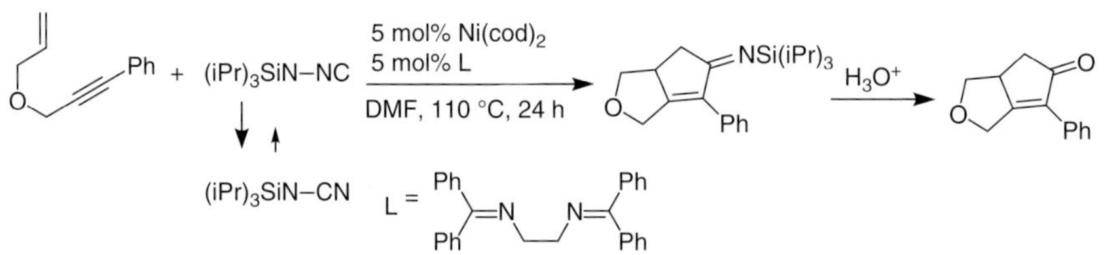

Figure 1.13 The formal nickel-catalyzed Pauson–Khand reaction.

aza-Pauson–Kahnd reaction has expanded asymmetric reaction. For the asymmetric reaction, phosphoramidite was an effective ligand. The ligand screening of phosphoramidite for the reaction was carried out (Figure 1.12) [22].

The formal nickel-catalyzed Pauson–Khand reaction has been reported in 1996 in which triisopropylsilylnitrile was employed instead of carbon monoxide by following hydration to afford the Pauson–Khand reaction product (Figure 1.13) [23].

The first nickel-catalyzed Pauson–Khand reaction has been reported in 2019 (Figure 1.14). In the presence of a catalytic amount of $Ni(CO)_3$(IPr) in toluene at 175 °C, intramolecular Pauson–Khand reaction proceeded to give the cyclic

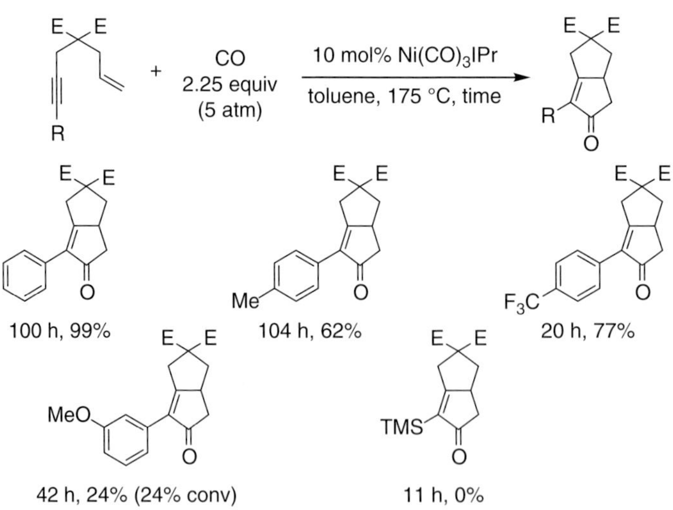

Figure 1.14 Nickel-catalyzed Pauson–Khand reaction.

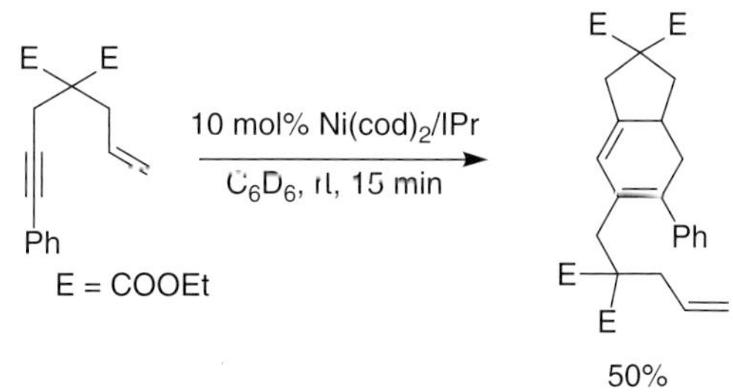

Figure 1.15 Nickel-catalyzed dimerization.

enones quantitatively. In this reaction, $Ni(CO)_3(IPr)$ is the most efficient catalyst. The higher reaction temperature is required for the dissociation of carbon monoxide to promote oxidative cyclization; however, in the presence of $Ni(cod)_2$ and IPr as a catalyst, no carbonylation product was observed. In fact, the reaction of enyne with $Ni(cod)_2$ and IPr gives the dimer compound of the enyne at room temperature in 15 minutes (Figure 1.15). Thus, in this reaction, the strong coordination of carbon monoxide to nickel is required to suppress the occurrence of dimerization of the enyne compound.

In contrast, in the presence of a catalytic amount of $Ni(cod)_2$ and either PCy_3 or IPr, the expected carbonylation product was not observed at all but only dimer of **1a**. This might be the reason why metal carbonyl complexes have been used as the catalyst for the Pauson–Khand reaction.

In this chapter, the formation of nickelacycles by the oxidative cyclization of unsaturated compounds with nickel(0) species and the reaction with carbon monoxide have been introduced. The key to the success of the construction of nickel-catalyzed cyclic carbonylation reaction is the control of the two key steps. One is the oxidative cyclization of unsaturated compounds with nickel(0) complex and the other is the control of the concentration of carbon monoxide in the solution.

References

1 Weiss, E., Merenyi, R.G., and Hubel, W. Cyclopentadienone iron carbonyls. *Chem. Ind.* 1960 (15): 407–408.
2 Khand, I.U., Knox, G.R., Pauson, P.L. et al. (1973). Organocobalt complexes. 2. Reaction of acetylenehexacarbonyldicobalt complexes, $(R^1C_2R^2)CO_2(CO)_6$, with norbornene and its derivatives. *J. Chem. Soc., Perkin Trans.* 1 (9): 977–981.
3 Torres, R.R. (ed.) (2012, IISBN). *The Pauson–Khand Reaction*. Wiley. ISBN: 978-0-470-97076-8.
4 Hicks, F.A., Kablaoui, N.M., and Buchwald, S.L. (1996). Titanocene-catalyzed cyclocarbonylation of enynes to cyclopentenones. *J. Am. Chem. Soc.* 118 (39): 9450–9451.
5 Chatani, N., Morimoto, T., Fukumoto, Y., and Murai, S. (1998). Ru-3(CO)(12)-catalyzed cyclocarbonylation of yne-aldehydes to bicyclic α,β-unsaturated γ-butyrolactones. *J. Am. Chem. Soc.* 120 (21): 5335–5336.

6 Kamitani, A., Chatani, N., Morimoto, T., and Murai, S. (2000). Carbonylative 5+1 cycloaddition of cyclopropyl imines catalyzed by ruthenium carbonyl complex. *J. Org. Chem.* 65 (26): 9230–9233.

7 Tobisu, M., Chatani, N., Asaumi, T. et al. (2000). Ru-3(CO)(12)-catalyzed intermolecular cyclocoupling of ketones, alkenes or alkynes, and carbon monoxide. 2+2+1 cycloaddition strategy for the synthesis of functionalized gamma-butyrolactones. *J. Am. Chem. Soc.* 122 (51): 12663–12674.

8 Chatani, N., Amako, K., Tobisu, M. et al. (2003). Ruthenium-catalyzed carbonylative cycloaddition of α-keto lactones with alkenes or alkynes: The participation of an ester-carbonyl group in cycloaddition reactions as the two-atom assembling unit. *J. Org. Chem.* 68 (4): 1591–1593.

9 Ogoshi, S. (2017). Highly atom economical molecular transformation via hetero-nickelacycle. *Bull. Chem. Soc. Jpn.* 90 (12): 1401–1406.

10 Ohashi, M., Hoshimoto, Y., and Ogoshi, S. (2015). Aza-nickelacycle key intermediate in nickel(0)-catalyzed transformation reactions. *Dalton Trans.* 44 (27): 12060–12073.

11 Hoshimoto, Y., Ohashi, M., and Ogoshi, S. (2015). Catalytic transformation of aldehydes with nickel complexes through η^2 coordination and oxidative cyclization. *Acc. Chem. Res.* 48 (6): 1746–1755.

12 Tamaki, T., Nagata, M., Ohashi, M., and Ogoshi, S. (2009). Synthesis and reactivity of six-membered oxa-nickelacycles: a ring-opening reaction of cyclopropyl ketones. *Chem. Eur. J.* 15 (39): 10083–10091.

13 Ogoshi, S., Tonomori, K., Oka, M., and Kurosawa, H. (2006). Reversible carbon–carbon bond formation between 1,3-dienes and aldehyde or ketone on nickel(0). *J. Am. Chem. Soc.* 128 (21): 7077–7086.

14 Ogoshi, S., Nagata, M., and Kurosawa, H. (2006). Formation of nickeladihydropyran by oxidative addition of cyclopropyl ketone. Key intermediate in nickel-catalyzed cycloaddition. *J. Am. Chem. Soc.* 128 (16): 5350–5351.

15 Ogoshi, S., Ikeda, H., and Kurosawa, H. (2007). Formation of an aza-nickelacycle by reaction of an imine and an alkyne with nickel(0): oxidative cyclization, insertion, and reductive elimination. *Angew. Chem. Int. Ed.* 46 (26): 4930–4932.

16 Morimoto, T. and Kakiuchi, K. (2004). Evolution of carbonylation catalysis: no need for carbon monoxide. *Angew. Chem. Int. Ed.* 43 (42): 5580–5588.

17 Friis, S.D., Lindhardt, A.T., and Skrydstrup, T. (2016). The development and application of two-chamber reactors and carbon monoxide precursors for safe carbonylation reactions. *Acc. Chem. Res.* 49 (4): 594–605.

18 Hoshimoto, Y., Ohata, T., Sasaoka, Y. et al. (2014). Nickel(0)-catalyzed 2+2+1 carbonylative cycloaddition of imines and alkynes or norbornene leading to γ-lactams. *J. Am. Chem. Soc.* 136 (45): 15877–15880.

19 Ueda, T., Konishi, H., and Manabe, K. (2012). Palladium-catalyzed carbonylation of aryl, alkenyl, and allyl halides with phenyl formate. *Org. Lett.* 14 (12): 3100–3103.

20 Fujihara, T., Hosoki, T., Katafuchi, Y. et al. (2012). Palladium-catalyzed esterification of aryl halides using aryl formates without the use of external carbon monoxide. *Chem. Commun.* 48 (64): 8012–8014.

21 Screpanti, C., Fonne-Pfister, R., Lumbroso, A. et al. (2016). Strigolactone derivatives for potential crop enhancement applications. *Bioorg. Med. Chem. Lett.* 26 (10): 2392–2400.
22 Hoshimoto, Y., Ashida, K., Sasaoka, Y. et al. (2017). Efficient synthesis of polycyclic γ-lactams by catalytic carbonylation of ene-imines via nickelacycle intermediates. *Angew. Chem. Int. Ed.* 56 (28): 8206–8210.
23 Zhang, M.H. and Buchwald, S.L. (1996). A nickel(0)-catalyzed process for the transformation of enynes to bicyclic cyclopentenones. *J. Org. Chem.* 61 (14): 4498–4499.

2

Transformation of Aldehydes via Nickelacycles

Yoichi Hoshimoto

Osaka University, Department of Applied Chemistry, 2-1 Yamadaoka, Suita, Osaka 565-0871, Japan

2.1 Introduction and Scope of This Chapter

Aldehydes are fundamental and common reagents in organic synthesis, as a number of methods are available to activate their formyl groups [1]. As a particularly important method, an η^1 coordination of the formyl oxygen to Lewis acidic metals has been well known (Figure 2.1), which promotes the reaction between aldehydes and a variety of nucleophiles through an enhancement of electrophilicity at the formyl carbon [1]. The oxidative addition of the formyl C—H bond to a metal center, which is a key elementary step in the transformation of aldehydes, has often been proposed to be initiated by the η^1 coordination [1, 2]. An η^2 coordination of aldehydes to the low-valent transition metals has been also well studied [1, 3–4]. Interestingly, a three-membered oxametallacycle complex can be considered as one of the contributing forms in the resonance structures, especially when the strong back-donation is operative from the dπ orbital in the metal center to the pπ^* orbital in the formyl ligand (Figure 2.1b). When the back-donation is dominant rather than the donation from the η^2-aldehyde ligand to the metal center, it should be reasonable to consider the η^2-aldehyde complex as the three-membered oxametallacycle complex.

So far, a variety of transition metal complexes bearing η^2-aldehyde ligands have been isolated in the field of coordination chemistry [3, 4]; however, these complexes have been rarely proposed and experimentally demonstrated as an intermediate in homogeneous transition metal catalysis. Both the formyl oxygen and the formyl carbon in the η^2-aldehyde ligand demonstrated a nucleophilic reactivity [5], which should further expand the use of aldehydes in organic synthesis.

The combination of an electron-rich transition metal and a strong electron-donating ligand would be ideal to generate an η^2-aldehyde intermediate in transition metal catalysis, as the strength of back-donation should play a role in increasing the contribution of the three-membered oxametallacycle form. In this aspect, the combination of electron-rich nickel(0) and strong σ-donating ligands should lead to the formation of a promising catalyst [4i], and the isolation of (η^2-aldehyde)Ni $(PR_3)_2$ complexes was indeed reported [4].

Nickel Catalysis in Organic Synthesis: Methods and Reactions,
First Edition. Edited by Sensuke Ogoshi.

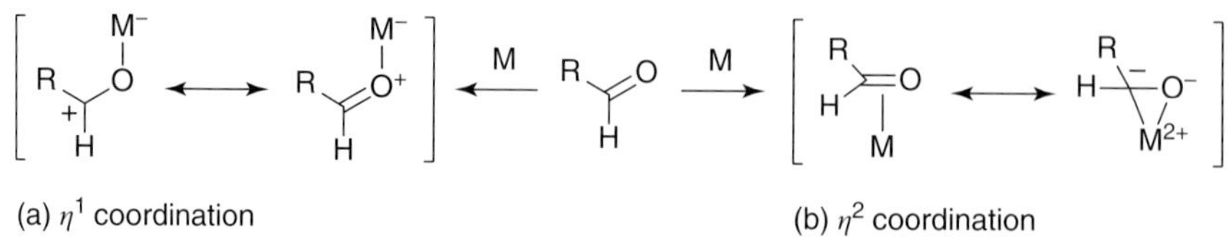

Figure 2.1 Activation of aldehydes through (a) η^1 coordination to Lewis acidic metals and (b) η^2 coordination to low-valent transition metals.

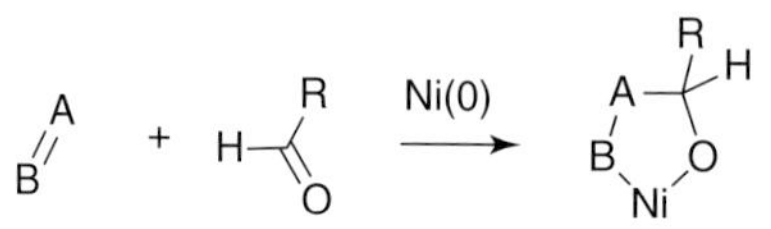

Figure 2.2 Formation of five-membered oxanickelacycle complexes through oxidative cyclization of aldehydes and other π-components (showing as A = B) on Ni(0).

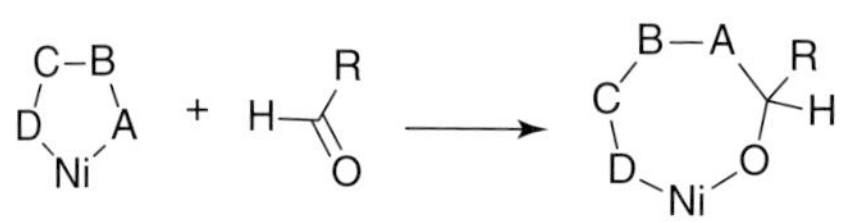

Figure 2.3 Formation of seven-membered oxanickelacycle complexes.

These complexes were proposed as a key intermediate in the nickel-catalyzed 1,2-addition of organometallic reagents to aldehydes, which is summarized in this chapter.

Furthermore, intriguingly, nickel(0) reacts with aldehydes and other π-components to afford five-membered oxanickelacycle complexes through oxidative cyclization (Figure 2.2). These oxanickelacycle complexes have been proposed as key intermediates for the catalytic transformation of aldehydes proceeding through the β-hydride elimination that is summarized in this chapter, as well as for the well-developed, nickel-catalyzed multicomponent coupling reactions [6].

Insertion of aldehydes to the five-membered nickelacycle complexes can furnish seven-membered complexes, which were also confirmed as intermediates in a catalytic transformation of aldehydes (Figure 2.3).

Based on this background, this chapter summarizes the catalytic transformation of aldehydes, which was proposed to proceed through oxanickelacycle intermediates. Stoichiometric transformations of oxanickelacycle complexes are referred when the references are critical to discuss the mechanism for the corresponding catalytic transformation of aldehydes.

2.2 Catalytic Transformation of Aldehydes Through Three-Membered Oxanickelacycle Complexes

Although reactive organometallic reagents such as organolithium, and Grignard reagents often react with aldehydes to yield alcohols as a 1,2-additon product without catalysts [7], the transition-metal-catalyzed 1,2-addition reactions with aldehydes have been extensively studied, as these procedures can expand the scope of organometallic reagents to less reactive organoboron and organosilicon compounds [8]. Moreover, an enantioselective process affording enantiopure alcohols can be developed in the presence of a catalytic amount of transition metals bearing chiral ligands.

Nickel is the catalyst of choice for such C—C bond formation reactions [9]. In general, a plausible reaction mechanism has been proposed to proceed through the formation of a three-membered oxanickelacycle complex (or η^2-aldehyde complex) followed by transmetalation with organometallic reagents (R′–M) and reductive elimination (Figure 2.4) [9a–n]. This mechanism was supported by results of some stoichiometric experiments reported by Ogoshi et al., including the electrophilic addition of Me_3SiX to (η^2-PhCHO)Ni(PCy_3)$_2$ (**1**) affording (η^1:η^1-$Me_3SiOCH(Ph)$)Ni(PCy_3)(X) (**2a**, X = OTf, **2b**, X = Cl) as well as addition of Me_2AlOTf affording (η^2-PhCHO-$AlMe_2OTf$)Ni(PCy_3)$_2$ (**3**) (Figure 2.5) [4c,d,f].

An alternative mechanism can be proposed to proceed through the insertion of aldehydes to a nickel–R′ bond that is able to be formed by transmetalation

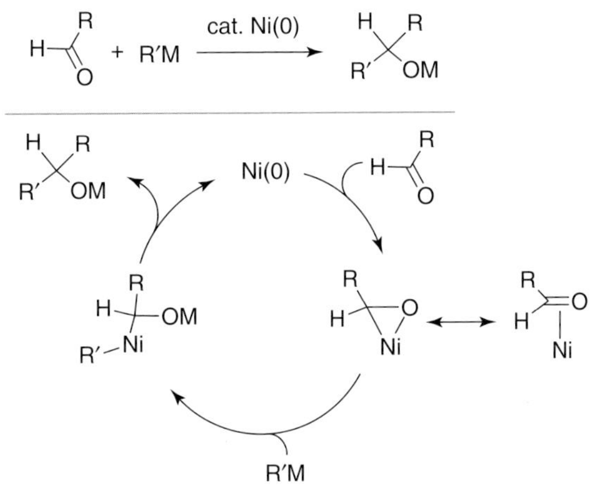

Figure 2.4 General reaction mechanism for nickel(0)-catalyzed 1,2-addition reaction of organometallic reagents to aldehydes.

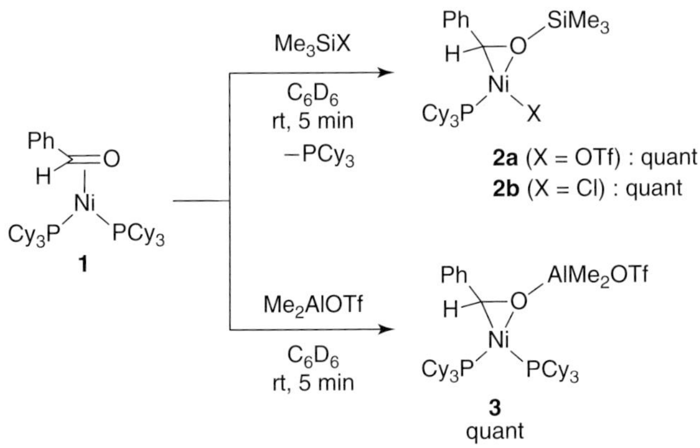

Figure 2.5 Electrophilic addition reactions of Me_3SiOTf (top) and Me_2AlOTf (below) to **1** [4c,d,f].

between R′–M and a nickel(II) complex, or by oxidative addition of nickel(0) to R′–X, where X is a (pseudo) halogen (Figure 2.6) [9o–q].

Representative examples of nickel(0)-catalyzed intermolecular 1,2-addition reactions between organometallic reagents and aldehydes are summarized in Figure 2.7 [9]. As organometallic reagents, organoaluminum [9a,c],

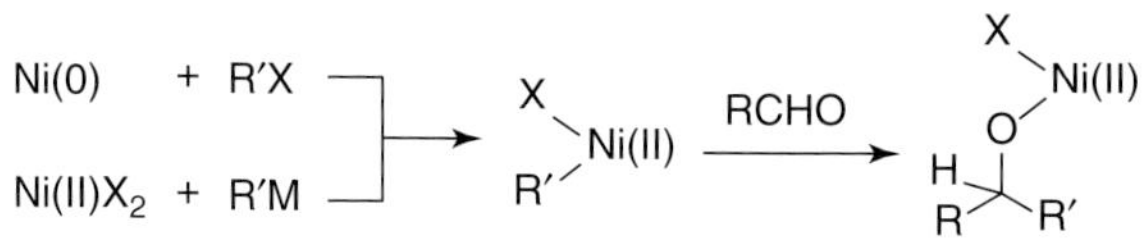

Figure 2.6 Plausible pathways for nickel(II)-catalyzed 1,2-addition reactions to aldehydes.

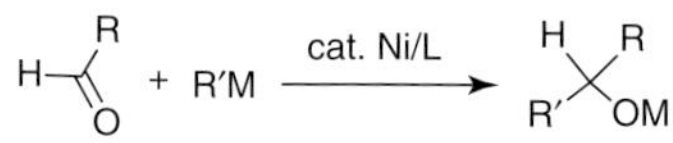

	R′M	cat. Ni/L	R
1998 Fujisawa	Me_3Al	Ni(acac)$_2$/2PPh$_3$	alkyl (1°, 2°), aryl, alkenyl
2003 Burguete Luis	Et_2Zn	Ni(OAc)$_2$/L1	Ph
2005 Woodward	(R′$_3$Al)$_2$·DABCO (R′ = Me, Et)	Ni(acac)$_2$/L2	alkyl (1°, 2°), aryl, alkenyl
2005 Shirakawa	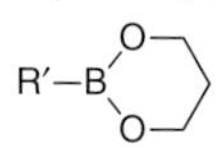 (R′ = aryl, alkenyl)	Ni(cod)$_2$/4-octyne	alkyl (1°), aryl
2005 Yorimitsu Oshima	R′$_3$B, R′-(9-BBN) (R′ = alkyl)	Ni(cod)$_2$/2P^{t}Bu$_3$	alkyl (1°), aryl
2006 Yorimitsu Oshima	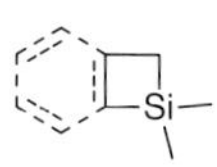	Ni(cod)$_2$ 2P^{n}Bu$_3$ or 2PPh$_2$Me	alkyl (1°, 2°), aryl, alkenyl
2007 Kondo Aoyama	(R′BO)$_3$ (R′ = aryl)	Ni(cod)$_2$/L3	aryl
2009 Itami	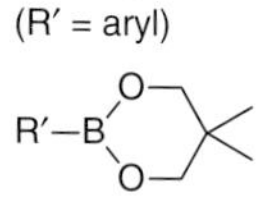 (R′ = aryl, alkenyl)	Ni(cod)$_2$/IPr	alkyl (1°, 2°), aryl
2014 Ogoshi	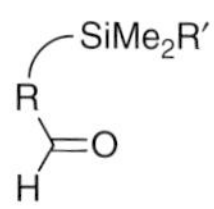 (R′ = aryl, vinyl, alkynyl)	Ni(cod)$_2$/IPr	aryl

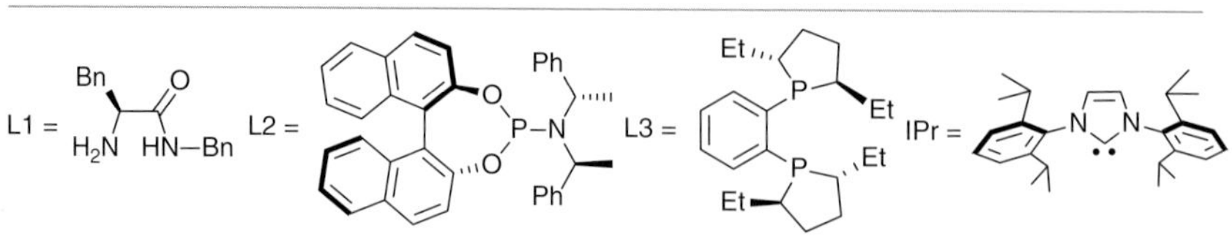

Figure 2.7 Selected examples on nickel(0)-catalyzed 1,2-addition reactions to aldehydes.

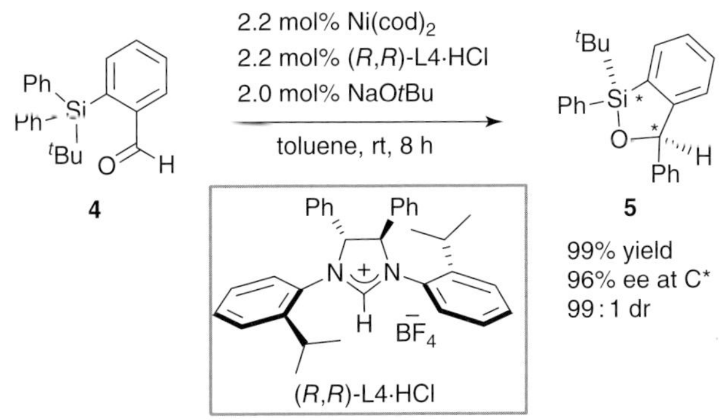

Figure 2.8 Nickel(0)-catalyzed asymmetric intramolecular 1,2-addition in **4** affording benzoxasilole **5**. Source: Kumar et al. 2015 [9n]. Reproduced with permission of American Chemical Society.

organozinc [9b], organoboron [9d,e,g–l], and organosilicon [9f,m,n] reagents have been used so far, and some asymmetric variants have been also developed [9b,c,h–j,n].

Among such asymmetric reactions, a variety of organoboron reagents have been used; however, organosilicon reagents have been less explored even though their use in organic synthesis should be of much interest owing to their high stability and availability. In 2015, Ogoshi et al. reported a formal intramolecular 1,2-addition of aryl-, vinyl-, and alkynyl-silanes to aldehydes, affording enantiopure benzoxasiloles in which the carbon as well as the silicon stereogenic centers were simultaneously constructed with excellent enantioselectivity (Figure 2.8) [9n]. A racemic variant of this reaction was proposed to proceed through an intermolecular R′ group transfer, which was clearly confirmed by a crossover experiment [9m], whereas the asymmetric process was confirmed to proceed through an intramolecular R′ transfer.

The other example of nickel(0)-catalyzed transformation of aldehydes was a reductive homocoupling of aryl aldehydes, proceeding through the three-membered oxanickelacycle complex (Figure 2.9) [4d]. The reaction was proposed to occur by the electrophilic addition of Me_3SiCl to (η^2-ArCHO)Ni(DPPF) intermediate (cf. Figure 2.5), followed by reduction of a Ni(II) intermediate by zinc dust to regenerate an active Ni(0) species.

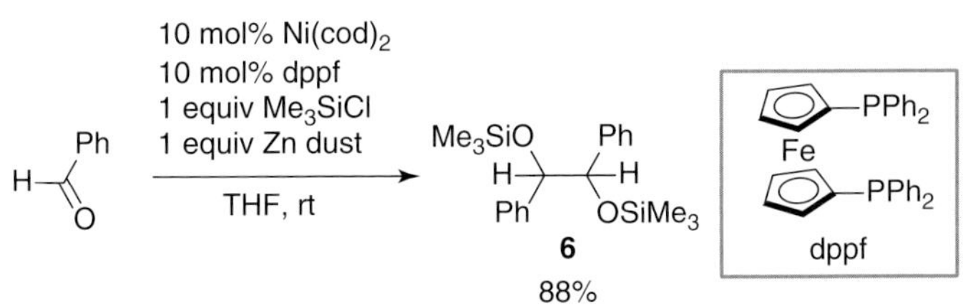

Figure 2.9 Nickel(0)-catalyzed homocoupling reaction of benzaldehyde to afford **6**. Source: Ogoshi et al. 2006 [4d]. Reproduced with permission of Elsevier.

2.3 Catalytic Transformation of Aldehydes Through Five-Membered Oxanickelacycle Complexes

So far, a variety of five-membered oxanickelacycle complexes have been proposed as a key intermediate for the multicomponent coupling reactions between aldehydes, π-components (alkynes, 1,3-dienes, allenes, and alkenes), and organometallic reagents [6]. Nevertheless, a different type of catalytic transformation of aldehydes has been rarely proposed to occur through five-membered oxanickelacycle complexes. One such transformation, β-H elimination and subsequent reductive elimination, can convert five-membered oxanickelacycle complexes to acyl derivatives (Figure 2.10).

The β-H elimination from oxanickelacycle complexes was proposed by Hillhouse and coworker, given that the thermolysis of seven-membered oxanickelacycle **8** afforded aldehyde **9** (Figure 2.11b) [10a], wherein they had prepared complex **8** from nickelacycle **7** and paraformaldehyde, followed by the procedure of Carmona et al. (Figure 2.11a) [10b].

In addition, Ogoshi et al. reported the formation of enone **12** (70%) from oxanickelacycle **10**, which was synthesized from benzaldehyde and 2-butyne through oxidative cyclization on nickel(0)/PCy_3 complex. A nickel(II)-hydride intermediate **11** was thus proposed as a possible intermediate (Figure 2.12) [4e].

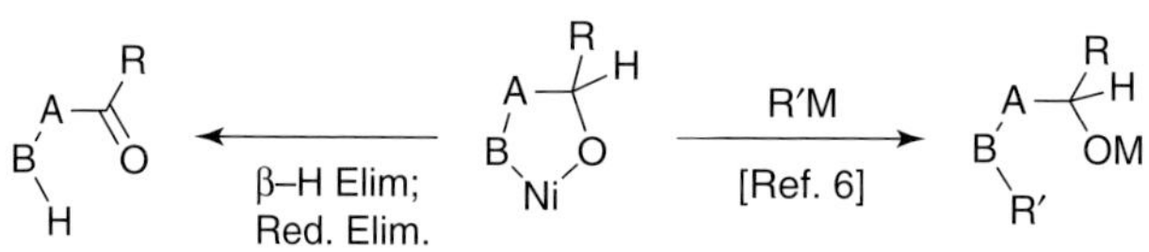

Figure 2.10 Transformation of five-membered oxanickelacycle complexes.

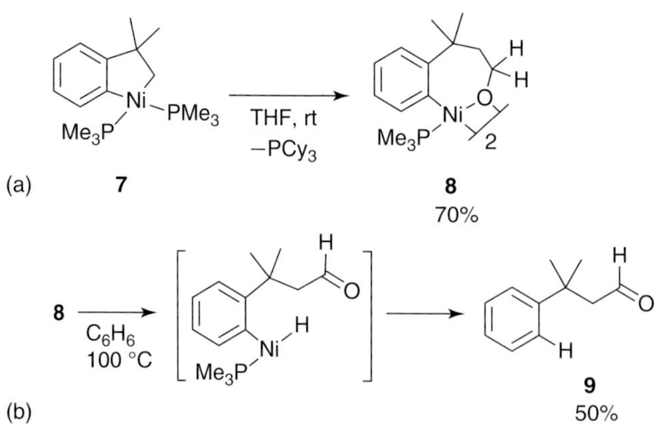

Figure 2.11 (a) Preparation of seven-membered oxanickelacycle **8**. Source: Carmona et al. 1989 [10b]. Reproduced with permission of American Chemical Society. (b) Observation of β-H elimination from **8** affording **9**. Source: Han and Hillhouse 1997 [10a]. Reproduced with permission of American Chemical Society.

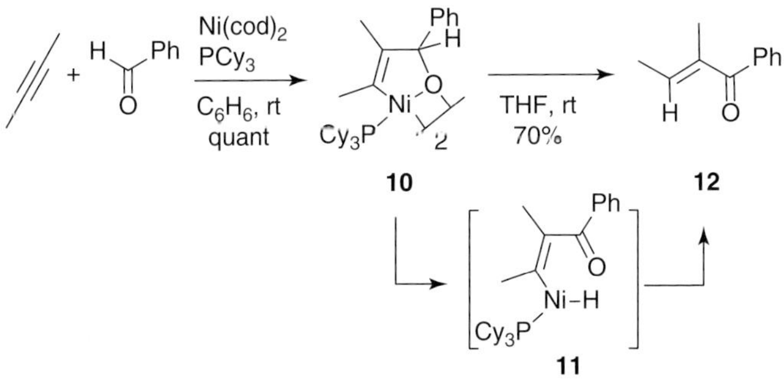

Figure 2.12 Synthesis and transformation of oxanickelacycle **10**. Source: Ogoshi et al. 2008 [4e]. Reproduced with permission of Royal Society of Chemistry.

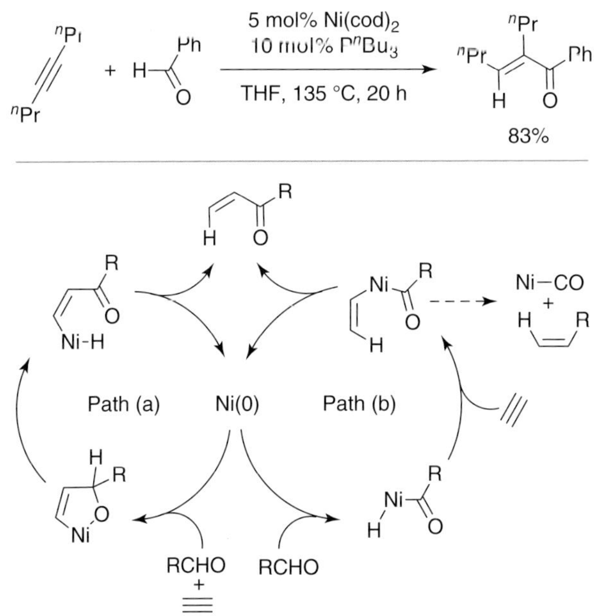

Figure 2.13 Nickel(0)-catalyzed alkyne hydroacylation of 4-octyne with benzaldehyde, and its possible mechanisms. Source: Tsuda et al. 1990 [11a]. Reproduced with permission of American Chemical Society.

These observations imply the participation of a five-membered oxanickelacycle as a possible intermediate in the nickel(0)-catalyzed alkyne hydroacylation, reported by Tsuda et al. (Figure 2.13) [11a]. In fact, Tsuda et al. discussed a possibility of the pathway (a) proceeding through the five-membered oxanickelacycle complex as well as the pathway (b) proceeding through an acyl nickel intermediate, and they concluded that the latter mechanism is favored based on the observation of decarbonylated olefins [11b].

After the report by Tsuda, Saegusa et al. [11], the nickel(0)-catalyzed hydroacylation reactions have been expanded to use alkenes; however, interestingly, both reaction mechanisms that proceed through an oxanickelacycle intermediate and through an acyl nickel intermediate have been proposed to be operative depending on the reaction system [12, 13]. Among these examples, Ogoshi and coworkers reported an intramolecular alkene hydroacylation that afforded a variety of 1-indanone and 1-tetralone derivatives catalyzed by nickel(0)/I^{t}Bu catalyst (Figure 2.14) [13a]. The reaction mechanism of this hydroacylation was proposed as shown in Figure 2.14, which was confirmed by (i) isolation and transformation of a dimeric five-membered oxanickelacycle complex shown as *cis*-(**18**)$_2$ in Figure 2.14, (ii) kinetics, and (iii) DFT calculations [13b,c]. The rate-limiting steps thus include oxidative cyclization from *anti*-**17** to *trans*-**18**, and a subsequent β-H elimination to yield a nickel hydride complex **19**.

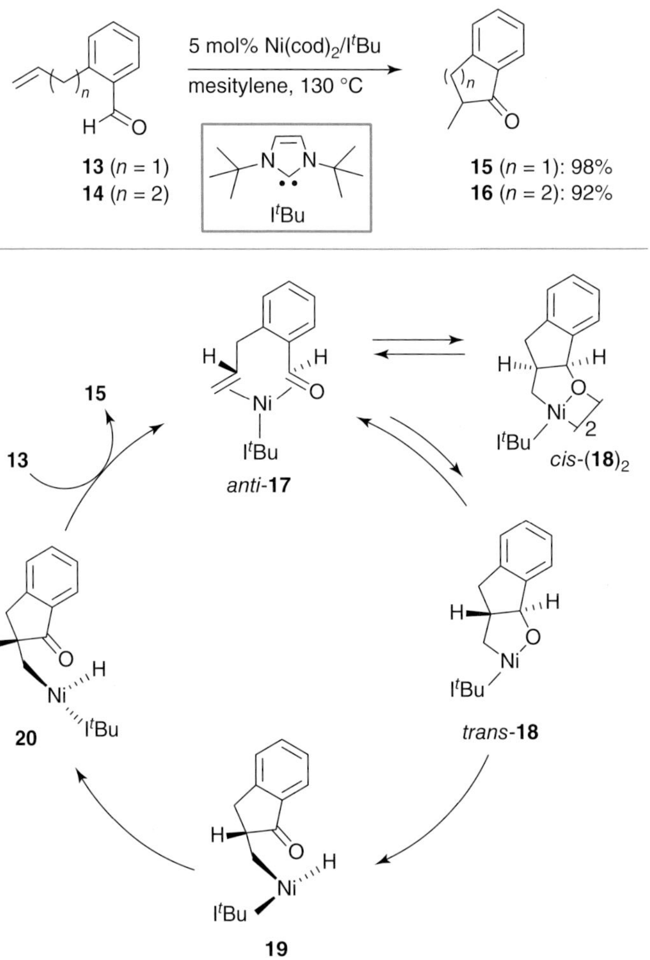

Figure 2.14 Nickel(0)-catalyzed intramolecular alkene hydroacylation and its reaction mechanism, wherein *anti-*, *cis-*, and *trans*-conformations refer to the relative stereochemistry between two formyl hydrogens or β-hydrogens. Source: Adapted from Hoshimoto et al. 2012 [13a] and 2017 [13b].

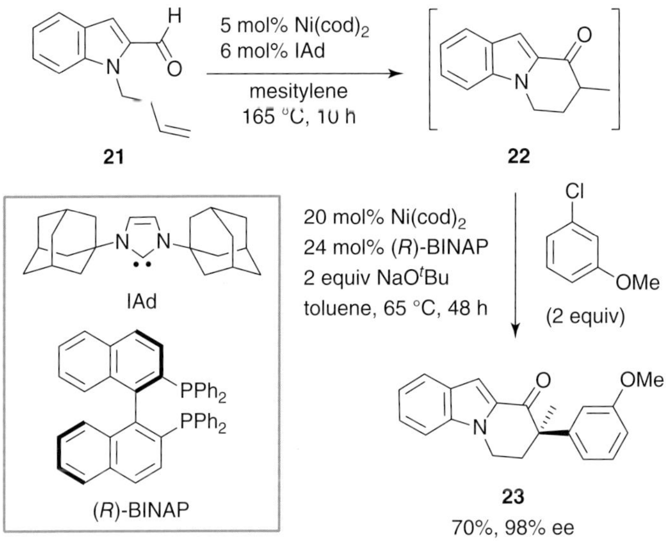

Figure 2.15 A one-pot synthesis of **23** through a sequential hydroacylation/α-arylation reaction. Source: Ghosh et al. 2016 [13d]. Reproduced with permission of American Chemical Society.

Later, Stanley and coworkers developed a one-pot synthesis of dihydropyridoindolone **23** with α-chiral quaternary stereogenic center from *N*-homoallylindole-2-carboxaldehyde (**21**) through a sequential protocol including nickel(0)/IAd-catalyzed hydroacylation and nickel(0)/(R)-BINAP-catalyzed α-arylation (Figure 2.15) [13d].

When aldehydes are used instead of alkynes or alkenes in nickel(0)-catalyzed hydroacylation, esters can be produced through a five-membered dioxanickelacycle intermediate. The homodimerization of aldehydes affording esters is known as the Tishchenko reaction, which is among the most important methods to prepare esters [14]. Ogoshi et al. reported the nickel(0)/IPrCl-catalyzed homodimerization of aldehydes bearing a wide range of aryl and alkyl substituents (Figure 2.16) [15a]. They observed the *in*

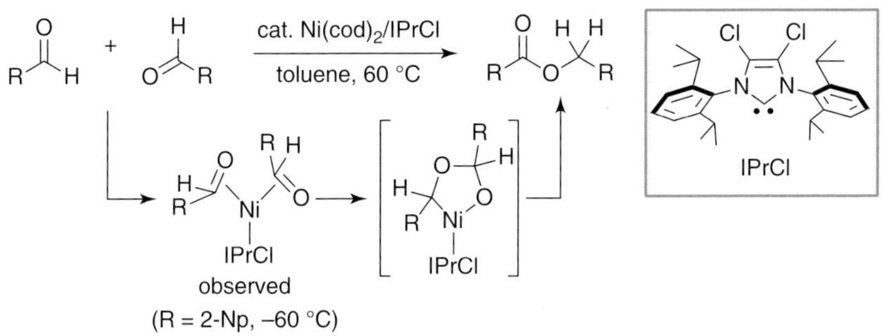

Figure 2.16 Nickel(0)-catalyzed homodimerization of aldehydes. Source: Ogoshi et al. 2010 [15a]. Reproduced with permission of Royal Society of Chemistry.

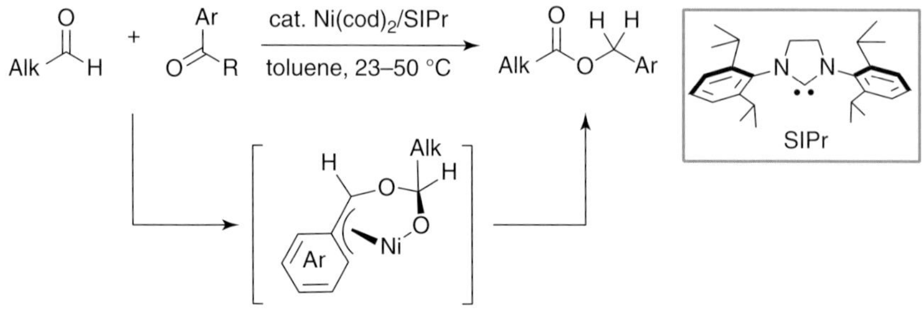

Figure 2.17 Nickel(0)-catalyzed cross-dimerization between alkyl (AlkCHO) and aryl (ArCHO) aldehydes. Source: Adapted from Hoshimoto et al. 2015 [4i] and 2011 [15b].

situ formation of bis(η^2-2-naphthaldehyde)Ni(IPrCl) and its conversion to the target esters, demonstrating the importance of simultaneous coordination of two aldehydes on the nickel(0) center.

Subsequently, Ogoshi and coworkers expanded the nickel(0)-catalyzed Tishchenko reaction to the selective cross-dimerization reaction between aryl and alkyl aldehydes, which afforded a single cross-coupled ester selectively from among the four possible esters including two cross-coupled and two homo-coupled products under nickel(0)/SIPr catalysis (Figure 2.17) [15b]. Although they did not give a decisive conclusion on the reaction mechanism, a possible reaction mechanism was proposed to take place through an oxanickelacycle intermediate including η^3-benzylnickel moiety, as shown in Figure 2.17, based on kinetic and deuterium-labeling experiments [4i, 15c].

2.4 Catalytic Transformation of Aldehydes Through Seven-Membered Oxanickelacycle Complexes

A further advanced derivatization of aldehydes can be achieved with more than two π-components by designing a catalyst system in which a seven-membered oxanickelacycle intermediate is involved.

Breit, Krische, and coworkers reported a regio- and stereoselective hydrohydroxymethylation of alkynes catalyzed by nickel(0)/PCy_3 (Figure 2.18) [16a]. They observed the formation of formate ester **27**, which can be converted to the product **28** by treatment with KOH, and they thus proposed a mechanism proceeding through β-H elimination from a seven-membered dioxanickelacycle intermediate **25**. The complex **25** was considered to be formed by insertion of formaldehyde to the Ni—O bond in a five-membered oxanickelacycle complex **24** that was formed by oxidative cyclization of an alkyne and an aldehyde on nickel(0)/PCy_3. Subsequently, the nickel(0)-catalyzed hydrohydroxymethylation protocol was expanded to the transformation of dienes by Breit, Krische, and coworkers, in which an oxanickelacycle intermediate bearing an η^3-allyl ligand was proposed [16b].

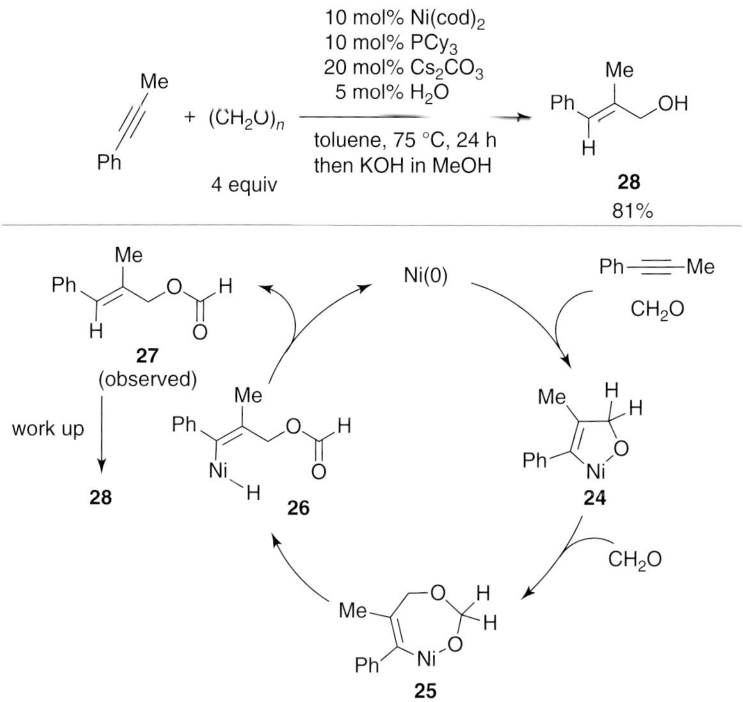

Figure 2.18 Nickel(0)-catalyzed hydrohydroxymethylation of alkyne. Source: Bausch et al. 2011 [16a]. Reproduced with permission of Wiley-VCH Verlag GmbH & Co. KGaA.

Later, Ohashi, Ogoshi, and coworkers reported a nickel(0)/IPr-catalyzed cross-trimerization reaction between tetrafluoroethylene, ethylene, and aldehydes (Figure 2.19) [17a]. A plausible mechanism was proposed to proceed through β-H elimination from **30**, which was afforded by the reaction between an aldehyde and 2,2,3,3,-tetrafluoronickelacyclopentane **29**. The reactivity of an analogous 2,2,3,3,-tetrafluoronickelacyclopentane complex was also explored by them [17b], which rationalizes a highly chemoselective pathway to furnish 4,4,5,5-tetrafluoro-1-pentanone derivatives.

2.5 Conclusion and Outlook

In this chapter, the catalytic transformations of aldehydes are summarized from the viewpoint of oxanickelacycle intermediates. Three-, five-, and seven-membered oxanickelacycle complexes have been proposed as key intermediates, and some of them were experimentally confirmed by isolation, spectroscopic and kinetic experiments, and theoretical approaches.

Five-membered and larger oxanickelacycle complexes can become a potential intermediate in a nickel-catalyzed bond forming reaction between aldehydes and multiple π-components. In this context, fundamental research is of importance to develop a method to generate more

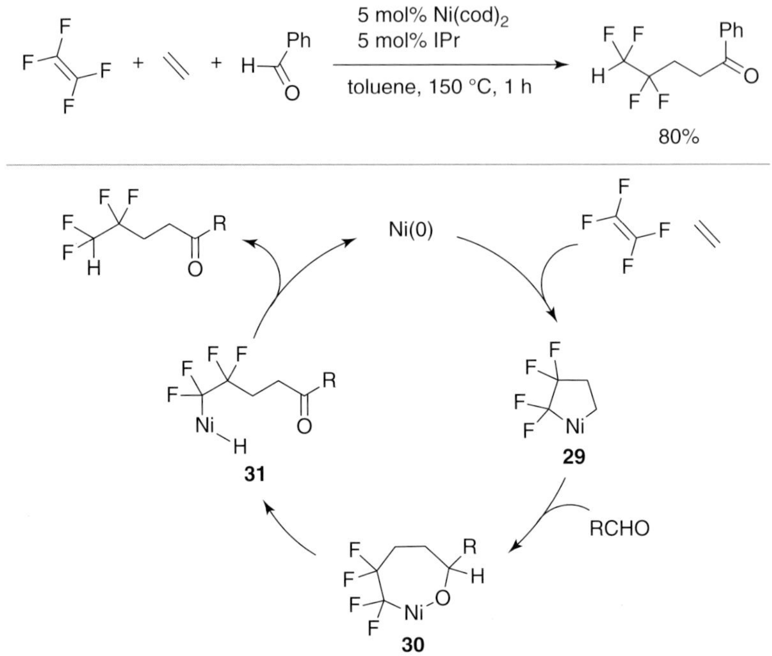

Figure 2.19 Nickel(0)-catalyzed cross-trimerization between tetrafluoroethylene, ethylene, and aldehydes. Source: Ohashi et al. 2015 [17a]. Reproduced with permission of American Chemical Society.

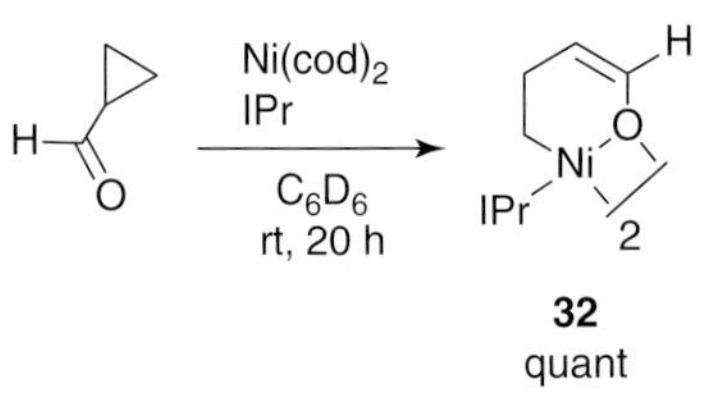

Figure 2.20 Formation of **32** through oxidative addition. Source: Tamaki et al. 2009 [18]. Reproduced with permission of John Wiley & Sons.

than five-membered oxanickelacycle complexes. For examples, Ogoshi and coworkers reported the isolation of a dimeric six-membered oxanickelacycle complex **32**, which proceeded through the oxidative addition of nickel(0) to cyclopropanecarboxaldehyde; however, so far this six-membered oxanickelacycle has not been involved in the catalytic transformation of aldehydes (Figure 2.20) [18]. In addition, Hillhouse and coworker suggested the *in situ* generation of 9- and 11-membered oxanickelacycle complexes, **33** and **35**, during the formation of a mixture of **34** and **36**, respectively, from the 7-membered oxanickelacycle **8** and paraformaldehyde (Figure 2.21) [10a].

How to expand these findings to an establishment of a catalyst system for the derivatization of aldehydes? This should be a big future challenge in the chemistry of nickel catalysis.

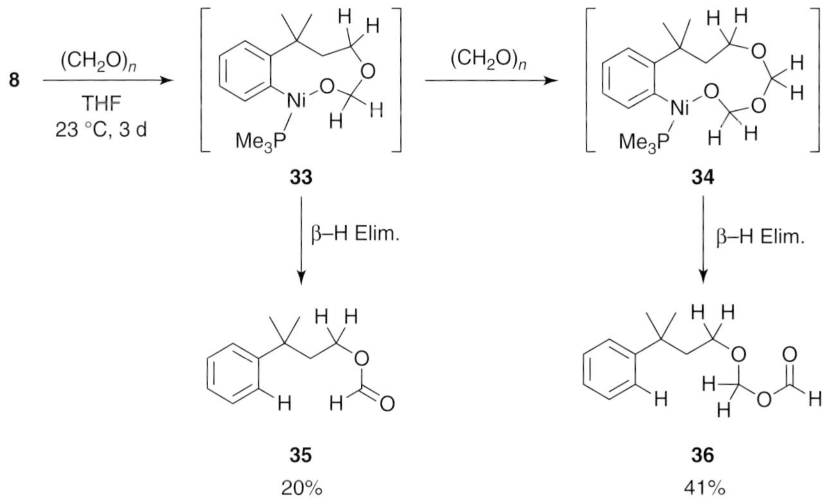

Figure 2.21 Generation of **33** and **34** in the reaction between **8** and $(CH_2O)_n$. Source: Han and Hillhouse 1997 [10a]. Reproduced with permission of American Chemical Society.

References

1 (a) Shambayati, S. and Schreiber, S.L. (1991). *Comprehensive Organic Synthesis* (eds. B.M. Trost and I. Fleming), 1. Oxford, UK: Pergamon Press, Vol, Chapter 1. (b) Crabtree, R.H. (2003). *The Organometallic Chemistry of the Transition Metals*, 4e. New York: Wiley-Interscience. (c) Collman, J.P., Hegedus, L.S., Norton, J.R., and Finke, R.G. (1987). *Principles and Applications of Organotransition-Metal Chemistry*. Mill Valley, CA: University Science Books. See also: (d) Huang, Y.-H. and Gladysz, J.A. (1988). *J. Chem. Educ.* 65: 298–303.

2 (a) Garralda, M.A. (2009). *Dalton Trans.*: 3635–3645. (b) Fristrup, P., Kreis, M., Palmelund, A. et al. (2008). *J. Am. Chem. Soc.* 130: 5206–5215.

3 For selected examples, see: Ti: (a) Li, L., Kristian, K.E., Han, A. et al. (2012). *Organometallics* 31: 8218–8224. V: (b) Gambarotta, S., Floriani, C., Chiesi-Villa, A., and Guastini, C. (1986). *Organometallics* 5: 2425–2433. Fe: (c) Lei, H., Royer, A.M., Rauchfuss, T.B., and Gray, D. (2012). *Organometallics* 31: 6408–6414. Co: (d) Lenges, C.P., Brookhart, M., and White, P.S. (1999). *Angew. Chem. Int. Ed.* 38: 552–555. (e) Schneider, J.J., Wolf, D., Bläser, D., and Boese, R. (2000). *Eur. J. Inorg. Chem.*: 713–718. (f) Gambarotta, S., Floriani, C., Chiesi-Villa, A., and Guastini, C. (1983). *J. Am. Chem. Soc.* 105: 1690–1691. Mo: (g) Brunner, H., Wachter, J., Bernal, I., and Creawick, M. (1979). *Angew. Chem. Int. Ed.* 18: 861–862. (h) Chen, C.S., Lin, C.S., and Yeh, W.Y. (2011). *J. Organomet. Chem.* 696: 1474–1478. (i) Jiménez, T., Barea, E., Oltra, J.E. et al. (2010). *J. Org. Chem.* 75: 7022–7025. W: (j) Yeh, W.Y. and Lin, C.S. (2004). *Organometallics* 23: 917–920. Re: (k) Buhro, W.E., Patton, A.T., Strouse, C.E. et al. (1983). *J. Am. Chem. Soc.* 105: 1056–1058. (l) Helberg, L.E., Gunnoe, T.B., Brooks, B.C. et al. (1999). *Organometallics* 18: 573–581.

Os: (m) Brown, K.L., Clark, G.R., Headford, C.E.L. et al. (1979). *J. Am. Chem. Soc.* 101: 503–505. (n) Spera, M.L., Chen, H., Moody, M.W. et al. (1997). *J. Am. Chem. Soc.* 119: 12772–12778. Pt: (o) Head, R.A. (1982). *J. Chem. Soc., Dalton Trans.*: 1637–1639.

4 For η^2-aldehyde nickel(0) complexes, see: (a) Walther, D. (1980). *J. Organomet. Chem.* 190: 393–401. (b) Kaiser, J., Sieler, J., Walther, D. et al. (1982). *Acta Cryst.* B38: 1584–1586. (c) Ogoshi, S., Oka, M.-A., and Kurosawa, H. (2004). *J. Am. Chem. Soc.* 126: 11802–11803. (d) Ogoshi, S., Kamada, H., and Kurosawa, H. (2006). *Tetrahedron* 62: 7583–7588. (e) Ogoshi, S., Arai, T., Ohashi, M., and Kurosawa, H. (2008). *Chem. Commun.*: 1347–1349. (f) Ohashi, M., Saijo, H., and Ogoshi, S. (2010). *Organometallics* 29: 6534–6540. (g) Sakaki, S., Kitaura, K., Maruoka, K., and Ohkubo, K. (1983). *Inorg. Chem.* 22: 104–108. (h) Delbecq, F. and Sautet, P. (1992). *J. Am. Chem. Soc.* 114: 2446–2455. (i) Hoshimoto, Y., Ohashi, M., and Ogoshi, S. (2015). *Acc. Chem. Res.* 48: 1746–1755. (j) Ogoshi, S. (2017). *Bull. Chem. Soc. Jpn.* 90: 1401–1406.

5 Thiyagarajan, B., Kerr, M.E., Bollinger, J.C. et al. (1997). *Organometallics* 16: 1331–1334.

6 Selected reviews, see: (a) Montgomery, J. (2004). *J. Angew. Chem. Int. Ed.* 43: 3890–3908. (b) Tasker, S.Z., Standley, E.A., and Jamison, T.F. (2014). *Nature* 509: 299–309. (c) Jackson, E.P., Malik, H.A., Sormunen, G.J. et al. (2015). *Acc. Chem. Res.* 48: 1736–1745. (d) Cai, Y., Li, F., Li, Y.-Q. et al. (2018). *Tetrahedron Lett.* 59: 1073–1079.

7 Panahi, F., Bahmani, M., and Iranpoor, N. (2015). *Adv. Synth. Catal.* 357: 1211–1220.

8 For recent reviews, see: (a) Mondal, S. and Panda, G. (2014). *RSC Adv.* 4: 28317–28358. (b) Sun, Y.-W., Zhu, P.-l., Xu, Q., and Shi, M. (2013). *RSC Adv.* 3: 3153–3168. (c) Partyka, D.V. (2011). *Chem. Rev.* 111: 1529–1595. (d) Miyaura, N. (2009). *Synlett* 2039–2050. (e) Tian, P., Dong, H.-Q., and Lin, G.-Q. (2012). *ACS Catal.* 2: 95, 95–119. (f) Pellissier, H. (2015). *Tetrahedron* 71: 2487–2524.

9 (a) Ichiyanagi, T., Kuniyama, S., Shimizu, M., and Fujisawa, T. (1998). *Chem. Lett.*: 1033–1034. (b) Burguete, M.I., Collado, M., Escorihuela, J. et al. (2003). *Tetrahedron Lett.* 44: 6891–6894. (c) Biswas, K., Prieto, O., Goldsmith, P.J., and Woodward, S. (2005). *Angew. Chem. Int. Ed.* 44: 2232–2234. (d) Takahashi, G., Shirakawa, E., Tsuchimoto, T., and Kawakami, Y. (2005). *Chem. Commun.*: 1459–1461. (e) Hirano, K., Yorimitsu, H., and Oshima, K. (2005). *Org. Lett.* 7: 4689–4691. (f) Hirano, K., Yorimitsu, H., and Oshima, K. (2006). *Org. Lett.* 8: 483–485. (g) Hirano, K., Yorimitsu, H., and Oshima, K. (2006). *Adv. Synth. Catal.* 348: 1543–1546. (h) Arao, T., Kondo, K., and Aoyama, T. (2007). *Tetrahedron Lett.* 48: 4115–4117. (i) Yamamoto, K., Tsurumi, K., Sakurai, F. et al. (2008). *Synthesis* 22: 3585–3590. (j) Sakurai, F., Kondo, K., and Aoyama, T. (2009). *Tetrahedron Lett.* 50: 6001–6003. (k) Bouffard, J. and Itami, K. (2009). *Org. Lett.* 11: 4410–4413. (l) Maekawa, T., Sekizawa, H., and Itami, K. (2011). *Angew. Chem. Int. Ed.* 50: 7022–7026. (m) Hoshimoto, Y., Yabuki, H., Kumar, R. et al. (2014). *J. Am. Chem. Soc.* 136: 16752–16755. (n) Kumar, R., Hoshimoto, Y., Yabuki, H. et al. (2015). *J. Am. Chem. Soc.* 137: 11838–11845. For recent examples in which a Ni(II)-catalyzed mechanism

was proposed, see: (o) Zhou, L., Du, X., He, R. et al. (2009). *Tetrahedron Lett.* 50: 406–408. (p) Xing, C.-H. and Hu, Q.-S. (2010). *Tetrahedron Lett.* 51: 924–927. (q) Chen, W., Baghbanzadh, M., and Kappe, C.O. (2011). *Tetrahedron Lett.* 52: 1677–1679.

10 (a) Han, R. and Hillhouse, G.L. (1997). *J. Am. Chem. Soc.* 119: 8135–8136. (b) Carmona, E., Gutiérrez-Puebla, E., Marín, J.M. et al. (1989). *J. Am. Chem. Soc.* 111: 2883–2891.

11 (a) Tsuda, T., Kiyoi, T., and Saegusa, T. (1990). *J. Org. Chem.* 55: 2554–2558. (b) Meng, Q., Shen, W., He, R., and Li, M. (2011). *Transition Met. Chem.* 36: 793–799. (c) Yang, F., Jin, T., and Yamamoto, Y. (2012). *Tetrahedron* 68: 5223–5228.

12 (a) Taniguchi, H., Ohmura, T., and Suginome, M. (2009). *J. Am. Chem. Soc.* 131: 11298–11299. (b) Xiao, L.-J., Fu, X.-N., Zhou, M.-J. et al. (2013). *J. Am. Chem. Soc.* 138: 2957–2960.

13 (a) Hoshimoto, Y., Hayashi, Y., Suzuki, H. et al. (2012). *Angew. Chem. Int. Ed.* 51: 10812–10815. (b) Hoshimoto, Y., Hayashi, Y., Ohashi, M., and Ogoshi, S. (2017). *Chem. Asian J.* 12: 278–282. (c) Meng, Q. and Wang, F. (2017). *J. Mol. Model* 23: 11. (d) Ghosh, A., Walker, J.A. Jr., Ellern, A., and Stanley, L.M. (2016). *ACS Catal.* 6: 2673–2680.

14 For reviews, see: (a) Törmäkangas, O.P. and Koskinen, A.M.P. (2001). *Recent Res. Devel. Organic Chem.* 5: 225. (b) Seki, T., Nakajo, T., and Onaka, M. (2006). *Chem. Lett.* 35: 824–829. (c) Dzik, W.I. and Gooβen, L.J. (2011). *Angew. Chem. Int. Ed.* 50: 11047–11049.

15 (a) Ogoshi, S., Hoshimoto, Y., and Ohashi, M. (2010). *Chem. Commun.* 46: 3354–3356. (b) Hoshimoto, Y., Ohashi, M., and Ogoshi, S. (2011). *J. Am. Chem. Soc.* 133: 4668–4671. (c) Yu, H. and Fu, Y. (2012). *Chem. Eur. J.* 18: 16765–16773.

16 (a) Bausch, C.C., Patman, R.L., Breit, B., and Krische, M.J. (2011). *Angew. Chem. Int. Ed.* 50: 5687–5690. (b) Köpfer, A., Sam, B., Breit, B., and Krische, M.J. (2013). *Chem. Sci.* 4: 1876–1880.

17 (a) Ohashi, M., Shirataki, H., Kikushima, K., and Ogoshi, S. (2015). *J. Am. Chem. Soc.* 137: 6496–6499. (b) Ohashi, M., Kawashima, T., Taniguchi, T. et al. (2015). *Organometallics* 34: 1604–1607.

18 Tamaki, T., Nagata, M., Ohashi, M., and Ogoshi, S. (2009). *Chem. Eur. J.* 15: 10083–10091.

3

Transformation of Imines via Nickelacycles

Masato Ohashi

Osaka Prefecture University, Department of Chemistry, Graduate School of Science, 1-1 Gakuen-cho, Naka-ku, Sakai, Osaka 599-8531, Japan

3.1 Introduction

This chapter focuses on the transformation of imines via five-membered aza-nickelacycle intermediates generated via the oxidative cyclization of an imine and an unsaturated compound such as alkynes with Ni(0). Such aza-nickelacycles are much rarer than the related oxanickelacycles generated via the oxidative cyclization of an aldehyde with Ni(0), since imines are generally weaker electrophiles than aldehydes [1]. Therefore, it is logical to assume that electron-withdrawing substituents on the nitrogen of the imine are required for oxidative cyclization by promoting back-donation from Ni(0) to imines. Furthermore, the generation of a five-membered aza-nickelacycle is efficiently facilitated by the chelate coordination of a donor atom on the *N*-substituent group to a vacant coordination site on the nickel center [2a,b]. Given such background, the first isolation of a five-membered aza-nickelacycle was achieved via the oxidative cyclization of *N*-sulfonylimine and an alkyne [2c, 3].

3.2 [2 + 2 + 1] Carbonylative Cycloaddition of an Imine and Either an Alkyne or an Alkene Leading to γ-Lactams

The transition-metal-catalyzed carbonylative cycloaddition of an imine, either an alkyne or an alkene, and CO (aza-Pauson–Khand reaction) is known as a straightforward method for constructing a γ-lactam skeleton, whereas it has historically been limited to some extent in the well-established Pauson–Khand reaction [4, 5]. Although a five-membered aza-nickelacycle (**3aa**), which was first isolated via the oxidative cyclization of *N*-(benzenesulfonyl)benzaldimine (**1a**) and diphenylacetylene (**2a**) [2c], definitely reacted with CO (5 atm) to afford the corresponding α,β-unsaturated γ-lactam (**4aa**; Schemes 3.1), such nickel-catalyzed transformation reactions are entirely hampered under a CO atmosphere due to the formation of catalytically unreactive nickel carbonyl complexes such as $Ni(CO)_3L$. Two strategies were expected to be the key to the

Nickel Catalysis in Organic Synthesis: Methods and Reactions,
First Edition. Edited by Sensuke Ogoshi.

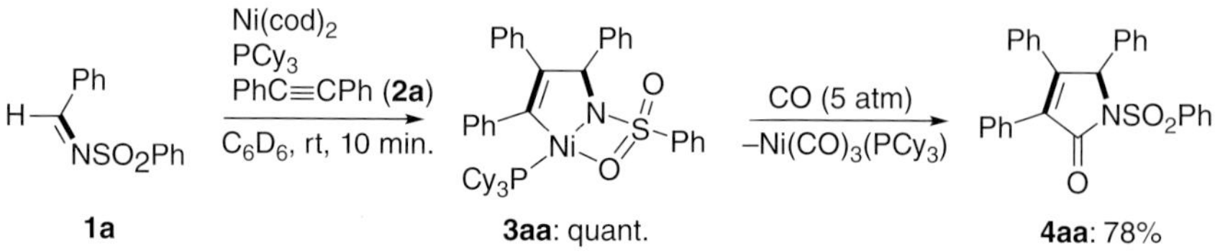

Scheme 3.1 Formation of five-membered aza-nickelacycle **3aa** and its reactivity toward carbon monoxide. Yields, determined by ^{1}H NMR spectroscopy, are given.

success in establishing a Ni(0)-catalyzed carbonylative cycloaddition leading to γ-lactams; (i) to regulate the CO concentration strictly, as it has to be sufficiently high for the reaction with the aza-nickelacycle intermediate, yet also sufficiently low to ensure the formation of $Ni(CO)_3L$ and (ii) to construct a carbonylation system that can effectively regenerate an active Ni(0) species from Ni(0) carbonyl complexes under a CO atmosphere.

Based on strategy (i), the Ogoshi research group reported that phenyl formate was employed as a CO source, since it has an interesting reactivity that allows it to generate CO *in situ* in the presence of organic/inorganic bases, and Tsuji and coworkers [6] and Manabe and coworkers [7] have independently reported its application to transformations involving carbonylation using CO generated *in situ*. Thus, in the presence of a catalytic amount of $Ni(cod)_2$ and PCy_3 (10 and 20 mol%, respectively) as well as a mixture of phenyl formate and NEt_3 (1.5 and 2.0 equiv, respectively), the desired γ-lactam **4aa** was obtained from the reaction of *N*-(benzenesulfonyl)benzaldimine **1a** with 2 equiv of diphenyacetylene **2a** at 70 °C (Scheme 3.2) [8]. This Ni(0)-catalyzed [2 + 2 + 1] carbonylative cycloaddition of imines and alkynes in the presence of phenyl formate can allow us to prepare a variety of α,β-unsaturated γ-lactams. In addition, the carbonylative [2 + 2 + 1] cycloaddition of imines and norbornene also occurs to furnish bicyclic γ-lactams in good to high yields. It should be noted that proper choice of both the ligand and the base was crucial for the smooth formation of γ-lactams; the use of other tertiary phosphine or NHC ligands significantly diminished the yield of γ-lactams, and other bases, such as DBU, DMAP, and quinuclidine, were not suitable under the presented conditions. As was expected, however, the direct use of CO gas in the present system remained unsuccessful because of the rapid generation of $Ni(CO)_3(PCy_3)$.

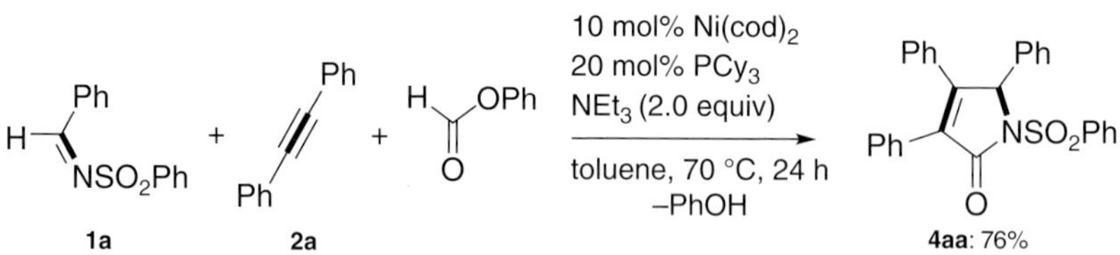

Scheme 3.2 Ni(0)/PCy_3-catalyzed [2 + 2 + 1] carbonylative cycloaddition reaction of imine **1a** and alkyne **2a** with phenyl formate.

On the other hand, Ogoshi and coworkers have recently demonstrated that CO gas itself can be utilized in a Ni(0)-catalyzed [2 + 2 + 1] carbonylative cycloaddition reaction with 1,5-ene-imines, which was employed on the basis their previous findings that 1,5-enals efficiently coordinate to nickel(0) in an $\eta^2:\eta^2$ manner [9], and thus was consistent with strategy (ii). In the presence of a catalytic amount of $Ni(cod)_2$ and PCy_3 (10 mol% each), the desired [2 + 2 + 1] carbonylative cycloaddition of *N*-(*o*-allylbenzylidene)-*p*-toluenesulfonamide (**5a**) took place in the presence of 0.5 atm of CO (1.3 equiv) at room temperature to furnish tricyclic γ-lactam (**6a**) in quantitative yield (Scheme 3.3). It should be worth mentioning that stoichiometric treatment of $Ni(CO)_3(PCy_3)$ with 3 equiv of **5a** even in the presence of 5.0 atm of CO gave the tricyclic γ-lactam **6a** in 79% yield, while this reaction was also conducted in a pressure-tight NMR tube without stirring the reaction mixture. In contrast, when the catalytic reaction was carried out under vigorous stirring (1350 rpm), a dramatic change to colorless reaction mixture within 15 minutes indicated the rapid formation of $Ni(CO)_3(PCy_3)$, and the yield of the target γ-lactam **6a** was only 12%. Continuous stirring did not increase the conversion of the 1,5-ene-imine **5a** at room temperature, and even after stopping stirring, the yield of **6a** did not increase substantially at either room temperature or at 80 °C. Thus, the second and important key to the success in the utilization of CO gas was that the reaction mixture was allowed to stand without stirring; otherwise, the nickel species quantitatively changed to $Ni(CO)_3(PCy_3)$ within a remarkably short time.

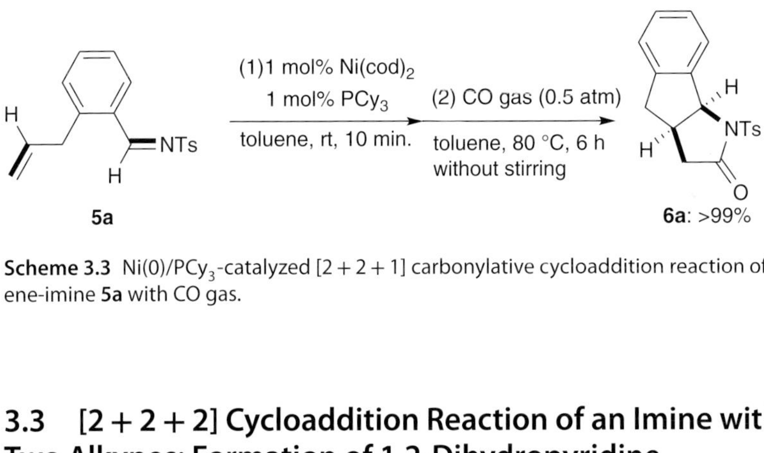

Scheme 3.3 $Ni(0)/PCy_3$-catalyzed [2 + 2 + 1] carbonylative cycloaddition reaction of ene-imine **5a** with CO gas.

3.3 [2 + 2 + 2] Cycloaddition Reaction of an Imine with Two Alkynes: Formation of 1,2-Dihydropyridine Derivatives

The five-membered aza-nickelacycle, which was generated via the oxidative cyclization of an imine and an alkyne, was allowed to react with another molecule of alkyne, thus leading to a seven-membered aza-nickelacycle. For example, the treatment of **3aa** with an additional equimolar amount of **2a** gave a seven-membered nickelacycle (**7aa**) in quantitative yield (Scheme 3.4) [2c]. The insertion of 2-butyne (**2b**) into **3aa** proceeded much faster than that of **2a**, and furnished the corresponding seven-membered nickelacycle (**7az**). Such a seven-membered nickelacycle (**7ab**) was also prepared via the reaction of **1a** with 2 equiv of **2b** in the presence of $Ni(cod)_2$ and PCy_3 at room temperature for

10 minutes (Scheme 3.4). Heating the seven-membered nickelacycles at 100 °C facilitated reductive elimination to yield the corresponding 1,2-dihydropyridine derivatives (**8aa, 8ab, 8az**), respectively (Scheme 3.4) [2c].

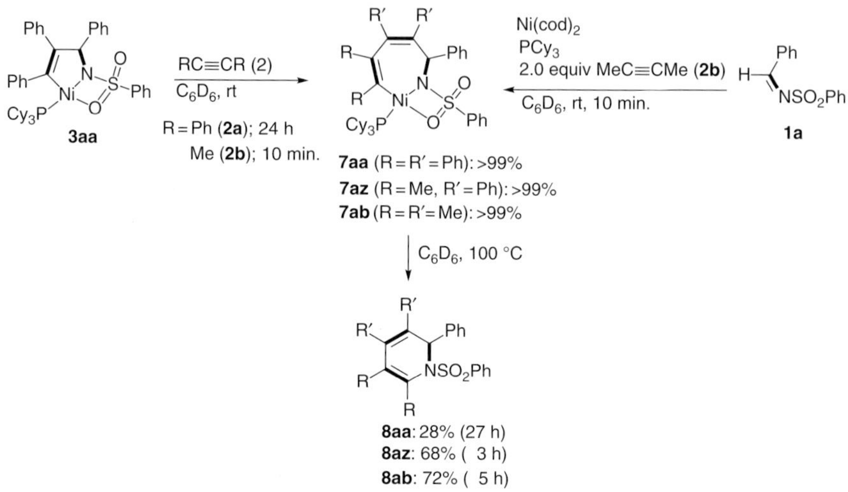

Scheme 3.4 Formation of seven-membered aza-nickelacycles **7** and reductive elimination from **7** leading to 1,2-dihydropyridines **8**.

As a corollary of the aforementioned stoichiometric reactions of seven-membered nickelacycles **7**, the intermolecular [2 + 2 + 2] cycloaddition reaction of *N*-sulfonyl imine **1a** and **2b** in the presence of a catalytic amount of $Ni(cod)_2$ and PMe^tBu_2 at 100 °C afforded 1,2-dihydropyridine **8ab** in 87 % yield (Table 3.1, entry 1) [2c]. Employing either 3-hexyne (**2c**) or trimethylsilylacetylene (**2d**) also resulted in the formation of the corresponding 1,2-dihydropyridines (**8ac** and **8ad**), respectively (entries 2 and 4), whereas the use of PCy_3 in place of PMe^tBu_2 as an auxiliary ligand led to worse results (entry 3). In contrast, when *N*-phosphinyl imine (**1b**; R = $POPh_2$) was employed in the [2 + 2 + 2] cycloaddition reaction with **2b** in the presence of $Ni(cod)_2$ and PCy_3 (10 and 20 mol%, respectively), the corresponding 1,2-dihydropyridine (**8bb**) was obtained in 64 % yield (Table 3.1, entry 5) [10].

It is logical to think that employing NHC ligands with a strong σ-donating ability would facilitate oxidative cyclization of imines, even in the absence of electron-withdrawing substituents on their nitrogen atom, by promoting back-donation from Ni(0) to imines, thus expanding the substrate scope of imines in the [2 + 2 + 2] cycloaddition reaction leading to 1,2-dihydropyridines. Indeed, Ni(0)/NHC-catalyzed [2 + 2 + 2] cycloaddition reactions of *N*-aryl imines with alkynes proceeded efficiently under relatively mild reaction conditions (40 °C) [10]. For instance, in the presence of $Ni(cod)_2$ and IMes (each of 2 mol%, respectively; IMes = 1,3-bis(2,4,6-trimethylphenyl)imidazole-2-ylidene), the reaction of *N*-benzylidene-(trifluoromethyl)aniline derivatives (**1c,d**) with 4-octyne (**2e**) afforded the corresponding 1,2-dihydropyridines (**8ce, 8de**) in 83% and 86% yield, respectively (Scheme 3.5). Although *N*-benzylidene-2-(trifluoromethyl)aniline

Table 3.1 Ni(0)/phosphine-catalyzed [2 + 2 + 2] cycloaddition reaction of *N*-sulfonyl or *N*-phosphinyl imines **1** with alkynes **2**.

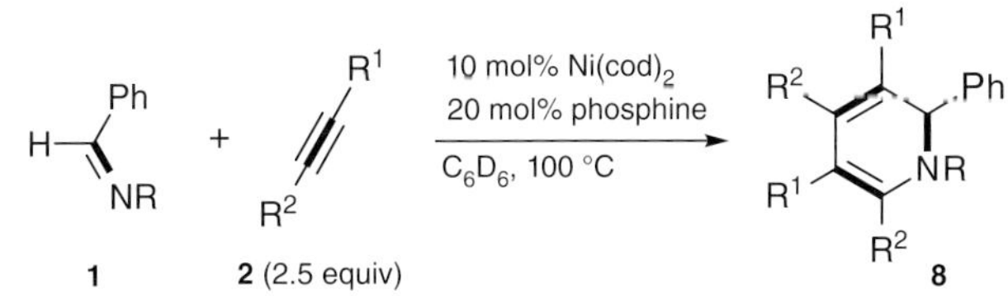

Entry	Imine 1	Alkyne 2	Phosphine	Time (h)	8 Yield(%)[a)]
1	**1a**	**2b** ($R^1 = R^2 = Me$)	P^tBu_2Me	48	**8ab** 87(50)
2		**2c** ($R^1 = R^2 = Et$)	P^tBu_2Me	70	**8ac** 64(55)
3			PCy_3	24	**8ac** 26
4		**2d** ($R^1 = SiMe_3$, $R^2 = H$)	P^tBu_2Me	18	**8ad** 58(38)
5[b)]	**1b**	**2b**	PCy_3	3	**8bb** —(64)

General conditions: **1** (0.10 mmol), **2** (0.25 mmol), $Ni(cod)_2$ (0.01 mmol), and phosphine ligand (0.02 mmol) were reacted in C_6D_6 (0.5 ml) at 100 °C.

a) Yields, determined by ^1H NMR spectroscopy, are given. The values in parentheses are of isolated yield.

b) $Ni(cod)_2$ (0.04 mmol), PCy_3 (0.08 mmol), **1b** (0.40 mmol), and **2b** (1.20 mmol) were employed in toluene (1.0 ml).

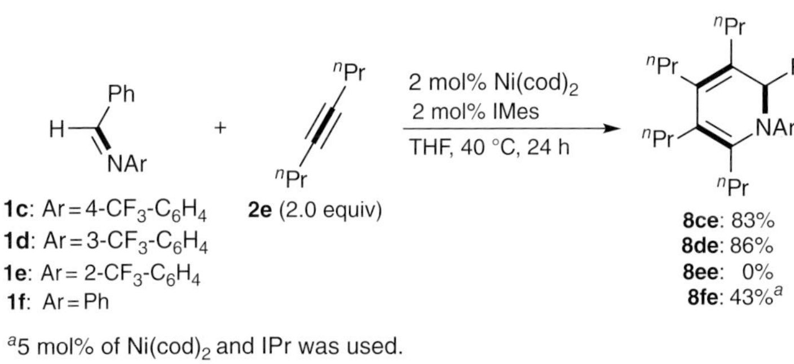

Scheme 3.5 Ni(0)/NHC-catalyzed [2 + 2 + 2] cycloaddition reaction of *N*-aryl imines **1c–f** and 4-octyne **2e**.

(**1e**) did not participate in the cycloaddition reaction under the same reaction conditions, this Ni(0)/NHC system allowed us to employ a simple *N*-phenyl imine (**1f**), thus affording **8fe** in 43 % isolated yield.

Yoshikai and coworkers also reported a related nickel-catalyzed [2 + 2 + 2] cycloaddition of two alkynes and an aldimine bearing a 3-methyl-2-pyridyl directing group on the nitrogen atom to furnish 1,2-dihydropyridines (Scheme 3.6) [11]. Their DFT calculation of the model compound revealed that intramolecular coordination of the pyridyl ring to the nickel center stabilized the five-membered aza-nickelacycle intermediate regardless of the steric strain caused by the four-membered chelation structure. A series of Ni(0)-catalyzed

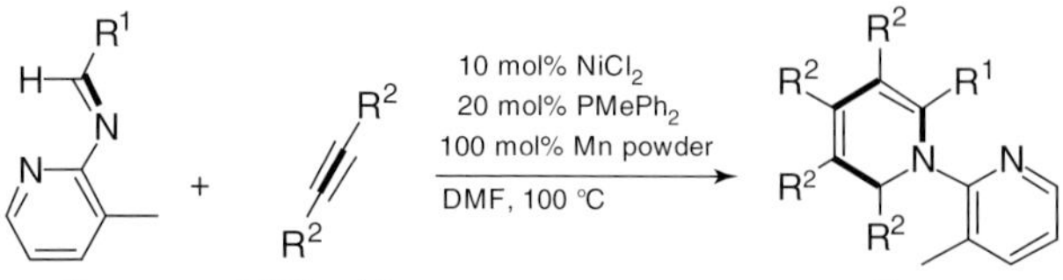

Scheme 3.6 Ni(0)-catalyzed [2 + 2 + 2] cycloaddition reaction of *N*-pyridyl imines **1g** and alkynes **2**.

[2 + 2 + 2] cycloaddition of imines with alkynes proceeded as follows [2c, 10, 11]: (i) oxidative cyclization of an imine and an alkyne with nickel(0), generating a five-membered aza-nickelacycle intermediate; (ii) insertion of a second alkyne, forming a seven-membered aza-nickelacycle intermediate; and (iii) reductive elimination from the seven-membered aza-nickelacycle, producing a 1,2-dihydropyridine with the concomitant regeneration of nickel(0).

3.4 Three-Component Coupling and Cyclocondensation Reactions of an Imine, an Alkyne, and Alkylmetal Reagents

The five-membered aza-nickelacycle generated via the oxidative cyclization of an imine and an alkyne is a key intermediate in the nickel-catalyzed three-component coupling reaction of an imine, an alkyne, and organometallic reagents. Jamison and coworker reported the first nickel-catalyzed three-component coupling reaction of an imine, an alkyne, and triethylborane, and the five-membered aza-nickelacycles were proposed as a key reaction intermediate in this catalytic reaction (Scheme 3.7) [12]. Ogoshi and coworkers have then demonstrated that treatment of aza-nickelacycle **3aa** with dimethylzinc in toluene at room temperature furnishes the expected methylzincamido (**11aa**) in 74% isolated yield (Scheme 3.8a) [13]. This stoichiometric reaction was successfully expanded to a nickel-catalyzed three-component coupling reaction of **1a**, **2a**, and dimethylzinc, thus leading to **11aa** in moderate yield (Scheme 3.8b) [13]. It should be noted that **3aa** does not react with triethylborane even when heated at 60 °C, which is consistent with the fact that *N*-tosyl imines cannot participate in the nickel-catalyzed three-component coupling of an alkyne, an imine, and triethylborane [12]. Zhou and coworkers reported the related

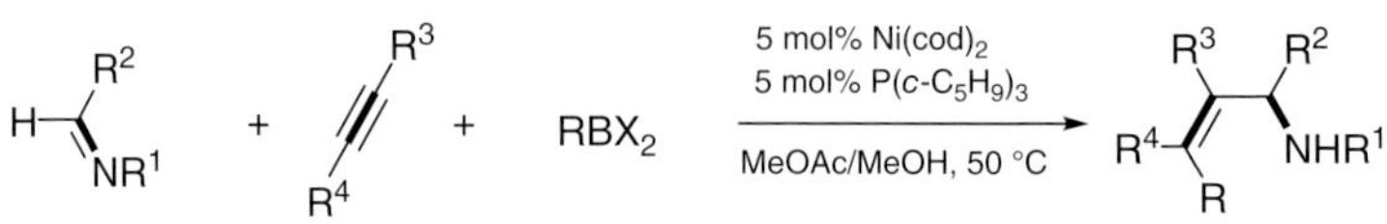

Scheme 3.7 Ni(0)-catalyzed three-component coupling reaction of alkynes, imines, and organoboron reagents.

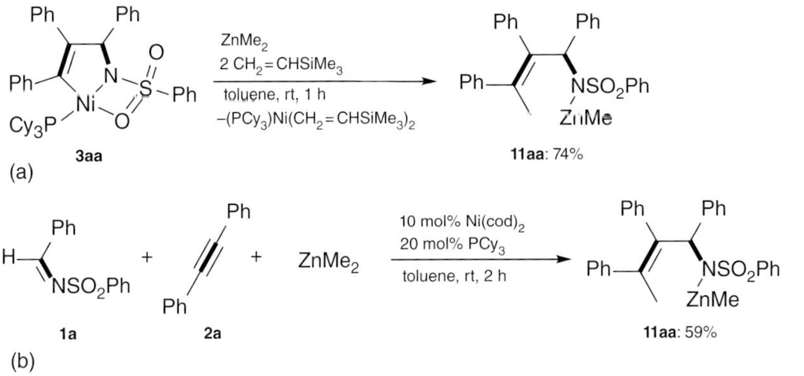

Scheme 3.8 (a) Reaction of five-membered aza-nickelacycle **3aa** with dimethylzinc. (b) Ni(0)-catalyzed three-component coupling reaction of **1a**, **2a**, and dimethylzinc.

nickel-catalyzed reductive coupling of imines and alkynes with $ZnEt_2$ as a reductant, providing allylic amines (Scheme 3.9) [14].

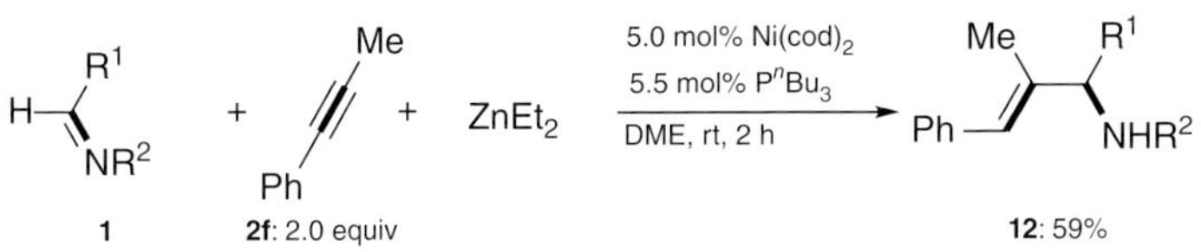

Scheme 3.9 Ni(0)-catalyzed reductive coupling reaction of imines, 1-phenyl-1-propyne (**2f**), and diethylzinc.

On the other hand, Ogoshi and coworkers found that treating **3aa** with trimethylaluminum in place of dimethylzinc resulted in the unexpected formation of a five-membered aza-aluminacycle (**13aa**) in 69% yield (Scheme 3.10) [13]. Unlike the three-component coupling product **11aa**, the aza-aluminacycle **13aa** is an organometallic reagent, in which the Al—C bond can react with a variety of electrophiles. This stoichiometric reaction was also successfully applied to a nickel-catalyzed cyclocondensation of *N*-sulfonyl imine **1a**, alkynes **2**, and trimethylaluminum to yield a variety of aza-aluminacycle (Table 3.2) [13]. The key to the success of this cyclocondensation reaction was the slow addition of $AlMe_3$. Without this slow addition, the yield of **13** was significantly

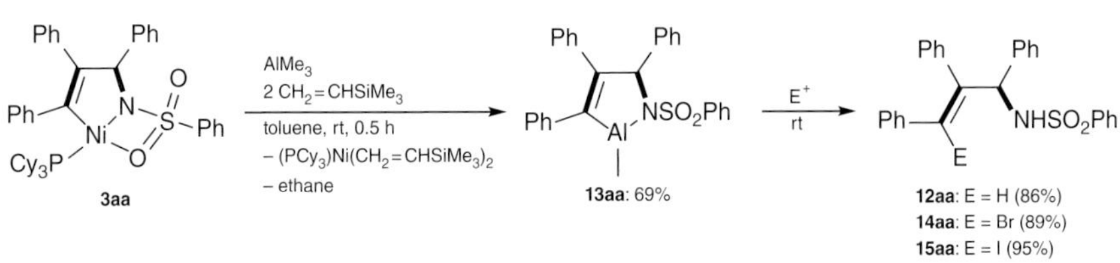

Scheme 3.10 Reaction of five-membered aza-nickelacycle **3aa** with trimethylaluminum. Reaction conditions: for **12aa**, MeOH/toluene, instant; for **14aa**, NBS (2 equiv) in CH_3CN, 0.5 hour; for **15aa**, I_2 (excess) in CH_3CN, 2 hour, then HCl.

Table 3.2 Ni(0)/PCy_3-catalyzed three-component cyclocondensation of *N*-sulfonyl imine **1a**, alkynes **2**, and trimethylaluminum.

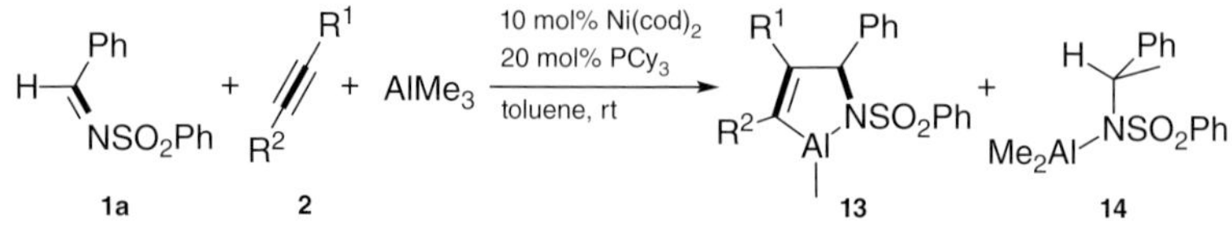

Entry	Alkyne 2	Time (h)	Product yield (%)[a]	Yield of 14 (%)[b]
1	**2a** ($R^1 = R^2 = Ph$)	1.0	**13aa** 86(71)	—
2[c]	**2f** ($R^1 = Me$, $R^2 = Ph$)	3.0	**13af**[d] 65(44)	13
3	**2g** ($R^1 = R^2 = p$-Me-C_6H_4)	6.0	**13ag** 85(82)	—
4	**2h** ($R^1 = R^2 = p$-CF_3-C_6H_4)	3.0	**13ah** 90(99)	—
5[e]	**2i** ($R^1 = SiMe_3$, $R^2 = Ph$)	3.0	**13ai** 85(73)	7
6[f]	**2c** ($R^1 = R^2 = Et$)	2.5	**13aa** 27(—)	59

General conditions: **1a**, **2** (0.30 mmol each), and $Ni(cod)_2/PCy_3$ (0.03 mmol) were reacted in toluene (10.0 ml) at room temperature. After dropwise addition of $AlMe_3$, the reaction mixture was stirred until the color derived from aza-nickelacycle **3** (typically purple) disappeared.

a) Isolated yield as allylamines **12** after protolysis. The values in parentheses are of isolated **13**.
b) Cited yields, determined by ^1H NMR, were of the corresponding protonated products.
c) The reaction was conducted with concomitant addition of $AlMe_3$ and **2f**.
d) The minor regioisomer (11%) was also obtained.
e) The reaction was conducted using 1.5 mmol of **2i**.
f) Formation of a 1,2-dihydropyridine derivative was observed.

decreased as a result of the direct reaction of **1a** with $AlMe_3$ to afford the side reaction product **14**. This is the first example of the catalytic formation of aza-aluminacyclopentenes, although cycloalumination of either olefins or acetylenes mediated by Cp_2Zr derivatives has been used in the preparation of organoaluminum compounds [15]. It should be noted that Montgomery and coworker developed a related nickel-catalyzed cyclocondensation reaction of aldehydes, alkynes, and dialkylsilanes to furnish oxasilacyclopentenes (Scheme 3.11) [16].

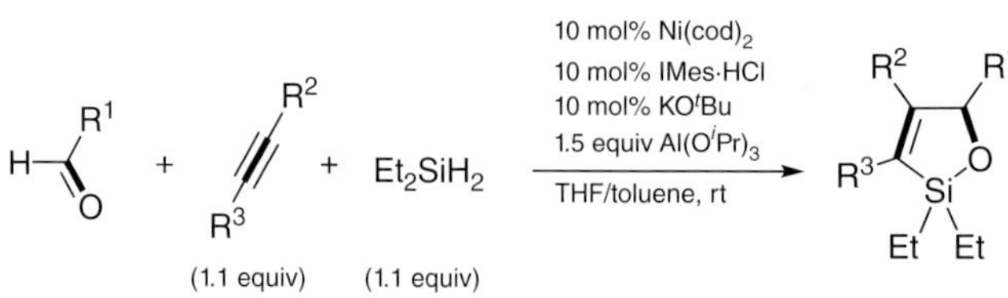

Scheme 3.11 Ni(0)-catalyzed cyclocondensation reaction of aldehydes, alkynes, and Et_2SiH_2.

Although nickel-catalyzed C—C bond forming reactions between alkenes and imines remain challenging given that the simultaneous coordination of alkenes and imines, followed by oxidative cyclization, is difficult, Cheng and coworkers developed a nickel-catalyzed reductive coupling reaction of a conjugated alkene and an aldimine using zinc powder as the reducing agent and water as the proton source (Scheme 3.12) [17]. Chuang and coworker applied this alkene/imine reductive coupling reaction to [60]fullerene [18]. In addition, Ogoshi and coworkers have recently developed a nickel-catalyzed reductive coupling reaction of tetrafluoroethylene, *N*-sulfonyl-substituted imines, and diethylsilane (Scheme 3.13) [19]. The *N*-sulfonyl group on the imines is crucial for this catalytic reaction by enhancing the coordination ability of the imine to Ni(0) and stabilizing the resulting aza-nickelacycle intermediates.

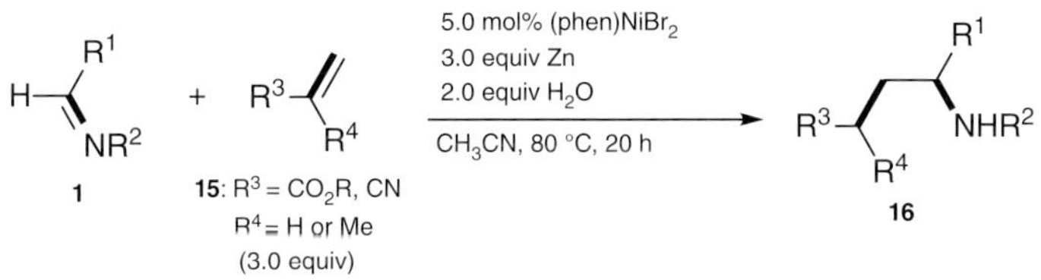

Scheme 3.12 Ni(0)-catalyzed reductive coupling reaction of imines and conjugated alkenes.

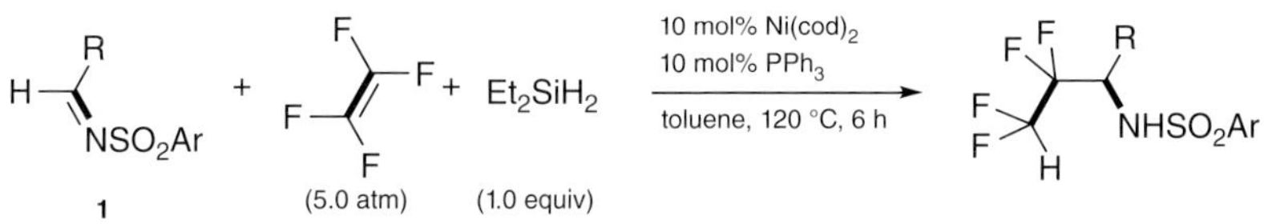

Scheme 3.13 Ni(0)-catalyzed cyclocondensation reaction of aldehydes, alkynes, and Et_2SiH_2.

References

1 For reviews on the nickel-catalyzed reactions via (hetero)nickelacycle intermediate, see: (a) Tamaru, Y. (ed.) (2005). *Modern Organo Nickel Chemistry.* Weinheim, Germany: Willey-VCH. (b) Ikeda, S.-I. (2003). *Angew. Chem. Int. Ed.* 42: 5120. (c) Montgomery, J. (2004). *Angew. Chem. Int. Ed.* 43: 3890. (d) Moslin, R.M., Miller-Moslin, K., and Jamison, T.F. (2007). *Chem. Commun.*: 4441. (e) Ng, S.-S., Ho, C.-Y., Schleicher, K.D., and Jamison, T.F. (2008). *Pure Appl. Chem.* 80: 929. (f) Tanaka, K. and Tajima, Y. (2012). *Eur. J. Org. Chem.* 2012: 3715. https://onlinelibrary.wiley.com/doi/full/10.1002/ejoc.201200098. (g) Montgomery, J. (2013). Organonickel chemistry. In: *Organometallics in Synthesis: Fourth Manual* (ed. B.H. Lipshutz), 319–428. Hoboken, NJ: Wiley. (h) Ogoshi, S. (2013). *Yuki Gosei Kagaku Kyokaishi* 71: 14. (i) Tasker, S.Z., Standley, E.A., and Jamison, T.F. (2014). *Nature* 509: 299. (j) Hoshimoto, Y., Ohashi, M., and Ogoshi, S. (2015). *Acc. Chem. Res.* 48: 1746.

2 For examples, see: (a) Ogoshi, S., Oka, M., and Kurosawa, H. (2004). *J. Am. Chem. Soc.* 126: 11802. (b) Ogoshi, S., Ueta, M., Arai, T., and Kurosawa, H. (2005). *J. Am. Chem. Soc.* 127: 12810. (c) Ogoshi, S., Ikeda, H., and Kurosawa, H. (2007). *Angew. Chem. Int. Ed.* 46: 4930.

3 Seminal studies on isolation of nickelalactams generated from an alkyne, an isocyanate, and nickel(0), see: (a) Hoberg, H. and Oster, B.W. (1982). *J. Organomet. Chem.* 234: C35. (b) Hoberg, H. and Oster, B.W. (1983). *J. Organomet. Chem.* 252: 359. a seminal study on nickel-catalyzed reaction of two alkyne and a carbodiimide to give a 2-imino-1,2-dihydropyridine, see: (c) Hoberg, H. and Burkhart, G. (1979). *Synthesis* 1979: 525.

4 For reviews on preparation of γ-lactams via hetero-Pauson–Khand (or aza-Pauson–Khand) reactions, see: (a) Torres, R.R. (2012). *The Pauson–Khand Reaction: Scope, Variations and Applications.* UK: Wiley. (b) Omae, I. (2011). *Coord. Chem. Rev.* 255: 139. (c) Park, J.H., Chang, K.-M., and Chung, Y.K. (2009). *Coord. Chem. Rev.* 253: 2461. (d) Chatani, N. (2008). *Chem. Rec.* 8: 201.

5 For selected recent reports on the synthesis of lactams via carbonylative cycloaddition reaction, see: (a) Wu, X.-F. and Neumann, H. (2012). *ChemCatChem* 4: 447. (b) Iwata, T., Inagaki, F., and Mukai, C. (2013). *Angew. Chem. Int. Ed.* 52: 11138. and references therein. (c) Fukuyama, T., Nakashima, N., Okada, T., and Ryu, I. (2013). *J. Am. Chem. Soc.* 135: 1006. and references therein.

6 Fujihara, T., Hosoki, T., Katafuchi, Y. et al. (2012). *Chem. Commun.* 48: 8012.

7 (a) Ueda, T., Konishi, H., and Manabe, K. (2012). *Org. Lett.* 14: 3100. (b) Ueda, T., Konishi, H., and Manabe, K. (2013). *Org. Lett.* 15: 5370. (c) Ueda, T., Konishi, H., and Manabe, K. (2013). *Angew. Chem. Int. Ed.* 52: 8611.

8 Hoshimoto, Y., Ohata, T., Sasaoka, Y. et al. (2014). *J. Am. Chem. Soc.* 136: 15877.

9 Hoshimoto, Y., Ashida, K., Sasaoka, Y. et al. (2017). *Angew. Chem. Int. Ed.* 56: 8206.

10 Hoshimoto, Y., Ohata, T., Ohashi, M., and Ogoshi, S. (2014). *Chem. Eur. J.* 20: 4105.

11 Adak, L., Chan, W.C., and Yoshikai, N. (2011). *Chem. Asian J.* 6: 359.

12 (a) Patel, S.J. and Jamison, T.F. (2003). *Angew. Chem. Int. Ed.* 42: 1364. (b) Patel, S.J. and Jamison, T.F. (2004). *Angew. Chem. Int. Ed.* 43: 3941.

13 Ohashi, M., Kishizaki, O., Ikeda, H., and Ogoshi, S. (2009). *J. Am. Chem. Soc.* 131: 9160.

14 Zhou, C.-Y., Zhu, S.-F., Wang, L.-X., and Zhou, Q.-L. (2010). *J. Am. Chem. Soc.* 132: 10955.

15 Selected reviews, see: (a) Dzhemilev, U.M. and Ibragimov, A.G. (2000). *Russ. Chem. Rev.* 69: 121. (b) Dzhemilev, U.M. and Ibragimov, A.G. (1994). *J. Organomet. Chem.* 466: 1.

16 Baxter, R.D. and Montgomery, J. (2008). *J. Am. Chem. Soc.* 130: 9662.

17 Yeh, C.-H., Korivi, R.P., and Cheng, C., H. (2008). *Angew. Chem. Int. Ed.* 47: 4892.

18 Rajeshkumar, V. and Chuang, S.-C. (2012). *Eur. J. Org. Chem.* 2012: 3795.

19 Shirataki, H., Ono, T., Ohashi, M., and Ogoshi, S. (2019). *Org. Lett.* 21: 851.

4

Asymmetric C—C Bond Formation Reactions via Nickelacycles

Ravindra Kumar[1] *and Sensuke Ogoshi*[2]

[1] *CSIR-Central Drug Research Institute (CDRI), Medicinal and Process Chemistry Division, BS-10/1, Sector-10, Jankipuram Extension, Lucknow 226031, India*
[2] *Osaka University, Department of Applied Chemistry, Suita, Osaka 565-0871, Japan*

4.1 Introduction

Nickelacycles, generated by oxidative cyclization of two unsaturated compounds of diverse structural scaffolds on nickel(0), are key intermediates in many molecular transformations through C—C and C—X bond(s) formation. Despite a wide spectrum of organic transformations derived from nickelacycles, asymmetric variants of these transformations, which enable one-step synthesis of chiral molecular scaffolds with high selectivity, are remarkably less explored. Chiral ligands are the key controlling element in asymmetric catalysis, and enantioselection can be tuned by adjusting the ligand (either phosphine or *N*-heterocyclic carbene [NHC]) sphere around the nickel center. In such reactions, enantioselectivity might emerge from chiral ligands by face-selective coordination of coupling partners to the nickel-chiral ligand complex.

This chapter describes the development of catalytic asymmetric transformations via nickelacycles, generated by oxidative cyclization of various π-components of diverse structural patterns such as alkynes, alkenes, carbonyls, and imines. Overall asymmetric molecular transformations involve either simple coupling of two π-components or multiple components in the presence of other π-components or organometallic reagents. It also includes some examples of reactive σ-components such as cyclopropyl substrates as one of the coupling partners.

4.2 Enantioselective Reactions Involving Nickelacycles

4.2.1 Nickel-Catalyzed Asymmetric Coupling of Alkynes and Aldehydes

In the arena of nickel catalysis via nickelacycle, organic transformations involving oxidative cyclization of an alkyne and an aldehyde are most studied in racemic

Nickel Catalysis in Organic Synthesis: Methods and Reactions,
First Edition. Edited by Sensuke Ogoshi.

as well as their asymmetric variants. Two types of couplings are reported involving these two substrates: (i) in the presence of reducing agents such as triethylborane or silanes, called reductive coupling and (ii) in the presence of some organometallic alkyl coupling partners, such as dialkylzinc reagents, called alkylative coupling. These reactions constitute one C—C σ-bond at the nickelacycle formation step and C—H or C—C bond with a coupling partner, either from reducing or alkylating agents, respectively.

4.2.1.1 Nickel-Catalyzed Asymmetric Reductive Coupling of Alkynes and Aldehydes

The first enantioselective method for catalytic reductive coupling of alkynes (**1**) and aldehydes (**2**) was reported by Jamison and coworkers in 2003 employing triethylborane (Et_3B) as a reducing agent (Scheme 4.1) [1]. The chiral monophosphine ligand, (+)-(neomenthyl)diphenylphosphine (NMDPP; **4**), was an optimal ligand for this transformation to give chiral allylic alcohols (**3**) in good yield, regioselectivity (>95 : 5), and enantioselectivity (42–96% ee). Unlike the aliphatic aldehyde, the aromatic benzaldehyde gave relatively inferior regioselectivity and enantioselectivity (91 : 9 rr and 73% ee). The scope of this asymmetric reaction is limited to alkyne substrates containing at least one aromatic substituent whereas dialkyl

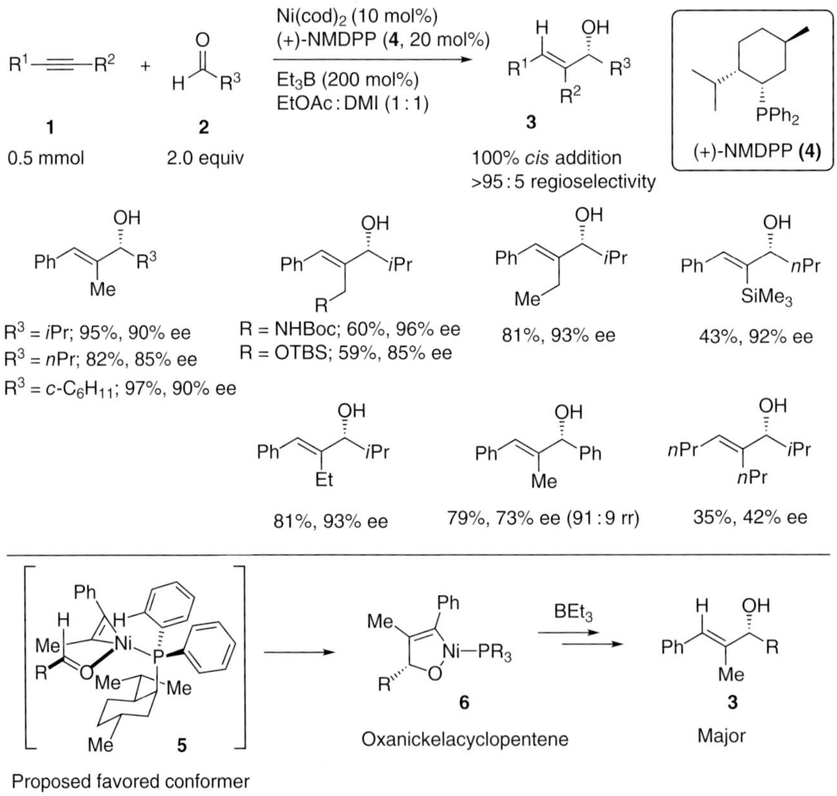

Scheme 4.1 Catalytic enantioselective reductive coupling of alkynes and aldehydes.

alkynes resulted in low yield (ethylation was the side reaction derived from direct coupling with BEt_3) and enantioselectivity (35% yield and 42% ee for 4-octyne).

The high selectivity was hypothesized to originate from the prerequisite arrangement of components **1** and **2** in a highly ordered transition state explained by a cooperative effect between steric properties of the ligand and electronic differences of the alkyne substituents. Enantioinduction (for major enantiomer) might emerge from the coordination of the aldehyde to the less encumbered side, keeping the aldehydic "R" group away from the nickel center to minimize steric interactions as in model conformer **5**. Formation of the favored oxanickelacycle (**6**) followed by reaction with BEt_3 gives the chiral allylic alcohol **3** (Scheme 4.1).

Later in 2003, the same group developed *P*-chiral monodentate ferrocenyl phosphines (**7** and its analogues) that can be applicable to dialkyl-alkynes, and afforded moderate regioselectivity and enantioselectivity [2]. Chiral allylic alcohols are useful starting materials and are prevalent in complex natural products; the strategy was employed to readily access this motif and rapidly assemble (−)-terpestacin (**11**) (Scheme 4.2) [3]. The enantioselective reductive coupling of dialkyl-alkyne (**8**) and aliphatic aldehyde (**9**) in the presence of *P*-chiral ferrocenyl phosphine ligand (**7**) favored the desired diastereomer (2 : 1), and regioisomers (2.6 : 1) of **10** served as a required starting material to accomplish the total synthesis in 10 steps.

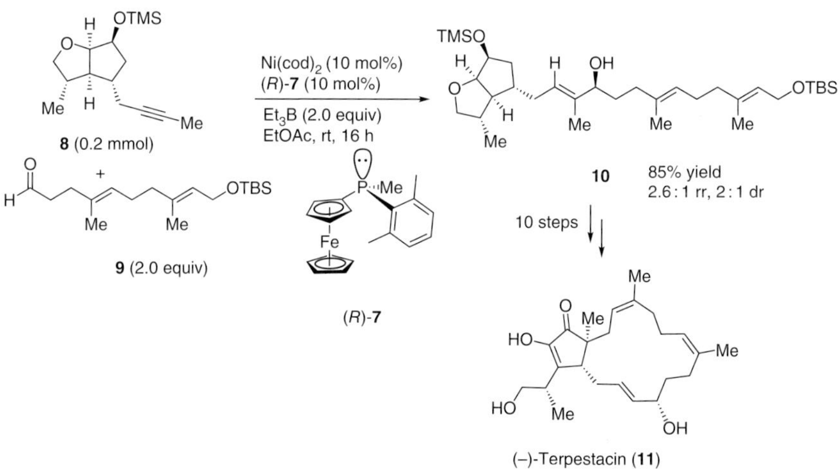

Scheme 4.2 Total synthesis of (−)-terpestacin via reductive coupling of alkynes and aldehydes.

Montgomery and coworkers explored various chiral NHC ligands to address the major challenge inherent in regioselectivity with the use of electronically biased coupling components, such as aryl–alkyl alkynes. The advantage of using NHC over phosphine ligands is the ease of fine tuning of substituents, which plays a key controlling role in altering the regio- and stereoselectivities. In 2007, this group has reported the first use of chiral NHCs in nickel-catalyzed reductive coupling reactions in the presence of silanes as a reducing agent (Scheme 4.3) [4]. The important features of this method include the broad scope of aldehydes (**12**, both aliphatic and aromatic) and alkynes (**13**, both internal and terminal alkynes), and the direct access to silyl group protected allylic alcohols **14**. Among various C_2-symmetric NHCs, *ortho, ortho* (at 2 and 6 positions)-disubstituted *N*-aryl was optimal from the standpoint of chemical yield, whereas steric differentiation of the two *ortho* substituents was optimal from the standpoint of enantioselectivity. The best result in terms of enantioselectivities (ranging from 65% to 85% ee) was reported with NHC **15**.

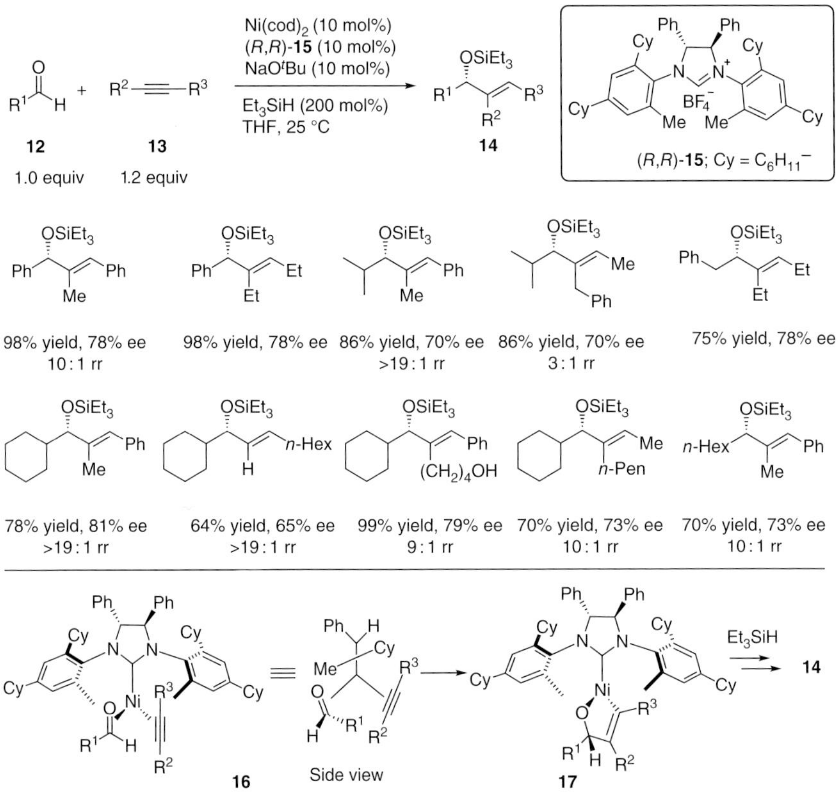

Scheme 4.3 Ni^0/NHC-catalyzed asymmetric reductive coupling of alkynes and aldehydes.

The reaction was hypothesized to proceed via generation of a three coordinate Ni^0/**15** complex and tilting of the *N*-aryl ring of NHC in such a way that bulky ortho substituents (cyclohexyl) would position anti to the imidazole backbone phenyl group and distal to nickel, as depicted in orientation model **16** (Scheme 4.3). The ortho methyl substituent syn to the backbone phenyl group and proximal to nickel thus dictates the selectivity of aldehyde binding. This model **16** leads to oxanickelacycle **17**, which on reaction with Et_3SiH gives silyl protected allylic alcohol **14** as an observed major enantiomer.

An intramolecular version of this reaction was also reported employing the same NHC **15**, giving 13- and 14-membered macrocycles (**19** and **20**) in 76% total yield with 86 : 14 regioselectivity from alkynyl-aldehyde substrate **18** (Scheme 4.4) [4]. The present methodology is important as it gives rapid entry to macrocyclic substructures found in many bioactive natural products.

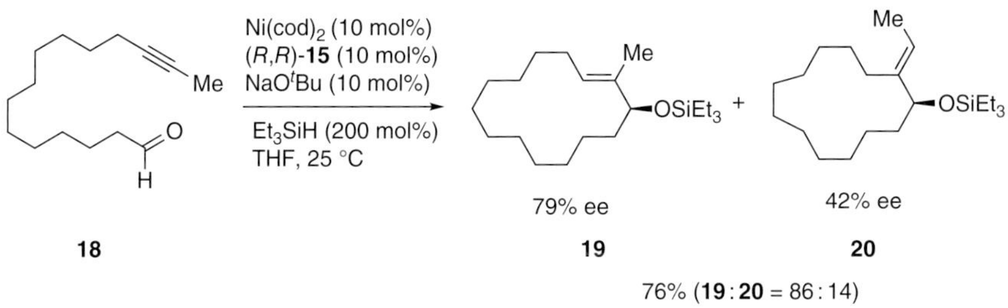

Scheme 4.4 Intramolecular asymmetric reductive coupling of alkynyl-aldehyde.

4.2.1.2 Nickel-Catalyzed Asymmetric Alkylative Coupling of Alkynes and Aldehydes

Besides reductive coupling reactions, nickel-catalyzed alkylative couplings have also been reported in the presence of dialkylzinc reagents. Such reactions are more atom-efficient processes compared to their reductive counterparts and commonly produce allylic alcohols with a tetrasubstituted olefin. The asymmetric reaction was first reported by Zhou and coworkers in the presence of Me_2Zn as a methylating agent and 6,6′-disubstituted spiro phosphoramidite as a chiral ligand (Scheme 4.5) [5]. Thus, a variety of chiral allylic alcohols **23** were

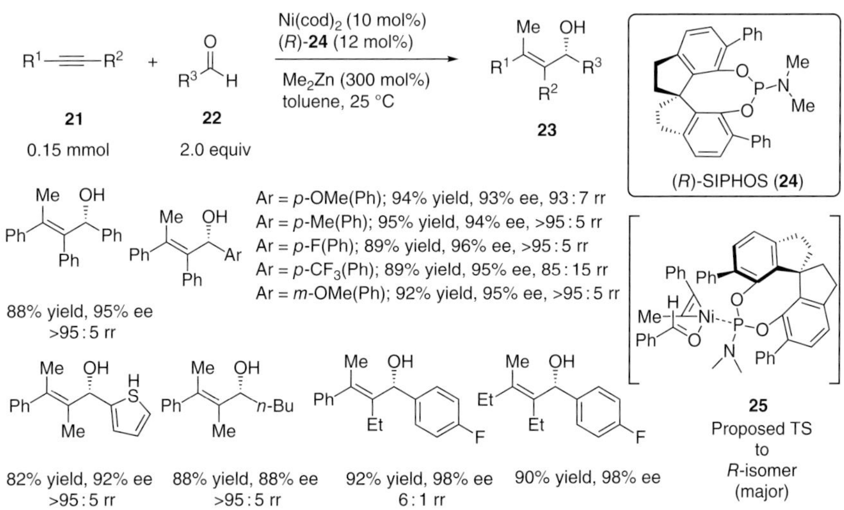

Scheme 4.5 Catalytic enantioselective alkylative coupling of alkynes and aldehydes.

obtained from a broad substrate scope (aliphatic and aromatic internal alkynes **21** and aldehydes **22**) in high yields (77–95%), regioselectivity (up to >99 : 5), and enantioselectivity (86–99% ee) using 10 mol% of $Ni(cod)_2$ and 12 mol% (*S*)-SIPHOS (**24**). This enantioselective alkylative coupling is mainly limited to Me_2Zn as another alkylating agent, Et_2Zn, gives a mixture of reductive coupling (33%, 57% ee) and alkylative coupling (50%, 71% ee) products, whereas Ph_2Zn fetches no reaction under these reaction conditions.

A model transition state **25** has been proposed to rationalize the formation of the observed absolute configuration of the reaction product. The coordination of aldehyde is favored in such a way that the phenyl group of the aldehyde is oriented away from the phenyl group on the 6,6′-position of the ligand to eliminate steric congestion, and thus the nickel-activated alkyne approaches the aldehyde from its "*Si*" face to give (*R*)-allylic alcohol.

4.2.2 Nickel-Catalyzed Asymmetric Coupling of Alkynes and Imines

Like aldehydes, imines have been investigated as a coupling partner in alkylative coupling with alkynes to afford allylic amines with tetrasubstituted olefins. In 2004, Jamison and coworkers reported the first nickel-catalyzed enantioselective synthesis of allylic amines **28** from alkynes **26**, protected imines **27**, and organoboranes (Et_3B) (Scheme 4.6) [6]. *N*-(*tert*-Butyldimethylsilyloxy)ethyl

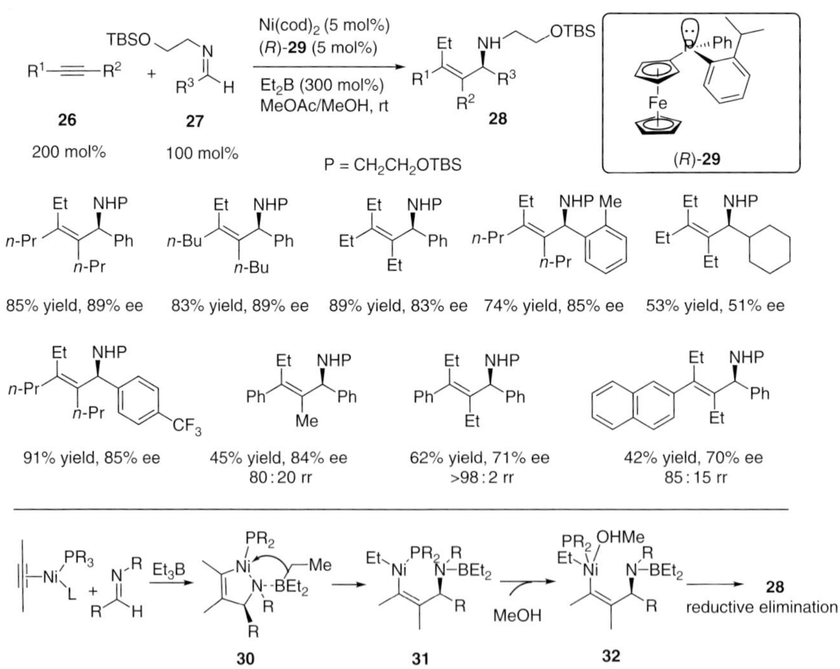

Scheme 4.6 Catalytic enantioselective alkylative coupling of alkyne and Imines.

($TBSOCH_2CH_2$) protected imine **27** was employed to maximize its reactivity and selectivity. Under optimized conditions, using hydroxylic solvent, methanol in particular gave high selectivity for the alkylative coupling product over the reductive coupling product. This reaction was promoted by the nickel-phosphine complex of chiral monodentate *P*-chiral ferrocenyl phosphanes, and high chemical yield (up to 95%) and enantioselectivity (up to 89% ee) were reported for tetrasubstituted allylic amines **28** with 5 mol% of $Ni(cod)_2$ and 2-isopropylphenyl substituted *P*-chiral ferrocenyl phosphane **29**. The highest yield and enantioselectivity have been observed with specific experimental conditions by portion-wise addition of the alkyne and Et_3B. A range of alkynes (includes aromatic and aliphatic) and aldimines were investigated and unsymmetrical alkynes showed good to excellent regioselectivity (up to >98 : 2). Removal of the (*tert*-butyldimethylsilyloxy) ethyl protecting group with TBAF provided direct access to versatile primary allylic amines, demonstrating the practical utility of this strategy.

A plausible reaction mechanism can be proposed via aza-nickelacycle **30**. Ethyl group transfer from B to the Ni center followed by reductive elimination, which is promoted by methanol, gives allylic amines **28**. In the absence of methanol, reductive coupling product formation is dominated via β-hydride elimination from **31**. Since both products give the same enantioselectivity, enantioinduction might arise at the oxidative cyclization step [7].

4.2.3 Nickel-Catalyzed Asymmetric Coupling of 1,3-Enynes and Aldehydes

Instead of simple alkynes, conjugated alkenyl-alkynes (1,3-enynes) have also been explored in coupling reactions with aldehydes and ketones. The pendant alkene in alkyne enhances both the reactivity and selectivity. In 2005, Jamison and coworkers reported the first enantioselective catalytic reductive coupling reactions of 1,3-enynes with aldehyde in the presence of triethylborane (Scheme 4.7) [8]. This asymmetric reaction is promoted by a nickel complex derived from 1 equiv of $Ni(cod)_2$ and 2 equiv of monodentate *P*-chiral ferrocenyl phosphanes. The same chiral ligand **29** was optimal for this reaction; however, a 10 mol% catalyst (prepared from $Ni(cod)_2$ and **29** in 1 : 2 ratio) provided excellent regioselectivity (>95 : 5) and modest enantioselectivity with aromatic aldehydes. Electronically diverse and functionalized aromatic aldehydes **34** and 1,3-enynes **33** were investigated to produce enantioenriched diene-ols **35** of diverse substitution pattern. Particularly, 1-(trimethylsilyl)-substituted enyne provided alkenylsilanes; however, in poor enantioselectivity (28% ee). This method provides a straightforward tool to access chiral 2,4-dienes, a synthetically useful scaffold employed in diastereoselective reactions such as Diels–Alder reactions. Moreover, selective reduction of diene-1-ol (less substituted olefin) gives alkyl-substituted allylic alcohols (regioselectivity >95 : 5), which is otherwise

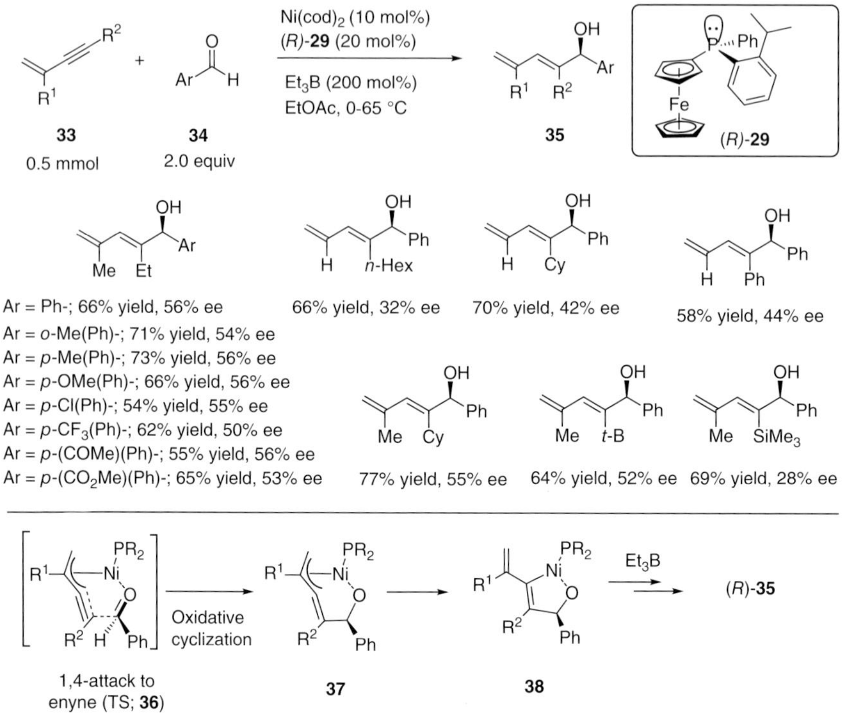

Scheme 4.7 Catalytic enantioselective coupling of 1,3-enynes and aromatic aldehydes.

difficult to achieve by direct coupling with dialkyl-alkynes under previous conditions [9].

The plausible reaction mechanism was explained by a slightly different oxidative cyclization transition state involving a formal 1,4-attack and a transient η^3-allylnickel-chiral phosphine ligand complex (**36**), which on oxidative cyclization gives an enantioenriched η^3-butadienyl intermediate **37** or **38**. The following reaction with BEt_3 gives the chiral diene-ols **35**.

4.2.4 Nickel-Catalyzed Asymmetric Coupling of 1,3-Enynes and Ketones

Unlike aldehydes, ketones have rarely been explored in any catalytic multicomponent coupling reactions due to their less reactivity, and Jamison's research group has further investigated with ketone variant. It gave enantioenriched allylic alcohols with formidable challenging tertiary alcohols. However, in this case, *P*-ortho-tolyl containing chiral ferrocenyl phosphine (*S*-**42**) was an optimized ligand to give products in moderate yield and enantioselectivity (Scheme 4.8) [10]. The electronically diverse aromatic and heteroaromatic ketone, however, limited to methyl ketones **40**, is compatible to give the corresponding allylic tertiary alcohols **41** in good yields and excellent regioselectivity (>95 : 5) with moderate enantioselectivity (up to 70% ee). The α,β-unsaturated ketone, 1-acetyl-1-cyclohexene, is also applicable in this methodology.

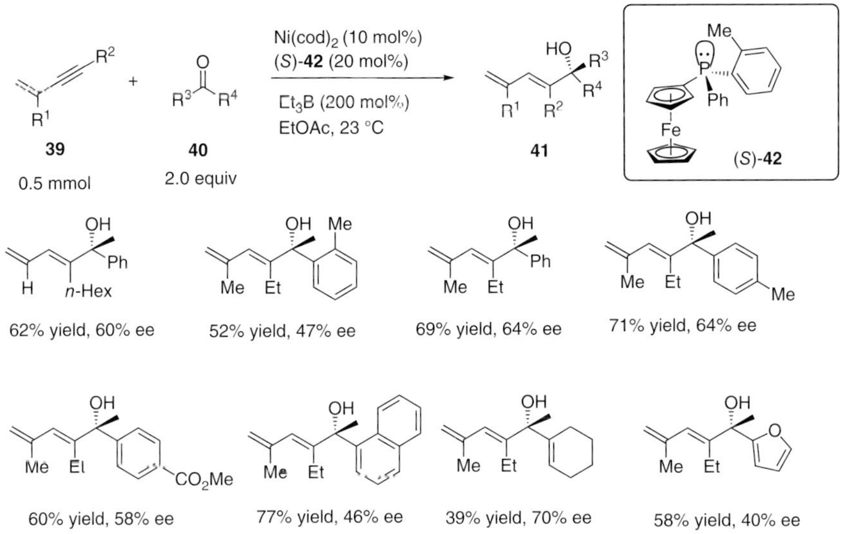

Scheme 4.8 Catalytic enantioselective coupling of 1,3-enynes and ketones.

4.2.5 Nickel-Catalyzed Asymmetric Coupling of 1,3-Dienes and Aldehydes

Unlike alkynes, alkenes have poor coordination ability to nickel. However, conjugated alkenes (1,3-dienes) can be employed due to π-allyl stabilization. In 2000, Mori and coworkers reported the nickel(0)-catalyzed asymmetric cyclization of 1,3-diene and a tethered aldehyde in the presence of silanes. Treatment of the prochiral substrate, dien-al **43**, with excess triethoxysilanes (5 equiv) in the presence of catalytic Ni(cod)$_2$ and chiral phosphine ligands gave five- or six-membered carbocycles or pyrrolidine derivatives **44** (Scheme 4.9)

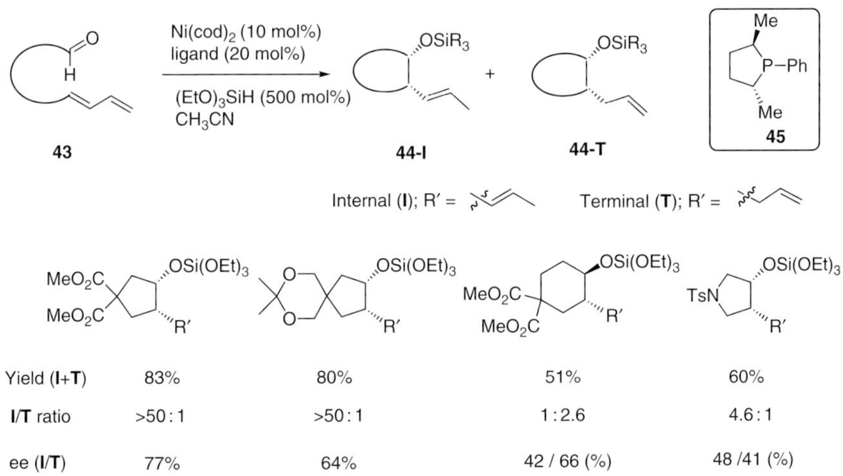

Scheme 4.9 Catalytic enantioselective cyclization of ω-formyl-1,3-dienes.

[11]. The best results are obtained with chiral phosphorane **45,** which gives cyclic compounds with up to 86% ee. Under the conditions, a mixture of regioisomers (internal [**44-I**] and terminal [**44-T**] olefins in the side chain) were obtained in moderate to good yields. It was reported that the ratio of regioisomers (internal vs. terminal olefin) and their enantiomeric excess varied depending upon the substrates and/or silanes. Regioselectivity was excellent in the case of cyclopentanes (up to **I/T** >50 : 1), whereas it was low with pyrrolidines (**I/T** 4.6 : 1). Cyclohexane products are obtained in reverse regioselectivity (**I/T** = 1 : 2.6). This method is quite important and indispensable in synthetic organic chemistry as it provides a one-step method to access cyclic compounds (carbocycles or heterocycles) in an optically active form, frequently occurring in many biologically active compounds.

The reaction course of this asymmetric cyclization of ω-formyl-1,3-dienes was explained by two possible mechanisms (Scheme 4.10). In path A the cyclization proceeded with the formation of cyclized product **44-I** via a π-allylnickel intermediate **46**, which is generated by the reaction of formyl-diene **43** and nickel hydride complex (generated in situ from $Ni(cod)_2$ and $HSiR_3$). Path B showed the cyclization process via intramolecular oxidative cyclization of formyl-diene **43** on nickel to nickelacycle **48** followed by a σ bond metathesis with silane. This bicyclic oxanickelacycle **49** produced a cyclized compound having a terminal olefin **44-T**. Formation of cyclized products **44-I** was also possible through path B via reductive elimination from silyloxy-cyclic nickel intermediate **50** via π-allylnickel intermediate. Since both the ratio and enantiomeric excess of the cyclized product varies with substrates and/or silanes, the authors speculated that both the mechanisms of path A and path B might be operational in this cyclization.

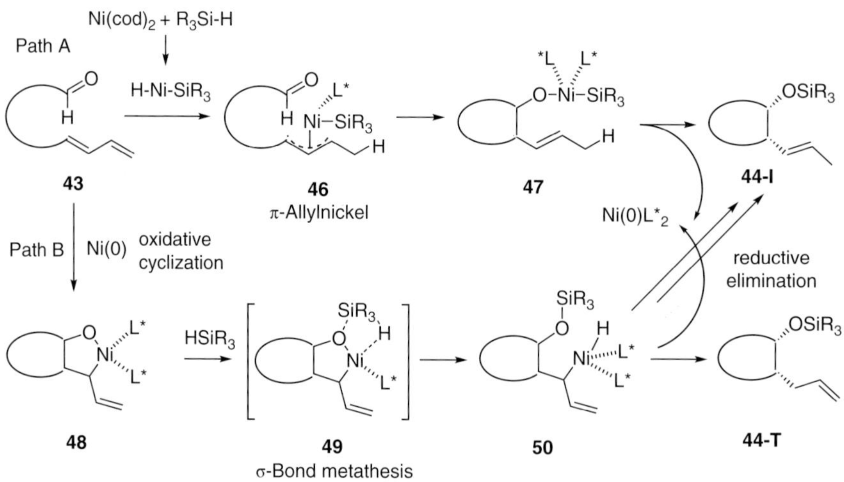

Scheme 4.10 Plausible mechanism for cyclization of ω-formyl-1,3-dienes.

Beside the formation of cyclized products via an intramolecular coupling of ω-formyl-1,3-dienes, reported by Mori et al., an intermolecular version was developed by Sato's research group in 2007 (Scheme 4.11) [12]. Various chiral NHC precursors have been examined by tuning their structures in the presence of base (Cs_2CO_3) to validate the intermolecular asymmetric coupling of 1,3-dienes **51**, aldehydes **52**, and Et_3SiH. The optimal *N*-1-(2,4,6-trimethylphenyl)propyl groups substituted NHC **54**, synthesized in this laboratory, gave various coupling products **53** in good yields (up to 91%) with high regioselectivities, diastereoselectivities (anti-selective in the case of the internal 1,3-diene), and enantioselectivities (up to 97% ee). The substrate scope included terminal as well as internal 1,3-dienes **51** and aliphatic as well as aromatic aldehydes **52**, and gave diverse structures of silyl-protected chiral homoallylic alcohols **53** with *Z*, and anti-selectivity. The diastereoselectivity in the case of internal 1,3-dienes was explained by a plausible mechanism, shown in the footnote of Scheme 4.11. The oxidative cyclization of aldehyde and one of the olefins (selectively with the olefin attached with the smaller group) of 1,3-diene forms oxa-nickelacycle **55** having all substituents in trans orientation to each other. This nickelacycle **55** is in equilibrium with the seven-membered nickelacycle **56** through insertion of pendent olefin. The σ-bond metathesis of the seven-membered oxa-nickelacycle with Et_3SiH followed by reductive elimination of nickel(0) gives the products with anti-(*Z*)-selectivity. The enantioselectivity might arise at the oxidative cyclization step.

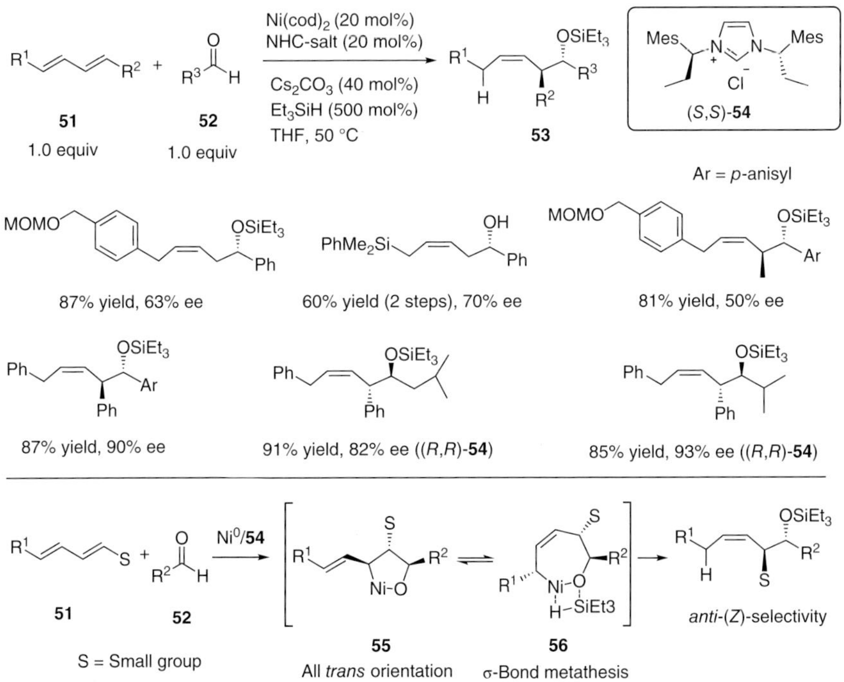

Scheme 4.11 Catalytic enantioselective intermolecular coupling of 1,3-dienes and aldehydes.

4.2.6 Nickel-Catalyzed Asymmetric Coupling of Enones and Alkynes

α,β-Unsaturated carbonyls (enons) have been employed extensively in nickel catalyzed asymmetric multicomponent reactions, either in the presence or in the absence of any organometallic coupling reagents.

4.2.6.1 Nickel-Catalyzed Asymmetric Alkylative Coupling of Enones and Alkynes

Ikeda and coworkers have reported the first enantioselective nickel-catalyzed tandem multicomponent couplings of enones, alkynes, and Me_2Zn in the presence of Lewis acid, chlorotrimethylsilane (TMSCl) (Scheme 4.12) [13]. In this study, classically used chiral bidentate ligands including bisphosphanes, bisoxazolines, phospanyloxazolines, and pyridinyloxazolines afforded poorer results both in efficiency and enantioselectivity. The optimal chiral ligand, (*S*)-4-(*tert*-butyl)-2-methyl-4,5-dihydrooxazole (**60**), afforded the coupling product enol-silyl ethers, which were converted into β-alkyenylketones **59** after treatment with aqueous acid. Yields and enantioselectivities of products varied with substrates, ranging from good to modest. The substrate scope includes five- and six-membered enones **57** and internal and terminal alkynes **58**. The present catalytic system is also applicable to crotonaldehyde.

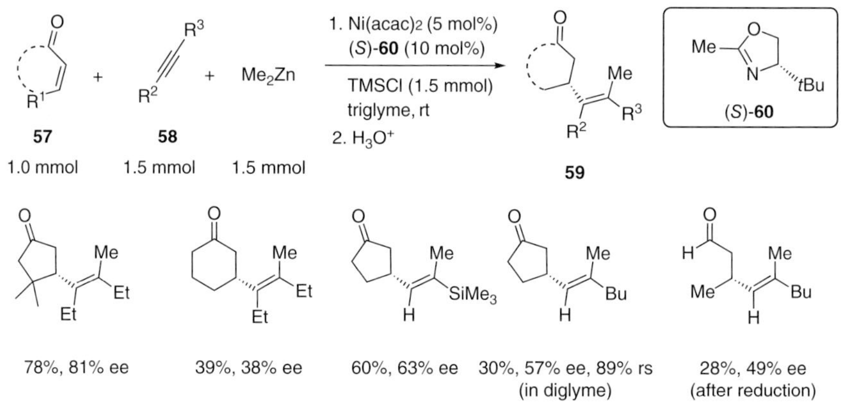

Scheme 4.12 Catalytic enantioselective coupling of alkynes, enones, and dimethylzinc.

Although the authors have not discussed any concrete mechanism, two plausible paths may be drawn as follows (Scheme 4.13) [14]. Path A was proposed via a π-allyl intermediate **61**, generated by the reaction of enones, nickel(0), and Lewis acid, TMSCl. Alkyne insertion to **61** may lead to alkenylnickel specie **65**, which undergoes transmetalation of the organozinc reagent, and finally reductive elimination affords the product **66**. However, because of lack of direct precedent for the conversion of π-allylnickel complex **62** into products and further because TMSCl is not an essential additive in most variants, nickelacycle-based mechanism via oxidative cyclization of an enone and an alkyne on nickel is probably more reasonable (path B). The oxa-allylnickelacycle **65** opens up to give alkenylnickel complex **66** in the presence of TMSCl.

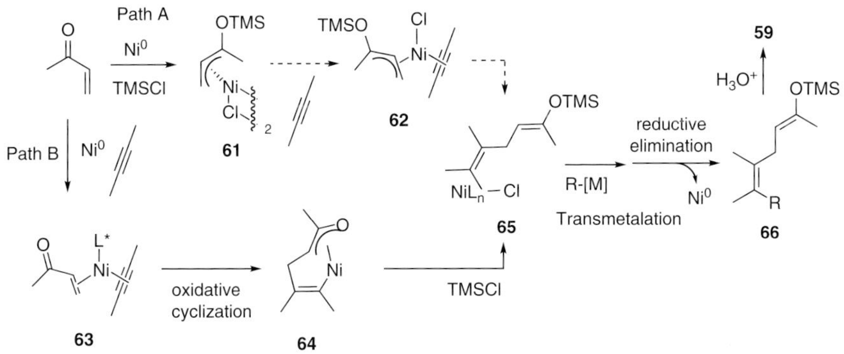

Scheme 4.13 Plausible mechanism for coupling of alkynes, enones, and dimethylzinc.

Transmetalation with organozinc reagent Me_2Zn and finally reductive elimination leads to β-alkyenyl-silyl-enol-ether **67**, which on acidic hydrolysis gives β-alkyenylketones **59**.

4.2.6.2 Nickel-Catalyzed Asymmetric Coupling of Enones and Alkynes

Beside the use of organometallic coupling reagents, such as reducing agents (R_3SiH or Et_3B) or alkylating partners (R_2Zn Et_3B) in multicomponent coupling reactions, research has also been devoted to the reaction of nickelacycle without these reagents. Such multicomponent molecular transformations are highly atom economical. Ogoshi and coworkers reported the isolation of nickelacycles, formed by the oxidative cyclization of nickel(0) complexes with an alkyne and enone, and their reactions in racemic as well as asymmetric versions.

In 2015, nickel-catalyzed asymmetric and fully intermolecular [2+2+2] cycloaddition of two enones and an alkyne was reported by Ogoshi et al. (Scheme 4.14) [15]. This asymmetric reaction was promoted by nickel-NHCs complex, whereas phosphine ligands failed to give any products. The reaction proceeds in an excellent diastereoselective (single out of possible eight), chemoselective, and regioselective manner with an optimal *N*-2-biphenyl substituted C_2-symmetric NHC precursor **70** in the presence of KO^tBu at 30 °C. A variety of acyclic enones **67** and internal alkynes **68** are included in this study to afford the respective cyclohexenes **69** in moderate to good yields and enantioselectivities ranging from 44% to 95% ee. The unsymmetrical alkyne, 1-phenyl-1-propyne, gave the corresponding cycloadduct in 40% yield with a high degree of regioselectivity (98 : 2). This transformation is synthetically important as the resulting multicenter cyclohexane scaffold is often found in pharmaceutical and natural products such as terpenoids, steroids, and alkaloids.

The same research group reported that conjugated alkyne (1,3-enyne) majorly favors [2+2]-cycloaddition with enone to afford cyclobutenes, instead of the formation of cyclohexenes via [2+2+2]-cycloaddition, and an asymmetric version of this reaction was reported in 2016. Various C_1 and C_2 symmetric chiral NHCs were examined in this study to afford enantioenriched cyclobutenes, among which 2,6-disubstituted (*bis-ortho*) *N*-phenyl substituted NHCs were found suitable for such transformations from the stand point of chemical yield

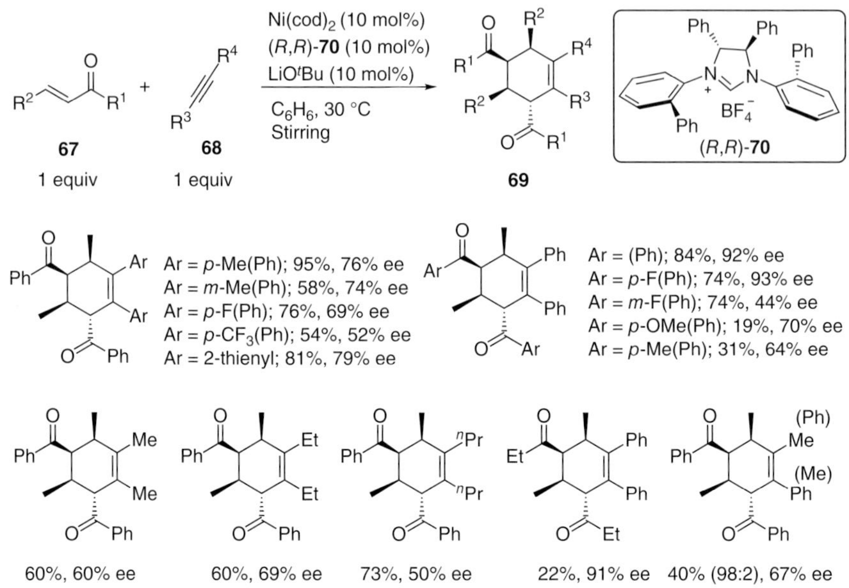

Scheme 4.14 Catalytic enantioselective [2+2+2] cycloaddition of two enones and an alkyne.

of the product. Moderate yield and enantioselectivity (ranged from 45% to 89% ee) are obtained with bulky *N*-(2,7-dicyclohexylnaphthyl)-substituted NHC **74** (Scheme 4.15) [16]. A broad range of functionalized α,β-unsaturated carbonyls **71** are included in this study, such as conjugated amides, esters, and phosphates, cyclohexenenones, and cyclopentenones. *N*-*tert*-Butylmaleimide was the best among them with good yield and enantioselectivity (89%).

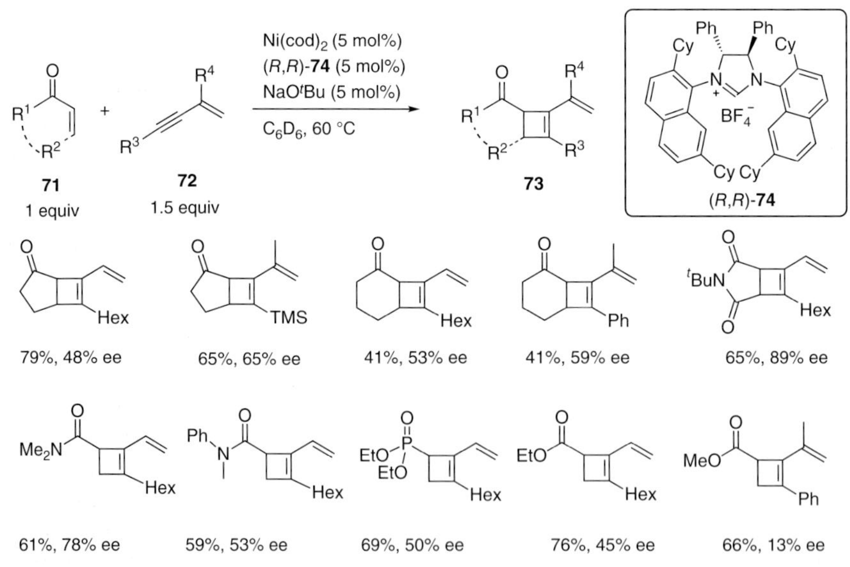

Scheme 4.15 Catalytic enantioselective [2+2] cycloaddition of an enone and an alkyne.

The plausible reaction mechanism for these transformations was explained based on their previous reports in non-chiral variants (Scheme 4.16). The oxidative cyclization of an enone and an alkyne with a nickel(0)/NHC species gives an enantioenriched η^1-*O*-nickelenolate intermediate **76**, which reacts with the second enone in a diastereoselective manner to generate a seven-membered nickelacycle intermediate or its isomer 77, which on reductive elimination gives a [2+2+2] cycloaddition product **69** and generates a total of four stereogenic centers. Furthermore, the formation of cyclobutenes is explained by the unique reactivity of 1,3-enyne that prohibits the insertion of another enone moiety to η^3-butadienyl intermediate **78**, thus allowing reductive elimination at higher temperature to give cyclohexene. The origin of enantioselectivity has not been explained in these asymmetric reactions; however, it possibly arises at the oxidative cyclization step due to face-selective coordination of enone and alkyne in the presence of chiral NHC.

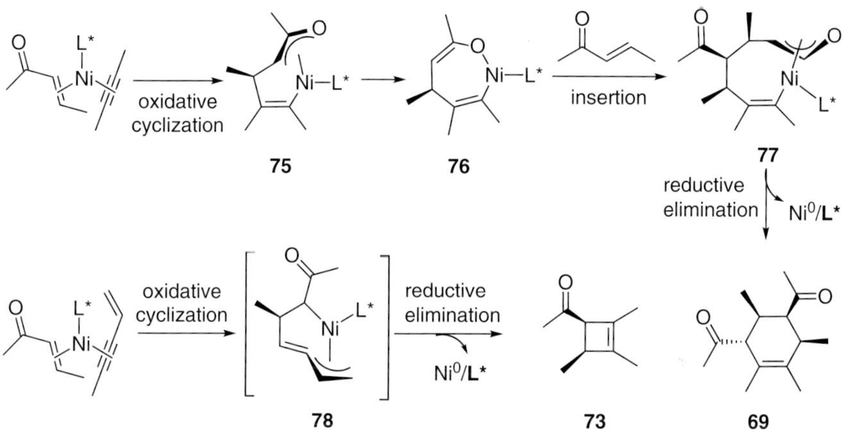

Scheme 4.16 Plausible mechanism for multicomponent reactions of enones and alkynes.

These asymmetric coupling reactions involved an intermolecular oxidative cyclization of α,β-unsaturated carbonyls and alkynes to yield various carbocycles such as cyclohexenes **69** and cyclobutenes **73** with the generation of four and two chiral centers, respectively. Ogoshi and coworkers further elaborated their strategy to construct fused tricyclic scaffolds via intramolecular oxidative cyclization of symmetrical cyclohexadienones with alkyne tethered in the same molecule (**79**) [17]. Intramolecular oxidative cyclization of **79** on nickel(0) gave a tricyclic nickelacycle **80** through desymmetrization, which on reaction with another enone **81** afforded a fused tricyclic scaffold with five chiral centers **82**. This transformation was very effective with NHC and a model reaction was demonstrated with racemic NHC, i.e. 1,3-bis-(2,6-diisopropylphenyl)imidazol-2-ylidene (IPr). Stoichiometric reaction was also reported by the reaction of **79** with IPr and isolation of *O*-allyl-nickelacycle **80** to propose the model transformation (Scheme 4.17). Alkynyl-cyclohexadienones **79** was synthesized in one step by oxidative dearomatization with hyperiodine reagents, phenyliodine bis(acetate)

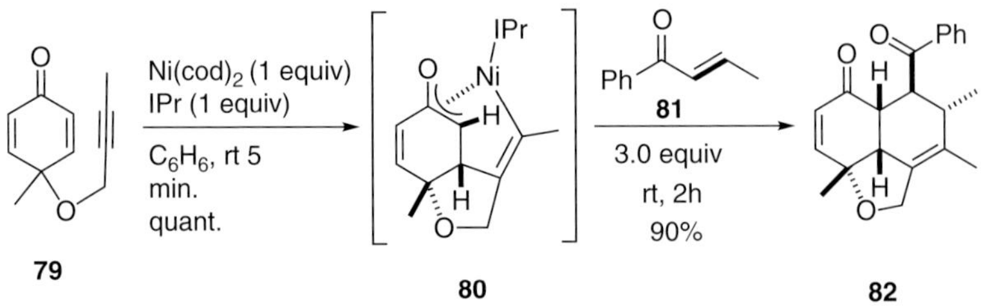

Scheme 4.17 Stoichiometric reaction of alkynyl-cyclohexadienones **79** with Ni^0/IPr.

(PIDA), or phenyliodine bis(trifluoroacetate) (PIFA) in the presence of propargyl alcohol.

Catalytic asymmetric reaction was evaluated with chiral NHCs, and good chemical yield and enantioselectivity were reported with 2,4,6-trisubstituted-*N*-phenyl NHC **86**. A single diastereomer was formed out of 16 possible isomers. A variety of substituted alkynyl-cyclohexadienones **83** and acyclic enones **84** are included in this study to achieve the respective tricyclic scaffolds **85** in good yields with excellent enantioselectivity (ranged from 95% to 99% ee) (Scheme 4.18). The high enantioselectivity in this transformation possibly arose at the desymmetrization step by oxidative cyclization of symmetrical dione with alkyne on the nickel-chiral ligand complex. The developed strategy is attractive as it provides a two-step synthesis (from phenols) of chiral hydronaphtho[1,8-*bc*]furan ring structures containing five contiguous chiral centers, a tricyclic scaffold frequently found in biologically active natural products.

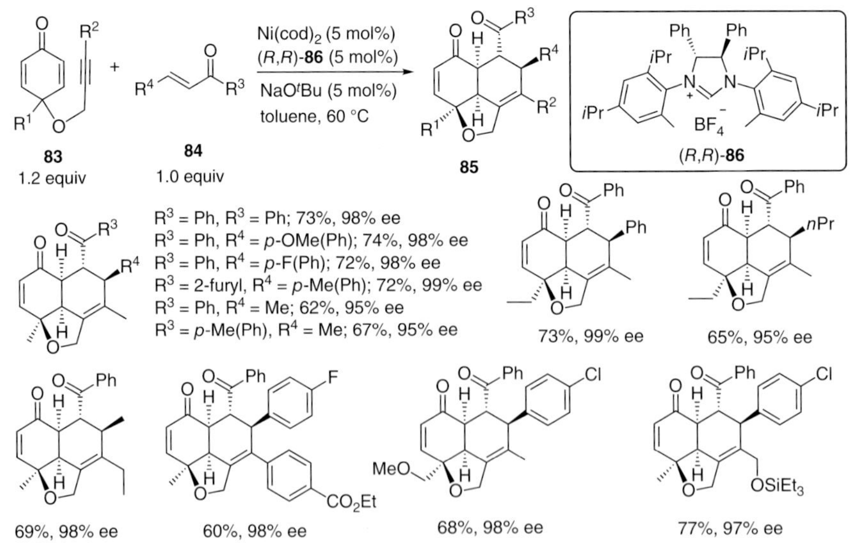

Scheme 4.18 Catalytic enantioselective synthesis of fused tricyclic scaffolds via [2+2+2] cycloaddition of an enone and an alkyne.

4.2.7 Nickel-Catalyzed Asymmetric Coupling of Arylenoates and Alkynes

Cramer and coworkers reported a nickel-catalyzed asymmetric synthesis of cyclopentenones via reductive [3+2] cycloaddition of α,β-unsaturated aromatic esters (acrylates) and alkynes in the presence of BEt_3 and *t*BuOH (Scheme 4.19) [18]. Chiral bulky C_1-symmetric NHC ligands are suitable for efficient transformation to cyclopentenones **89** from mesityl enoates **87**. High enantioselectivity (ranging from 85% to 96% ee) and regioselective incorporation of unsymmetrically substituted alkynes are obtained using an optimal mixed-designed C_1-symmetric NHC ligand (*S*)-**90**. Different aryl-substitution patterns at the β-position of acrylic ester **87** were investigated. Internal alkynes **88**, including dialkyl or aryl-alkyl alkynes, are suitable under these conditions, whereas diaryl alkynes failed to provide the cyclopentenones.

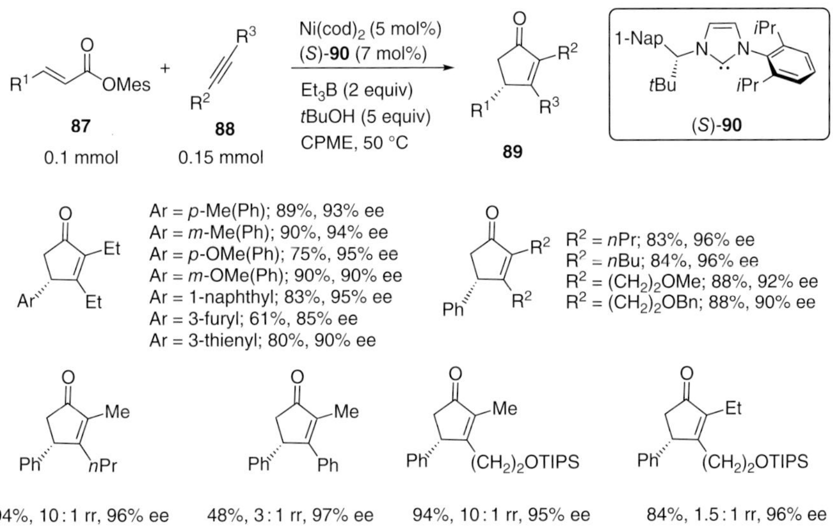

Scheme 4.19 Catalytic enantioselective annulations of alkynes and arylenoate.

The following plausible mechanistic scenario was proposed involving coordination, followed by oxidative cyclization of the arylenoate **87** and alkyne **88** substrates on the chiral Ni^0/L^* species to give oxanickelacyclic intermediate **92** (Scheme 4.20). Cyclization and subsequent β-alkoxide elimination releases cyclopentenone **89**. Transmetalation with triethylborane followed by β-hydride elimination gives a nickel hydride species **97**, which on reductive elimination regenerates Ni(0). The authors hypothesized that the enantioselectivity determining step of the process is the oxidative cyclization and the high enantioselectivity emerged by the involvement of a single chiral side chain of the NHC in the enantioselection, presumably during the face-selective coordination and incorporation of the arylenoate.

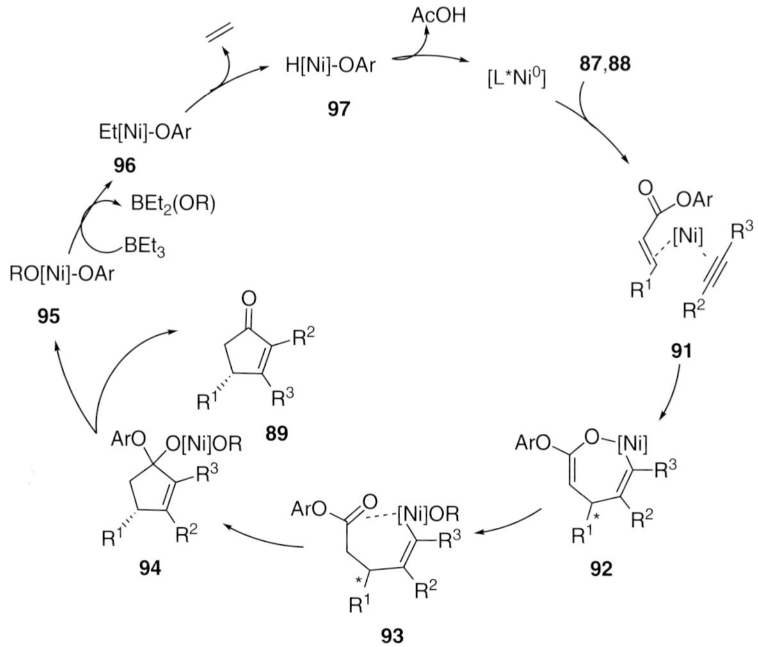

Scheme 4.20 Plausible mechanism for annulations of alkynes and arylenoates.

4.2.8 Nickel-Catalyzed Asymmetric Coupling of Diynes with Ketenes

Besides alkynes, dialkynes are also investigated in multicomponent reactions to construct bicyclic compounds. Louie et al. reported a nickel-catalyzed cycloaddition between diynes and ketenes to afford 2,4-cyclohexadienones, a bicyclic compound containing one quaternary stereocenter in good yields [19]. In this report, an example with substrates **98** and **99** was investigated to achieve a formidable, challenging chiral quaternary stereocenter and by employing (*R*)-BINAP, the corresponding enantioenriched product **100** was formed in excellent enantioselectivity (99% ee), albeit in moderate yield (38%) due to low reactivity (Scheme 4.21). A higher yield was obtained when the reaction temperature was increased to 100 °C.

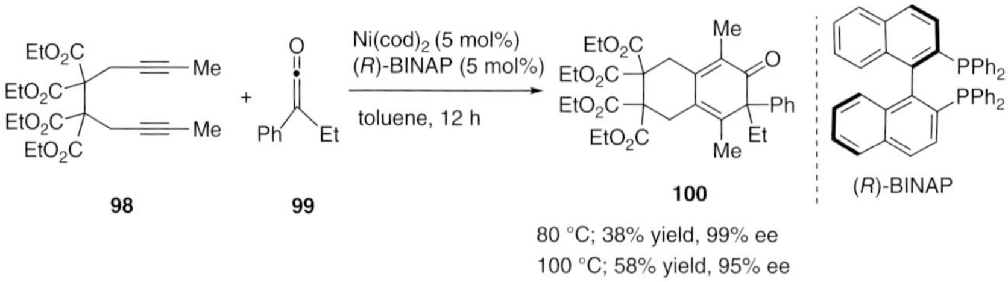

Scheme 4.21 Example of catalytic enantioselective coupling of diyne with ketene.

4.2.9 Nickel-Catalyzed Asymmetric Coupling of Allenes, Aldehydes, and Silanes

Unlike alkynes, allenes are relatively less explored in nickel-catalyzed reductive coupling with aldehydes. In 2005, Jamison's group developed the first example of reductive coupling of allenes with aldehydes in the presence of silane [20]. This transformation was completely site selective (a double bond reacts preferentially) and highly enantioselective in terms of transfer of the axial chirality of the allene to the product, *Z*-trisubstituted allylic alcohol **103**, with very high fidelity (>95%). It also represents the unusual reactivity of an allene, that is, the addition of an electrophile (formyl carbon) **102** to the least nucleophilic central, *sp*-hybridized carbon of an unactivated allene **101**. The use of an imidazolinyl carbene ligand NHC-IPr is suitable for complete transfer of axial chirality (95–98% ee) to central chirality of allylic alcohol (95% ee), whereas the nickel-phosphine ligand (tricyclopentylphosphine, Cyp_3P) tends toward erosion of enantiomeric purity (95–62% ee). Thus, a variety of silyl-protected *Z* allylic alcohols **103** are synthesized from chiral allenes (95–98% ee) and aromatic aldehydes in the presence of Et_3SiH in good yields and selectivity using employing Ni^0-IPr complex (Scheme 4.22).

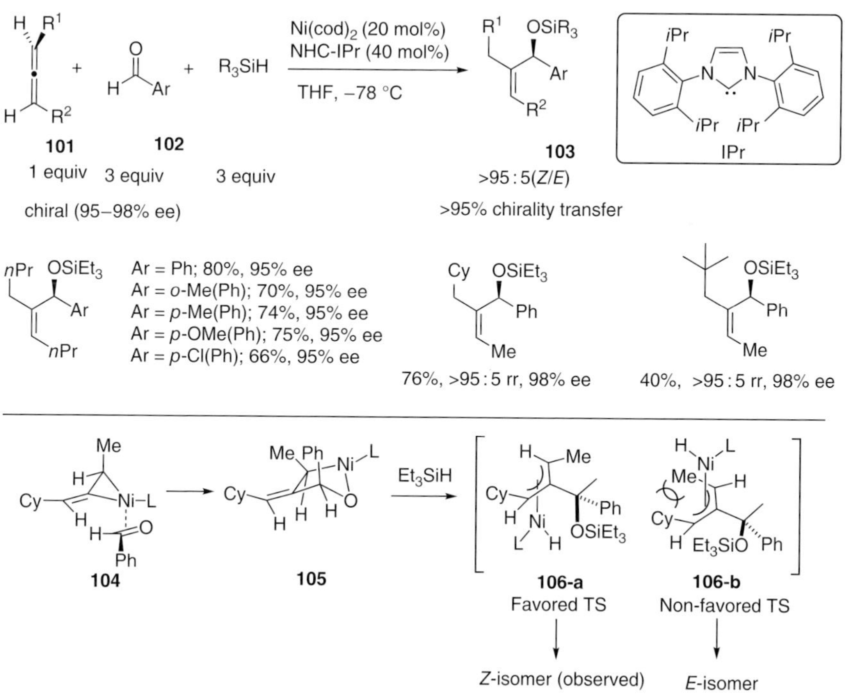

Scheme 4.22 Catalytic enantioselective coupling of chiral allenes, aldehydes, and silanes.

A plausible mechanism was proposed based on the result of deuterium labeling experiments. Coordination of one of the double bonds of allene and back-bonding of NHC-IPr induces significant deformation from linearity of the chiral allene, and of the four isomeric 1 : 1 : 1 complexes; the favored model conformer **104** places the large Ni-IPr complex on the less hindered allene face and less substituted double bond. The sense of enantioinduction was explained by the coordination of benzaldehyde that was away from the methyl group of allene with the Ph group placed between IPr and (cyclohexyl)-methylidene, reorganization to allow overlap between a C—Ni bond and π^*, and oxidative addition to give metalacycle **105**. σ-Bond metathesis between **105** and Et_3SiH affords η^3-allyl-Ni complex **106-a** or **106-b**, which on reductive elimination with retention leads to allylic alcohol. The favored nickel complex **106-a** (no steric congestion between cyclohexyl and methyl group on allene) forms the observed *Z*-alkene and *R* configuration.

Since, IPr has shown high efficiency in allene–aldehyde reductive coupling reactions with complete transfer of axial chirality to the product, a few chiral NHC ligands were evaluated by the same group to achieve chiral allylic alcohols from achiral (terminal) allenes [21]. In an initial study, as an example, cyclohexylallene **107** was investigated with benzaldehyde (**108**) and *tert*-butyldimethylsilane. A maximum of 24% ee was achieved with a chiral NHC (*R,R*)-**110** for the corresponding silyl-protected allylic alcohol **109**, however in good chemical yield (86%) (Scheme 4.23).

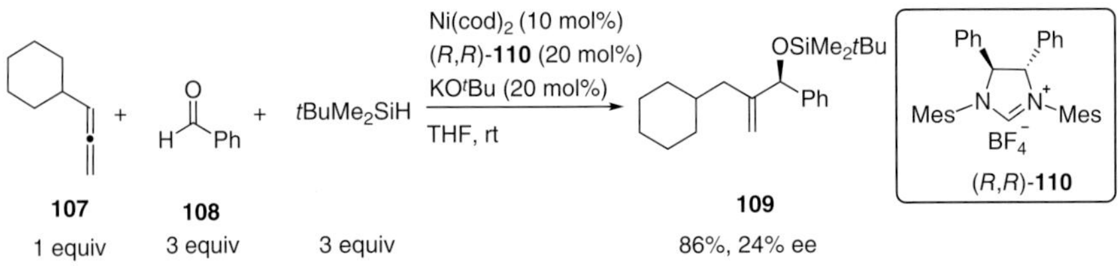

Scheme 4.23 Catalytic enantioselective reductive coupling of terminal allene, aldehyde, and silane.

4.2.10 Nickel-Catalyzed Asymmetric Coupling of Allenes and Isocyanates

In 2010, Murakami and coworkers employed isocyanate and allenes in nickel-catalyzed coupling reactions to form dihydropyrimidine-2,4-diones **113** via [2+2+2]-cycloaddition of 1 equiv of allene **111** and 2 equiv of isocyanates **112** [22]. High levels of enantioselectivity (up to 99%) and selectivity (**112** verses **113**; >20 : 1) were reported with an optimized 10 mol% of Ni-chiral monophosphine ligand complex (prepared from $Ni(cod)_2$ and (*S,S*)-*i*-Pr-FOXAP **115** in 1 : 2 ratio) (Scheme 4.24). Benzyl isocyanate gives low yield, whereas other alkyl isocyanates including hexyl-, cyclohexyl-, and *tert*-butyl-isocyanates all failed to undergo the cycloaddition reaction.

A plausible mechanism for this transformation was proposed through the coordination of a hetero pair of an allene (less substituted olefinic site) and isocyanate (**116**) followed by oxidative cyclization on nickel/L*. Subsequent insertion of another molecule of isocyanate into the nickel–nitrogen bond of the resulting five-membered aza-nickelacyclic intermediate **117** leads to expansion

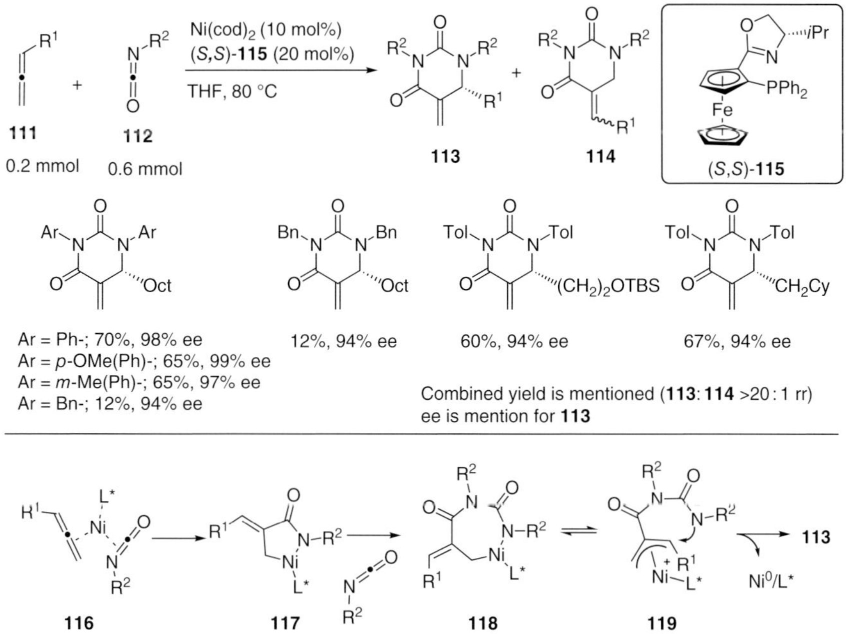

Scheme 4.24 Nickel-catalyzed [2+2+2] cycloaddition of isocyanates and allenes.

to the seven-membered ring aza-nickelacycle **118**, which is in equilibrium with zwitterionic π-allylnickel species **119**. Finally, intramolecular recombination occurs at the more substituted carbon of the allyl moiety affording the required [2+2+2] cycloaddition product **113** with the generation of nickel/L* species. Enantioselectivity might emerge from the chiral environment of π-allylnickel intermediate **119**.

4.2.11 Nickel-Catalyzed Asymmetric Coupling of Alkenes, Aldehydes, and Silanes

Unlike alkynes, alkenes are relatively less explored due to their less reactivity toward nickel. Cramer et al. have employed reactive alkenes, norbornenes **120**, in the enantioselective reductive three-component coupling with aromatic aldehydes **121** and silanes [23]. A variety of indanol derivatives **122** are obtained in high enantioselectivity and diastereoselectivity using the newly synthesized bulky chiral C_2-symmetric NHC ligand **123** based on the 1,2-di(napthalen-1-yl)ethylene diamine backbone (Scheme 4.25). Different substitution patterns in aromatic aldehydes and norbornene derivatives were included in this study; between exo and endo methylenealkoxy-substituted norbornene derivatives, exo-norbornene showed a substantially higher reactivity and high enantioselectivity. Unstrained alkenes such as cyclopentene failed to react with an aldehyde under these optimized reaction conditions. Other silanes, such as TES, TBS, and TBDPS, worked almost with equal efficiency.

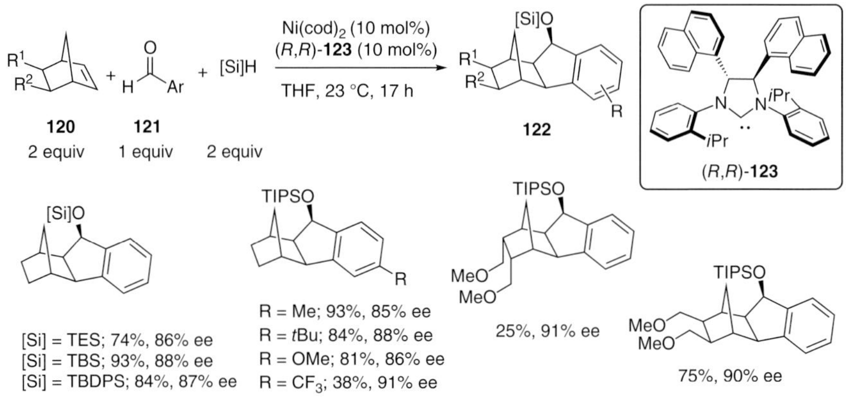

Scheme 4.25 Catalytic enantioselective couplings of norbornenes and aldehydes and silanes.

The mechanistic pathway was proposed through enantio-determining oxidative cyclization on Ni^0/L^* formed oxanickelacycle **125** (Scheme 4.26). High enanatioselectivity might emerge from face-selective coordination of the aldehyde and norbornene to Ni^0/L^* complex with the aryl group of the aldehyde placed away from the methylene bridge, avoiding steric congestion. Two orientations were proposed (**124-a** and **124-b**) and the favored **124-a** led to the desired major enantiomer of the product. Subsequently, the silane reacts with the oxanickelacycle **125**, forming nickel-hydride intermediate **126**. The direct reductive elimination from **126** gave non-cyclized product **127**, which

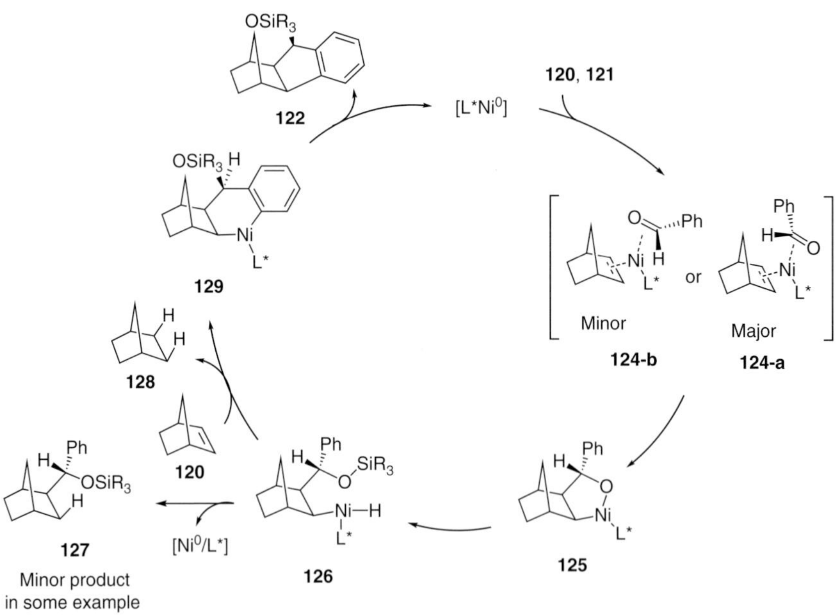

Scheme 4.26 Plausible mechanism for coupling of alkenes, aldehydes, and silanes.

was reported as a minor product in some examples. On the other hand, when another molecule of norbornene **120** is involved in mechanism, it undergoes hydrometallation with the subsequent C(sp^2)–H_{Ortho} bond activation of Ph-group. This step leads to formation of cyclometalated species **129** with the release of norbornane **128**. Finally, reductive elimination gave the required indanol derivatives **122**. Here, excess of norbornene acted as a hydrogen acceptor.

Besides the highly reactive alkenes such as norbornenes, Ogoshi et al. reported in 2016 an intramolecular reductive coupling reaction of simple alkenes and aldehydes with hydrosilanes, catalyzed by (η^6-toluene)Ni(SIPr) (TNSI) [24]. A variety of silyl-protected 1-indanol derivatives have been synthesized in a diastereoselective manner. An example is demonstrated in its asymmetric version to achieve enantioenriched indanol (**131**, 98% yield, 36% ee) using chiral NHC (*R*,*R*)-**86** (Scheme 4.27). This reaction analogously proceeded via an intramolecular oxidative cyclization of an aldehyde and an alkene, promoted by a Lewis acidic silane reagent.

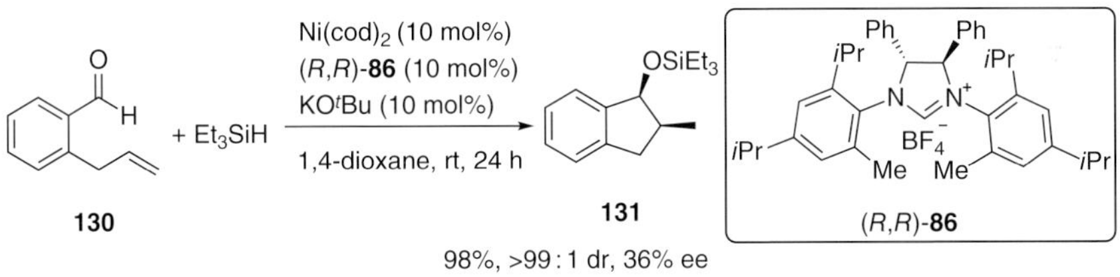

Scheme 4.27 Catalytic enantioselective intramolecular reductive coupling of enal **130** with Et_3SiH.

4.2.12 Nickel-Catalyzed Asymmetric Coupling of Formamide and Alkene

Unlike hydroacylations, hydrocarbamoylations to alkene (addition of formamide to alkene) are less studied due to the intrinsic less reactivity of formamide C—H bonds toward transition metals. In 2013, Cramer et al. developed a novel class of chiral diaminophosphineoxide ligands that enabled asymmetric hydrocarbamoylations of homoallylic formamides by a heterobimetallic activation mode (Scheme 4.28) [25].

The best reactivity was achieved when a combination of 5 mol% of chiral secondary phosphine oxide (SPO) **134,** $Ni(cod)_2$, and PPh_3 was employed in the presence of 40 mol% of Lewis acid, $AlMe_3$. A series of cyclic amides **133** such as γ-lactams, bicyclic pyrrolidines, and larger ring size lactams were synthesized from homoallylic formamides **132** in good yields (75–98%), diastereoselectivities (4 : 1 to >20 : 1), and enantioselectivities (ranged from 79% to 95% ee). The scope of the substrate included a wide range of substitution pattern on nitrogen and the method was also suitable to constitute larger ring sizes (five-, six-, and seven-membered lactams).

The proposed reaction mechanism was explained in Scheme 4.29 through the activation of amide **132** with *in situ* generated chiral heterobimetallic Ni–Al

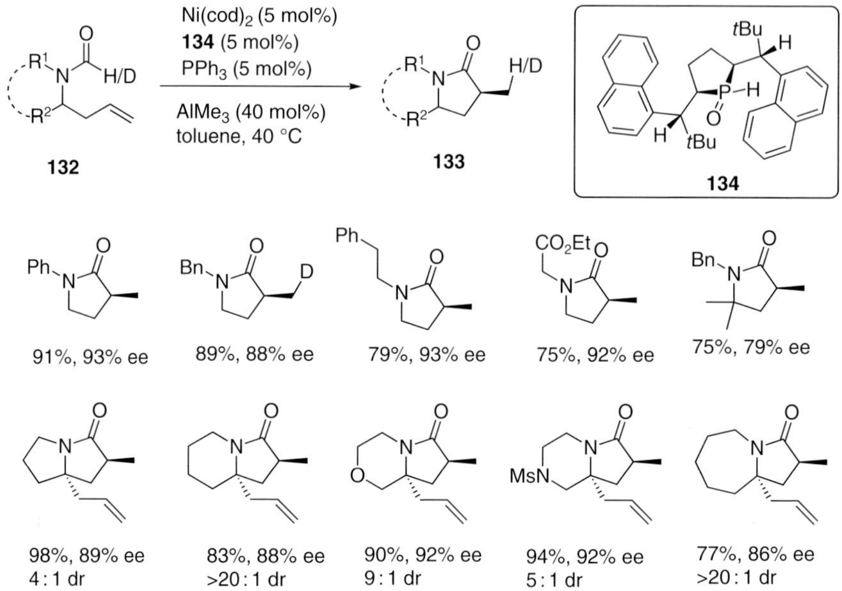

Scheme 4.28 Nickel-catalyzed asymmetric hydrocarbamoylations of alkenes.

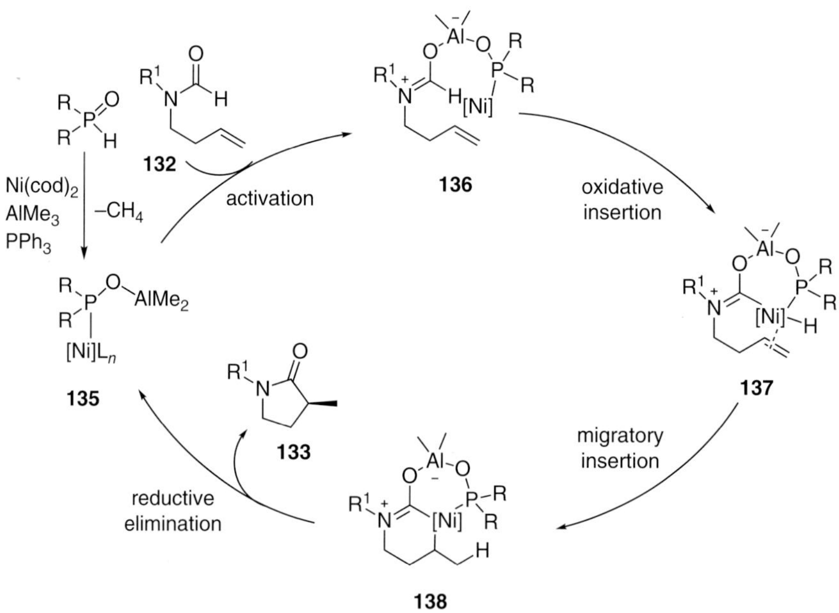

Scheme 4.29 Proposed mechanism for hydrocarbamoylations of alkenes.

complex **135** and subsequent oxidative insertion of nickel center into the now activated formamide C—H bond in a favorable intramolecular manner forms **137**. Migratory insertion of olefin to the Ni-H leads to nickelacycle **138**, which in turn reductively eliminates bimetallic catalyst **135** and delivers the cyclic amide **133** through the formation of an asymmetric C—C bond.

4.2.13 Nickel-Catalyzed Asymmetric Coupling of Alkynes and Cyclopropyl Carboxamide

In addition to various π-coupling partners, the oxidative addition of cyclopropyl ketones and imines also participates in nickel catalyzed multicomponent reactions via the corresponding hetero-nickelacycles. Among various functionalized cyclopropanes used in cycloaddition reactions, cyclopropyl carboxylates (acids, esters, and amides) are attractive classes of substrates because the carboxylate groups allow further organic transformations. Based on Cramer's concept of use of bifunctional SPO and heterometallic Ni–Al complex to activate formamide [25], Ye et al. reported such SPO complexes for the cycloaddition reaction of cyclopropyl carboxamide with alkyne in 2017 (Scheme 4.30) [26]. An asymmetric version of this transformation was also investigated to give enantioenriched cyclopentenyl carboxamides in excellent yield and enantioselectivity with TADDOL-derived SPO chiral ligand **142**. A series of enantioenriched cyclopentenyl carboxamides **141** were synthesized in up to 99% yield and 94% ee from various *N*-substituted cyclopropyl carboxamides **139** and internal alkynes **140**.

The mechanism for this transformation was drawn based on Ogoshi's mechanism reported previously for the [3+2] cycloaddition reaction of cyclopropyl ketones with alkynes (Scheme 4.31). Through coordination and activation of cyclopropyl carboxamide **139** with *in situ* generated Ni–Al heterometallic complex **143** forms the intermediate **144**. A subsequent PR_3-assisted oxidative addition of nickel with the C—C bond of cyclopropane provided the

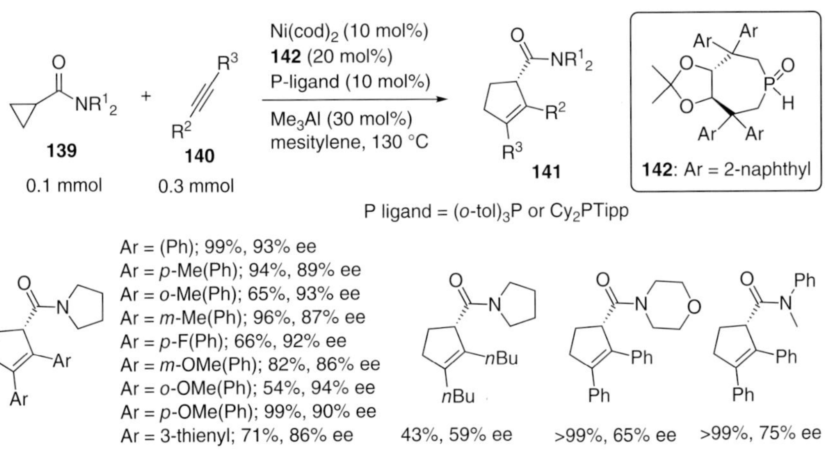

Scheme 4.30 Catalytic enantioselective coupling of cyclopropyl carboxamides and alkynes.

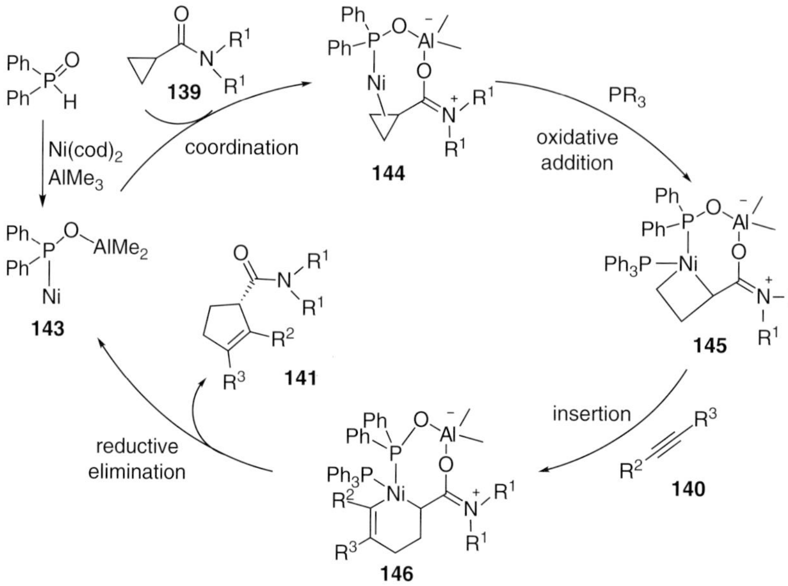

Scheme 4.31 Plausible mechanism for coupling of alkynes and cyclopropyl carboxamides.

four-membered nickelacycle **145**, which inserted into alkyne to afford the six-membered nickelacycle **146**. A final reductive elimination afforded the desired cyclopentenyl carboxamide **141** and regenerated the bimetallic catalytic species **143**.

4.3 Miscellaneous

4.3.1 Nickel-Catalyzed Asymmetric Annulation of Pyridones via Hydroarylation to Alkenes

In 2015, Cramers and coworker reported nickel-catalyzed intramolecular hydroarylation to alkenes forming 1,6-annulated 2-pyridones [27]. This is an important structural core in indolizidine and quinolizidine alkaloids. It was reported that the regioselectivity, either exo or endo in cyclization during the olefin hydrometalation (five or six-membered nickelacycles; **148** vs. **150**), is fully controlled by ligand participation. The reaction in the presence of cyclooctadiene (**L2** = COD) exclusively delivers exo-cyclization products, whereas bulky NHC ligands (**L1**) selectively promote the endo-cyclization mode. The asymmetric version of this transformation was also demonstrated with a few examples of **152** and **154** employing chiral bulky NHC **156**, which gave enantioenriched bi- and tricyclic pyridones **153** and **155**, respectively with 57% ee in each case (Scheme 4.32). The authors proposed a C6–H nickel(0) insertion/hydroarylation mechanism based on deuterium experiments and the absence of any deuterium scrambling indicates an irreversible hydroarylation step.

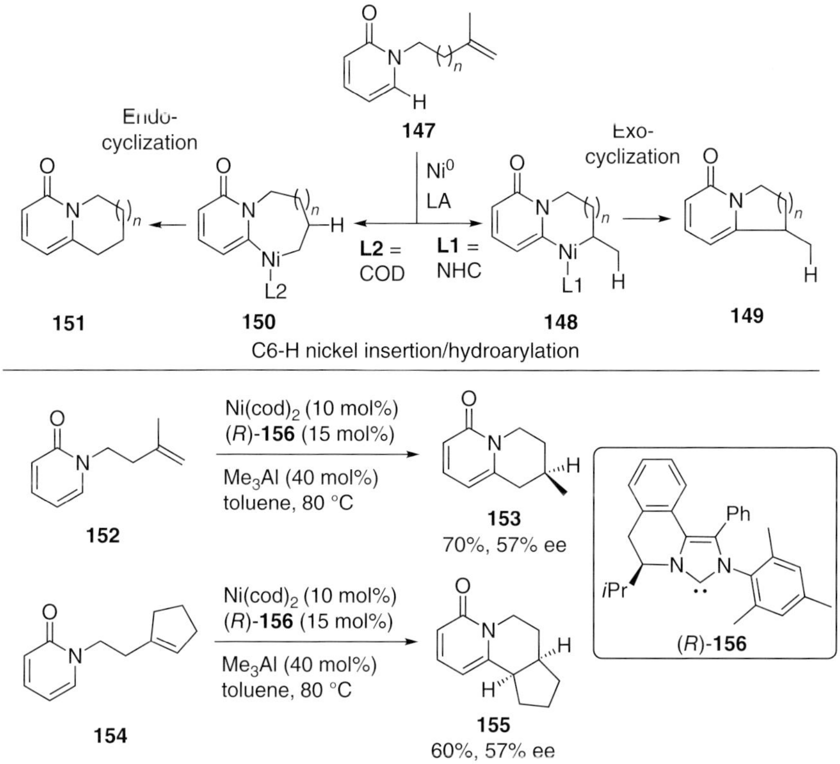

Scheme 4.32 Nickel-catalyzed annulation of pyridones via hydroarylation to alkenes.

4.3.2 Nickel-Catalyzed Asymmetric Synthesis of Benzoxasilole

Apart from nickel-catalyzed multiple component reactions using various π-components including carbonyls, and σ-component such as cyclopropyl carbonyls, Ogoshi and coworkers demonstrated a unique reactivity of carbonyl-nickel complex for the activation of silicons [28]. The reaction involved the formation of oxanickelacycle with an aldehyde followed by transfer of aryl group from the silicon center to the carbonyl carbon. This reaction is highly efficient to form benzoxasiloles **158** in the presence of NHC such as IPr. An asymmetric version of this reaction was also reported in the presence of chiral NHC **159** to give enantioenriched benzoxasiloles **158** (Scheme 4.33; up to 99% yield and ee) [29]. The authors reported that there is a switching of the reaction pathway with the change of ligand from IPr to chiral NHC **159**. In the presence of IPr, the racemic reaction proceeded with intermolecular aryl transfer, whereas the aryl group transferred in an intramolecular manner when **159** was employed. This switching of mechanism was explained by DFT calculation and suggested that in the case of chiral NHC (*R*,*R*)-**159**, there is formation of a unique η^2:η^2 coordination of nickel/**159** with formyl group and aryl group (**161**), which facilitated an intramolecular aryl transfer from the silicon atom

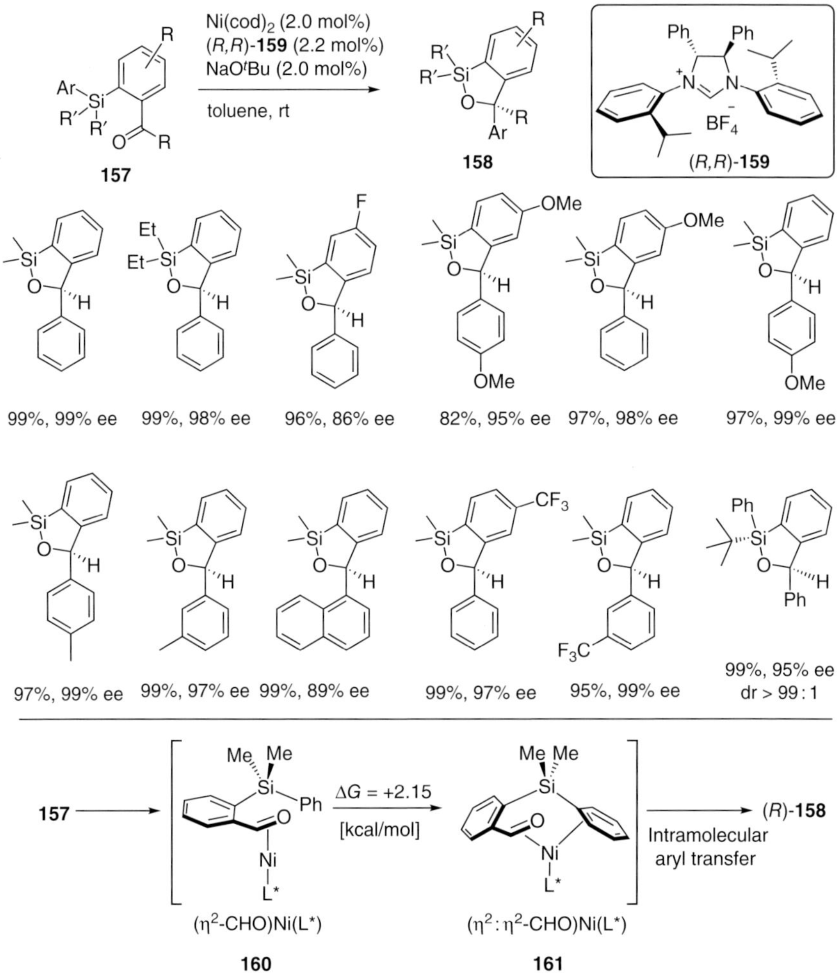

Scheme 4.33 Nickel-catalyzed enantioselective synthesis of benzoxasiloles.

to the formyl group giving (*R*)-**158,** whereas such a nickel complex was not observed with IPr and thus η^2-Ni(CHO)(IPr) (**160**) favored intermolecular aryl transfer.

4.4 Overview and Future Perspective

A variety of molecular transformations are reported to date from nickelacycle in the racemic form. However, unlike palladium and other metals, asymmetric transformations using nickel are still under development stage. Thus, nickel catalysis in asymmetric transformations has the potential to be a productive field of study into the future to address even more challenging substrates and the rapid assembly of chiral complex molecules.

References

1 Miller, K.M., Huang, W.-S., and Jamison, T.F. (2003). *J. Am. Chem. Soc.* 125: 3442–3443.
2 Colby, E.A. and Jamison, T.F. (2003). *J. Org. Chem.* 68: 156–166.
3 Chan, J. and Jamison, T.F. (2003). *J. Am. Chem. Soc.* 125: 11514–11515.
4 Chaulagain, M.R., Sormunen, G.J., and Montgomery, J. (2007). *J. Am. Chem. Soc.* 129: 9568–9569.
5 Yang, Y., Zhu, S.-F., Zhou, C.-Y., and Zhou, Q.-L. (2008). *J. Am. Chem. Soc.* 130: 14052–14053.
6 Patel, S.J. and Jamison, T.F. (2004). *Angew. Chem., Int. Ed.* 43: 3941–3944.
7 Patel, S.J. and Jamison, T.F. (2003). *Angew. Chem. Int. Ed.* 42: 1364–1367.
8 Miller, K.M., Colby, E.A., Woodin, K.S., and Jamison, T.F. (2005). *Adv. Synth. Catal.* 347: 1533–1536.
9 Miller, K.M., Luanphaisarnnont, T., Molinaro, C., and Jamison, T.F. (2004). *J. Am. Chem. Soc.* 126: 4130–4131.
10 Miller, K.M. and Jamison, T.F. (2005). *Org. Lett.* 7: 3077–3080.
11 (a) Sato, Y., Saito, N., and Mori, M. (2000). *J. Am. Chem. Soc.* 122: 2371–2371. (b) Sato, Y., Saito, N., and Mori, M. (2002). *J. Org. Chem.* 67: 9310–9317.
12 Sato, Y., Hinata, Y., Seki, R. et al. (2007). *Org. Lett.* 9: 5597–5599.
13 Ikeda, S., Cui, D.-M., and Sato, Y. (1999). *J. Am. Chem. Soc.* 121: 4712–4713.
14 Ikeda, S., Kondo, K., and Sato, Y. (1996). *J. Org. Chem.* 61: 8248–8255.
15 Kumar, R., Tokura, H., Nishimura, A. et al. (2015). *Org. Lett.* 17: 6018–6021.
16 Kumar, R., Tamai, E., Ohnishi, A. et al. (2016). *Synthesis* 48: 2789–2794.
17 Kumar, R., Hoshimoto, Y., Tamai, E. et al. (2017). *Nat. Commun.* 8: 32.
18 Ahlin, J.S.E., Donets, P.A., and Cramer, N. (2014). *Angew. Chem., Int. Ed.* 53: 13229–13233.
19 Kumar, P., Troast, D.M., Cella, R., and Louie, J. (2011). *J. Am. Chem. Soc.* 133: 7719–7721.
20 Ng, S.-S. and Jamison, T.F. (2005). *J. Am. Chem. Soc.* 127: 7320–7321.
21 Ng, S.-S. and Jamison, T.F. (2006). *Tetrahedron* 62: 11350–11359.
22 Miura, T., Morimoto, M., and Murakami, M. (2010). *J. Am. Chem. Soc.* 132: 5836–15838.
23 Ahlin, J.S.E. and Cramer, N. (2016). *Org. Lett.* 18: 3242–3245.
24 Hayashi, Y., Hoshimoto, Y., Kumar, R. et al. (2016). *Chem. Commun.* 52: 6237–6240.
25 Donets, P.A. and Cramer, N. (2013). *J. Am. Chem. Soc.* 135: 11772–11775.
26 Liu, Q.-S., Wang, D.-Y., Yang, Z.-J. et al. (2017). *J. Am. Chem. Soc.* 139: 18150–18153.
27 Donets, P.A. and Cramer, N. (2015). *Angew. Chem., Int. Ed.* 54: 633–637.
28 Hoshimoto, Y., Yabuki, H., Kumar, R. et al. (2014). *J. Am. Chem. Soc.* 136: 16752–16755.
29 Kumar, R., Hoshimoto, Y., Yabuki, H. et al. (2015). *J. Am. Chem. Soc.* 137: 11838–11845.

Part II

Functionalization of Unreactive Bonds

5

Recent Advances in Ni-Catalyzed Chelation-Assisted Direct Functionalization of Inert C—H Bonds

Yan-Hua Liu[1], *Fang Hu*[2], *and Bing-Feng Shi*[1]

[1] *Zhejiang University, Department of Chemistry, 38 Zheda Rd., Hangzhou 310027, China*
[2] *Zhejiang University, College of Pharmaceutical Sciences, 866 Yuhangtang Rd., Hangzhou 310058, China*

5.1 Introduction

Transition-metal-catalyzed (TM-catalyzed) chelation-assisted C–H activation strategy has witnessed rapid progress during the last decade [1]. This synthetic method obviates the prefunctionalization of starting materials and represents one of the most straightforward routes in organic synthesis. Generally, directing groups (DGs) are introduced to the substrate to facilitate the efficient and selective cleavage of the proximal C—H bond [1].

While the second or third row TMs (e.g. Rh, Ir, Ru, Pd) keep leading the field of inert C—H bond activation, the more sustainable variant dominated by first row TMs (e.g. Fe, Co, Ni, Cu) has witnessed remarkable progress [2]. This can be partially attributed to their nonprecious, low toxic, and earth-abundant nature as well as their new reactivity profiles. Among these, Ni-catalyzed reactions have received much attention, owing to the high catalyst flexibility and chemical versatility caused by the wide oxidation state of the Ni element (from Ni(0), Ni(I), Ni(II), Ni(III) to Ni(IV)) [3]. In this chapter, we will describe the recent advances in nickel-catalyzed chelation-assisted inert C—H bond functionalization. Early elegant examples on Ni-catalyzed functionalization of electronically activated (hetero)aryl C—H bonds, or C—H bonds activated via hyperconjugations, have been well documented and will not be discussed in this review [4]. For clarity, the content will be classified by the mono- or bidentate DGs adopted and further classified by newly formed chemical bonds.

5.2 Ni-Catalyzed Functionalization of Inert C—H Bonds Assisted by Bidentate Directing Groups

The bidentate DGs adopted in these works generally consist of a monoanionic coordinate site, such as the deprotonated anionic N atom of the amide substrates, and a neutral nitrogen coordinating atom on quinoline (e.g. 8-aminoquinoline, AQ) [5] or pyridine (2-pyridinylisopropyl amine, PIP-amine) [6]. In some cases,

Nickel Catalysis in Organic Synthesis: Methods and Reactions,
First Edition. Edited by Sensuke Ogoshi.

the neutral coordinating atom was extended to the O atom and N atom of amino acid derivatives. The reactions developed via these DGs were applied to the aliphatic or aromatic carboxylic acid substrates [5].

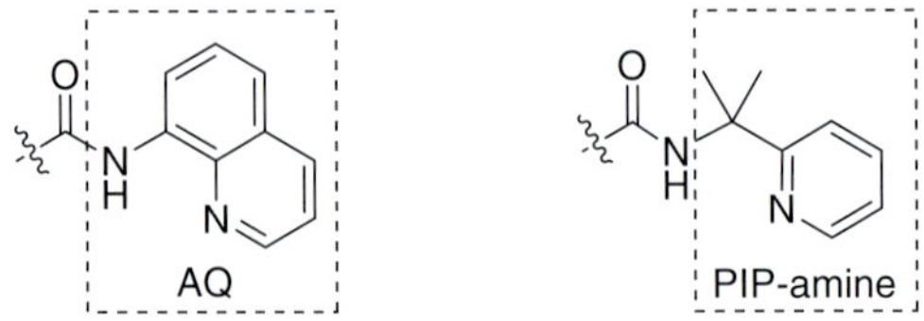

5.2.1 Arylation

In 2014, Chatani and coworker first reported the arylation of unactivated C(sp^3)—H bond catalyzed by $Ni(OTf)_2$ using AQ as the DG (Scheme 5.1) [7a]. The reaction of **1** with aryl iodides with $Ni(OTf)_2$ as a catalyst, Na_2CO_3 as base, and MesCOOH as ligand in DMA at 140 °C for 24 hours gave the β-arylation product **2**. Both Ni(II) and Ni(0) ($Ni(COD)_2$) show high catalytic activity. The addition of a sterically bulky carboxylic acid, such as 2,4,6-trimethylbenzoic acid (MesCOOH) or 2,6-$Ph_2C_6H_3COOH$ as ligand, could improve the yield. The starting material with a quaternary carbon at the α-position reacted smoothly to afford the desired arylated product while substrates with a tertiary or a secondary carbon atom showed less reactivity. Other aryl sources such as PhBr and PhOTf were proved unreactive under this condition. Deuterium-labeling experiments suggest that C–H activation is reversible.

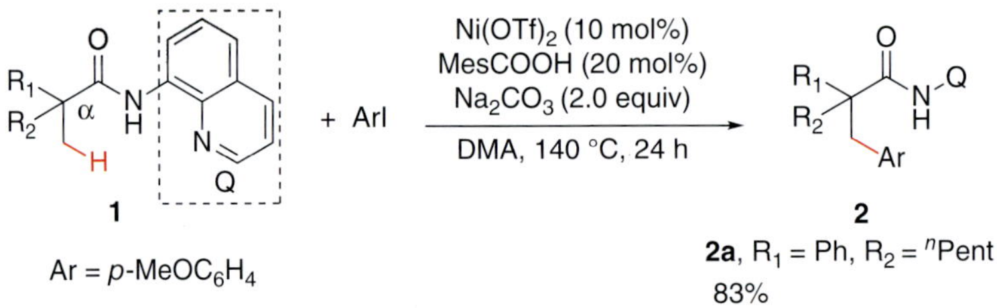

Scheme 5.1 Ni-catalyzed arylation of C(sp^3)—H bond.

Based on the mechanistic investigations, the author proposed a Ni(II)/Ni(IV) catalytic cycle (Scheme 5.2). Coordination and deprotonation of the amide by Ni(II) gives the intermediate **INT 1**, which undergoes reversible C(sp^3)–H activation via a CMD pathway delivering the cyclonickel(II) species **INT 2**. Subsequent oxidative addition of PhI yields the Ni(IV) species **INT 3**, which undergoes reductive elimination to give intermediate **INT 4**. A protonation step releases the product **2** and regenerates the Ni(II) catalyst. They have also reported Ni(II)-catalyzed arylation of (hetero)aryl C(sp^2)—H bonds with aryl iodides using AQ as the DG [7b].

In 2014, Chatani and coworkers successfully applied diaryliodonium salts as arylation reagents in Ni-catalyzed C–H arylation reaction (Scheme 5.3a) [7c].

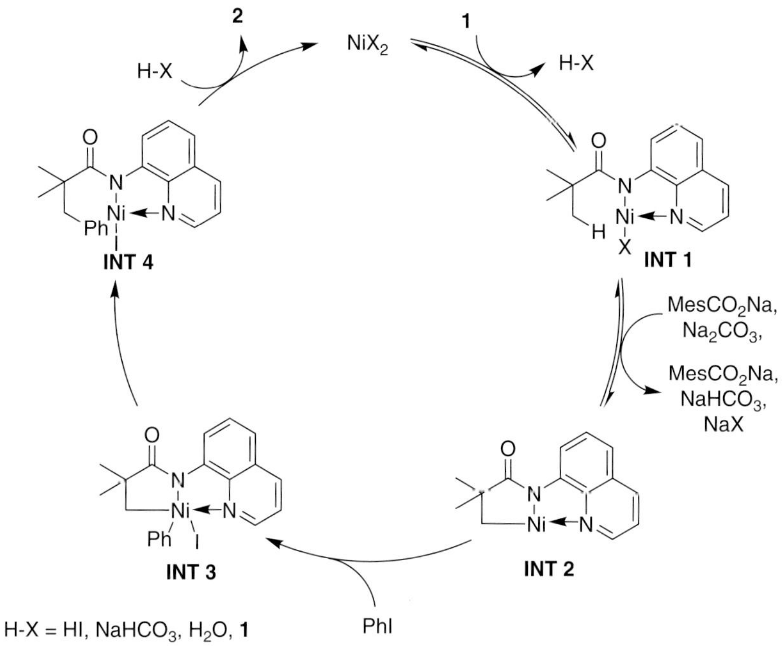

Scheme 5.2 Proposed mechanism for Ni-catalyzed C(sp^3)–H arylation.

To exclude the possible two-step pathway, which involves the decomposition of diaryliodonium salt into aryl iodide and aryl triflate followed by the arylation with aryl iodide as the actual aryl reagent, they have conducted a series of control experiments (Scheme 5.3b). The reaction with aryl iodide gave the arylated product **2b** in 37% yield and no desired product was observed in the reaction with aryl triflate, while the reaction with diaryliodonium salt gave the arylated product **2b** in 76% yield. These results clearly suggest that the main pathway to **2b** involves direct arylation with the diaryliodonium salt. Besides, the effect of the counter anions of the diaryliodonium salt was also examined, indicating that the coordinating OTf and Cl anions were competent while the non-coordinating BF_4 and PF_6 were completely unreactive.

Mechanistically, the author proposed that the diaryliodonium salt underwent oxidative addition to the cyclonickel species **INT 2** leading to the Ni(IV) species **INT 3**, and that a following reductive elimination and protonation step released **2** and the Ni(II) catalyst (Scheme 5.4).

In 2014, You and coworkers independently reported a Ni-catalyzed C(sp^3)–H arylation with the assistance of AQ auxiliary (Scheme 5.5) [8]. Impressively, aryl bromides could also be used as the coupling partners.

In 2015, the Qiu group reported the Ni(II)-catalyzed heteroarylation of C(sp^3)—H bonds with 2-bromoheteroarenes assisted by AQ auxiliary (Scheme 5.6) [9].

The Shi group has developed a removable PIP bidentate DG, which has shown versatile reactivity in Pd-catalyzed C–H functionalization [6]. In 2016, Shi and

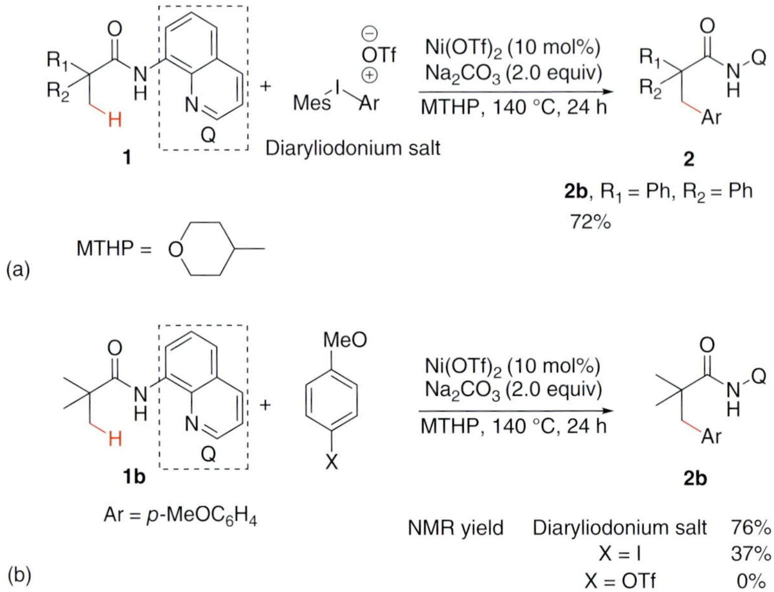

Scheme 5.3 (a) Ni-catalyzed arylation of C(sp^3)—H bond using diaryliodonium salt. (b) Control experiments.

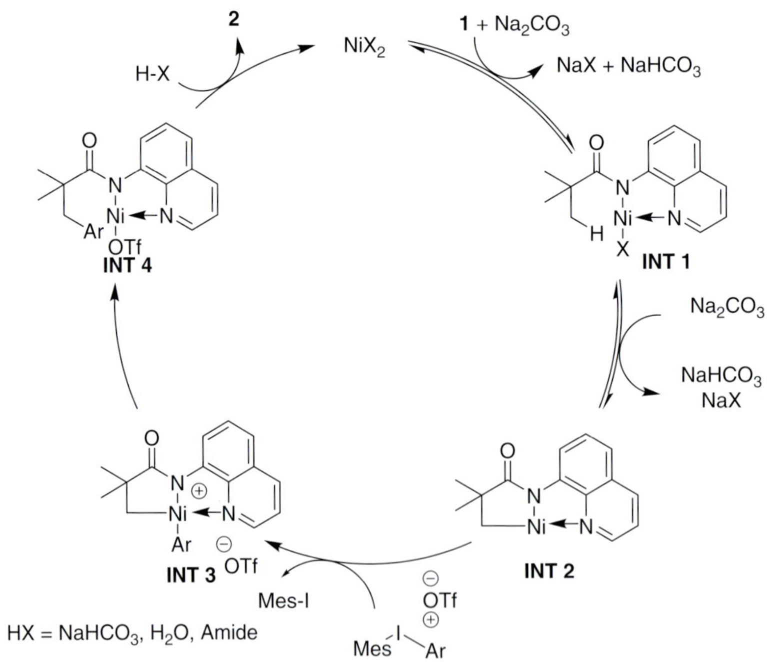

Scheme 5.4 Proposed mechanism for Ni-catalyzed arylation of C(sp^3)—H bond.

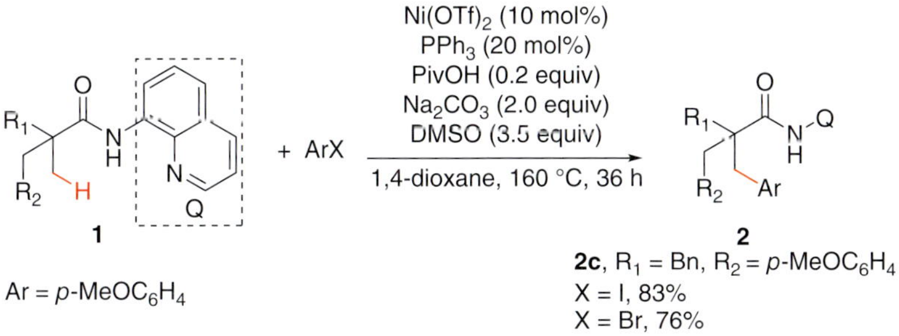

Scheme 5.5 Ni-catalyzed arylation of $C(sp^3)$—H bond.

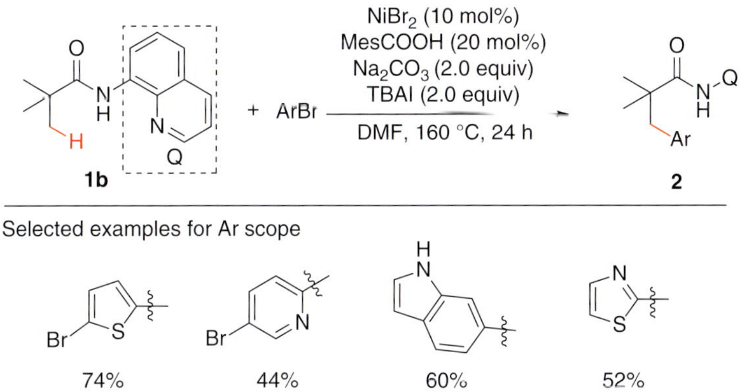

Scheme 5.6 Ni-catalyzed heteroarylation of $C(sp^3)$—H bonds.

coworkers reported the nickel-catalyzed arylation of $C(sp^2)$—H bonds with aryl-boron reagents assisted by PIP auxiliary (Scheme 5.7a) [10]. Ag_3PO_4 was used as oxidant. Phosphine ligands such as PPh_3, PCy_3, dppp, BINAP, and $P(o\text{-}MeOC_6H_4)_3$ were crucial to promote the reaction and no desired product was observed in the absence of these ligands. Replacement of arylboronic acid pinacol esters by arylboronic acid led to a trace amount of the desired product. The PIP DG could be readily removed via a nitrosylation/hydrolysis sequence, affording the corresponding 2-arylbenzoic acid **5** in 90% yield (Scheme 5.7b).

The complete inhibition of this reaction by 1.0 equiv of TEMPO, hydroquinone, or 2,6-di-*tert*-butyl-4-methylphenol indicated that this reaction may proceed via an SET pathway. Accordingly, the author proposed a Ni(I)/Ni(II)/Ni(III) catalytic cycle (Scheme 5.8).

In the same year, Shi and coworkers reported the Ni(II)-catalyzed oxidative arylation of $C(sp^2)$—H bond using triethoxy(aryl)silanes as arylation reagents (Scheme 5.9) [11]. KF was used in this system to facilitate the transmetalation process.

In 2016, You and coworkers reported an oxidative $C(sp^2)$–H/$C(sp^2)$–H cross-coupling reaction of (hetero) arenes assisted by AQ to synthesize biheteroaryls (Scheme 5.10) [12]. The use of Ag_2CO_3, PPh_3, and PivCOOH was crucial for the efficiency of this reaction. Thiazoles, benzothiazoles, benzoxazole,

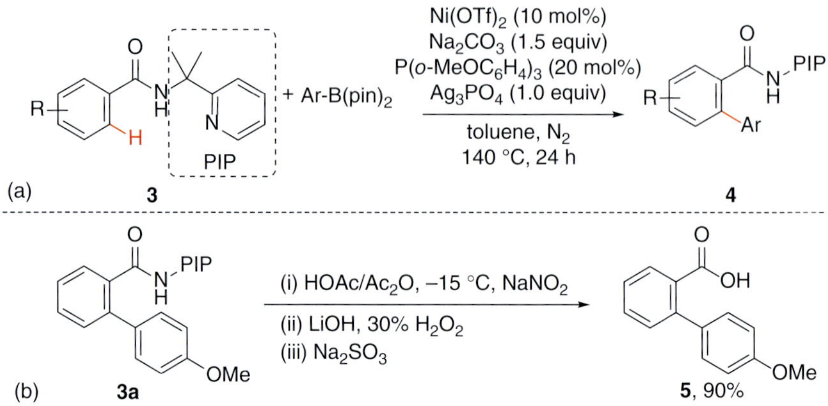

Scheme 5.7 Ni-catalyzed arylation of C(sp^2)—H bonds assisted by PIP auxiliary.

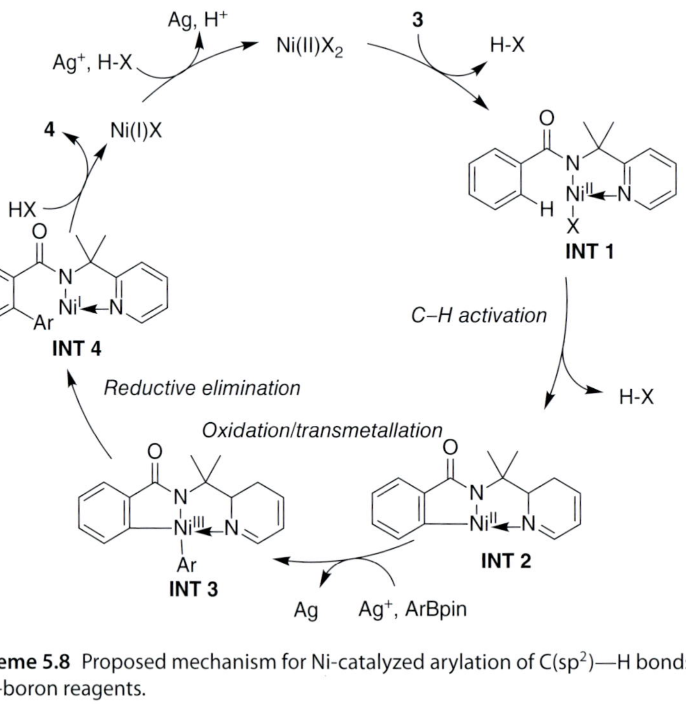

Scheme 5.8 Proposed mechanism for Ni-catalyzed arylation of C(sp^2)—H bonds with aryl-boron reagents.

purine, and caffeine can be used as coupling partners. Various synthetically useful functional groups, such as aldehyde, cyano, ester, nitro, acyl, bromo, and hydroxyl, were well tolerated.

5.2.2 Alkylation

In 2013, Chatani and coworker reported the first example of Ni(II)-catalyzed C(sp^2)–H alkylation of benzamides and acrylamides assisted by AQ auxiliary

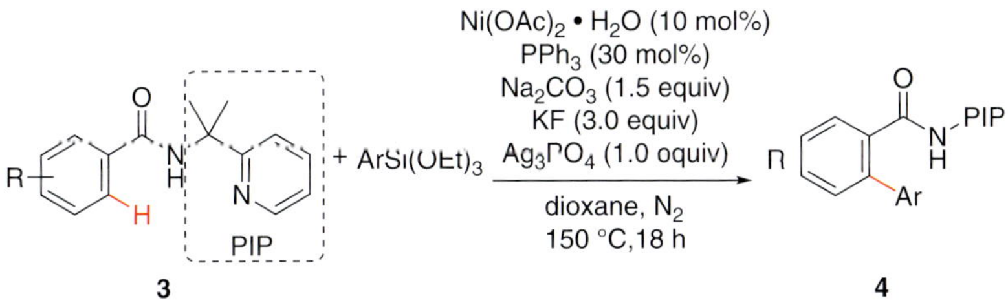

Scheme 5.9 Ni-catalyzed oxidative arylation of C(sp^2)—H bonds.

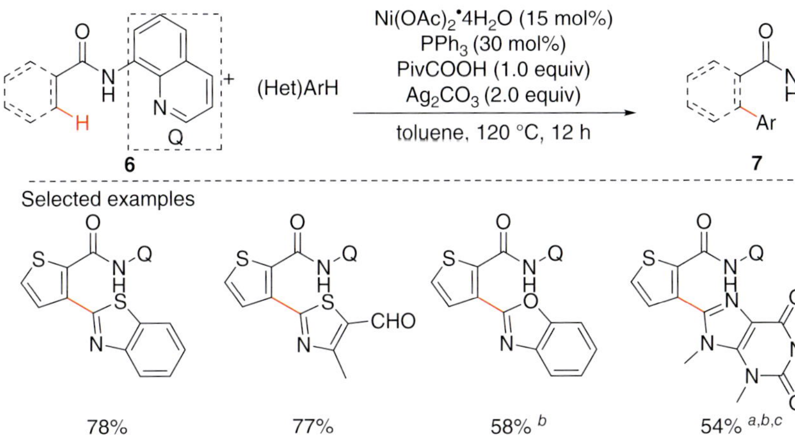

a 24 h. b PivCOOH (2.0 equiv), *o*-xylene. c 140 °C.

Scheme 5.10 Ni-catalyzed heteroarylation of C(sp^2)—H bonds.

(Scheme 5.11) [13]. Alkyl bromides were used as alkylation reagent. In some cases, the addition of 2.0 equiv of NaI was needed to improve the yield, especially when using alkyl chlorides as alkyl source. This alkylation reaction was also applicable to *α*,β-unsaturated amides. For these cases, the presence of a substituent at the *α*-carbon atom was critical for success. Generally, a broad range of primary alkyl halides were compatible for this protocol; however, secondary halides proved completely ineffective.

In 2014, Ni(II)-catalyzed secondary alkylation and trifluoroethylation of C(sp^2)–H assisted by AQ auxiliary was reported by the Ackermann group (Scheme 5.12) [14]. [(DME)NiCl$_2$] was employed as catalyst and BDMAE was used as ligand. Intriguingly, *t*-BuOLi was proved to be crucial for the reaction since other potential alternatives, such as Na$_2$CO$_3$, Cs$_2$CO$_3$, and NaOAc, did not give any desired product under otherwise identical conditions.

In 2014, Ni-catalyzed alkylation of primary C(sp^3)—H bonds with alkyl halides directed by AQ was reported by Ge and coworkers (Scheme 5.13a) [15]. The phosphine-based ligand was essential for high efficiency. Notably, both alkyl bromide and alkyl chloride could also be used as alkyl sources when CsI was added.

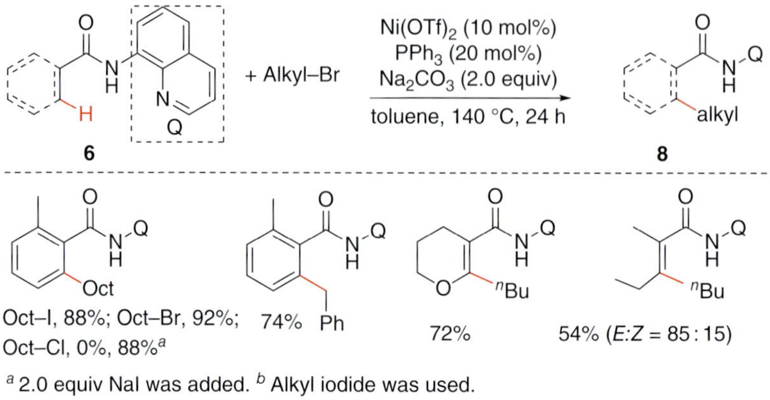

Scheme 5.11 Ni-catalyzed alkylation of C(sp^2)—H bonds with primary alkyl halides.

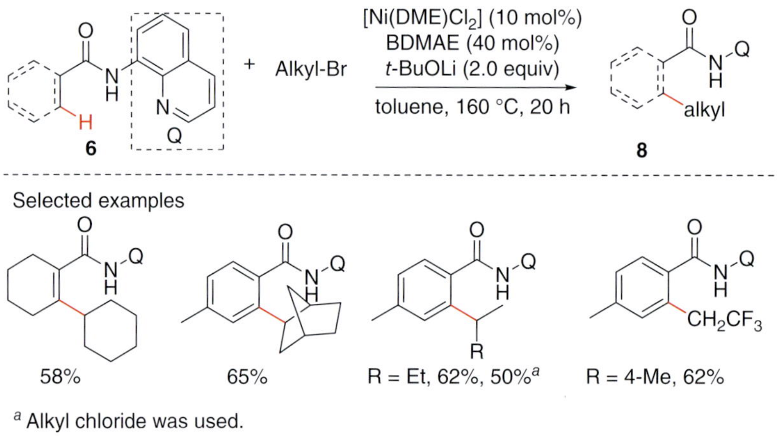

Scheme 5.12 Ni-catalyzed secondary alkylation of C(sp^2)—H bonds with secondary alkyl bromides.

TEMPO trapping experiments resulted in decreased yield as well as the isolation of radical trapped compound 2,2,6,6-tetramethyl-1-(pentyloxy)piperidine **10** (Scheme 5.13b), which could not be isolated without the Ni(II) catalyst. Accordingly, the author proposed a N(II)/Ni(III)/Ni(I) catalytic cycle (Scheme 5.14). However, a Ni(II)/Ni(IV) catalytic cycle could not be ruled out either.

Apart from alkyl halides, various other reagents could also be used as alkylation reagents. In 2014, Chatani and coworkers reported the benzylation of C(sp^2)—H bonds assisted by AQ via cross-dehydrogenative coupling of C(sp^2)–H in benzamides with benzylic C(sp^3)–H in toluene derivatives using $^iC_3F_7I$ as a radical type oxidant (Scheme 5.15) [16]. To avoid the addition of $^iC_3F_7I$ to quinoline group, the 5-position of 8-AQ was blocked by substitutions, such as chloro or methoxy. α,β-Unsaturated amides were also comparable with this reaction.

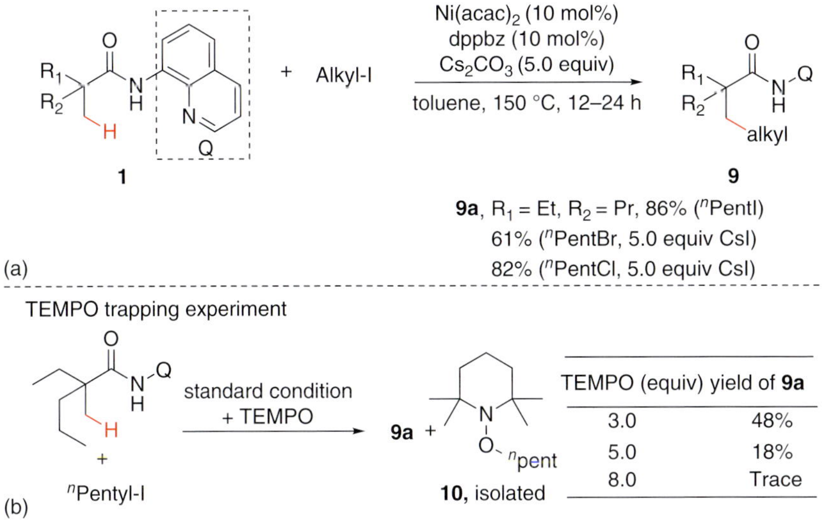

TEMPO (equiv)	yield of **9a**
3.0	48%
5.0	18%
8.0	Trace

Scheme 5.13 Ni-catalyzed alkylation of C(sp^3)—H bonds with alkyl halides.

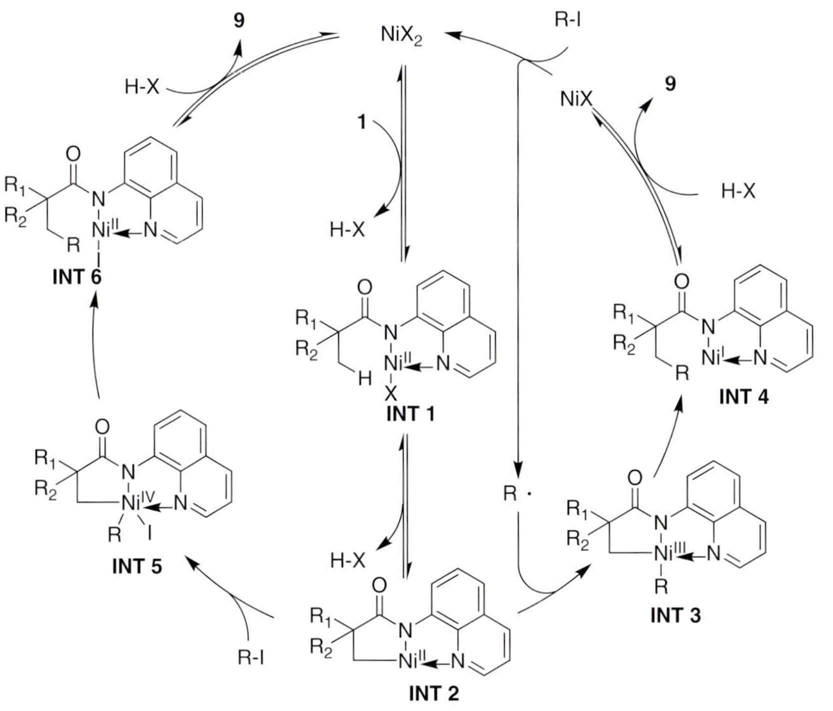

Scheme 5.14 Proposed mechanism for Ni-catalyzed alkylation of C(sp^3)—H bonds.

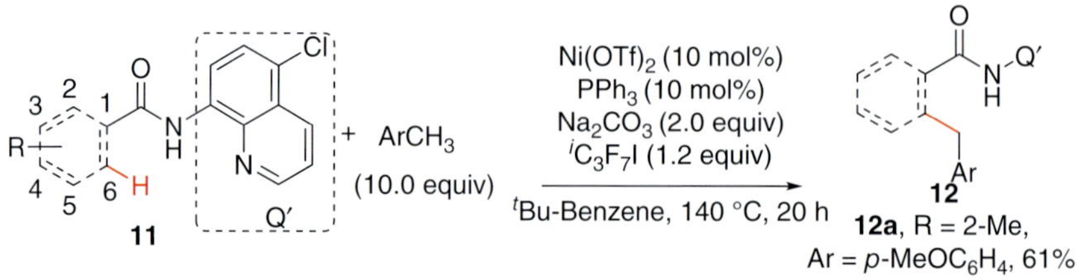

Scheme 5.15 Ni-catalyzed benzylation of C(sp^2)—H bonds with toluene derivatives.

In the proposed Ni(II)/Ni(III)/Ni(I) catalytic cycle (Scheme 5.16), base-promoted single-electron transfer (SET) of R_f-I yielded the $R_f\cdot$ radical; subsequent hydrogen abstraction from toluene generated a benzyl radical and R_f-H. The two species (benzyl radical and $R_f\cdot$ radical) acted as critical single electron oxidant participating in the generation of cyclo-Ni(III) species **INT 3** and regeneration of Ni(II) species from cyclonickel(I) species. However, the Ni(II)/Ni(IV) catalytic cycle was also viable.

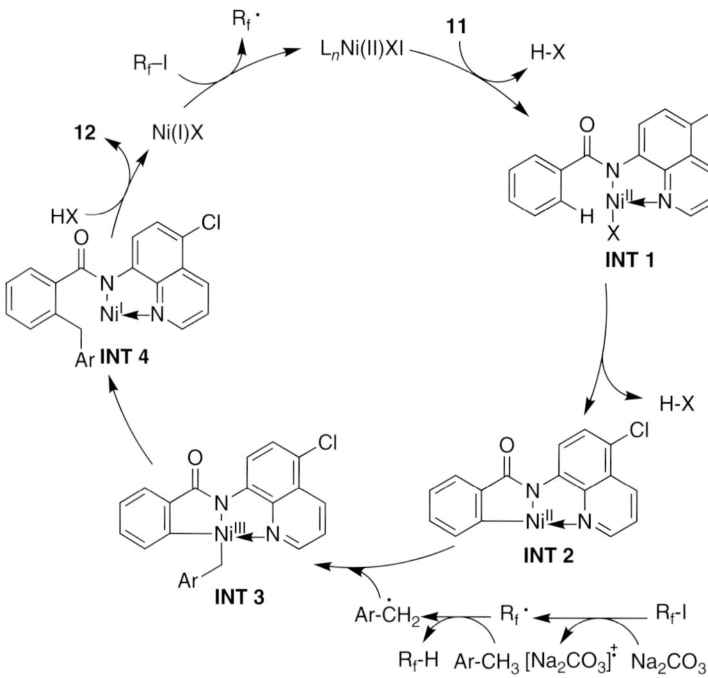

Scheme 5.16 Proposed mechanism for benzylation of C(sp^2)—H bonds.

In 2015, Maiti and coworkers reported the Ni(II)-catalyzed alkylation of C(sp^3)–H using electron-deficient olefins assisted by AQ (Scheme 5.17) [17]. Various activated alkenes, such as acrylates and acrylonitriles, could act as alkylation reagents, giving the desired alkylation products in synthetically useful yields.

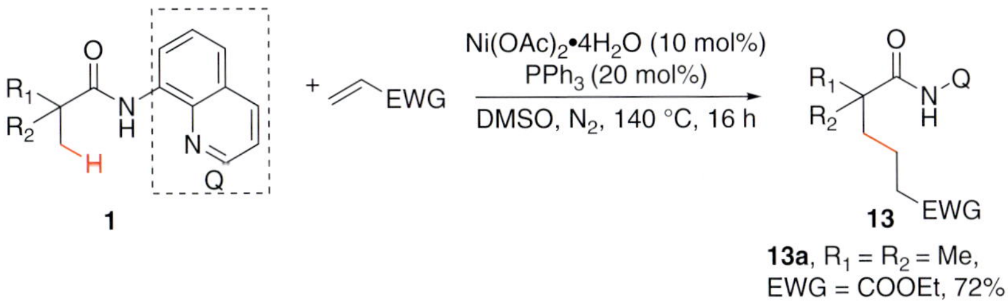

Scheme 5.17 Ni-catalyzed alkylation of C(sp^3)—H bonds with alkenes.

In 2014, Zeng and coworkers reported the Ni-catalyzed allylic alkylation of C(sp^2)—H bonds using allyl phosphates as coupling partners (Scheme 5.18) [18]. Either Ni(COD)$_2$ or Ni(OTf)$_2$ could catalyze the alkylation reaction.

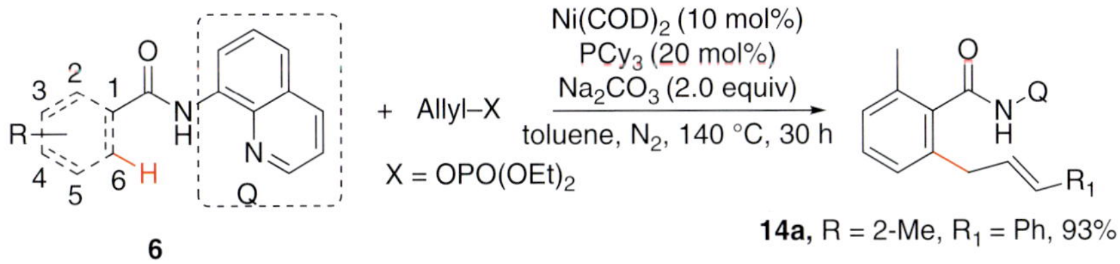

Scheme 5.18 Ni-catalyzed allylic allylation of C(sp^2)—H bonds with allyl phosphates.

In 2015, Sundararaju and coworkers reported the Ni-catalyzed allylic alkylation of benzamides using allylic bromides as allylation reagents (Scheme 5.19) [19]. A sterically bulky acid could significantly promote the reactivity.

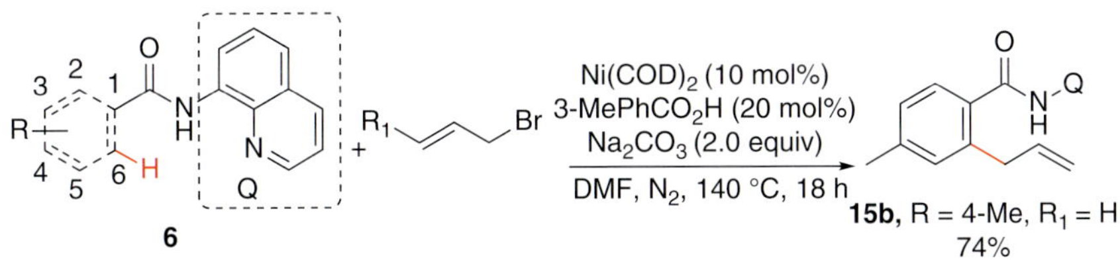

Scheme 5.19 Ni-catalyzed allylic alkylation of C(sp^2)—H bonds with allylic bromides.

In 2015, Ni-catalyzed direct benzylation, allylation, and alkylation of benzamides with various alkyl halides was reported by the Chatani group [20]. Since C–H methylation is extremely challenging, they embarked on the development of novel methylation reactions. They found that the methylation product **8a** could be obtained in 56% yield when 4.0 equiv of MeI was used. However, the *N*-methylation product was also formed in 47% yield. When MeOTs were used as the sole methylation reagent, only trace amount of **8a** was afforded. Intriguingly, the combination of 2.0 equiv of MeOTs and 2.0 equiv of NaI resulted in dramatically increasing yield (91%, Scheme 5.20).

In 2016, Chatani and coworkers reported the Ni(II)-catalyzed C(sp^2)–H methylation using $PhMe_3N^+I^-$ as methyl source (Scheme 5.21) [21]. Various

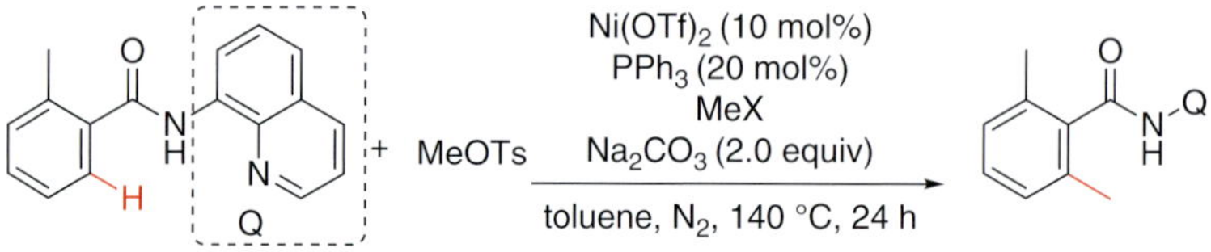

6a **8a**

MeI (4 equiv)	56% (NMR yield)
MeOTs (4 equiv)	Trace
MeOTs (2 equiv), NaI (2 equiv)	91%

Scheme 5.20 Ni-catalyzed methylation of C(sp^2)—H bonds with MeOTs and NaI.

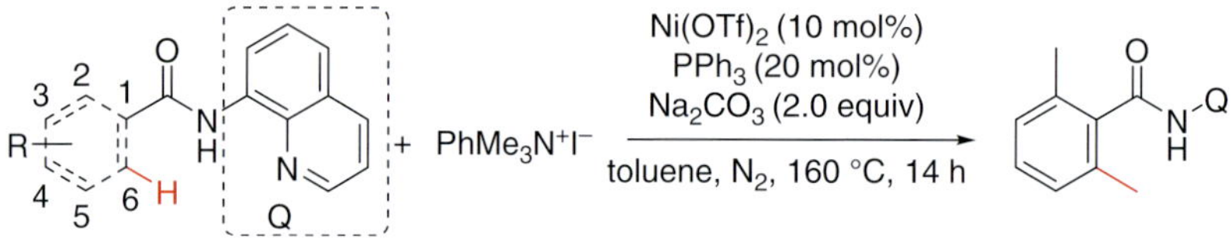

6 **8a,** R = 2-Me, 88%

Scheme 5.21 Ni-catalyzed methylation of C(sp^2)—H bonds with $PhMe_3N^+I^-$.

other phenyltrimethylammonium salts were tested and $PhMe_3N^+Br^-$ showed comparable reactivity. However, $PhMe_3N^+PF_6^-$ and $Me_4N^+I^-$ showed no reactivity. Examples of treating these ammonium salts as electrophilic arylation reagents instead of the widely used aryl halides or their pseudohalides in Ni(0)-catalyzed cross-coupling reactions have been reported; however, these ammonium salts have never been used as methylation sources. In this transformation, no arylation product was detected, indicating that the Ni(II)-catalyzed methylation might be mechanistically different from the previous Ni(0)-catalyzed arylation reaction. Accordingly, they suspected that owing to the insufficient nucleophilicity, Ni(II) species did not participate in the oxidative addition of $PhMe_3N^+I^-$ to cleave the Ph—N bond via a similar way as the Ni(0)-catalyzed reactions, but instead, a Me—N bond undergoes oxidative addition through an S_N2-type mechanism, leading to the methylation product. A Ni(II)/Ni(IV) catalytic pathway was proposed to be involved.

In 2016, another elegant work on Ni(II)-catalyzed methylation using DCP (dicumyl peroxide) as methylation reagent was reported by Chatani and coworker, (Scheme 5.22) [22]. Peroxides such as di-*tert*-butyl peroxide (DTBP)

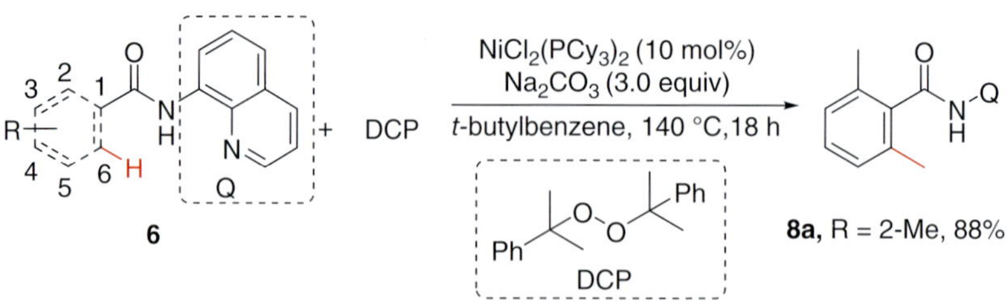

6 **8a,** R = 2-Me, 88%

Scheme 5.22 Ni-catalyzed methylation of C(sp^2)—H bonds.

and DCP undergo thermal decomposition and β-scission to generate a methyl radical, which functions as the methylating reagent. This reaction could be completely inhibited by 3.0 equiv of TEMPO, 1,4-cyclohexadiene, or α-methylstyrene under otherwise identical reaction conditions, indicating the involvement of a radical pathway.

In 2018, Wei and coworkers reported the benzylation of arenes using benzyltrimethylammonium salts as alkylation reagent (Scheme 5.23) [23]. Changing the anion to chloride and iodide slightly weakened the efficiency, while PF_6^- and OTf^- entirely shut down the reaction.

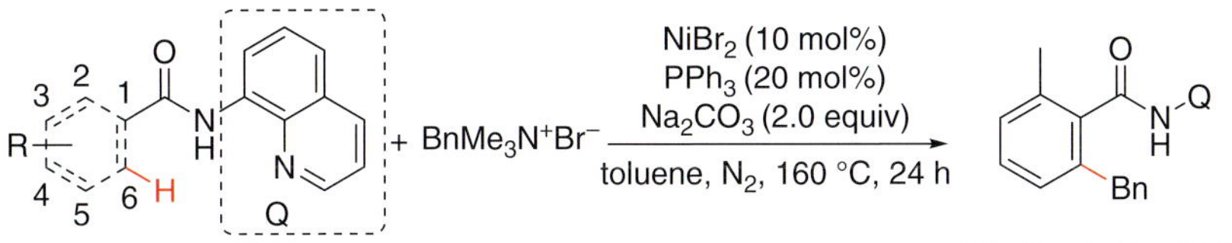

6 **12b**, R = 2-Me, 94%

Scheme 5.23 Ni-catalyzed benzylation of $C(sp^2)$—H bonds.

5.2.3 Alkenylation

In 2011, Chatani and coworkers reported the first example of Ni-catalyzed chelation-assisted functionalization of C—H bonds using an *N*,*N*′-bidentate DG [24]. In a seminal work, isoquinolones were synthesized via Ni(0)-catalyzed $C(sp^2)$–H alkenylation/annulation assisted by 2-pyridinylmethylamine (Scheme 5.24). Various other DGs have also been tested. AQ was unreactive, while 2-pyridinylethylamine could also give the desired product in 66% yield. The internal alkynes acted as coupling partner as well as hydrogen acceptor. The C–H cleavage in this reaction was believed to proceed via σ-bond metathesis mechanism and a Ni(0)/Ni(II) catalytic cycle was proposed.

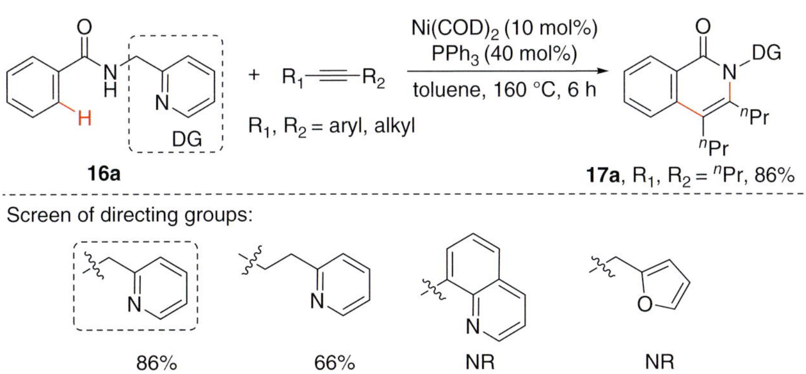

Scheme 5.24 Ni-catalyzed $C(sp^2)$–H alkenylation/annulation.

In 2015, Shi and coworkers reported the first Ni(II)-catalyzed $C(sp^3)$–H alkenylation with vinyl iodides assisted by AQ DG (Scheme 5.25) [25]. The ligand effect

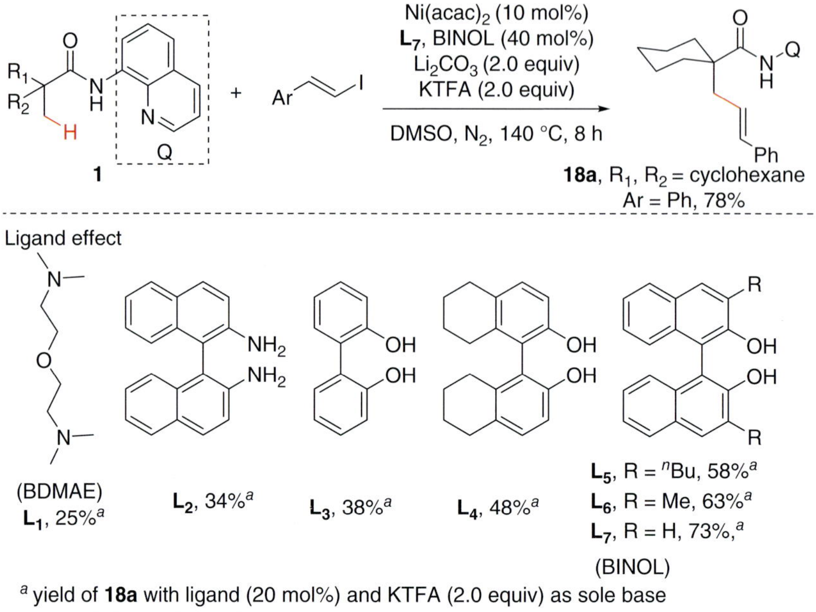

Scheme 5.25 Ni-catalyzed C(sp^3)–H alkenylation with vinyl iodides.

was extensively investigated, which showed the superior effect of diamine ligands ($\mathbf{L_1}$, $\mathbf{L_2}$) and diol-based ligands ($\mathbf{L_3}$–$\mathbf{L_7}$) toward this reaction when compared to phosphine-based ligands (PPh_3, BINAP, dppb, less than 15% yield). Finally, BINOL($\mathbf{L_7}$) proved to be the optimal ligand.

In 2015, You and coworkers reported a redox-neutral Ni(II)-catalyzed C(sp^3)–H alkenylation using internal alkynes assisted by AQ (Scheme 5.26) [26]. A similar result was independently reported by Maiti and coworkers [17].

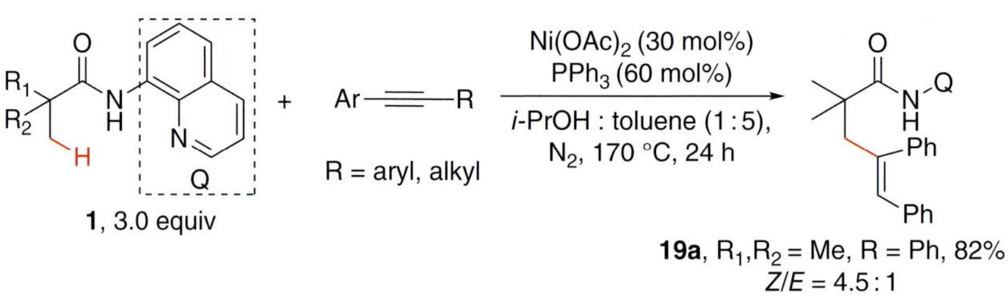

Scheme 5.26 Ni-catalyzed C(sp^3)–H alkenylation with internal alkynes.

In 2017, the Zhang group extended the alkyne insertion reaction to terminal alkynes (Scheme 5.27) [27]. In addition to aryl acetylenes, the aliphatic alkynes were also tolerated.

In 2016, Chatani and coworkers reported the direct homologation of aromatic amides with internal alkynes through a Ni-catalyzed double C(sp^2)–H activation assisted by AQ group (Scheme 5.28a) [28]. NHC ligands such as ICy·HCl, SICy·HBF4, I*t*Bu·HCl, IMes·HCl, and SIMes·HCl all showed superior reactivity

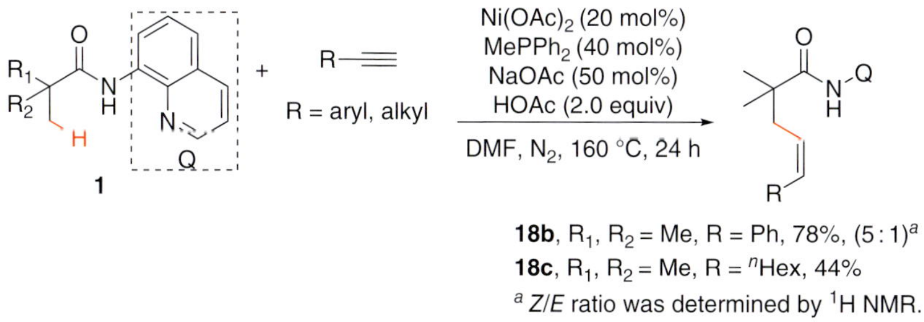

Scheme 5.27 Ni-catalyzed C(sp^3)–H alkenylation with terminal alkynes.

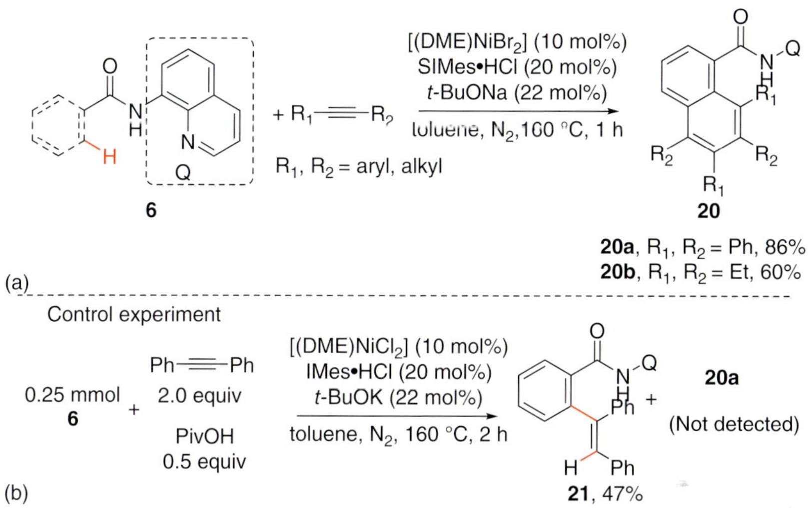

Scheme 5.28 Ni-catalyzed aromatic homologation reaction via C(sp^2)–H activation/oxidative annulation.

than PCy_3 ligand. In addition, when PivOH was added, the reaction could be quenched prior to the annulation step to afford alkene **21**, and no annulation product **20a** was obtained (Scheme 5.28b).

Shortly after, the Huang group realized the same [2+2+2]-type annulation through a substrate-assisted ligand exchange pathway (Scheme 5.29, path a) [29]. They found that a bulky, electron-rich phosphine ligand can revert the reaction to the [4+2]-annulation due to the facile C–N reductive elimination, affording isoquinolones predominantly (Scheme 5.29, path b).

5.2.4 Alkynylation

In 2015, Shi and coworkers reported the first example of Ni(II)-catalyzed C(sp^2)–H alkynylation using alkynyl bromides with the assistance of PIP (Scheme 5.30) [30]. Impressively, the reaction did not employ expensive ligands or metal oxidants. Furthermore, the catalyst loading could be reduced to as low as 0.5% (TON up to 196). A variety of bromoalkynes were well tolerated.

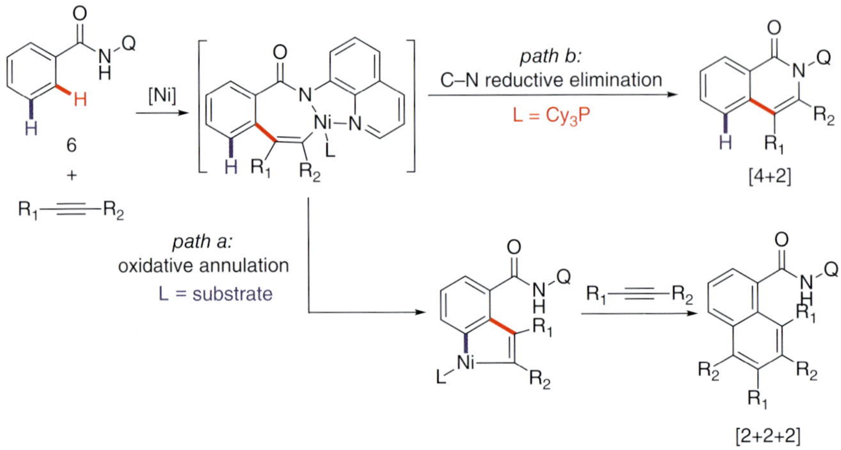

Scheme 5.29 Ni-catalyzed divergent C–H annulation of aromatic amides.

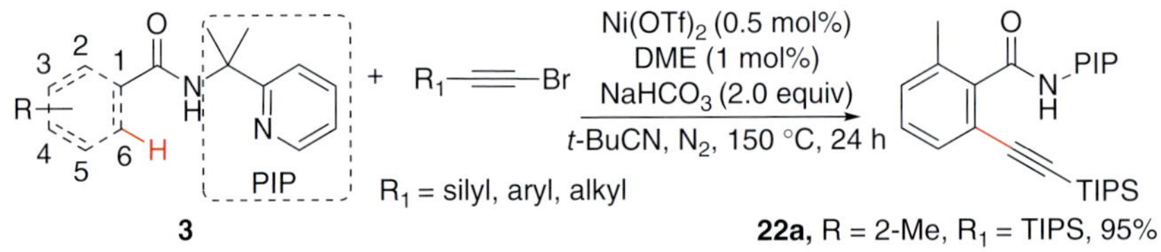

Scheme 5.30 Ni-catalyzed C(sp^2)–H alkynylation with alkynyl bromides assisted by PIP auxiliary.

Li and coworkers also reported a Ni-catalyzed C–H alkynylation using BDMAE as ligand with the assistance of AQ group (Scheme 5.31) [31].

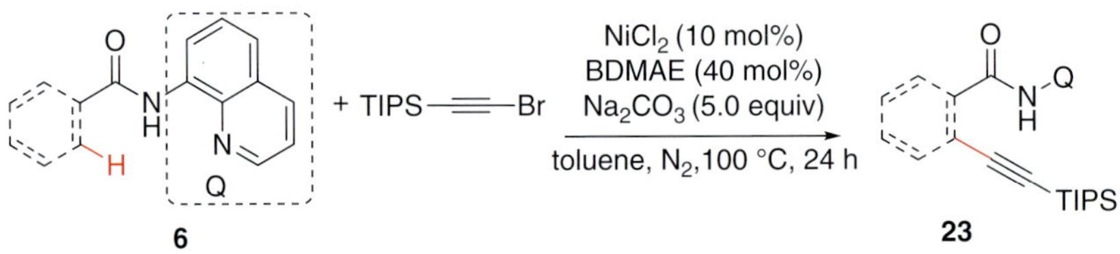

Scheme 5.31 Ni-catalyzed C(sp^2)–H alkynylation with alkynyl bromides assisted by AQ group.

Shortly after, a Ni(II)-catalyzed dehydrative alkynylation of unactivated C(sp^2)–H with terminal alkynes under atmospheric pressure of oxygen assisted by PIP auxiliary was reported by the Shi group (Scheme 5.32) [32]. The reaction featured the use of inexpensive, environmentally benign O_2 as the sole oxidant, terminal alkynes as alkylation reagents, and a simple catalytic system (no need for ligand, bases, or metal co-oxidants).

In 2016, Zhang and coworkers reported the Ni(II)-catalyzed oxidative alkynylation of C(sp^3)—H bonds using terminal alkynes with stoichiometric amount of Ag_2CO_3 as oxidant (Scheme 5.33a) [33]. Various aryl acetylenes, aliphatic

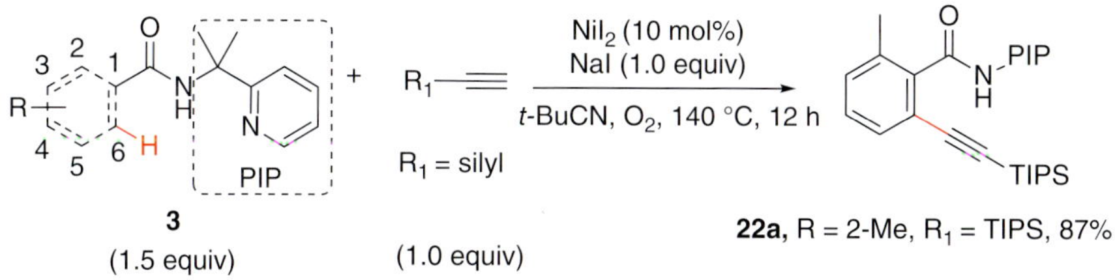

Scheme 5.32 Ni-catalyzed dehydrative alkynylation of C(sp²)–H bonds with terminal alkynes assisted by PIP auxiliary.

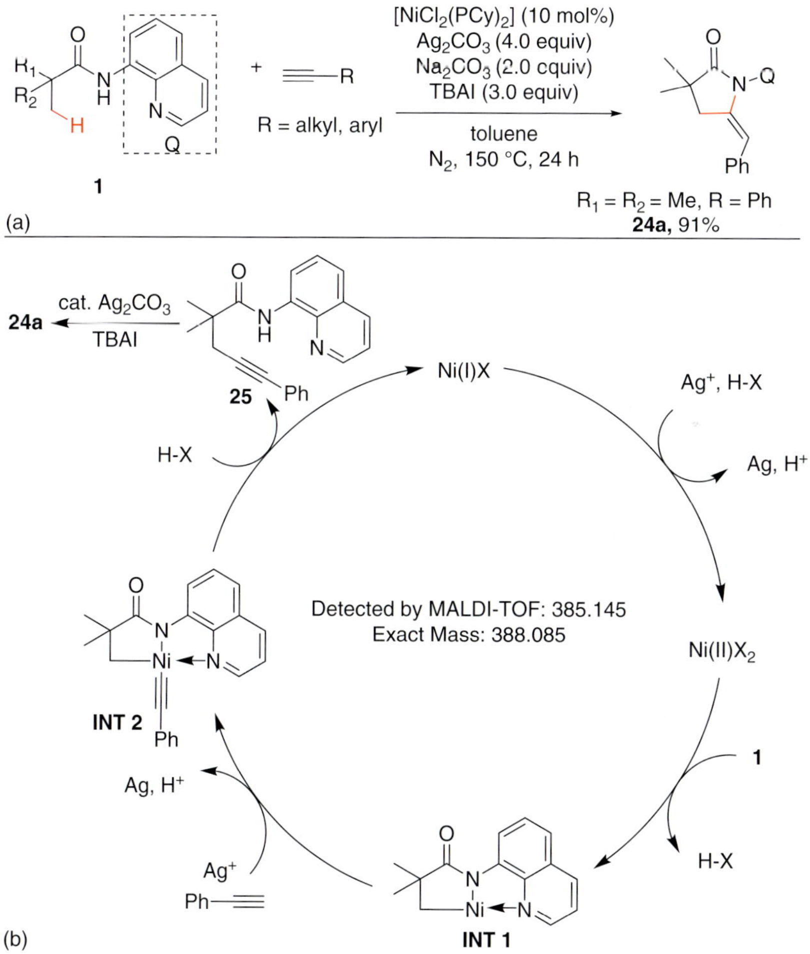

Scheme 5.33 Ni-catalyzed C–H alkynylation/tandem annulation.

alkynes, and enynes could be well tolerated. A study of the mechanism revealed that the annulation step from **25** to **24a** was mediated by Ag_2CO_3 with the assistance of TBAI. Accordingly, a Ni(II)/Ni(III)/Ni(I) catalytic cycle was proposed (Scheme 5.33b).

In 2016, Shi and coworkers reported the Ni(II) and Cu(I)-cocatalyzed C(sp^3)–H alkynylation with the assistance of the AQ group (Scheme 5.34) [34]. Stoichiometric amount of $Cu(OAc)_2$ was selected as the single-electron metal oxidant and bulky ethynyl triisopropylsilane proved to be the sole reactive alkyne coupling partner.

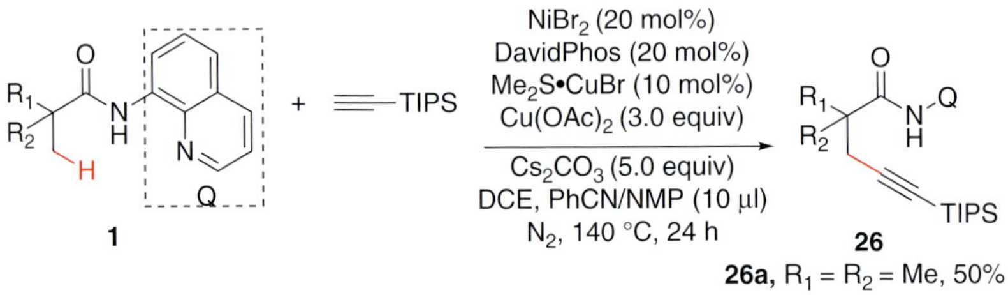

Scheme 5.34 Ni/Cu-cocatalyzed C(sp^3)–H alkynylation with bulky terminal alkyne.

In 2016, Song and coworkers reported a Ni(II)-catalyzed C(sp^2)–H alkynylation/annulation cascade reaction using O_2 as external oxidant assisted by 2-aminopyridine-1-oxide moiety (PyO) as an *N*,*O*-bidentate DG (Scheme 5.35) [35]. Similar to Zhang's work [33], this reaction proceeded through a C(sp^2)–H alkylation, followed by the intramolecular hydroamination of the resulting alkyne.

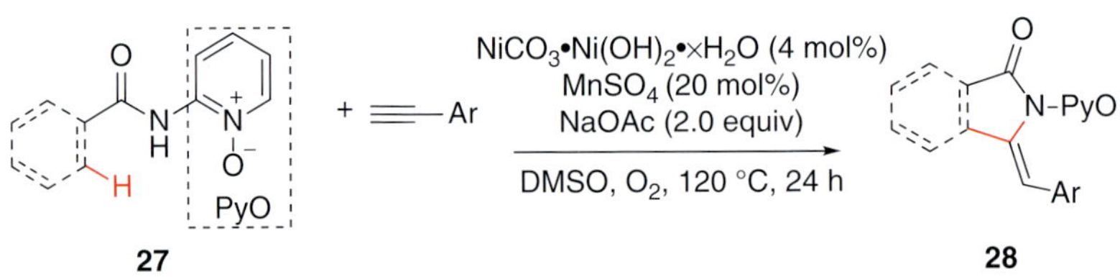

Scheme 5.35 Ni-catalyzed C(sp^2)–H alkynylation/annulation.

5.2.5 Other C—C Bond Formation Reactions Directed by Bidentate Directing Group

In 2015, Ge and coworkers reported the Ni(II)/Cu(II) co-catalyzed carbonylation of C(sp^3) and C(sp^2)—H bond using DMF as the carbonyl source under atmospheric O_2 (Scheme 5.36) [36]. Mechanistic investigations indicated that the incorporated carbonyl group mainly comes from the methyl group in DMF.

In 2016, Lei and coworkers reported the Ni(II)-catalyzed isocyanide insertion reaction of C(sp^2)—H bond using DTBP as oxidant (Scheme 5.37) [37]. A Ni(II)/Ni(III)/Ni(I) catalytic cycle was proposed to rationalize this transformation where DTBP served as a single-electron oxidant via homolytic decomposition.

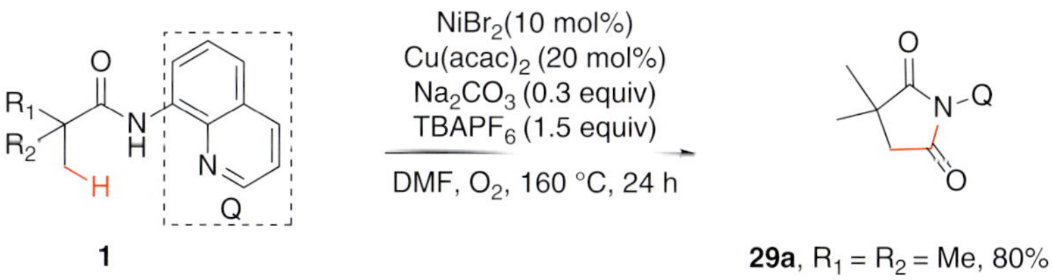

Scheme 5.36 Ni-catalyzed C–H carbonylation with DMF.

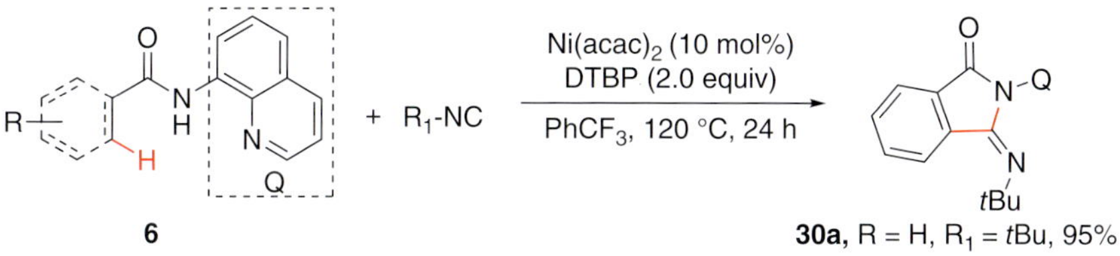

Scheme 5.37 Ni-catalyzed isocyanide insertion reaction of $C(sp^2)$—H bonds.

5.2.6 C—N Bond Formation

In 2014, Ge and coworkers reported the Ni(II)-catalyzed intramolecular oxidative amination of $C(sp^3)$–H using TEMPO as oxidant (Scheme 5.38a) [38]. A Ni(II)/Ni(III)/Ni(I) catalytic cycle was proposed to rationalize this transformation (Scheme 5.38b).

In 2015, the Zhang group reported the Ni(II)-catalyzed intermolecular $C(sp^2)$–H amination with secondary amines using Ag_2CO_3 as oxidant (Scheme 5.39) [39]. O_2 or air was also needed to further improve the yield. A tentative Ni(II)/Ni(III)/Ni(I) catalytic cycle was proposed.

In 2018, Gui and Tan group reported a Ni(II)-mediated $C(sp^2)$–H amination using sodium azide as the nitrogen source assisted by AQ (Scheme 5.40) [40]. The reaction was conducted by employing 2 equiv of $NiCl_2$ as TM mediator, delivering the primary anilines in moderate to good yields.

5.2.7 C–Chalcogen (Chalcogen = O, S, Se) Bond Formation

In 2017, the You group reported a Ni(II)-catalyzed $C(sp^3)$–H aminoxylation using stable nitroxyl radical as coupling partner (Scheme 5.41) [41]. In contrast to Ge's work [38], the cyclonickel(III) species **INT 3** (Scheme 5.38) underwent a C—O bond reductive elimination rather than a C—N bond reductive elimination.

In 2015, Shi and coworkers [42] and Yin and coworkers [43] independently reported the Ni(II)-catalyzed $C(sp^3)$–H thiolation with disulfides assisted by AQ group (Scheme 5.42).

In 2015, the Ni(II)-catalyzed $C(sp^2)$–H thiolation with disulfides assisted by PIP auxiliary was reported by Shi and coworkers [44] and Lu and coworkers [45] independently (Scheme 5.43). BINOL was screened as an effective ligand in Shi's protocol while a sterically bulky acid was used in Lu's work. Ag_2CO_3 turned out to be crucial in Lu's work, while in Shi's work the solvent DMSO itself proved to be a viable oxidant to regenerate diaryl disulfides from arylthiol to enter the

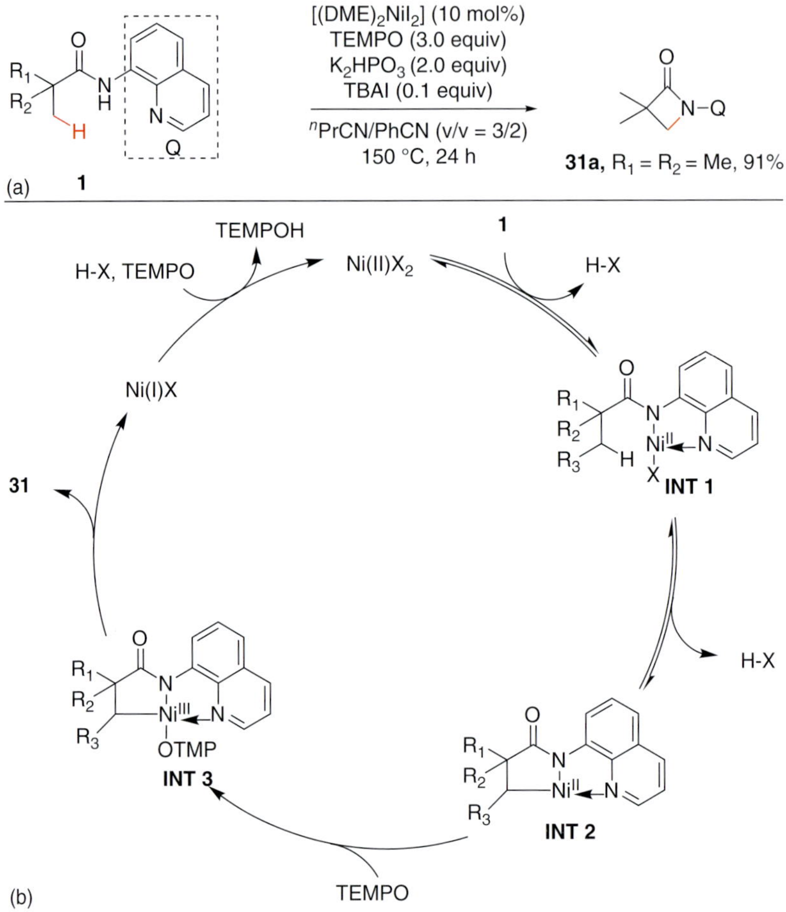

Scheme 5.38 Ni-catalyzed intramolecular C(sp^3)–H amination.

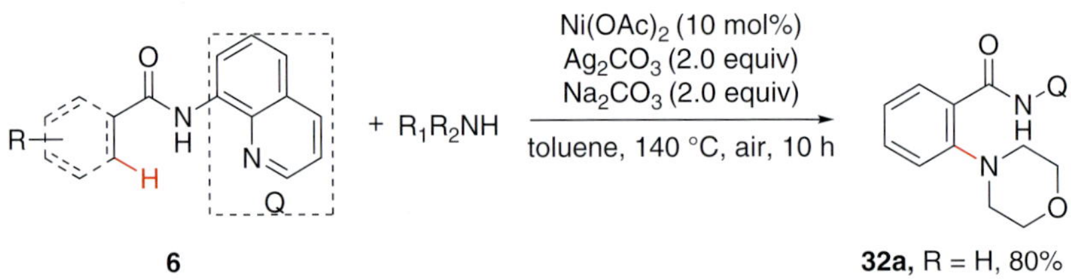

Scheme 5.39 Ni-catalyzed intermolecular C(sp^2)–H amination with secondary amines.

catalytic cycle. An analogous C(sp^2)–H thiolation assisted by AQ group was disclosed by the Kambe group [46]. Meanwhile, the Ni-catalyzed C–H thiolation with disulfides assisted by AQ group was also independently reported by Zhang and Shi groups [47].

In 2015, Chatani and coworker reported the first example of Ni(II)-catalyzed C(sp^2)–H sulfonylation with arylsulfonyl chlorides assisted by 5-chloro-8-

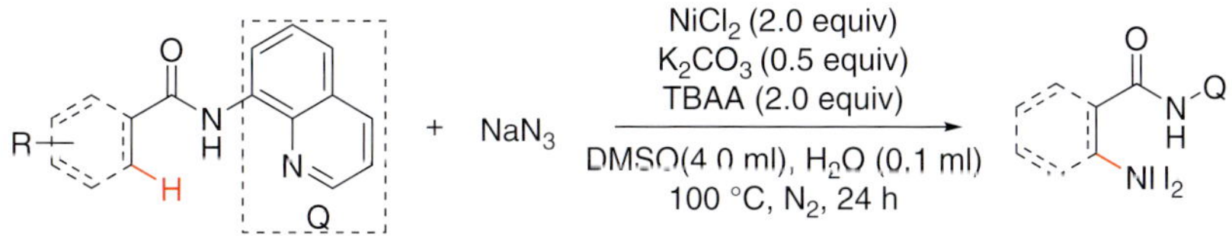

6 **33a,** R = H, 72%

Scheme 5.40 Ni-catalyzed C(sp^2)–H amination with sodium azide.

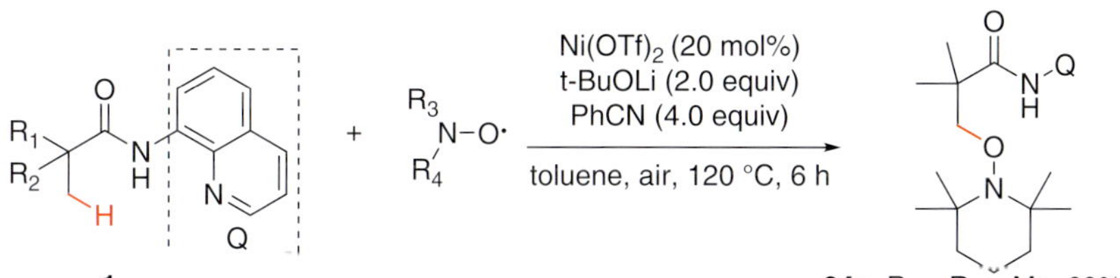

1 **34a,** $R_1 = R_2$ = Me, 69%

Scheme 5.41 Ni-catalyzed C(sp^3)–H aminoxylation.

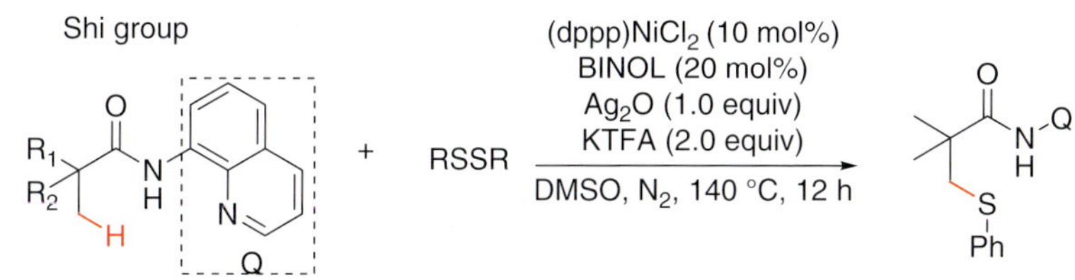

1 **35a,** R = Ph, $R_1 = R_2$ = Et, 47%

(a)

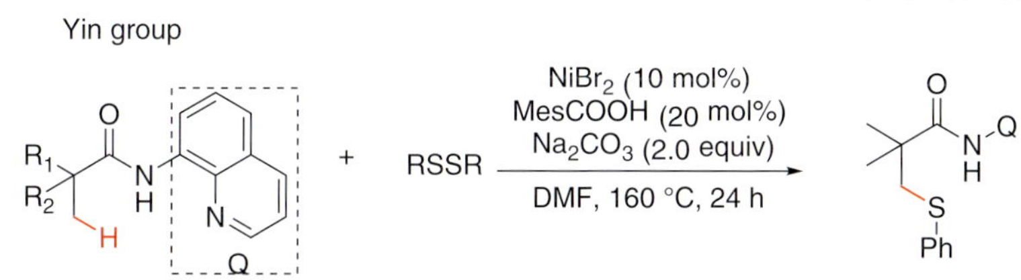

1 **35b,** R = Ph, $R_1 = R_2$ = Me, 75%

(b)

Scheme 5.42 Ni-catalyzed C(sp^3)–H thiolation assisted by AQ group.

aminoquinoline (8-AQ') auxiliary (Scheme 5.44) [48]. A block substituent at the 5-position of 8-aminoquinoline was needed to avoid the sulfonylation occurring at the 5-position.

In 2016, Liu and coworkers reported the Ni(II)-catalyzed thiolation and selenation of C(sp^2)—H bonds assisted by α-amino acid derived DG (Scheme 5.45) [49]. Substrates bearing α-amino acid moieties, such as alanine, valine, leucine, and phenylalanine, all worked well under the standard conditions. Dipeptide derivatives can also be used, albeit with decreased reactivity.

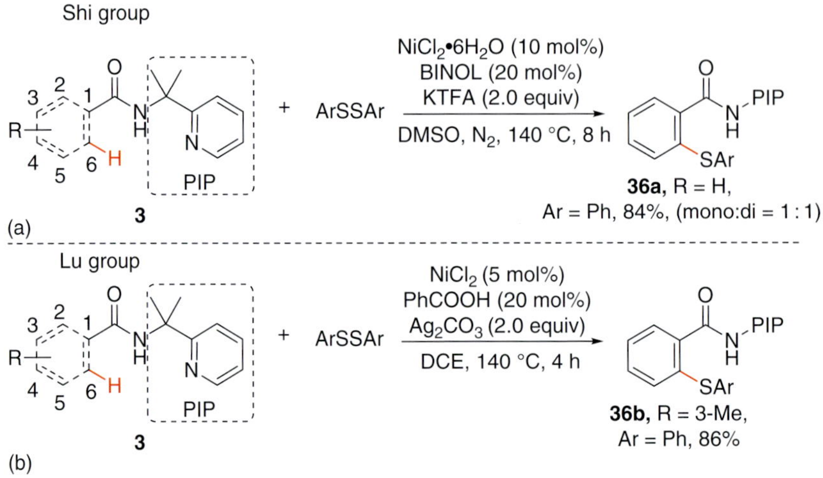

Scheme 5.43 Ni-catalyzed C(sp^2)–H thiolation assisted by PIP auxiliary.

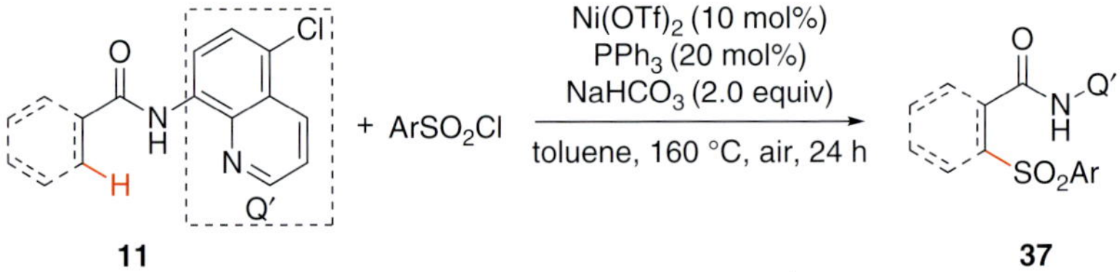

Scheme 5.44 Ni-catalyzed C(sp^2)–H sulfonylation.

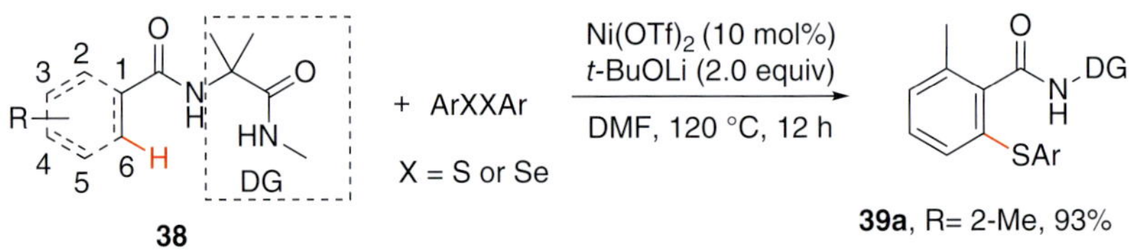

Scheme 5.45 Ni-catalyzed chalcogenation of C(sp^2)—H bonds.

In 2017, the Nishihara group reported the Ni(II)-catalyzed oxidative selenation and tandem annulation of benzamides with elemental selenium to construct isoselenazolones using O_2 as the sole oxidant (Scheme 5.46) [50]. ^{n}Bu$_4$NCl could improve the yield significantly, and probably acted as phase transfer reagent to help dissolve element selenium.

5.2.8 C–Halogen Bond Formation

In 2016, Shi and coworkers reported the Ni(II)-catalyzed C(sp^2)–H halogenation with readily available lithium halides (LiX, X = Br, I, Cl) assisted by PIP auxiliary (Scheme 5.47) [51]. $KMnO_4$ turned out to be a facile oxidant while the generally

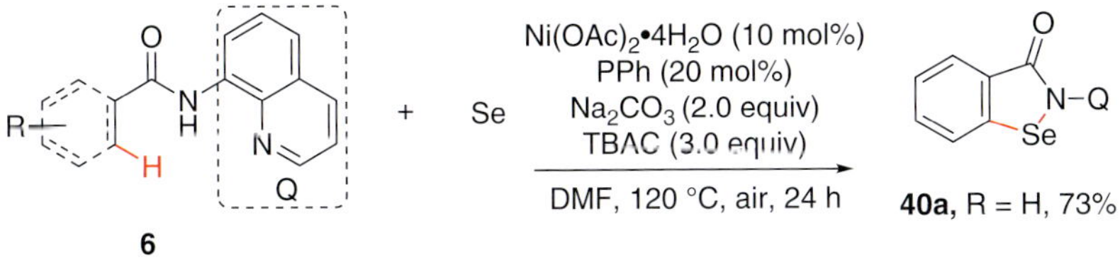

Scheme 5.46 Ni-catalyzed oxidative selenation of C(sp^2)—H bonds.

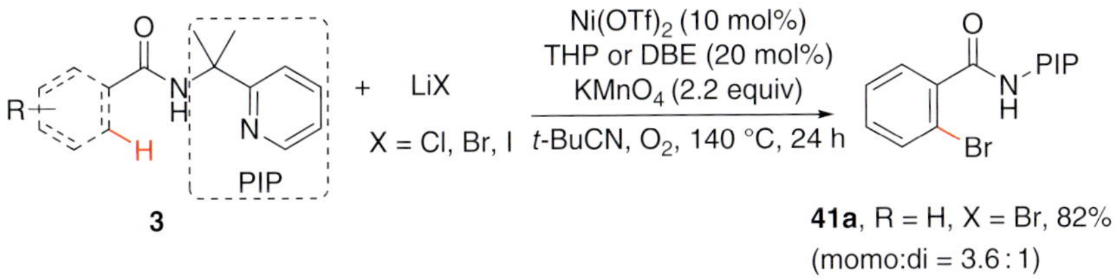

Scheme 5.47 Ni-catalyzed C(sp^2)–H halogenation with lithium halides.

used Ag_2CO_3 was ineffective. The current catalytic system could be applied to iodination, bromination as well as chlorination with slightly modified conditions.

In 2016, Chatani and coworker reported the Ni(II)-catalyzed iodination of C(sp^2)—H bonds using molecular iodine assisted by 8-amino-5-choloroquinoline (8-AQ′) (Scheme 5.48a) [52]. When 8-AQ was used as DG, the iodination occurred at the 5-position of 8-AQ. Intriguingly, when catalytic amount of Ag_2CO_3 was added and the solvent changed to DMF, this system could be applied to the C(sp^3)–H iodination. However, expected β-iodinated products were not obtained, but instead β-lactams were isolated (Scheme 5.48b). Control experiment convinced that the final product was formed via a stepwise β-iodination and subsequent intramolecular nucleophilic attack of the resulting C—I bond. Shortly after, Koley and coworkers reported a similar work using AQ as the DG [53].

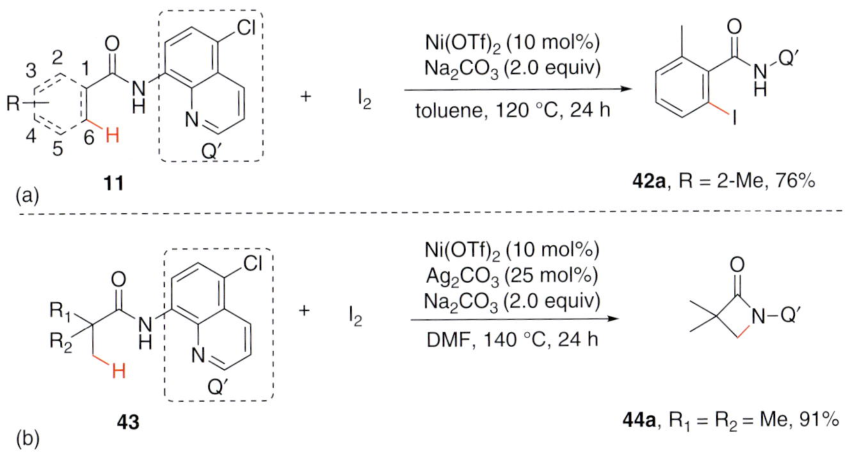

Scheme 5.48 Ni-catalyzed iodination (tandem-amination) of C—H bond.

5.3 Ni-Catalyzed Functionalization of Inert C—H Bonds Assisted by Monodentate Directing Groups

In contrast to the bidentate DG, the monodentate DGs used in these reactions contain only one coordinating site, which was commonly a neutral N atom of the pyridine or pyrimidine moiety. Example of using purine bases as DGs in the Ni-catalyzed functionalization of inert C—H bond was also reported. The substrates used in these reactions include anilines, indoles, 2-pyridones, etc.

5.3.1 Alkylation

In 2016, Ackermann and coworkers reported the Ni(II)-catalyzed C(sp^2)–H alkylation of aniline derivatives using 2-pyrimidine as the DG (Scheme 5.49a) [54].

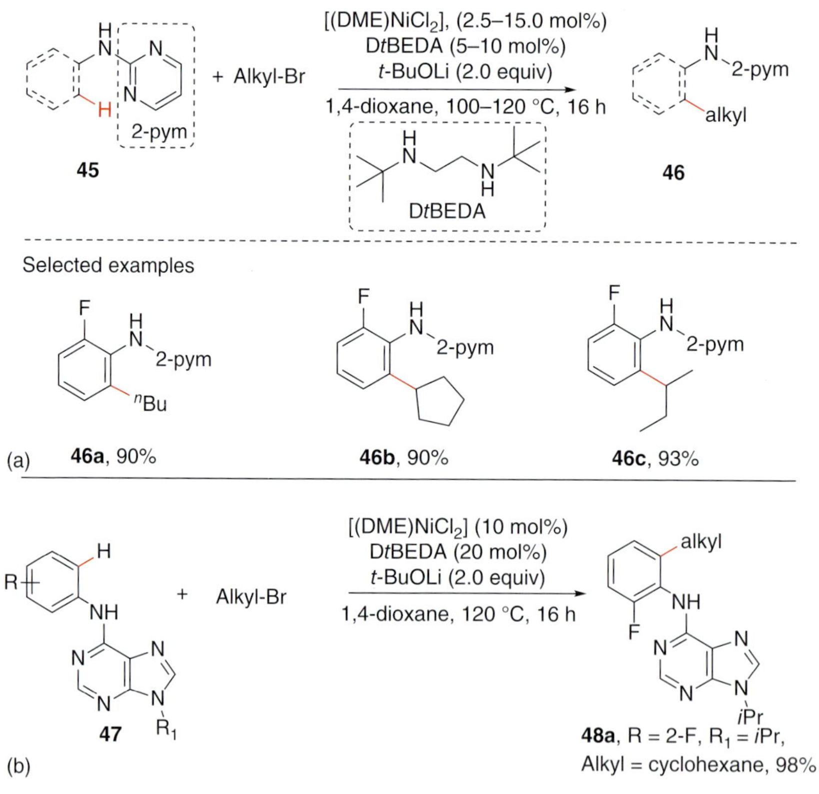

Scheme 5.49 Ni-catalyzed C(sp^2)–H alkylation of anilines.

The reaction was compatible with both primary and secondary alkyl bromides. The key factor for the highly efficient transformation was the choice of a vicinal secondary diamine (D*t*BEDA) as ligand. Other ligands such as NHCs, PPh_3 as well as other disubstituted vicinal diamines were far less efficient. One year later, this protocol was further applied to the C(sp^2)–H alkylation of anilines using purine bases as DG by the same group (Scheme 5.49b) [55].

In 2016, Punji and coworkers reported primary or secondary alkylation at C2 position of indoles assisted by monodentate pyridine or pyrimidine DG using a well-defined (quinolinyl)amido-nickel catalyst (Scheme 5.50) [56]. LiHMDS was critical to this transformation. 20 mol% LiHMDS in conjugation with 2.0 equiv of *t*-BuOLi was also effective; however, replacing *t*-BuOLi with other mild bases (e.g. Na_2CO_3, K_2CO_3, Cs_2CO_3, NaOAc) led to substantially reduced yield. In 2017, the same group achieved the dehydrative benzylation of indole derivatives using 2-iodobutane as oxidant under an analogous catalytic system [57].

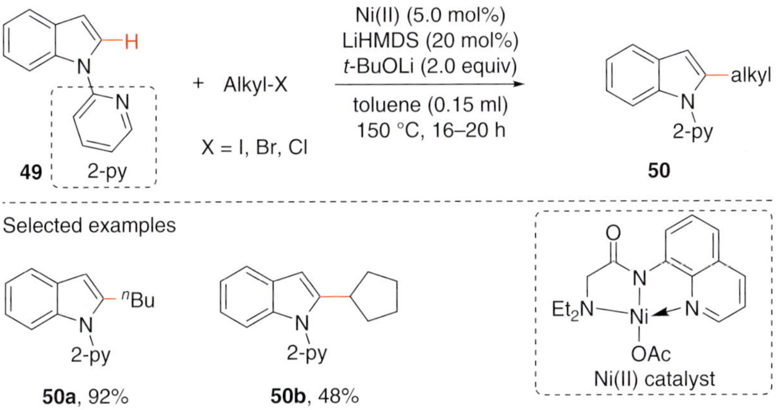

Scheme 5.50 Ni-catalyzed C(sp^2)–H alkylation of indoles.

In 2017, Miura and coworkers reported the pyridine-directed Ni(0)-catalyzed C6-selective alkylation of 2-pyridones with dienes and activated alkenes (Scheme 5.51) [58].

In 2017, Shi and coworkers reported a highly selective nickel-catalyzed C–H trifluoroethylation of heteroarenes (Scheme 5.52) [59]. This protocol provides an efficient access to various trifluoroethyl substituted heteroarenes, including indoles, pyrroles, furans, and thiophenes, with commercially available CF_3CH_2I as an alkylation reagent.

5.3.2 Alkenylation

In 2013, Ackermann and coworker reported the Ni(0)-catalyzed C(sp^2)–H alkenylation/annulation of aniline derivatives assisted by 2-pyrimidyl DG (Scheme 5.53) [60]. A variety of internal alkynes were compatible with this protocol, providing C2, C3-disubstituted indoles in good yields.

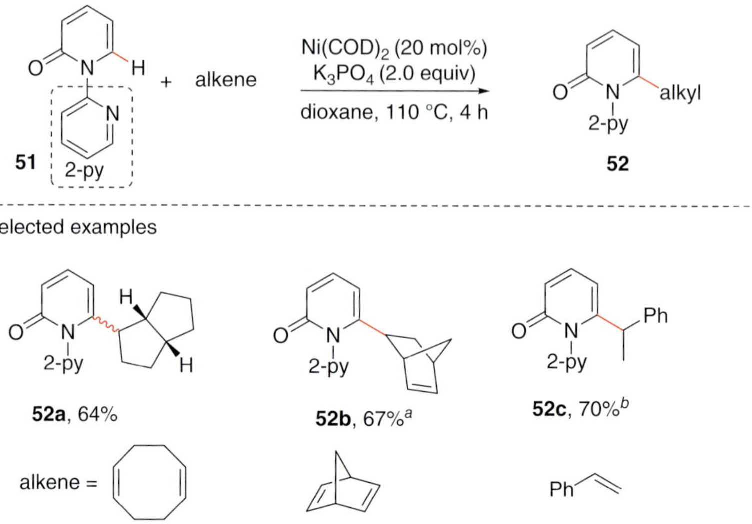

a K_3PO_4 1.0 equiv. b 4.0 equiv H_2O was added.

Scheme 5.51 Ni-catalyzed C(sp^2)–H alkylation of 2-pyridones.

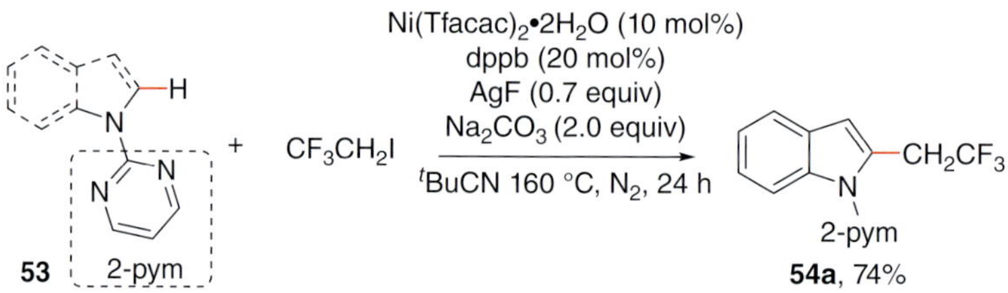

Scheme 5.52 Ni-catalyzed trifluoroethylation of C(sp^2)—H bonds.

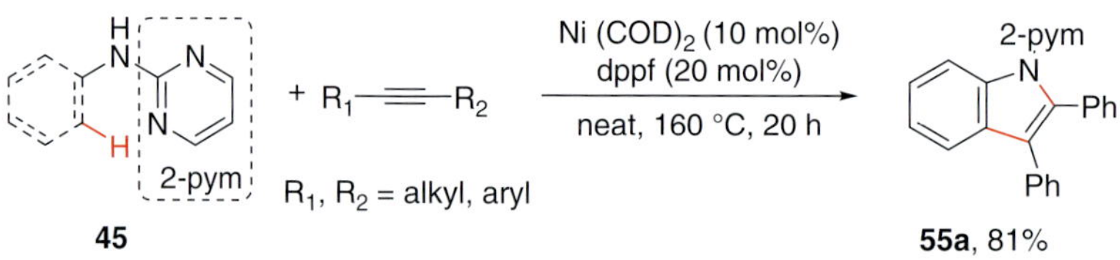

Scheme 5.53 Ni-catalyzed C(sp^2)–H alkenylation/annulation of anilines.

5.3.3 Alkynylation

In 2016, Ackermann and coworkers reported the C(sp^2)–H alkynylation of aniline derivatives using bulky silyl protected alkynyl bromides with the assistance of 2-primidyl or purine nucleobase as monodentate DG (Scheme 5.54) [61]. Among a variety of ligands, such as PPh_3, PCy_3, dppe, X-phos, bisoxazoline, BINOL, and various diamine ligands, D*t*BEDA was identified as the optimal. *t*BuOLi was also necessary and other bases, such as Na_2CO_3, K_2CO_3, LiHMDS, KHMDS, and tBuOK, were not effective.

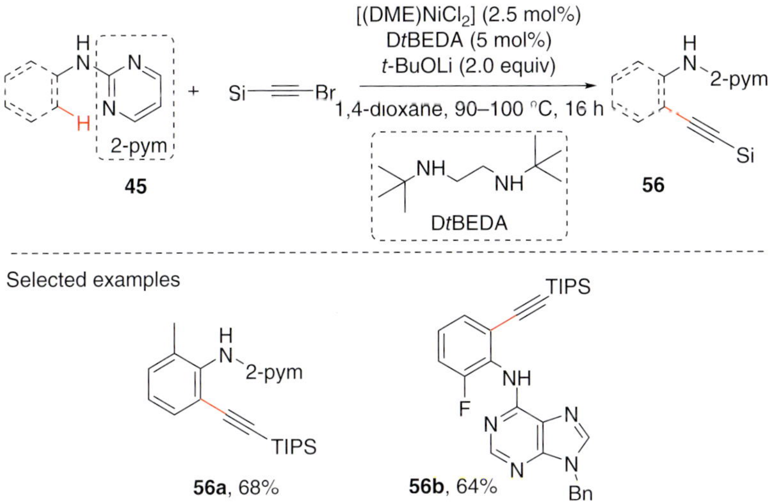

Scheme 5.54 Ni-catalyzed C(sp^2)–H alkynylation of anilines.

In 2016, Punji and coworkers reported the C(sp^2)–H alkynylation at the C–2 position of indoles, pyrroles, imidazoles, and pyrazoles with alkynyl bromides assisted by pyridine DG (Scheme 5.55) [62]. *t*BuOLi was essential for this reaction. 1,10-Phenanthroline (Phen) was identified to be the optimal ligand. [(Phen)$_3$Ni]NiBr$_4$ (**Cat. 1**) was successfully isolated and was found to be as effective as (THF)$_2$NiBr$_2$/Phen.

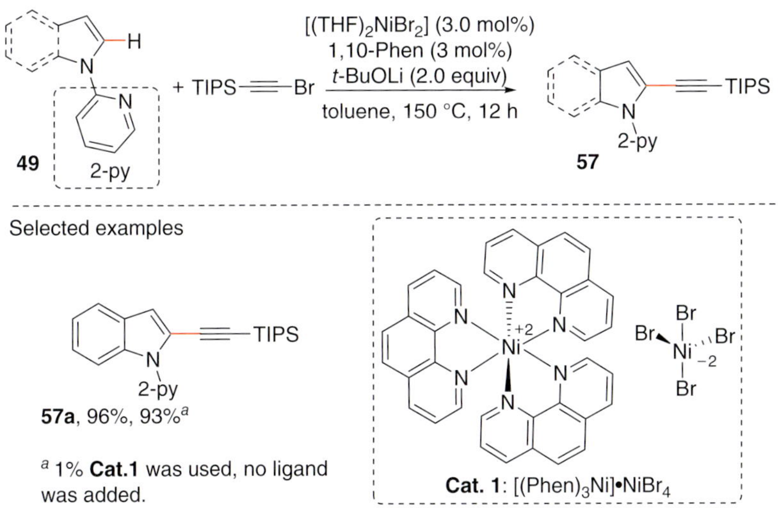

Scheme 5.55 Ni-catalyzed C(sp^2)–H alkynylation of indoles.

5.3.4 C–Calcogen Bond Formation

In 2016, Ackermann and coworkers reported the Ni(II)-catalyzed C(sp^2)–H thiolation of aniline derivatives using pyrimidine as DG (Scheme 5.56) [63].

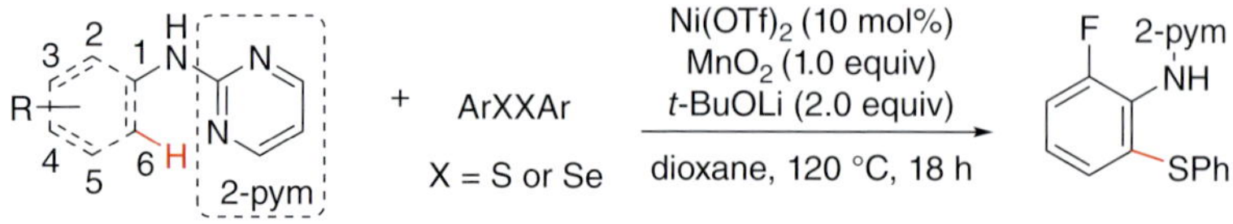

45

58a, 96%

Scheme 5.56 Ni-catalyzed C(sp^2)–H thiolation and selenylation of anilines.

This transformation proceeded smoothly without any external ligand. However, additional oxidant (MnO_2) was critical to dramatically promote the yield while Ag_2CO_3 was far less effective. Furthermore, the present protocol could be applied to the C–H selenylation under otherwise identical reaction conditions.

5.4 Summary

This chapter summarized the recent advances in nickel-catalyzed chelation-assisted inert C–H functionalization reactions. Ever since the pioneer work of Chatani, this field has witnessed significant progress during the past decade, highlighting the reactional flexibility and functional diversity of nickel catalyst. Despite these elegant advancements, there are still several challenges that need to be overcome: for example, hush conditions (higher than 130 °C or using strong bases, such as *t*-BuOLi); the necessity of a α-tertiary carbon atom adjacent to the reactive methyl group in the C(sp^3)–H functionalization reaction; the challenge in activating the more inert methylene group in secondary alkanes. To overcome these problems, more structurally and functionally diverse DGs as well as ligands should be developed. Furthermore, enormous prospects still remain undeveloped in nickel-catalyzed C–H functionalization when compared to palladium (e.g. transient DGs [64], remote C–H activation [65], enantioselective C–H activation [66]), thus leaving huge opportunities for those who are highly involved in this area.

References

1 (a) Chu, J.C.K. and Rovis, T. (2018). *Angew. Chem. Int. Ed.* 57: 62. (b) Larsen, M.A. and Hartwig, J.F. (2016). *ACS Cent. Sci.* 2: 281. (c) Park, Y., Kim, Y., and Chang, S. (2017). *Chem. Rev.* 117: 9247. (d) Mishra, N.K., Sharma, S., Park, J. et al. (2017). *ACS Catal.* 7: 2821. (e) Zhu, R.-Y., Farmer, M.E., Chen, Y.-Q., and Yu, J.-Q. (2016). *Angew. Chem. Int. Ed.* 55: 10578. (f) Hummel, J.R., Boerth, J.A., and Ellman, J.A. (2017). *Chem. Rev.* 117: 9163. (g) Liu, B., Hu, F., and Shi, B.-F. (2015). *ACS Catal.* 5: 1863. (h) Chen, Z., Wang, B., Zhang, J. et al. (2015). *Org. Chem. Front.* 2: 1107. (i) Kuhl, N., Hopkinson, M.N., Wencel-Delord, J., and Glorius, F. (2012). *Angew. Chem. Int. Ed.* 51: 10236. (j) Song, G., Wang, F., and Li, X. (2012). *Chem. Soc. Rev.* 41: 3651. (k) Sun, C.-L., Li, B.-J., and Shi, Z.-J. (2011). *Chem. Rev.* 111: 1293. (l) Ackermann,

L. (2011). *Chem. Rev.* 111: 1315. (m) Lyons, T.W. and Sanford, M.S. (2010). *Chem. Rev.* 110: 1147.
2 (a) Cera, G. and Ackermann, L. (2016). *Top. Curr. Chem.* 374: 57. (b) Usman, M., Ren, Z.-H., Wang, Y.-Y., and Guan, Z.-H. (2017). *Synthesis* 49: 1419. (c) Moselage, M., Li, J., and Ackermann, L. (2016). *ACS Catal.* 6: 498. (d) Yamaguchi, J., Muto, K., and Itami, K. (2016). *Top. Curr. Chem.* 374: 55. (e) Castro, L.C.M. and Chatani, N. (2015). *Chem. Lett.* 44: 410. (f) Zhan, B.-B., Liu, B., and Shi, B.-F. (2015). *Chin. Sci. Bull.* 60: 2907. (g) Rao, W.-H. and Shi, B.-F. (2016). *Org. Chem. Front.* 3: 1028. (h) Liu, J., Chen, G., and Tan, Z. (2016). *Adv. Synth. Catal.* 358: 1174. (i) Liang, Y., Liang, Y.-F., and Jiao, N. (2015). *Org. Chem. Front.* 2: 403. (j) Daugulis, O., Do, H.-Q., and Shabashov, D. (2009). *Acc. Chem. Res.* 42: 1074.
3 Tasker, S.Z., Stan, E.A., and Jamison, T.F. (2014). *Nature* 509: 299.
4 (a)Yamaguchi, J., Muto, K., and Itami, K. (2013). *Eur. J. Org. Chem.*: 19. (b)Nakao, Y. (2011). *Chem. Rec.* 11: 242.
5 (a) Daugulis, O., Roane, J., and Tran, L.D. (2015). *Acc. Chem. Res.* 48: 1053. (b) Rouquet, G. and Chatani, N. (2013). *Angew. Chem. Int. Ed.* 52: 11726. (c)Zaitsev, V.G., Shabashov, D., and Daugulis, O. (2005). *J. Am. Chem. Soc.* 127: 13154.
6 (a) Chen, F.-J., Zhao, S., Hu, F. et al. (2013). *Chem. Sci.* 4: 4187. (b) Zhang, Q., Chen, K., Rao, W. et al. (2013). *Angew. Chem. Int. Ed.* 52: 13588. (c) Zhang, Q., Yin, X.-S., Chen, K. et al. (2015). *J. Am. Chem. Soc.* 137: 8219.
7 (a) Aihara, Y. and Chatani, N. (2014). *J. Am. Chem. Soc.* 136: 898. (b) Yokota, A., Aihara, Y., and Chatani, N. (2014). *J. Org. Chem.* 79: 11922. (c) Iyanaga, M., Aihara, Y., and Chatani, N. (2014). *J. Org. Chem.* 79: 11933.
8 Li, M., Dong, J., Huang, X. et al. (2014). *Chem. Commun.* 50: 3944.
9 Wang, X., Zhu, L., Chen, S. et al. (2015). *Org. Lett.* 17: 5228.
10 Liu, B., Zhang, Z.-Z., Li, X., and Shi, B.-F. (2016). *Org. Chem. Front.* 3: 897.
11 Zhao, S., Liu, B., Zhan, B.-B. et al. (2016). *Org. Lett.* 18: 4586.
12 Cheng, Y., Wu, Y., Tan, G., and You, J. (2016). *Angew. Chem. Int. Ed.* 55: 12275.
13 Aihara, Y. and Chatani, N. (2013). *J. Am. Chem. Soc.* 135: 5308.
14 Song, W., Lackner, S., and Ackermann, L. (2014). *Angew. Chem. Int. Ed.* 53: 2477.
15 Wu, X., Zhao, Y., and Ge, H. (2014). *J. Am. Chem. Soc.* 136: 1789.
16 Aihara, Y., Tobisu, M., Fukumoto, Y., and Chatani, N. (2014). *J. Am. Chem. Soc.* 136: 15509.
17 Maity, S., Agasti, S., Earsad, A.M. et al. (2015). *Chem. Eur. J.* 21: 11320.
18 Cong, X., Li, Y., Wei, Y., and Zeng, X. (2014). *Org. Lett.* 16: 3926.
19 Barsu, N., Kalsi, D., and Sundararaju, B. (2015). *Chem. Eur. J.* 21: 9364.
20 Aihara, Y., Wülbern, J., and Chatani, N. (2015). *Bull. Chem. Soc. Jpn.* 88: 438.
21 Uemura, T., Yamaguchi, M., and Chatani, N. (2016). *Angew. Chem., Int. Ed.* 55: 3162.
22 Kubo, T. and Chatani, N. (2016). *Org. Lett.* 18: 1698.
23 Li, J.-W., Zheng, Z.-J., Xiao, T.-T. et al. (2018). *Asian J. Org. Chem.* 7: 133.
24 Shiota, H., Ano, Y., Aihara, Y. et al. (2011). *J. Am. Chem. Soc.* 133: 14952.
25 Liu, Y.-J., Zhang, Z.-Z., Yan, S.-Y. et al. (2015). *Chem. Commun.* 51: 7899.

26 Li, M., Yang, Y., Zhou, D. et al. (2015). *Org. Lett.* 17: 2546.
27 Lin, C., Chen, Z., Liu, Z., and Zhang, Y. (2017). *Org. Lett.* 19: 850.
28 Misal Castro, L.C., Obata, A., Aihara, Y., and Chatani, N. (2016). *Chem. Eur. J.* 22: 1362.
29 He, Z. and Huang, Y. (2016). *ACS Catal.* 6: 7814.
30 Liu, Y.-J., Liu, Y.-H., Yan, S.-Y., and Shi, B.-F. (2015). *Chem. Commun.* 51: 6388.
31 Yi, J., Yang, L., Xia, C., and Li, F. (2015). *J. Org. Chem.* 80: 6213.
32 Liu, Y.-H., Liu, Y.-J., Yan, S.-Y., and Shi, B.-F. (2015). *Chem. Commun.* 51: 11650.
33 Lin, C., Zhang, J., Chen, Z. et al. (2016). *Adv. Synth. Catal.* 358: 1778.
34 Luo, F.-X., Cao, Z.-C., Zhao, H.-W. et al. (2017). *Organometallics* 36: 18.
35 Zheng, X.-X., Du, C., Zhao, X.-M. et al. (2016). *J. Org. Chem.* 81: 4002.
36 Wu, X., Zhao, Y., and Ge, H.-B. (2015). *J. Am. Chem. Soc.* 137: 4924.
37 Hao, W., Tian, J., Li, W. et al. (2016). *Chem. Asian J.* 11: 1664.
38 Wu, X., Zhao, Y., and Ge, H.-B. (2014). *Chem. Eur. J.* 20: 9530.
39 Yan, Q., Chen, Z., Yu, W. et al. (2015). *Org. Lett.* 17: 2482.
40 Yu, L., Chen, X., Liu, D. et al. (2018). *Adv. Synth. Catal.* 360: 1346.
41 Wang, C., Zhang, L., and You, J. (2017). *Org. Lett.* 19: 1690.
42 Yan, S.-Y., Liu, Y.-J., Liu, B. et al. (2015). *Chem. Commun.* 51: 7341.
43 Wang, X., Qiu, R., Yan, C. et al. (2015). *Org. Lett.* 17: 1970.
44 Yan, S.-Y., Liu, Y.-J., Liu, B. et al. (2015). *Chem. Commun.* 51: 4069.
45 Yang, K., Wang, Y., Chen, X. et al. (2015). *Chem. Commun.* 51: 3582.
46 Reddy, V.P., Qiu, R., Iwasaki, T., and Kambe, N. (2015). *Org. Biomol. Chem.* 13: 6803.
47 (a) Lin, C., Li, D., Wang, B. et al. (2015). *Org. Lett.* 17: 1328. (b) Ye, X., Petersen, J.L., and Shi, X. (2015). *Chem. Commun.* 51: 7863.
48 Yokota, A. and Chatani, N. (2015). *Chem. Lett.* 44: 902.
49 Gao, F., Zhu, W., Zhang, D. et al. (2016). *J. Org. Chem.* 81: 9122.
50 Iwasaki, M., Miki, N., Tsuchiya, Y. et al. (2017). *Org. Lett.* 19: 1092.
51 Zhan, B.-B., Liu, Y.-H., Hu, F., and Shi, B.-F. (2016). *Chem. Commun.* 52: 4934.
52 Aihara, Y. and Chatani, N. (2016). *ACS Catal.* 6: 4323.
53 Khan, B., Kant, R., and Koley, D. (2016). *Adv. Synth. Catal.* 358: 2352.
54 Ruan, Z., Lackner, S., and Ackermann, L. (2016). *Angew. Chem. Int. Ed.* 55: 3153.
55 Ruan, Z., Ghorai, D., Zanoni, G., and Ackermann, L. (2017). *Chem. Commun.* 53: 9113.
56 Soni, V., Jagtap, R.A., Gonnade, R.G., and Punji, B. (2016). *ACS Catal.* 6: 5666.
57 Soni, V., Khake, S.M., and Punji, B. (2017). *ACS Catal.* 7: 4202.
58 Miura, W., Hirano, K., and Miura, M. (2017). *J. Org. Chem.* 82: 5337.
59 Yan, S.-Y., Zhang, Z.-Z., and Shi, B.-F. (2017). *Chem. Commun.* 53: 10287.
60 Song, W. and Ackermann, L. (2013). *Chem. Commun.* 49: 6638.
61 Ruan, Z., Lackner, S., and Ackermann, L. (2016). *ACS Catal.* 6: 4690.
62 Khake, S.M., Soni, V., Gonnade, R.G., and Punji, B. (2017). *Chem. Eur. J.* 23: 2907.

63 Müller, T. and Ackermann, L. (2016). *Chem. Eur. J.* 22: 14151.
64 (a) Kim, D.-S., Park, W.-J., and Jun, C.-H. (2017). *Chem. Rev.* 117: 8977. (b) Gandeepan, P. and Ackermann, L. (2018). *Chem* 4: 199. (c) Bhattacharya, T., Pimparkar, S., and Maiti, D. (2018). *RSC Adv.* 8: 19456. (d) St John-Campbell, S. and Bull, J.A. (2008). *Org. Biomol. Chem.* 16: 4582.
65 (a) Mihai, M.T., Genov, G.R., and Phipps, R.J. (2018). *Chem. Soc. Rev.* 47: 149. (b) Dey, A., Agasti, S., and Maiti, D. (2016). *Org. Biomol. Chem.* 14: 5440.
66 (a) Saint-Denis, T.G., Zhu, R.-Y., Chen, G. et al. (2018). *Science* 359: 759. (b) Newton, C.G., Wang, S.-G., Oliveira, C.C., and Cramer, N. (2017). *Chem. Rev.* 117: 8908. (c) Zheng, C. and You, S.-L. (2014). *RSC Adv.* 4: 6173. (d) Wencel-Delord, J. and Colobert, F. (2013). *Chem. Eur. J.* 19: 14010.

6

C—C Bond Functionalization

Yoshiaki Nakao

Kyoto University, Department of Material Chemistry, Graduate School of Engineering, Katsura, Nishikyo-ku, Kyoto 615-8510, Japan

6.1 Introduction

C—C bond activation has gained increasing attention because it can be an elemental reaction to develop new and innovative synthetic transformations. In particular, the C—C bond activation and functionalization of readily available organic compounds can provide synthetic organic chemists with opportunities to have novel and unique ideas for retro-synthetic analysis to construct target molecules. Metal catalysis plays key and unique roles in C—C bond activation, and nickel catalysts have been one of the major metal catalysts enabling C—C bond activation of several different classes of organic molecules. In this chapter, examples of nickel-catalyzed C—C bond functionalization are introduced.

There are two major modes to activate C—C bonds by nickel catalysis. One is the oxidative addition of C—C bonds to nickel(0) complexes (Scheme 6.1). An elemental step has been proposed for the activation of C—CN bonds of nitriles and C—acyl bonds of some ketones. The other mode for C—C bond activation is β-carbon elimination (Scheme 6.2). These processes have been integrated with other common elemental reactions of nickel complexes to develop novel transformations involving C—C bond activation. Nickel-catalyzed C—C bond activation reactions through decarbonylation from acylnickel(II) intermediates [1] are not discussed in this chapter.

6.2 C—C Bond Functionalization of Three-Membered Rings

The rearrangement reaction of vinylcyclopropanes can be catalyzed by nickel(0) to give cyclopentenes at room temperature (Scheme 6.3) [2]. Nickel catalysis has been applied to the intramolecular reaction of vinylcyclopropanes with alkynes to give 1,3,6-triene products (Scheme 6.4) [3]. When a substituent at the alkyne terminus is sterically hindered, the reaction gives bicyclic products bearing a cycloheptadiene ring (Scheme 6.5). The reactions are proposed to proceed through an

Nickel Catalysis in Organic Synthesis: Methods and Reactions,
First Edition. Edited by Sensuke Ogoshi.

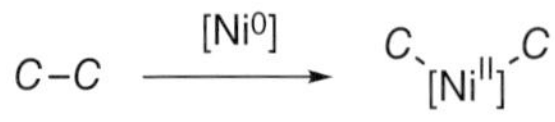

Scheme 6.1 Oxidative addition of C—C bonds to nickel(0) complexes.

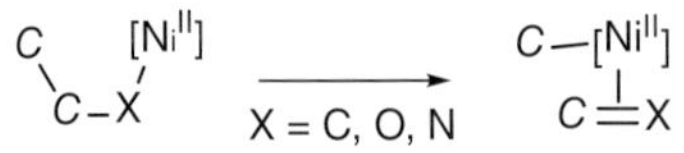

Scheme 6.2 β-Carbon elimination.

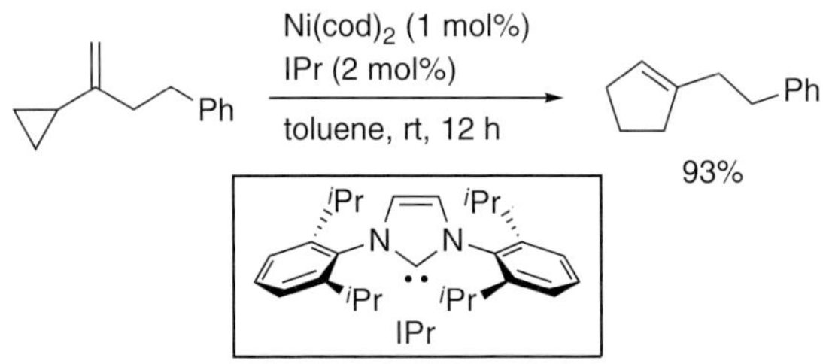

Scheme 6.3 Ni-catalyzed rearrangement of vinylcyclopropanes to cyclopenetenes.

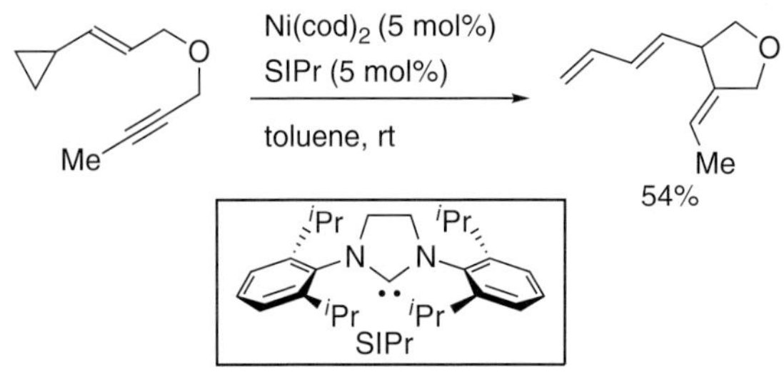

Scheme 6.4 Ni-catalyzed rearrangement of cyclopropylene-ynes to 1,3,6-trienes.

eight-membered nickelacycle intermediate derived from either the formation of nickelacyclopentene followed by β-carbon elimination or the oxidative addition of the C—C bond followed by insertion of alkynes (Scheme 6.6).

Cyclopropyl ketones undergo nickel-catalyzed reactions involving the cleavage of C—C bonds of the three-membered ring. Ring-opening coupling of cyclopropyl ketones with organometallic reagents was initially reported (Scheme 6.7) [4]. Cyclopropyl ketones also undergo cycloaddition reactions with C—C unsaturated bonds through C—C bond cleavage. The reaction with enones proceeds stereospecifically to give highly substituted cyclopentanes (Scheme 6.8) [5, 6], whereas the reaction across alkynes affords substituted cyclopentenes (Scheme 6.9) [7]. A proposed catalytic cycle is initiated by the formation of a six-membered nickeladihydropyran intermediate, which has been characterized and reacts with alkenes or alkynes to give the five-membered products upon reductive elimination (Scheme 6.10). Similar transformations using cyclopropyl imines (Scheme 6.11) [8] and cyclopropyl carboxamides (Scheme 6.12) [9] are also available. These reactions are accelerated by Lewis acid additives. Nickel

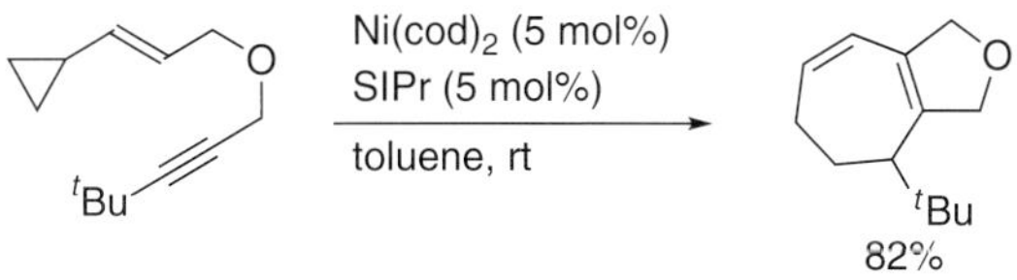

Scheme 6.5 Ni-catalyzed rearrangement of cyclopropylene-ynes to 1,4-cycloheptadienes.

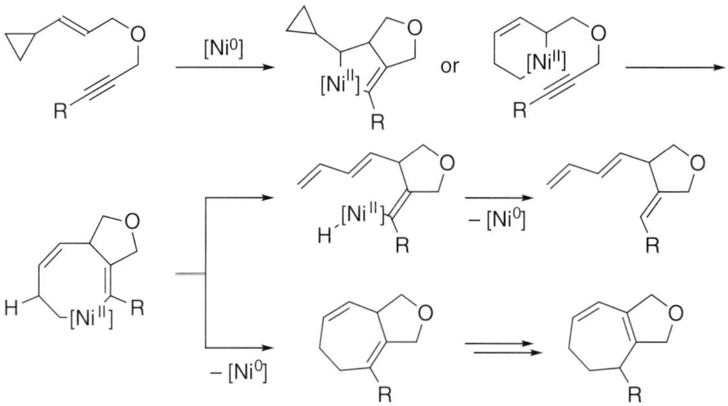

Scheme 6.6 A mechanism of Ni-catalyzed rearrangement of cyclopropylene-ynes.

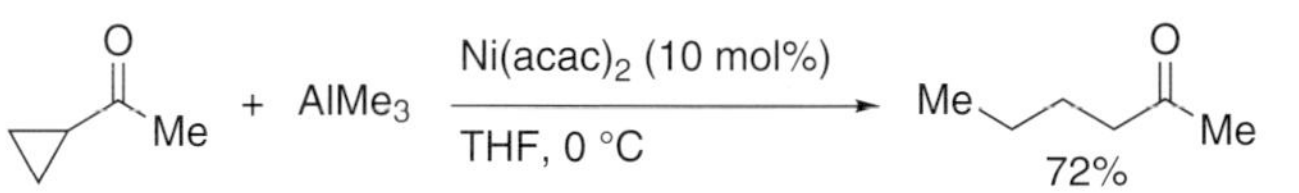

Scheme 6.7 Ni-catalyzed ring-opening coupling reaction of cyclopropyl ketones with $AlMe_3$.

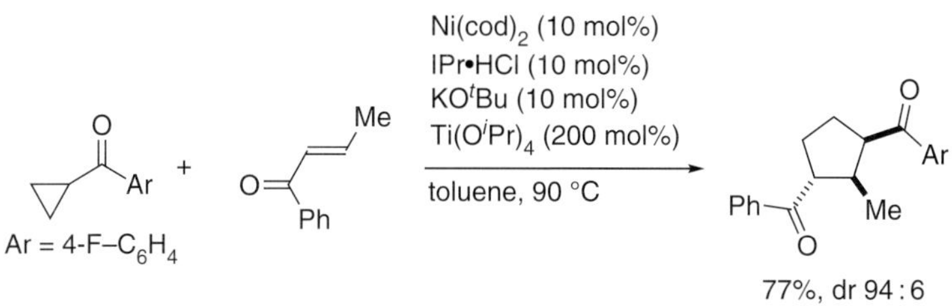

Scheme 6.8 Ni-catalyzed [3+2] cycloaddition reaction of cyclopropyl ketones with enones.

also catalyzes rearrangement of 1-acyl-2-vinylcyclopropanes to substituted dihydrofurans in a stereospecific manner (Scheme 6.13) [10].

Methylenecyclopropanes react with unsaturated compounds through C—C bond cleavage to give cyclic products by nickel catalysis in a diverse manner depending on the substituents and reaction conditions. Initial studies showed that they reacted with electron-poor olefins to give methylenecyclopentanes (Scheme 6.14) [11, 12]. Benzylidene-3-oxobutanoate also reacts with methylenecyclopropanes but gives substituted dihydrofurans (Scheme 6.15) [13]. A mechanism initiated by the oxidative addition of the C—C bond to

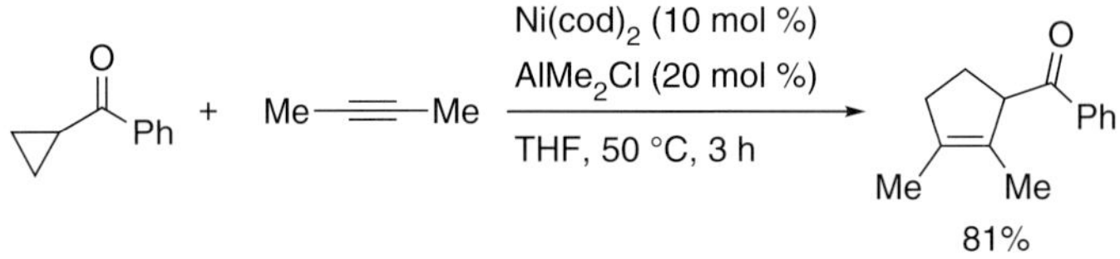

Scheme 6.9 Ni-catalyzed [3+2] cycloaddition reaction of cyclopropyl ketones with alkynes.

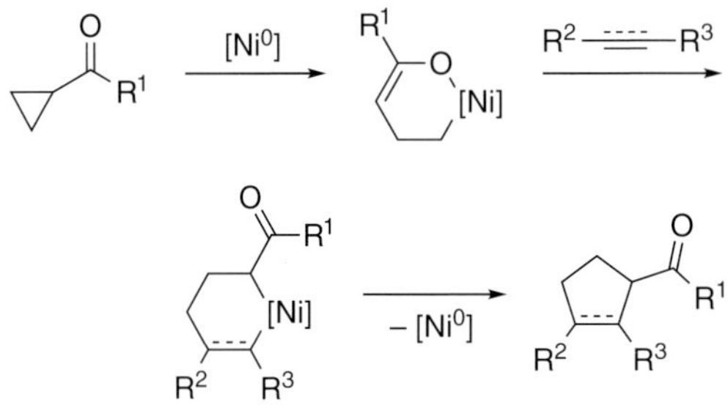

Scheme 6.10 A mechanism of Ni-catalyzed [3+2] cycloaddition reaction of cyclopropyl ketones with unsaturated compounds.

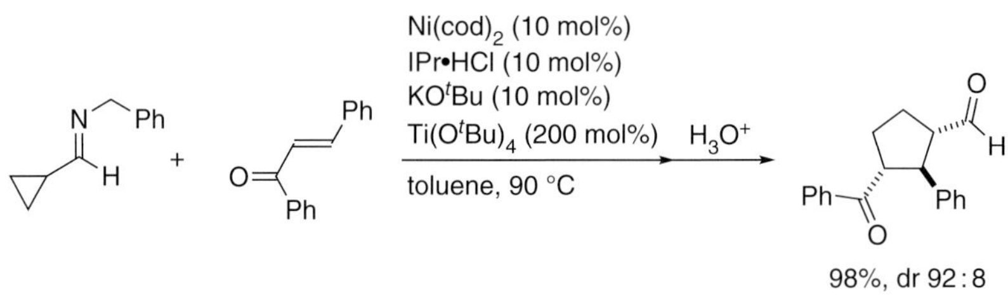

Scheme 6.11 Ni-catalyzed [3+2]-cycloaddition reaction of cyclopropyl imines with enones.

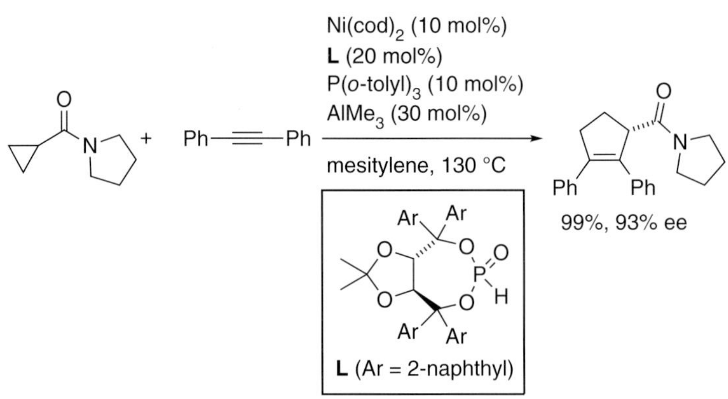

Scheme 6.12 Ni-catalyzed [3+2] cycloaddition reaction of cyclopropyl carboxamides with alkynes.

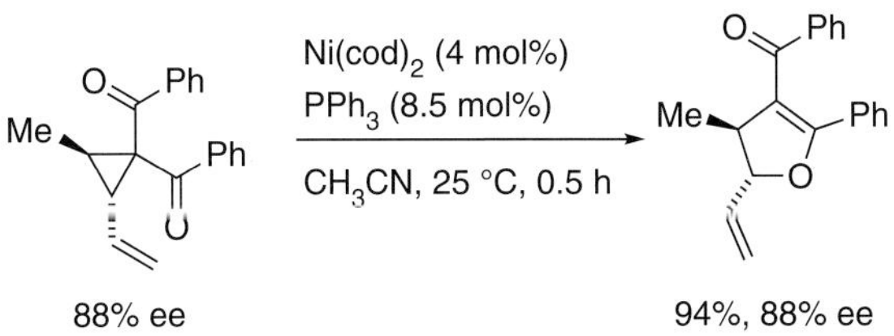

Scheme 6.13 Ni-catalyzed rearrangement of 1-acyl-2-vinylcyclopropanes to dihydrofurans.

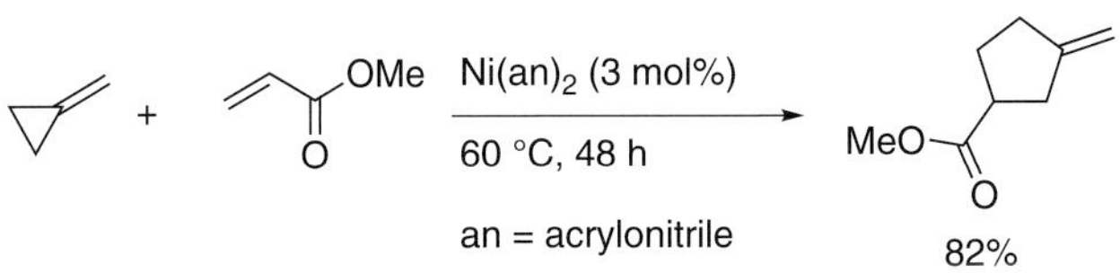

Scheme 6.14 Ni-catalyzed [3+2] cycloaddition reaction of methylenecyclopropane with methyl acrylate.

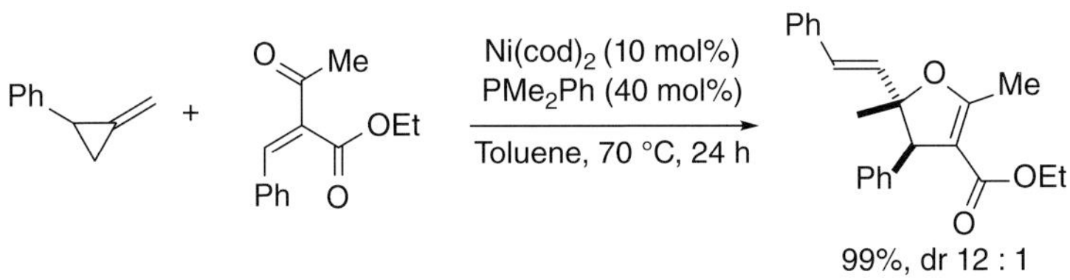

Scheme 6.15 Ni-catalyzed [3+2] cycloaddition reaction of methylenecyclopropane with methyl benzylidene-3-oxobutanoate.

nickel(0) followed by the reaction across alkenes is proposed for the former, whereas the initial formation of a nickeladihydrofuran intermediate, the C—Ni bond of which adds across methylenecyclopropanes, and subsequent β-carbon elimination followed by product-forming reductive elimination are suggested for the latter transformation. Ethyl cyclopropylideneacetate reacts with two molecules of alkynes through C—C bond cleavage to give seven-membered products by nickel catalysis (Scheme 6.16) [14]. Two alkyne substrates can be tethered to give bicyclic products (Scheme 6.17) [15] or differently to afford ethyl cycloheptylideneacetate with a variety of substituents (Scheme 6.18) [16]. 1,3-Dienes are also competent substrates for the cycloaddition reaction (Scheme 6.19) [17], and tethered dienynes can participate in the reaction to give

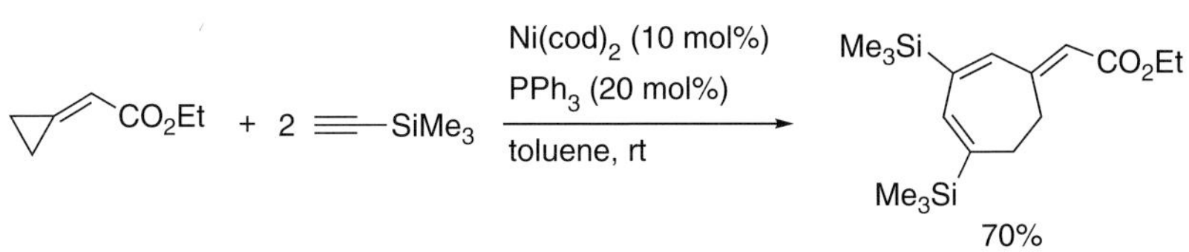

Scheme 6.16 Ni-catalyzed [3+2+2] cycloaddition reaction of ethyl cyclopropylideneacetate with alkynes.

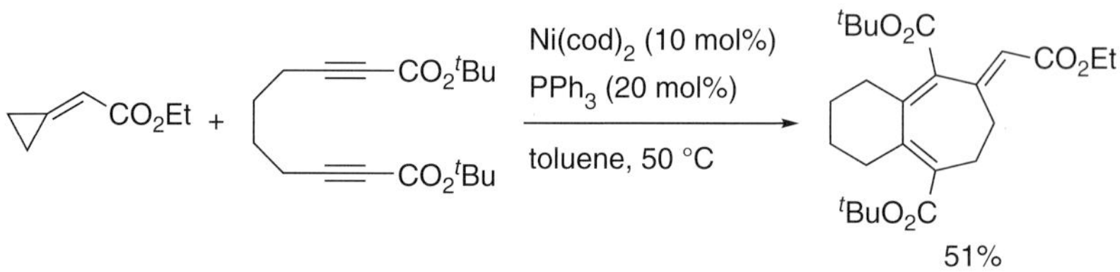

Scheme 6.17 Ni-catalyzed [3+2+2] cycloaddition reaction of ethyl cyclopropylideneacetate with diynes.

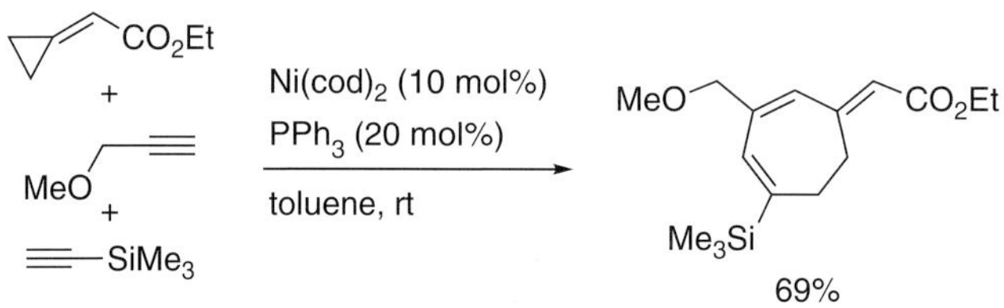

Scheme 6.18 Ni-catalyzed [3+2+2] cycloaddition reaction of ethyl cyclopropylideneacetate with two different alkynes.

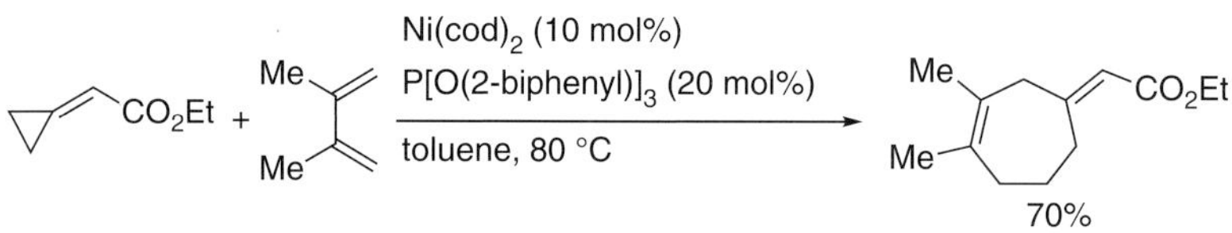

Scheme 6.19 Ni-catalyzed [4+3] cycloaddition reaction of ethyl cyclopropylideneacetate with 1,3-dienes.

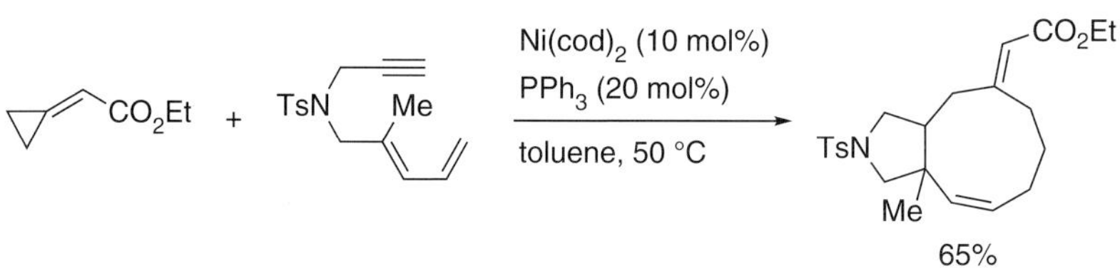

Scheme 6.20 Ni-catalyzed [4+3+2] cycloaddition reaction of ethyl cyclopropylideneacetate with dienynes.

bicyclic products bearing a nine-membered ring (Scheme 6.20) [18]. All of these reactions are likely initiated by the formation of nickelacycle species derived from nickel(0) and π-bonds of unsaturated substrates (Scheme 6.21) [19]. The double bond of ethyl cyclopropylideneacetate then inserts into the C—Ni bond. The nickelacycle intermediate could also be formed from cyclopropylideneacetate and relatively electron-rich alkynes according to theoretical calculations. It then reacts with another alkyne to expand the nickelacycle species. In either case, β-carbon elimination is proposed as a C—C bond activation step to afford the products upon C—C bond forming reductive elimination. Methyleneaziridines also react with diynes, in which only a single triple bond is incorporated to give substituted pyrroles through a mechanism (Scheme 6.22) [20].

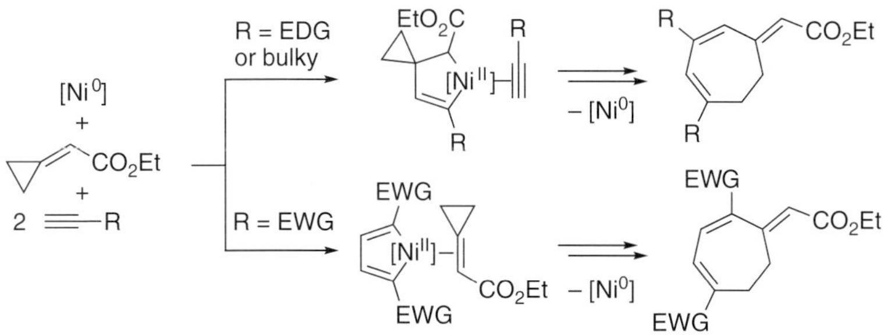

Scheme 6.21 Mechanisms of Ni-catalyzed cycloaddition reactions of ethyl cyclopropylideneacetate with alkynes.

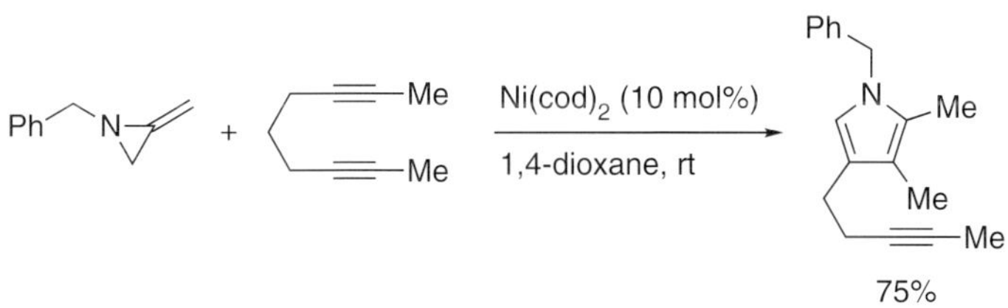

Scheme 6.22 Ni-catalyzed [3+2] cycloaddition reaction of methyleneaziridines with diynes.

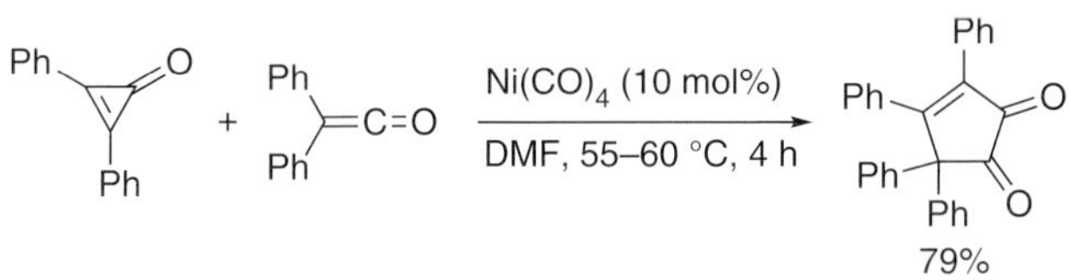

Scheme 6.23 Ni-catalyzed [3+2] cycloaddition reaction of diphenylcyclopropenone with diphenylketene.

There are reactions of other three-membered substrates through C—C bond cleavage by nickel catalysis. Cyclopropenone undergoes the reaction with ketenes to give cyclopentendiones (Scheme 6.23) [21]. Bicyclo[1.1.0]butane reacts with acrylates to give substituted cyclopropanes (Scheme 6.24) [22]. The oxidative addition of the C—C bonds to nickel(0) is proposed to initiate the respective catalytic cycles.

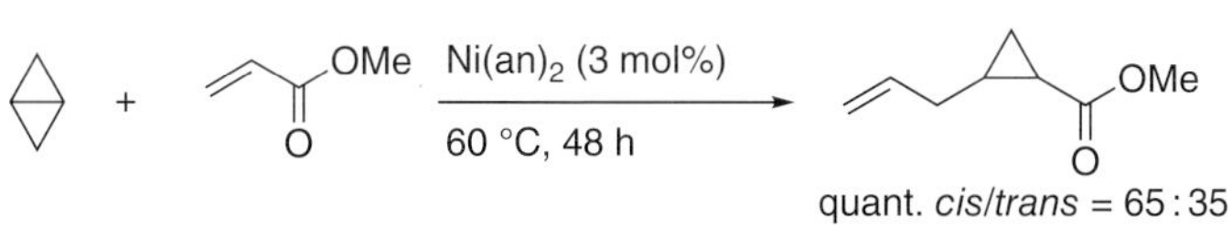

Scheme 6.24 Ni-catalyzed ring-opening reaction of bicyclo[1.1.0]butane with methyl acrylate.

6.3 C—C Bond Functionalization of Four- and Five-Membered Rings

Biphenylene undergoes the insertion of alkynes into the C—C bond to give phenanthrenes by nickel catalysis (Scheme 6.25) [23–27]. A catalytic cycle involving the oxidative addition of the C—C bond to nickel(0) followed by the insertion of alkynes and reductive elimination is proposed based on experiments and theoretical calculations. A similar catalytic cycle including the oxidative addition of C—C bonds has also been proposed for the reaction of cyclobutenones with alkynes to afford substituted phenols (Scheme 6.26) [28].

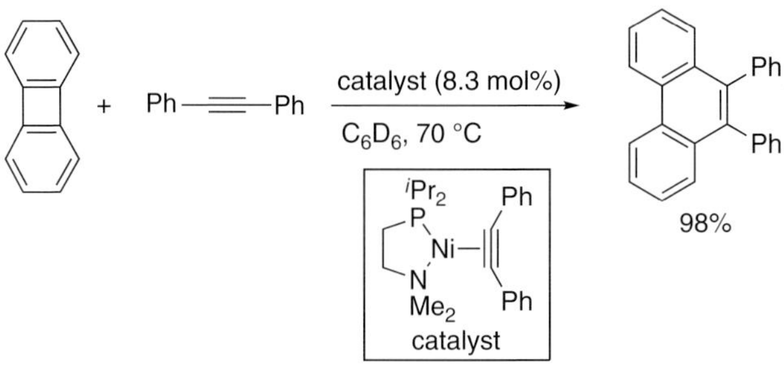

Scheme 6.25 Ni-catalyzed [4+2] cycloaddition reaction of biphenylene with diphenylacetylene.

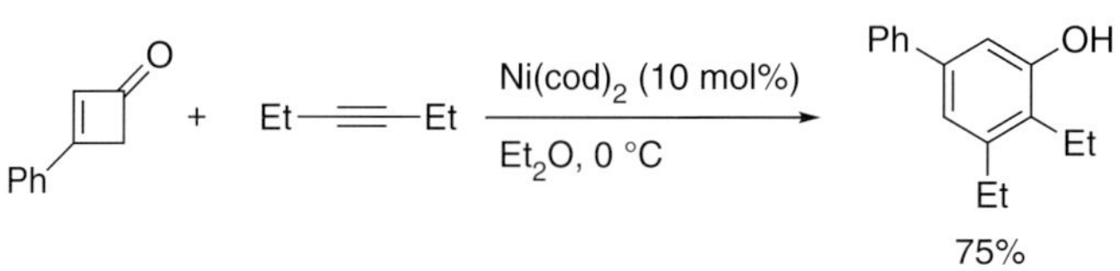

Scheme 6.26 Ni-catalyzed [4+2] cycloaddition reaction of cyclobutenones with alkynes.

The reaction of cyclobutanones with alkynes proceeds in a manner similar to those of biphenylenes and cyclobutenones (Scheme 6.27), but a distinct catalytic cycle is proposed to include the formation of oxanickelacyclopentene intermediates, which undergo β-carbon elimination to cleave the C—C bond (Scheme 6.28, path A) [29]. The use of diynes as a substrate allows the formation of eight-membered ketones through formal insertion of two alkynes into the C—C bond of the cyclobutanone moiety (Scheme 6.29) [30]. Likewise, 3-azetidinones and 3-oxetanones can be reacted with alkynes (Scheme 6.30) [31–33] and diynes (Scheme 6.31) [34] to give six- and eight-membered products bearing heteroatom functionalities by nickel catalysis. Furthermore, they react with 1,3-dienes to give another class of eight-membered products (Scheme 6.32) [35]. In the case of the reaction of 3-azetidinones with alkynes, a catalytic cycle through the oxidative addition of the C—C bond, rather than that through β-carbon elimination, has been proposed based on theoretical calculations

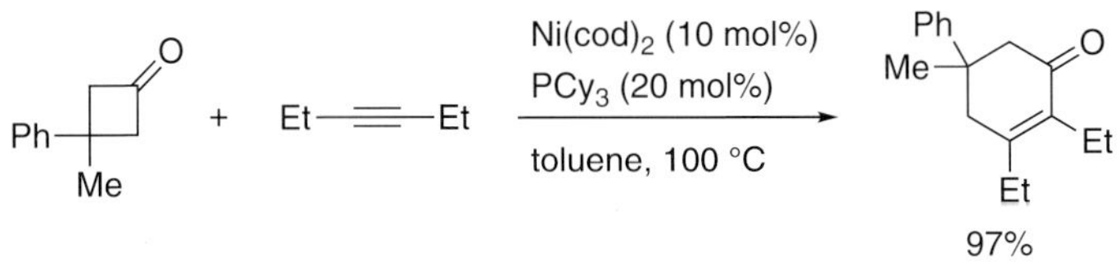

Scheme 6.27 Ni-catalyzed [4+2] cycloaddition reaction of cyclobutanones with alkynes.

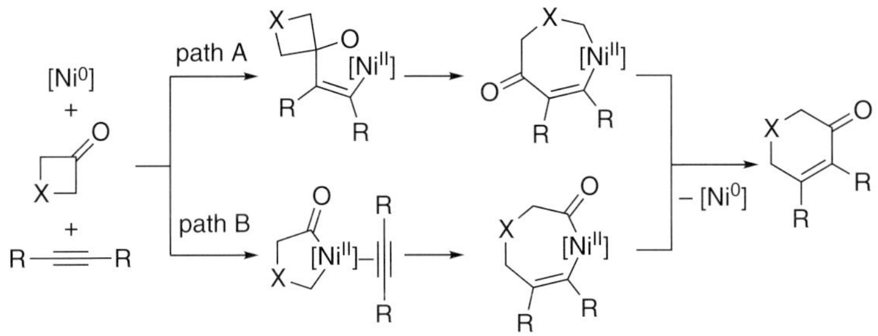

Scheme 6.28 A mechanism of Ni-catalyzed cycloaddition reactions of four-membered ketones with alkynes.

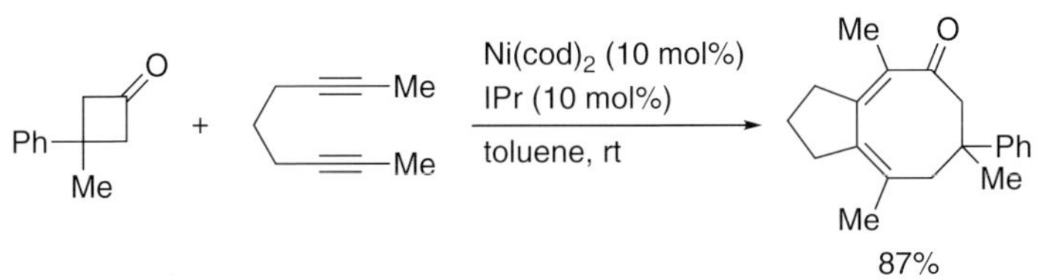

Scheme 6.29 Ni-catalyzed [4+2+2] cycloaddition reaction of cyclobutanones with diynes.

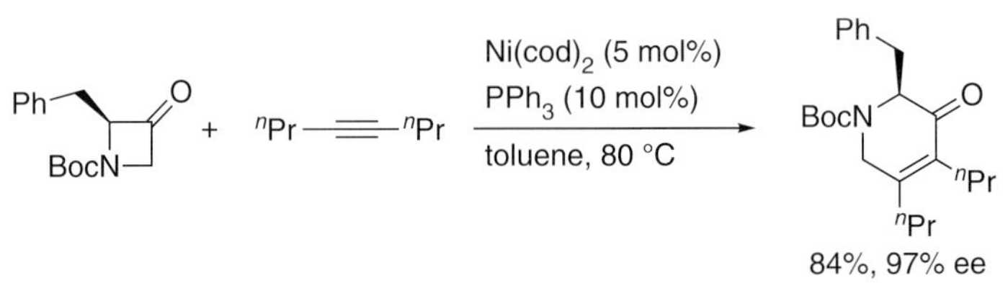

Scheme 6.30 Ni-catalyzed [4+2] cycloaddition reaction of 3-azetidinones with alkynes.

(Scheme 6.28, path B) [36]. An intramolecular insertion of vinylarenes into the C—C bond of cyclobutanones has been achieved by nickel catalysis to give a bicyclo[2.2.2]octan-2-one product (Scheme 6.33) [37, 38], whereas the same substrate gives a bicyclo[3.2.1]octan-3-one framework by rhodium catalysis most likely through the oxidative addition of the C—C bond [39], demonstrating that the mode of C—C bond activation can be complementary by different metal catalysts.

One other class of four-membered compounds that undergo C—C bond functionalization is bicyclo[2.1.0]pentane. It reacts with electron-deficient alkenes

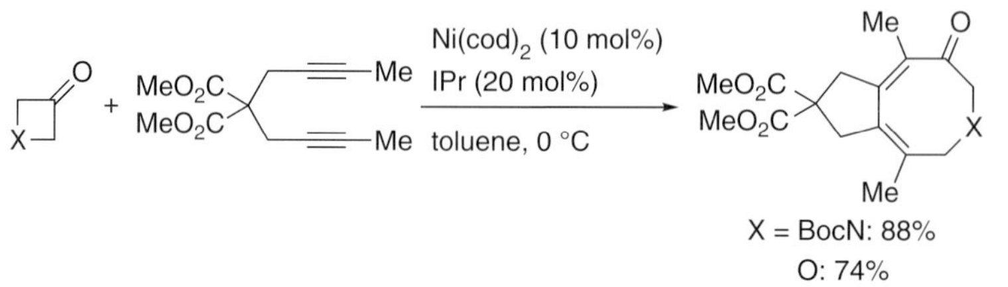

Scheme 6.31 Ni-catalyzed [4+2+2] cycloaddition reaction of 3-azetidinones and 3-oxetanones with diynes.

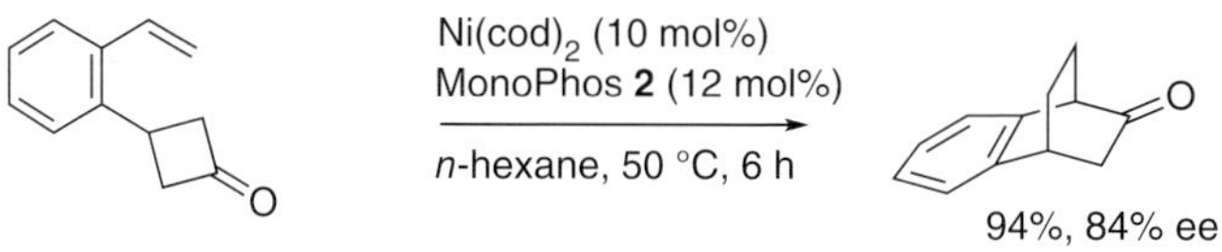

Scheme 6.32 Ni-catalyzed [4+2] cycloaddition reaction of 3-azetidinones and 3-oxetanones with 1,3-dienes.

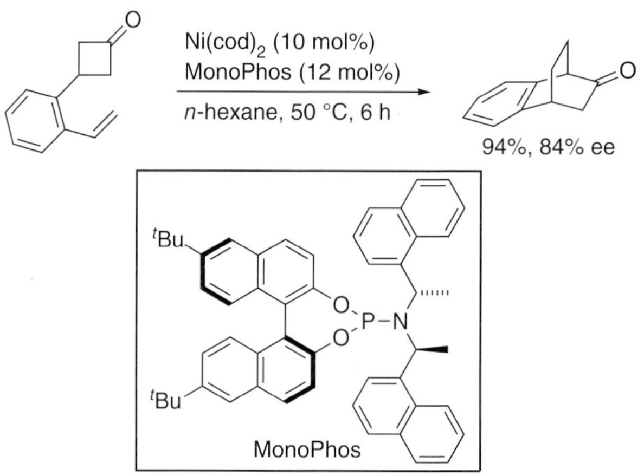

Scheme 6.33 Ni-catalyzed intramolecular [4+2] cycloaddition reaction of cyclobutanones with alkenes.

through their insertion selectively into the transannular C—C bond to give bicyclo[2.2.1]heptane products (Scheme 6.34) [40].

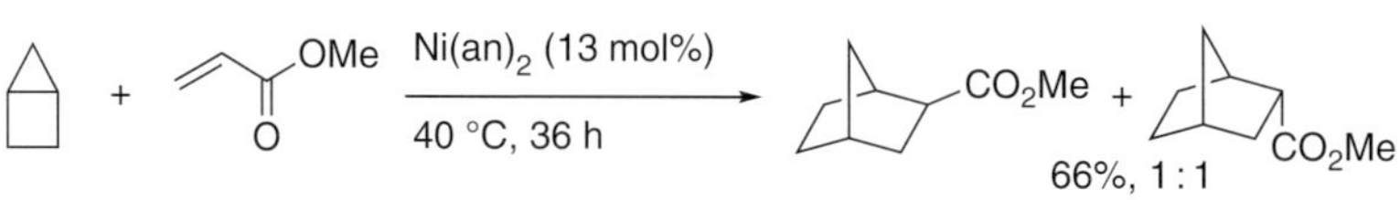

Scheme 6.34 Ni-catalyzed [4+2] cycloaddition reaction of bicycle[2.1.0]pentane with methyl acrylate.

Norbornadiene has a strain energy higher than that of cyclopropanes and cyclobutanes. The C—C bond of norbornadiene thus can be activated by a dinuclear nickel complex to react with carbon monoxide to give a bicyclo[3.3.0] product through the oxidative addition of the C—C bond across the di-nuclear nickel centers (Scheme 6.35) [41].

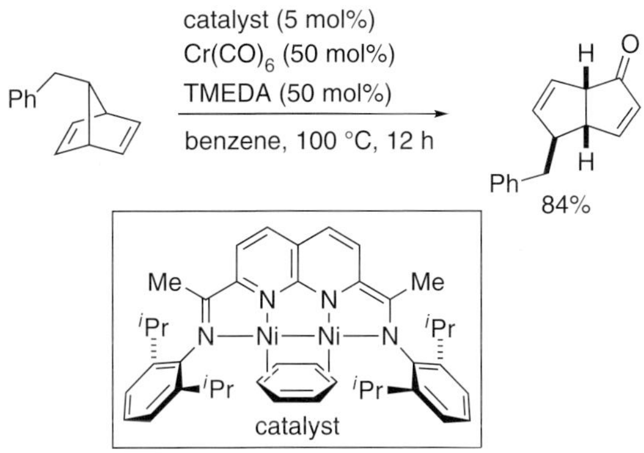

Scheme 6.35 Ni-catalyzed carbonylative rearrangement of norbornadiene.

6.4 C—C Bond Functionalization of Less Strained Molecules

Some less strained C—C bonds can also be activated by nickel catalysis. The isomerization of 1,4-pentadiene to isoprene proceeds by nickel catalysis through C—C bond cleavage (Scheme 6.36) [42]. A mechanism involving β-carbon elimination is proposed [43]. Catalytic reactions through cleavage of allylic C—C bonds also through β-carbon elimination have been reported by using nickel catalysts (Scheme 6.37) [44–46].

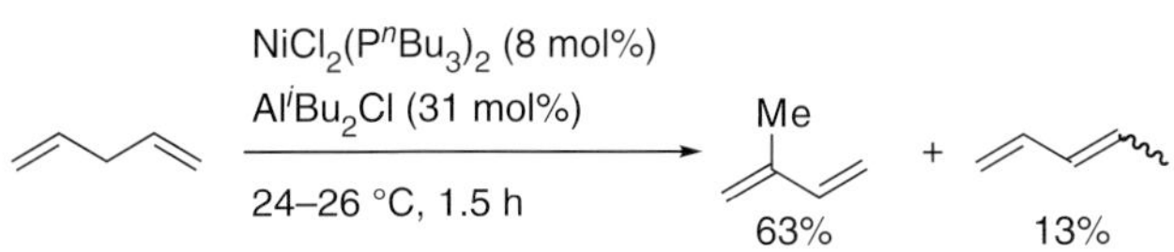

Scheme 6.36 Ni-catalyzed skeletal rearrangement of 1,4-pentadiene.

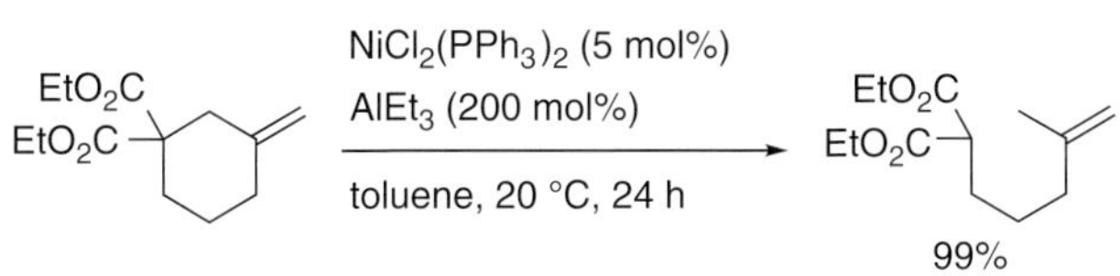

Scheme 6.37 Ni-catalyzed deallylation reaction.

Acyl C—C bonds can be activated by nickel catalysis. A cycloaddition reaction of 2-iodoacetophenone with norbornene proceeds through the cleavage of the acyl–Me bond (Scheme 6.38) [47]. β-Carbon elimination forms a nickel alkoxide intermediate, which forms through intramolecular nucleophilic addition of a *sec*-alkylnickel species derived from arylnickelation of norbornene, is proposed. On the other hand, a decarbonylation reaction of diarylketones to give biaryls through the Ar—C bond activation has been reported using a stoichiometric amount of a nickel(0) complex bearing an NHC ligand (Scheme 6.39) [48]. The use of diarylketones bearing an sp^2-nitrogen-based directing group allows the decarbonylation reaction to proceed in a catalytic manner (Scheme 6.40) [49]. Oxidative addition of the acyl–aryl bonds is proposed as a C—C bond activating step in these transformations.

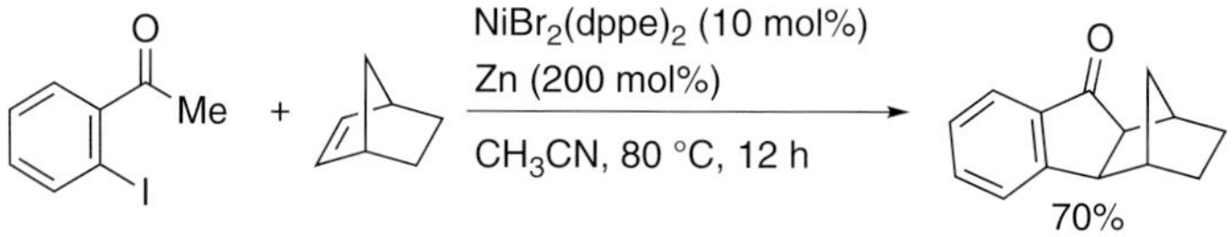

Scheme 6.38 Ni-catalyzed [3+2] cycloaddition reaction of 2-iodoacetophenone with norbornene.

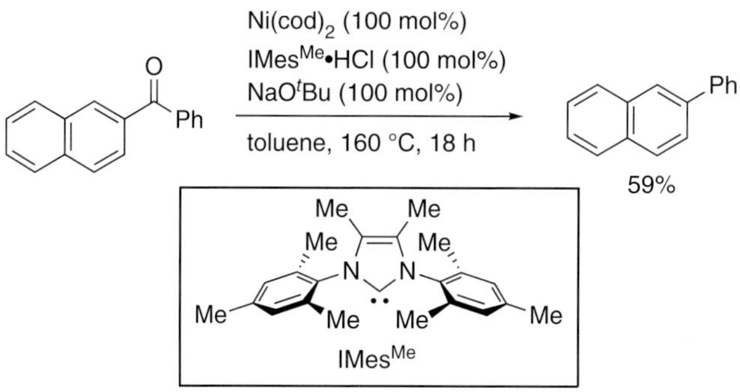

Scheme 6.39 Ni-mediated decarbonylation of diaryl ketones.

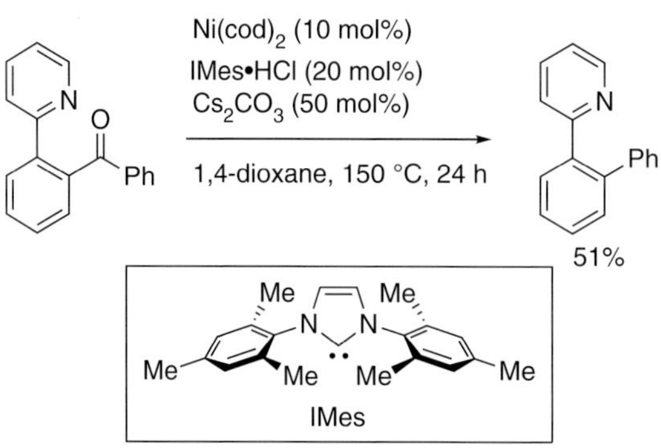

Scheme 6.40 Ni-catalyzed decarbonylation of diaryl ketones.

6.5 C—CN Bond Functionalization

One other type of C—C bond activation by nickel catalysis that has been applied to organic synthesis is the functionalization of C—CN bonds of nitriles [50]. All the nickel-catalyzed reactions involving the C—CN bond activation likely proceed through the oxidative addition of C—CN bonds to nickel(0) species. The elemental reaction was originally employed for the catalytic isomerization of 2-methyl-3-butenenitrile to pentenenitriles as a part of the DuPont's process to produce adiponitrile starting from 1,3-butadiene and HCN (Scheme 6.41) [51–59]. The activation of Ar—CN bonds has been applied to the cross-coupling reaction of aromatic nitriles with carbon (Scheme 6.42) [60, 61] and nitrogen nucleophiles (Scheme 6.43) [62]. Nickel(II) intermediates generated by the oxidative addition of C—CN bonds can also be reacted with unsaturated compounds to enable the carbocyanation reactions (Scheme 6.44) [63]. A wide variety of nitriles has been demonstrated to participate in the newly developed C—C bond forming addition reaction across alkynes and alkenes. The use of aluminum- and boron-based Lewis acid co-catalysts has been proved to be highly effective to promote oxidative addition. The C—CN bond activation in combination with the acyl–aryl bond cleavage mentioned in the previous Section 6.4 allows a cycloaddition reaction of 2-arylcarboxybenzonitrile with alkynes to give substituted 2*H*-chromen-2-ones (Scheme 6.45) [64–66]. The nickel/aluminum bimetallic catalysis also allows transfer hydrocyanation of alkenes through the alkyl—CN bond activation (Scheme 6.46) [67].

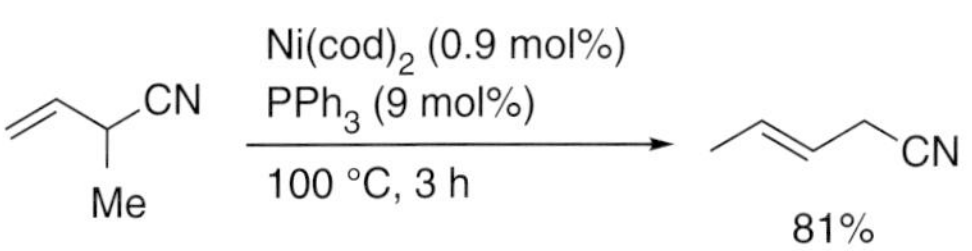

Scheme 6.41 Ni-catalyzed isomerization of 2-methyl-3-butenenitrile to 3-pentenenitrile.

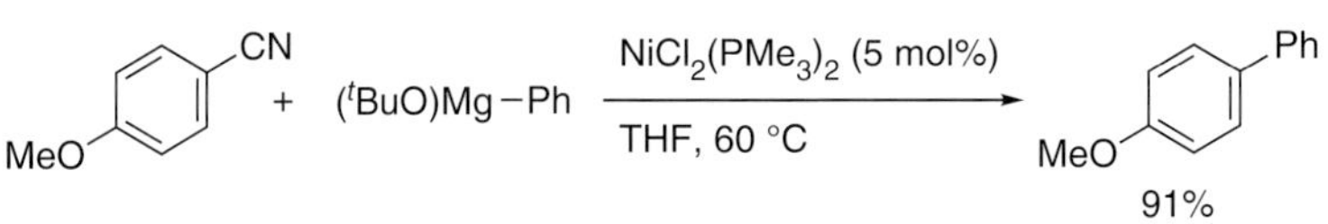

Scheme 6.42 Ni-catalyzed cross-coupling of aryl cyanides with Grignard reagents.

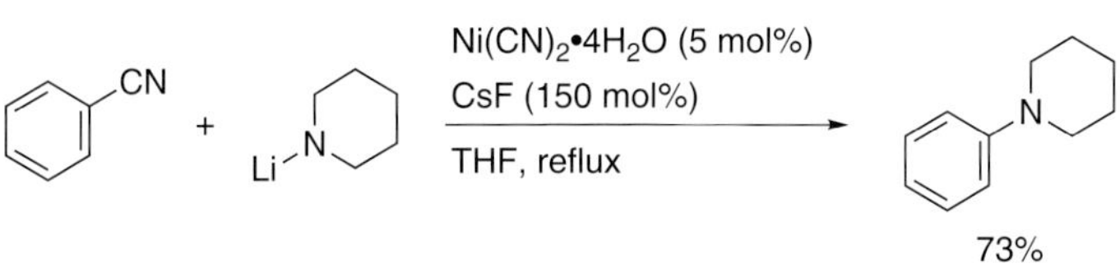

Scheme 6.43 Ni-catalyzed cross-coupling of aryl cyanides with lithium amides.

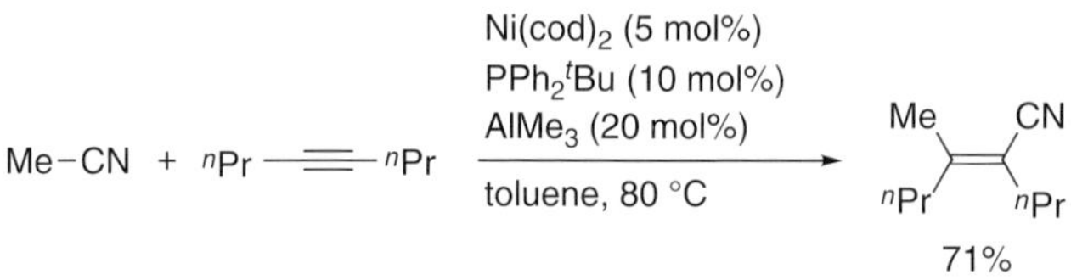

Scheme 6.44 Ni-catalyzed carbocyanation of alkynes.

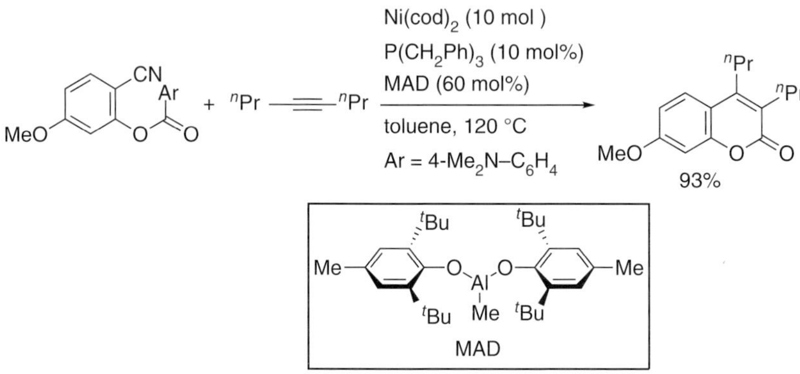

Scheme 6.45 Ni-catalyzed [4+2] cycloaddition reaction of 2-arylcarboxybenzonitrile with alkynes.

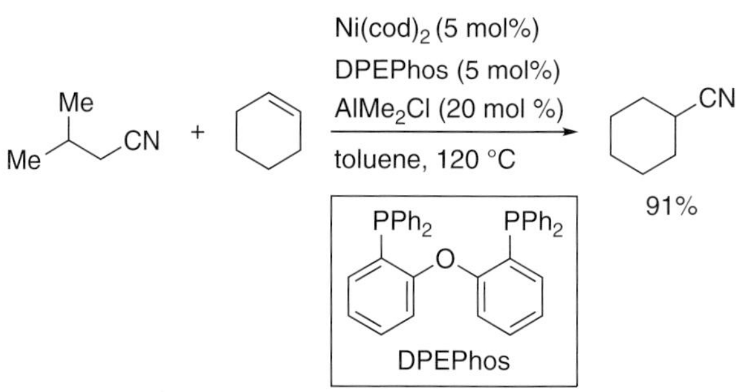

Scheme 6.46 Ni-catalyzed transfer hydrocyanation of alkenes.

6.6 Summary and Outlook

Synthetic transformations through C—C bond activation by nickel catalysis are surveyed in this chapter. Nickel catalysts have been shown to be effective for the activation of C—C bonds of not only strained molecules conventionally employed such as three- and four-membered rings but also less strained compounds such as ketones and nitriles in recent years. These transformations using "less activated C–C bonds" will be particularly attractive for synthetic organic chemists. Given a diverse range of reaction modes for C—C bond activation by nickel complexes, synthetically unique and useful transformations of such readily available compounds would be further envisioned. Development of new ligands as well as co-catalysts such as Lewis acid catalysts will be important to further develop novel C—C bond functionalizations by nickel catalysis.

References

1 Otsuka, S., Nakamura, A., Yoshida, T. et al. (1973). Chemistry of alkoxycarbonyl, acyl, and alkyl compounds of nickel(II) and palladium(II). *J. Am. Chem. Soc.* 95 (10): 3180–3188.

2 Zuo, G. and Louie, J. (2004). Highly active nickel catalysts for the isomerization of unactivated vinyl cyclopropanes to cyclopentenes. *Angew. Chem. Int. Ed.* 43 (17): 2277–2279.

3 Zuo, G. and Louie, J. (2005). Selectivity in nickel-catalyzed rearrangements of cyclopropylenynes. *J. Am. Chem. Soc.* 127 (16): 5798–5799.

4 Ichiyanagi, T., Kuniyama, S., Shimizu, M., and Fujisawa, T. (1997). Regioselective ring-opening of cyclopropyl ketones with organometallic reagents. *Chem. Lett.*: 1149–1150.

5 Liu, L. and Montgomery, J. (2006). Dimerization of cyclopropyl ketones and crossed reactions of cyclopropyl ketones with enones as an entry to five-membered rings. *J. Am. Chem. Soc.* 128 (16): 5348–5349.

6 Ogoshi, S., Nagata, M., and Kurosawa, H. (2006). Formation of nickeladihydropyran by oxidative addition of cyclopropyl ketone. Key intermediate in nickel-catalyzed cycloaddition. *J. Am. Chem. Soc.* 128 (16): 5350–5351.

7 Tamaki, T., Ohashi, M., and Ogoshi, S. (2011). [3+2] Cycloaddition reaction of cyclopropyl ketones with alkynes catalyzed by nickel/dimethylaluminum chloride. *Angew. Chem. Int. Ed.* 50 (50): 12067–12070.

8 Liu, L. and Montgomery, J. (2007). [3+2] Cycloaddition reactions of cyclopropyl imines with enones. *Org. Lett.* 9 (20): 3885–3887.

9 Liu, Q.S., Wang, D.Y., Yang, Z.J. et al. (2017). Ni–Al bimetallic catalyzed enantioselective cycloaddition of cyclopropyl carboxamide with alkyne. *J. Am. Chem. Soc.* 139 (50): 18150–18153.

10 Bowman, R.K. and Johnson, J.S. (2006). Nickel-catalyzed rearrangement of 1-acyl-2-vinylcyclopropanes. A mild synthesis of substituted dihydrofurans. *Org. Lett.* 8 (4): 573–576.

11 Noyori, R., Odagi, T., and Takaya, H. (1970). Nickel(0)-catalyzed reaction of methylenecyclopropanes with olefins. A novel [$\sigma 2 + \pi 2$] cycloaddition. *J. Am. Chem. Soc.* 92 (19): 5780–5781.

12 Noyori, R., Hayashi, N., and Kato, M. (1971). Reaction of methylenecyclopropanes with tetracyanoethylene. A new cycloaddition involving three- and two-carbon units. *J. Am. Chem. Soc.* 93 (19): 4948–4950.

13 Inami, T., Sako, S., Kurahashi, T., and Matsubara, S. (2011). Methylenecyclopropanes in [4 + 1] cycloaddition with enones. *Org. Lett.* 13 (15): 3837–3839.

14 Saito, S., Masuda, M., and Komagawa, S. (2004). Nickel-catalyzed intermolecular [3 + 2 + 2] cocyclization of ethyl cyclopropylideneacetate and alkynes. *J. Am. Chem. Soc.* 126 (34): 10540–10541.

15 Maeda, K. and Saito, S. (2007). Nickel-catalyzed [3 + 2 + 2] cycloaddition of ethyl cyclopropylideneacetate and diynes. Synthesis of 7,6- and 7,5-fused bicyclic compounds. *Tetrahedron Lett.* 48 (18): 3173–3176.

16 Komagawa, S. and Saito, S. (2006). Nickel-catalyzed three-component [3 + 2 + 2] cocyclization of ethyl cyclopropylideneacetate and alkynes-selective synthesis of multisubstituted cycloheptadienes. *Angew. Chem. Int. Ed.* 45 (15): 2446–2449.

17 Saito, S. and Takeuchi, K. (2007). Nickel-catalyzed [4 + 3] cycloaddition of ethyl cyclopropylideneacetate and 1,3-dienes. *Tetrahedron Lett.* 48 (4): 595–598.

18 Saito, S., Maeda, K., Yamasaki, R. et al. (2010). Synthesis of nine-membered carbocycles by the [4 + 3 + 2] cycloaddition reaction of ethyl cyclopropylideneacetate and dienynes. *Angew. Chem. Int. Ed.* 49 (10): 1830–1833.

19 Komagawa, S., Wang, C., Morokuma, K. et al. (2013). Mechanistic origin of chemo- and regioselectivity of nickel-catalyzed [3 + 2 + 2] cyclization reaction. *J. Am. Chem. Soc.* 135 (39): 14508–14511.

20 Pan, B., Wang, C., Wang, D. et al. (2013). Nickel-catalyzed [3 + 2] cycloaddition of diynes with methyleneaziridines via C–C bond cleavage. *Chem. Commun.* 49 (44): 5073–5075.

21 Baba, A., Ohshiro, Y., and Agawa, T. (1976). Reactions of diphenylcyclopropenone with ketenes in the presence of nickel tetracarbonyl. *J. Organomet. Chem.* 110: 121–127.

22 Noyori, R., Suzuki, T., Kumagai, Y., and Takaya, H. (1971). Nickel(0)-catalyzed reaction of Bicyclo[1.1.0]butanes with olefins. *J. Am. Chem. Soc.* 93 (22): 5894–5896.

23 Edelbach, B.L., Lachicotte, R.J., and Jones, W.D. (1999). Catalytic carbon–carbon bond activation and functionalization by nickel complexes. *Organometallics* 18 (20): 4040–4049.

24 Müller, C., Lachicotte, R.J., and Jones, W.D. (2002). Catalytic C–C bond activation in biphenylene and cyclotrimerization of alkynes: increased reactivity of P,N-versus P,P-substituted nickel complexes. *Organometallics* 21 (9): 1975–1981.

25 Schaub, T. and Radius, U. (2005). Efficient C–F and C–C activation by a novel *N*-heterocyclic carbene-nickel(0) complex. *Chem. Eur. J.* 11 (17): 5024–5030.

26 Schaub, T., Backes, M., and Radius, U. (2006). Nickel(0) complexes of *N*-alkyl-substituted *N*-heterocyclic carbenes and their use in the catalytic carbon–carbon bond activation of biphenylene. *Organometallics* 25 (17): 4196–4206.

27 Gu, Z., Boursalian, G.B., Gandon, V. et al. (2011). Activated phenacenes from phenylenes by nickel-catalyzed alkyne cycloadditions. *Angew. Chem. Int. Ed.* 50 (40): 9413–9417.

28 Huffman, M.A. and Liebeskind, L.S. (1991). Nickel(0)-catalyzed synthesis of substituted phenols from cyclobutanones and alkynes. *J. Am. Chem. Soc.* 113 (7): 2771–2772.

29 Murakami, M., Ashida, S., and Matsuda, T. (2005). Nickel-catalyzed intermolecular alkyne insertion into cyclobutanones. *J. Am. Chem. Soc.* 127 (19): 6932–6933.

30 Murakami, M., Ashida, S., and Matsuda, T. (2006). Eight-membered ring construction by [4 + 2 + 2] annulation involving β-carbon elimination. *J. Am. Chem. Soc.* 128 (7): 2166–2167.
31 Ho, K.Y. and Aïssa, C. (2012). Regioselective cycloaddition of 3-azetidinones and 3-oxetanones with alkynes through nickel-catalysed carbon-carbon bond activation. *Chem. Eur. J.* 18 (12): 3486–3489.
32 Kumar, P. and Louie, J. (2012). A single step approach to piperidines via Ni-catalyzed beta-carbon elimination. *Org. Lett.* 14 (8): 2026–2029.
33 Ishida, N., Yuhki, T., and Murakami, M. (2012). Synthesis of enantiopure dehydropiperidinones from alpha-amino acids and alkynes via azetidin-3-ones. *Org. Lett.* 14 (15): 3898–3901.
34 Kumar, P., Zhang, K., and Louie, J. (2012). An expeditious route to eight-membered heterocycles by nickel-catalyzed cycloaddition: low-temperature $C(sp^2)–C(sp^3)$ bond cleavage. *Angew. Chem. Int. Ed.* 51 (34): 8602–8606.
35 Thakur, A., Facer, M.E., and Louie, J. (2013). Nickel-catalyzed cycloaddition of 1,3-dienes with 3-azetidinones and 3-oxetanones. *Angew. Chem. Int. Ed.* 52 (46): 12161–12165.
36 Li, Y. and Lin, Z. (2013). Theoretical studies on nickel-catalyzed cycloaddition of 3-azetidinone with alkynes. *Organometallics* 32 (10): 3003–3011.
37 Murakami, M. and Ashida, S. (2006). Nickel-catalysed intramolecular alkene insertion into cyclobutanones. *Chem. Commun.* (44): 4599–4601.
38 Liu, L., Ishida, N., and Murakami, M. (2012). Atom- and step-economical pathway to chiral benzobicyclo[2.2.2]octenones through carbon–carbon bond cleavage. *Angew. Chem. Int. Ed.* 51 (10): 2485–2488.
39 Murakami, M., Itahashi, T., and Ito, Y. (2002). Catalyzed intramolecular olefin insertion into a carbon–carbon single bond. *J. Am. Chem. Soc.* 124 (47): 13976–13977.
40 Noyori, R., Suzuki, T., and Takaya, H. (1971). Nickel(0)-catalyzed reaction of bicyclo[2.1.0]pentane with olefins. *J. Am. Chem. Soc.* 93 (22): 5896–5897.
41 Hartline, D.R., Zeller, M., and Uyeda, C. (2017). Catalytic carbonylative rearrangement of norbornadiene via dinuclear carbon–carbon oxidative addition. *J. Am. Chem. Soc.* 139 (39): 13672–13675.
42 Miller, R.G., Pinke, P.A., Stauffer, R.D. et al. (1974). Nickel-catalyzed skeletal rearrangements of 1,4-dienes. *J. Am. Chem. Soc.* 96 (13): 4211–4220.
43 Pinke, P.A. and Miller, R.G. (1974). Mechanistic study of the nickel-promoted 1,4-pentadiene to isoprene rearrangement. Isomerization of deuterio-cis-1,4-hexadienes. *J. Am. Chem. Soc.* 96 (13): 4221–4229.
44 Bricout, H., Carpentier, J.-F., and Mortreux, A. (1997). Further developments in metal-catalysed C–C bond cleavage in allylic dimethyl malonate derivatives. *Tetrahedron Lett.* 38 (6): 1053–1056.
45 Necas, D. and Kotora, M. (2008). Ring opening of methylenecycloalkenes via the C–C bond cleavage. *Org. Lett.* 10 (22): 5261–5263.
46 Ambe-Suzuki, K., Ohyama, Y., Shirai, N., and Ikeda, S.-i. (2012). Nickel/zinc chloride-promoted domino reaction of enynes and enones including unstrained C–C bond cleavage. *Adv. Synth. Catal.* 354 (5): 879–888.

47 Shukla, P. and Cheng, C.H. (2006). Facile β-alkyl and β-hydride elimination in the nickel-catalyzed annulation of *o*-iodophenyl ketones and aldehydes with bicyclic alkenes. *Org. Lett.* 8 (13): 2867–2869.
48 Morioka, T., Nishizawa, A., Furukawa, T. et al. (2017). Nickel-mediated decarbonylation of simple unstrained ketones through the cleavage of carbon–carbon bonds. *J. Am. Chem. Soc.* 139 (4): 1416–1419.
49 Zhao, T.T., Xu, W.H., Zheng, Z.J. et al. (2018). Directed decarbonylation of unstrained aryl ketones via nickel-catalyzed C–C bond cleavage. *J. Am. Chem. Soc.* 140 (2): 586–589.
50 For a recent review, see:Nakao, Y. (2014). Catalytic C–CN bond activation. *Top. Curr. Chem.* 346: 33–58.
51 Tolman, C.A., McKinney, R.J., Seidel, W.C. et al. (1985). Homogeneous nickel-catalyzed olefin hydrocyanation. *Adv. Catal.* 33: 1–46.
52 Chaumonnot, A., Lamy, F., Sabo-Etienne, S. et al. (2004). Catalytic isomerization of cyanoolefins involved in the adiponitrile process. C–CN bond cleavage and structure of the nickel π-allyl cyanide complex Ni(η^3-1-Me-C_3H_4)(CN)(dppb). *Organometallics* 23 (14): 3363–3365.
53 van der Vlugt, J.I., Hewat, A.C., Neto, S. et al. (2004). Sterically demanding diphosphonite ligands–synthesis and application in nickel-catalyzed isomerization of 2-methyl-3-butenenitrile. *Adv. Synth. Catal.* 346 (8): 993–1003.
54 Wilting, J., Müller, C., Hewat, A.C. et al. (2005). Nickel-catalyzed isomerization of 2-methyl-3-butenenitrile. *Organometallics* 24 (1): 13–15.
55 Acosta-Ramírez, A., Muñoz-Hernández, M., Jones, W.D., and García, J.J. (2006). Isomerization of 2-methyl-3-butenenitrile with (bis-diphenylphosphinoferrocene)nickel compounds: catalytic and structural studies. *J. Organomet. Chem.* 691 (18): 3895–3901.
56 Acosta-Ramírez, A., Muñoz-Hernández, M., Jones, W.D., and García, J.J. (2007). Catalytic isomerization of 2-methyl-3-butenenitrile by nickel systems using bis-diphosphinoferrocene ligands: evidence for hemilability. *Organometallics* 26 (24): 5766–5769.
57 Acosta-Ramírez, A., Morales-Morales, D., Serrano-Becerra, J.M. et al. (2008). Study of the reactivity of 2-methyl-3-butenenitrile with Ni(0)-*N*-heterocyclic carbene complexes. *J. Mol. Catal. A: Chemical* 288 (1–2): 14–18.
58 Tauchert, M.E., Kaiser, T.R., Göthlich, A.P.V. et al. (2010). Phosphonite ligand design for nickel-catalyzed 2-methyl-3-butenenitrile isomerization and styrene hydrocyanation. *ChemCatChem* 2 (6): 674–682.
59 Li, T. and Jones, W.D. (2011). DFT calculations of the isomerization of 2-methyl-3-butenenitrile by [Ni(bisphosphine)] in relation to the DuPont adiponitrile process. *Organometallics* 30 (3): 547–555.
60 Miller, J.A. (2001). C–C bond activation with selective functionalization: preparation of unsymmetrical biaryls from benzonitriles. *Tetrahedron Lett.* 42 (40): 6991–6993.
61 Miller, J.A. and Dankwardt, J.W. (2003). Nickel catalyzed cross-coupling of modified alkyl and alkenyl Grignard reagents with aryl- and heteroaryl nitriles: activation of the C–CN bond. *Tetrahedron Lett.* 44 (9): 1907–1910.
62 Miller, J.A., Dankwardt, J.W., and Penney, J.M. (2003). Nickel catalyzed cross-coupling and amination reactions of aryl nitriles. *Synthesis*: 1643–1648.

63 For an account, see:Nakao, Y. (2012). Nickel/Lewis acid-catalyzed carbocyanation of unsaturated compounds. *Bull. Chem. Soc. Jpn.* 85 (7): 731–745.

64 Nakai, K., Kurahashi, T., and Matsubara, S. (2011). Nickel-catalyzed cycloaddition of *o*-arylcarboxybenzonitriles and alkynes via cleavage of two carbon–carbon sigma bonds. *J. Am. Chem. Soc.* 133 (29): 11066–11068.

65 Nakai, K., Kurahashi, T., and Matsubara, S. (2013). Synthesis of quinolones by nickel-catalyzed cycloaddition via elimination of nitrile. *Org. Lett.* 15 (4): 856–859.

66 Guan, W., Sakaki, S., Kurahashi, T., and Matsubara, S. (2014). Reasons two nonstrained C–C σ-bonds can be easily cleaved in decyanative [4 + 2] cycloaddition catalyzed by nickel(0)/Lewis acid systems. Theoretical insight. *ACS Catal.* 5 (1): 1–10.

67 Fang, X., Yu, P., and Morandi, B. (2016). Catalytic reversible alkene-nitrile interconversion through controllable transfer hydrocyanation. *Science* 351: 832–836.

7

C—O Bond Transformations

Mamoru Tobisu

Osaka University, Department of Applied Chemistry, Suita, Osaka 565-0871, Japan

7.1 Introduction

The ability of transition metal complexes to activate aromatic carbon-halogen [C(aryl)—X] bonds has led to the development of a series of catalytic transformation of aryl halides, as exemplified by the tremendous success of cross-coupling reactions [1]. Recent efforts have been directed to the use of phenol derivatives as less harmful and less costly alternatives to aryl halides in catalytic arylation reactions. Although it has long been known that the reactivity of aryl trifluoromethane sulfonates (triflates) is comparable to that of aryl halides, fluorine-based leaving groups are even more toxic and costly than aryl halides. Therefore, the use of non-fluorinated phenol derivatives, such as aryl methyl ethers, aryl esters, and carbamates, has significantly attracted the attention of synthetic chemists for use in catalytic arylation reactions. These phenol derivatives are unreactive under the typical conditions used for palladium-catalyzed transformation of aryl halides due to the inert nature of the C(aryl)—O bonds. During the past decades, a substantial progress has been made in the development of methods for the transformation of these inert phenol derivatives [2–6]. The key to the successful development is the use of nickel-based catalysts. Given its less electronegative character, nickel tends to promote an oxidative addition process more readily than palladium [7].

This chapter deals with nickel-catalyzed reactions that involve the cleavage of inert C—O bonds. The focus of the chapter is on three classes of inert C—O bonds: C(aryl)—O, C(benzyl)—O, and C(acyl)—O bonds. Among the numerous derivatives, this chapter overviews the nickel-catalyzed reactions of ethers, esters, carbamates, and non-protected phenols, which are common but challenging substrates for use in catalytic reactions. The choice of the ligand is crucial in developing catalytic C—O bond transformations. The ligands that are commonly used in palladium-catalyzed reactions of aryl halides are normally ineffective in many nickel-catalyzed C—O bond transformations. Ligands frequently appearing in this chapter are listed in Figure 7.1.

Nickel Catalysis in Organic Synthesis: Methods and Reactions,
First Edition. Edited by Sensuke Ogoshi.

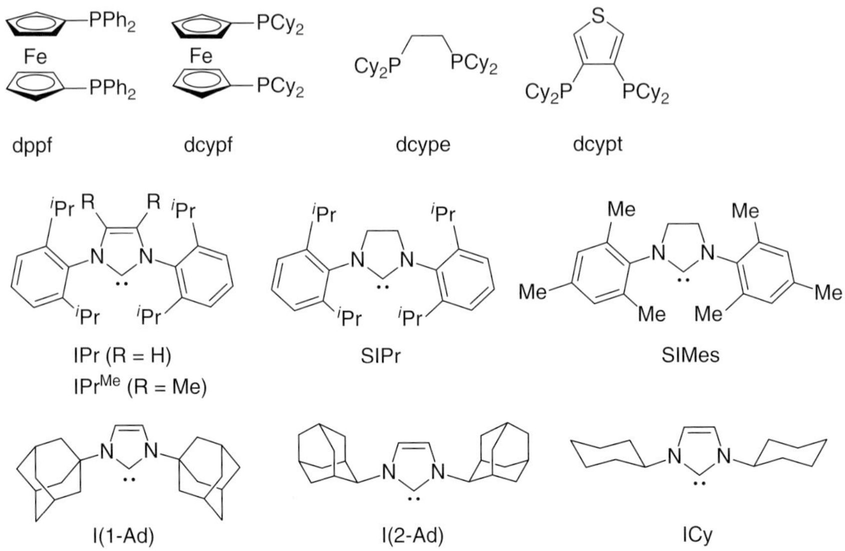

Figure 7.1 Structures and abbreviations of the ligands used in nickel-catalyzed C—O bond transformations.

7.2 C(aryl)—O Bond Cleavage

Because phenols are abundant in Nature, the use of their derivatives as arylating reagents in catalytic reactions is of great synthetic value. To achieve such processes, the activation of C(aryl)—O bonds in phenol derivatives is essential. The difficulty in activating these bonds can be estimated by comparing the calculated activation energies for the oxidative addition of a series of aryl electrophiles (Table 7.1) [8]. The activation barriers for the oxidative addition of phenyl acetate and anisole to $Pd(PMe_3)$ are higher than those of the commonly used bromobenzene by 31–39 kcal/mol. These values can be significantly lowered by switching the metal center to nickel. In fact, the vast majority of reactions involving the activation of C(aryl)—O bonds of inert phenol derivatives are catalyzed by nickel complexes. Once the oxidative addition of these phenol derivatives becomes viable, its application to cross-coupling with various nucleophiles and C—H bonds is possible as is the case for the catalytic transformation of aryl halides. This section summarizes the nickel-catalyzed transformation of aryl esters, carbamates, ethers, and phenols. It should be noted that most of the reactions described in this section can also be applied to the corresponding alkenyl esters and ethers, in which nickel complexes activate $C(sp^2)$—O bonds in a similar manner [9].

7.2.1 Aryl Esters, Carbamates, and Carbonates

In addition to the inertness of the C(aryl)—O bond, the selectivity between the cleavage of C(aryl)—O and C(acyl)—O bonds needs to be addressed in

Table 7.1 Calculated activation barriers for oxidative addition of various aryl electrophiles: nickel vs palladium.

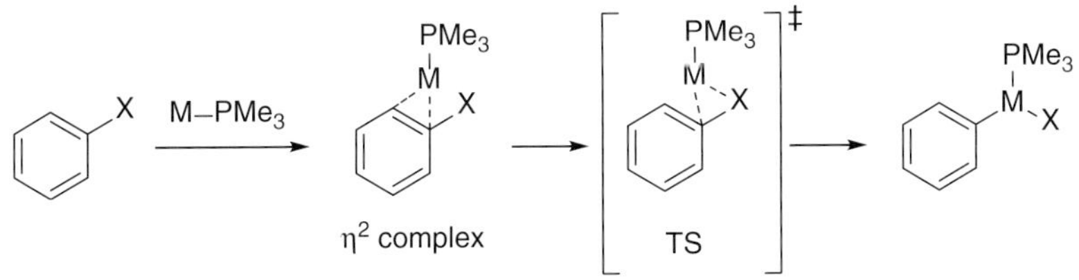

Entry	X	Activation barrier (kcal/mol) Pd	Ni
1	Br	3.4	3.3
2	Cl	14.8	9.6
3	OTf	25.9	4.5
4	OAc	34.0	26.4
5	OMe	42.0	29.5

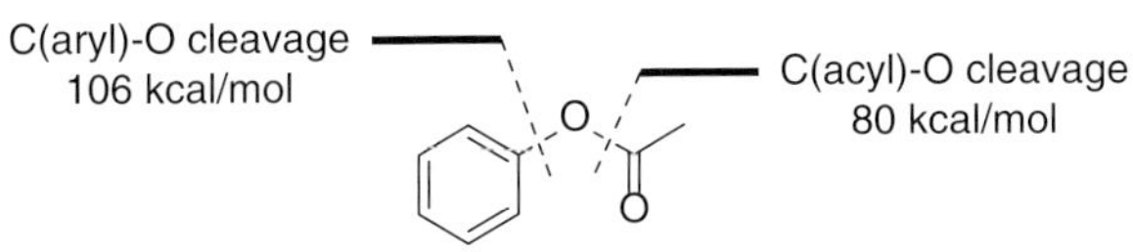

Figure 7.2 C(aryl)–O cleavage vs. C(acyl)–O cleavage.

developing catalytic methods for aryl ester substrates (Figure 7.2). C(acyl)—O bonds are more reactive than C(aryl)—O bonds in terms of bond dissociation energies [8]. However, overall selectivity is determined by many other factors such as the reversibility of the C—O bond cleavage and the relative rates of the subsequent reaction. The use of bulky acyl groups, such as a pivaloyl group, is also beneficial for achieving the selective transformation of C(aryl)—O bonds of aryl esters.

The nickel-catalyzed cross-coupling of aryl esters, carbamates, and carbonates with organometallic reagents is summarized in Table 7.2. Snieckus and coworkers reported the first Kumada–Tamao–Corriu type cross-coupling of aryl carbamates, in which phenyl, methyl, and trimethylsilylmethyl Grignard reagents are coupled under ligand-free conditions (Entry 1) [10]. Nakamura and coworkers reported that a phosphine/OH bidentate ligand generates a more active catalyst, which can be used to suppress the undesired homocoupling of aryl carbamates observed under ligand-free conditions (Entry 2). It has been proposed that this ligand allows for a facile oxidative addition of C(aryl)—O bonds via the formation of a highly ordered transition state involving substrate activation by a Lewis acidic magnesium alkoxide moiety (Figure 7.3) [11].

Table 7.2 Nickel-catalyzed cross-coupling of aryl esters, carbamates and carbonates with organometallic nucleophiles.

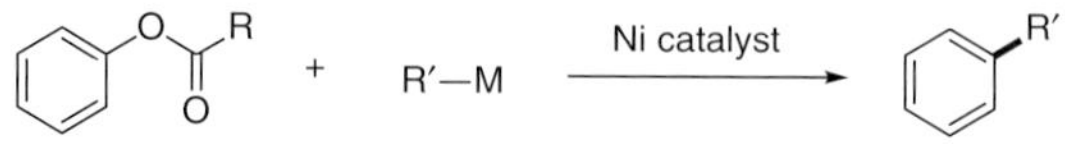

Entry	R	R′—M	Catalyst/ligand	References
1	NEt_2	PhMgX MeMgX Me_3SiCH_2MgX	$Ni(acac)_2$	[10]
2	NEt_2	Ar′MgX	$Ni(acac)_2$/ 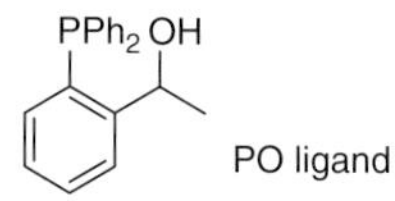	[11]
3	^tBu	Ar′ZnX	$NiCl_2(PCy_3)_2$	[12]
4	^tBu	Alkyl-ZnX	$Ni(cod)_2$/dcype	[13]
5	NEt_2	Ar′-Al^iBu_2 $^iPr_3Si$—≡—Al^iBu_2	$NiCl_2(PCy_3)_2$	[14]
6	^tBu	$Ar'B(OH)_2$	$NiCl_2(PCy_3)_2$	[16, 18]
7	Me, ^tBu, Ph, etc.	$(Ar'BO)_3$	$NiCl_2(PCy_3)_2$	[17]
8	^tBu	$Ar'BF_3K$	$Ni(cod)_2/PCy_3$	[19]
9	^tBu	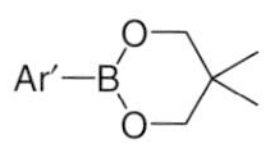	$Ni(cod)_2/PCy_3$	[20]
10	NEt_2 O^tBu	$Ar'B(OH)_2$	$NiCl_2(PCy_3)_2$	[21]
11	NEt_2	$Ar'B(OH)_2 + 0.1(Ar'BO)_3$	$NiCl_2(PCy_3)_2$	[22]
12	NEt_2	$(Ar'BO)_3$	$NiCl_2(PCy_3)_2$	[23]
13	OMe	$Ar'B(OH)_2$	$Ni(cod)_2$/dcypf	[24]
14	NEt_2	$Ar'Si(OEt)_3$	$Ni(cod)_2/dcype/CuF_2$	[25]

Cross-coupling of aryl esters is possible with less nucleophilic organometallic reagents. Aryl (Entry 3) and alkylzinc (Entry 4) reagents can serve as suitable coupling partners for aryl pivalate substrates when PCy_3 [12] and dcype [13] are used as ligands, respectively. Cross-coupling with organoaluminum reagents has also been reported (Entry 5) [14]. The Suzuki–Miyaura reaction arguably represents the most useful cross-coupling reaction because organoboron reagents are stable to moisture and air and compatible with most of the common functional groups [15]. Garg and coworkers [16] and Shi and coworkers [17] independently reported the first

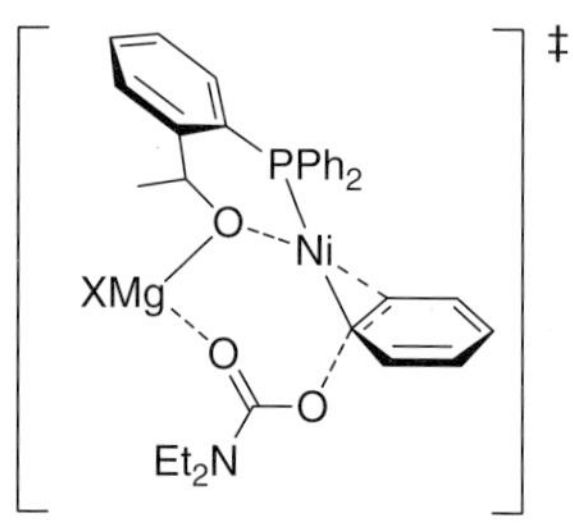

Figure 7.3 Transition state model for C—O bond cleavage in Kumada–Tamao–Corriu type cross-coupling using a PO ligand.

Suzuki–Miyaura type cross-coupling of aryl esters using a nickel catalyst ligated with PCy_3 (Entries 6 and 7) [18]. Although a slight modification in the reaction conditions was required depending on the protecting group of the boron reagents or when aryl carbamates are applied, PCy_3 uniformly was reported to be an effective ligand (Entries 8–12) [19–23]. Aryl carbonates are also suitable substrates for nickel-catalyzed cross-coupling with organoboron reagents (Entries 10 and 13) [21a, 24]. Organosilicon reagents represent another class of synthetically useful coupling partners, and an additional cost benefit would be expected because silicon is an earth-abundant element. However, there is only one successful example of the cross-coupling of inert phenol derivatives with organosilicon nucleophiles (Entry 14) [25]. Aryl carbamates can be coupled with arylsilanes under the influence of a nickel/copper combined catalyst system.

Replacing organometallic nucleophiles with simpler organic compounds in cross-coupling reactions and reacting at their C—H bonds represent a subject of intensive research. In this context, it would be valuable if inert phenol derivatives, instead of aryl halides, could be used as aryl donors in C–H arylation reactions. The nickel-catalyzed arylation of non-acidic arene C—H bonds with inert phenol derivatives has only been accomplished in an intramolecular setting. Thus, the reaction of biaryl ether bearing an ortho pivaloxyl group can be cyclized to form benzofuranes by a nickel catalyst (Eq. (7.1)) [26] The intermolecular cross-coupling of aryl esters with non-acidic C—H bonds has been accomplished via the use of cobalt [27] or rhodium [28] catalysts.

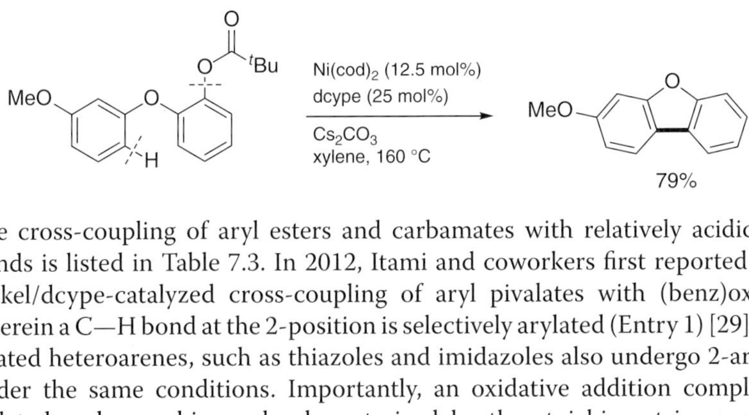

(7.1)

The cross-coupling of aryl esters and carbamates with relatively acidic C—H bonds is listed in Table 7.3. In 2012, Itami and coworkers first reported on the nickel/dcype-catalyzed cross-coupling of aryl pivalates with (benz)oxazoles, wherein a C—H bond at the 2-position is selectively arylated (Entry 1) [29]. Other related heteroarenes, such as thiazoles and imidazoles also undergo 2-arylation under the same conditions. Importantly, an oxidative addition complex was isolated and unambiguously characterized by the stoichiometric reaction of 2-naphthyl pivalates with $Ni(cod)_2$/dcype [30]. Although π-extended aryl esters, such as naphthyl pivalates, participate in this cross-coupling efficiently, the reactivity of phenyl pivalates is much lower under the nickel/dcype-catalyzed conditions. Iwai and Sawamura solved this problem by developing PS-dppbz, a new polymer-supported bisphosphine derivative, which can be used to promote the cross-coupling of phenyl pivalates with benzoxazoles [31]. Shi and coworkers reported that a C—H bond in penta- and tetrafluorobenzenes can be coupled with aryl carbamates with the aid of a nickel/copper dual catalytic system (Entry 2) [32]. A perfluoroarylcopper species is proposed to be involved as the nucleophile. The same catalyst system has also been used in the Sonogashira-type coupling of aryl carbamates using triisopropylsilylacetyle

Table 7.3 Nickel-catalyzed cross-coupling of aryl esters and carbamates with C—H bonds.

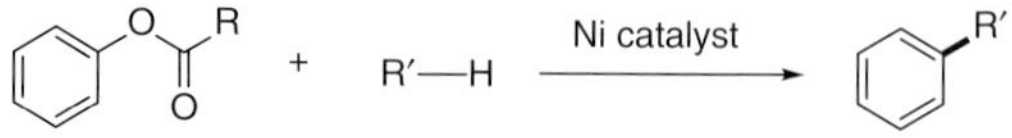

Entry	R	R′–H	Catalyst/ligand	References
1	^{t}Bu NEt_2 O^{t}Bu	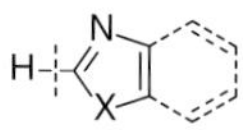 X = O, S, NMe	$Ni(cod)_2$ or $Ni(OTf)_2$/dcype $Ni(cod)_2$/PS-dppbz	[29–31]
2	NEt_2	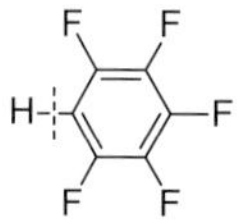	$Ni(cod)_2$/dcype/CuF/PPh_3	[32]
3	NEt_2	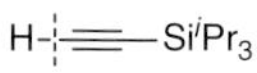	$Ni(cod)_2$/dcype/CuF/PPh_3	[32]
4	^{t}Bu NEt_2	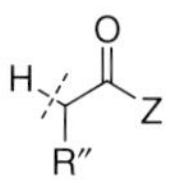 R″ = H, alkyl, aryl Z = aryl, ^{t}Bu, OMe, NMeAr	$Ni(cod)_2$/dcypt	[34]
5	^{t}Bu	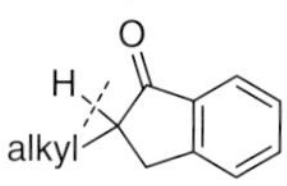	$Ni(cod)_2$ 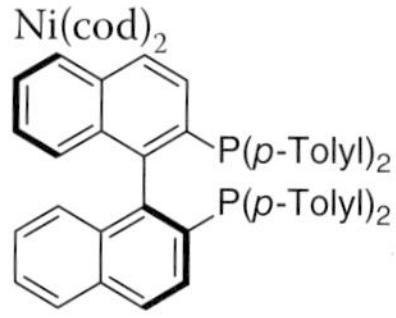	[35]
6	^{t}Bu	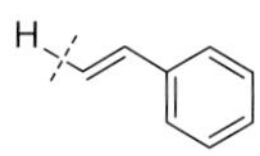	$Ni(cod)_2$/dppf	[36]

PS-dppbz =

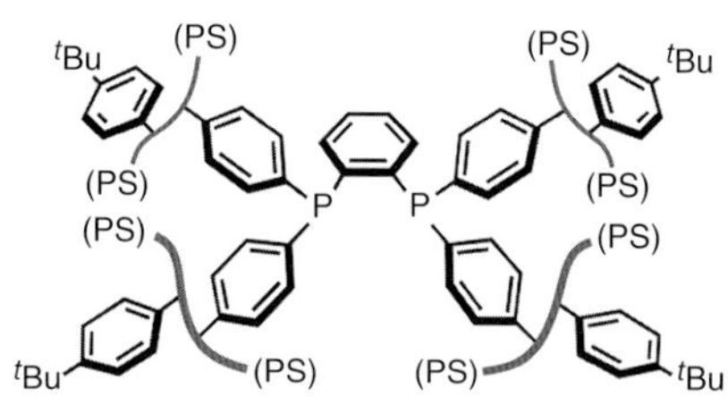

(Entry 3) [32]. The α-arylation of ketones, esters, and amides with aryl halides is a useful method for the synthesis of α-aryl carbonyl compounds [33]. Nickel catalysts allow aryl esters and carbamates to be used as arylating reagents in such α-arylation reactions. Itami and coworkers reported that dcypt, a new bisphosphine ligand based on a thiophene backbone (Figure 7.1), is a suitable ligand for this purpose (Entry 4) [34]. Martin and coworkers reported that nickel-catalyzed α-arylation can be applied to the enantioselective construction of a quaternary stereogenic center using cyclic ketone substrates (Entry 5) [35]. The Mizoroki–Heck reaction, in which aryl halides and alkenes are coupled via an insertion/β-hydrogen elimination sequence, is another class of C–H functionalizations. Watson and coworkers disclosed that the alkenylation of aryl pivalates can occur in the presence of a nickel/dppf catalyst (Entry 6) [36].

Catalytic cyanation of aryl halides is a useful method for the synthesis of aryl cyanides, which find widespread use in synthetic organic chemistry [37]. In such catalytic cyanation reactions, metal cyanides, such as $Zn(CN)_2$, are frequently used as a cyanating reagent. Itami and coworkers reported that aminoacetonitrile derivatives are a suitable cyano source in the nickel-catalyzed cyanation of aryl pivalates and carbamates (Eq. (7.2)) [38]. A cyanide ion generated in equilibrium with aminoacetonitrile is proposed to react with the oxidative addition complex.

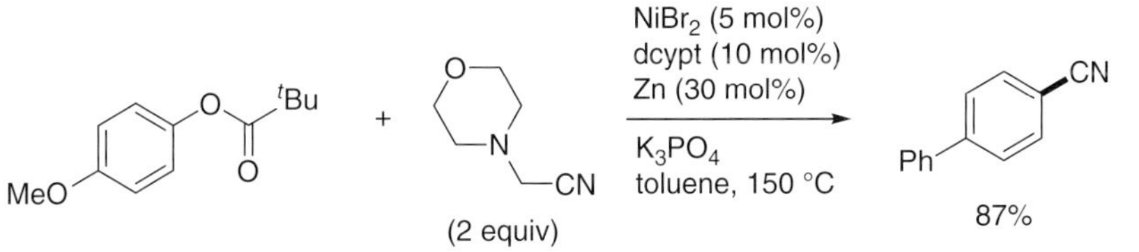

(7.2)

The cross-coupling of aryl halides with heteroatom-based nucleophiles has become a powerful method for the formation of carbon–heteroatom bonds. For example, the Buchwald–Hartwig amination is one of the most reliable methods for accessing aromatic amines [39]. Nickel-catalyzed C—O bond activation allows aryl esters and carbamates to be used in catalytic amination reactions. Tobisu and Chatani developed the first nickel-catalyzed amination of aryl esters [40], while Garg and coworkers reported on the amination of aryl carbamates [41] (Table 7.4, Entry 1). Both reactions employ an *N*-heterocyclic carbene ligand. Primary anilines can also be synthesized from aryl esters and carbamates by amination using a benzophenone imine, followed by acid hydrolysis of the resulting imines (Entry 2) [42]. The introduction of other heteroatoms to aryl esters and carbamates has also been accomplished. Shi and coworkers reported on the first borylation of inert phenol derivatives using a Ni/PCy_3 system (Entry 3) [43]. The use of a neopentylglycol-protected diboron reagent, not the corresponding pinacol-protected derivative, is essential for an efficient reaction. Martin and coworker reported that the use of a silylboron reagent leads to the formation of a silylated product with the aid of nickel/copper combined catalytic system (Entry 4) [44]. The role of a copper co-catalyst is likely to generate an active silylcopper species. Carbon-phosphorus bond formation was first accomplished by Han and coworkers, who developed a method for the cross-coupling of aryl esters with various organophosphorus reagents (Entry 5) [45]. Stannylation of aryl pivalates was also reported using a silylstannane reagent in conjunction with a nickel/dcype catalyst (Entry 6) [46].

The removal of functional groups often plays a key role in organic synthesis because it allows the electronic, steric, and coordinating properties of the

Table 7.4 Nickel-catalyzed cross-coupling of aryl esters and carbamates with heteroatom-based nucleophiles.

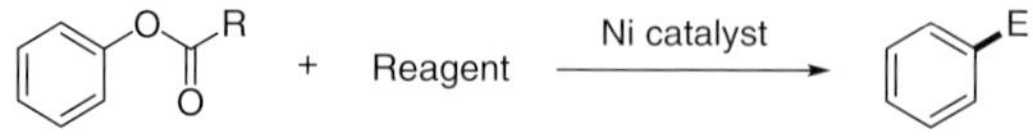

Entry	R	Reagent	Catalyst/ligand	Product	References
1	^{t}Bu NEt_2	HNR′R″	$Ni(cod)_2$/IPr or SIPr	Ar–NR′R″	[40, 41]
2	^{t}Bu NEt_2	HN=CPh₂ → $HN{=}CPh_2$	$Ni(cod)_2$/dcype	$Ar{-}NH_2$	[42]
3	NEt_2	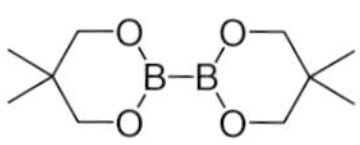	$NiCl_2(PCy_3)_2$	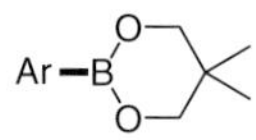	[43]
4	^{t}Bu	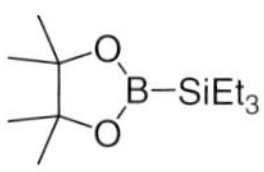	$Ni(cod)_2/PCy_3/CuF_2$	$Ar{-}SiEt_3$	[44]
5	^{t}Bu NEt_2	$H{-}P(O)R'_2$ R′ = Ph, alkyl, O^{i}Pr $HPPh_2$	$Ni(cod)_2$/dcype	$Ar{-}P(O)R'_2$ $Ar{-}PPh_2$	[45]
6	^{t}Bu	$Me_3Si{-}SnBu_3$	$Ni(cod)_2$/dcype	$Ar{-}SnBu_3$	[46]

functional group to be used in a temporary manner. The nickel-catalyzed activation of aryl esters and carbamates was successfully applied to the catalytic removal of these functional groups by performing the reaction in the presence of a suitable reducing agent (Table 7.5). Isopropylmagnesium halides were first reported to serve as a hydride donor in the nickel-catalyzed reduction of aryl carbamates (Entry 1) [10]. Later, methods using milder reducing agents, such

Table 7.5 Nickel-catalyzed reductive cleavage of aryl esters and carbamates.

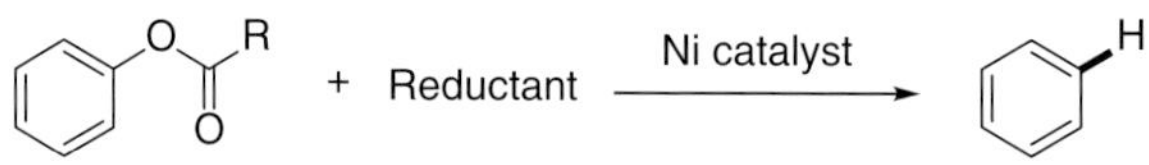

Entry	R	Reductant	Catalyst/ligand	References
1	NEt_2	^{i}PrMgX	$Ni(acac)_2$	[10]
2	^{t}Bu	$HSiMe(OMe)_2$	$Ni(cod)_2/PCy_3$	[47]
3	NEt_2	$HMe_2Si{-}O{-}SiMe_2H$	$NiCl_2(PCy_3)_2$	[48]
4	^{t}Bu	HCO_2Na	$Ni(cod)_2$/dcype	[49]

as hydrosilanes (Entries 2 and 3) [47, 48] and sodium formate (Entry 4) [49], were developed, allowing for the functional group tolerant modification of elaborate aromatic systems. For example, the sequential *ortho* and *para* C–H functionalization of aryl pivalate, followed by the removal of the pivaloxyl group, leads to the formation of a *meta*-terphenyl derivative (Scheme 7.1) [47]. In this synthetic scheme, the pivaloxy group can serve as a traceless handle to direct the initial bromination to occur at the para position and the subsequent palladium-catalyzed C–H arylation to occur at the ortho position.

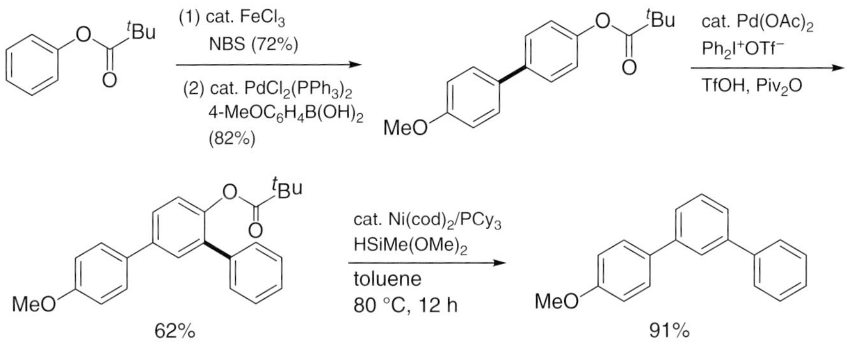

Scheme 7.1 Regioselective functionalization of aryl pivalate and catalytic removal of a pivaloxy group.

As a method for the synthesis of aromatic carboxylic acids, the catalytic carboxylation of aryl halides with CO_2 has emerged as a powerful and environmentally benign alternative to the nucleophilic addition reaction of organolithium and magnesium reagents to CO_2. Nickel is an excellent catalyst, allowing the carboxylation of aryl halides to occur under milder conditions using cheaper reducing agents, such as metallic zinc and manganese [50]. The key feature of the nickel system is the involvement of the Ar-Ni(I) species, which is capable of efficiently adding across CO_2. Another advantage of a nickel catalyst is that aryl pivalates can be used in place of aryl halides (Eq. (7.3)) [51]. The addition of aryl pivalates to isocyanates also proceeds, with the corresponding amides being produced under similar nickel-catalyzed conditions (Eq. (7.4)) [52].

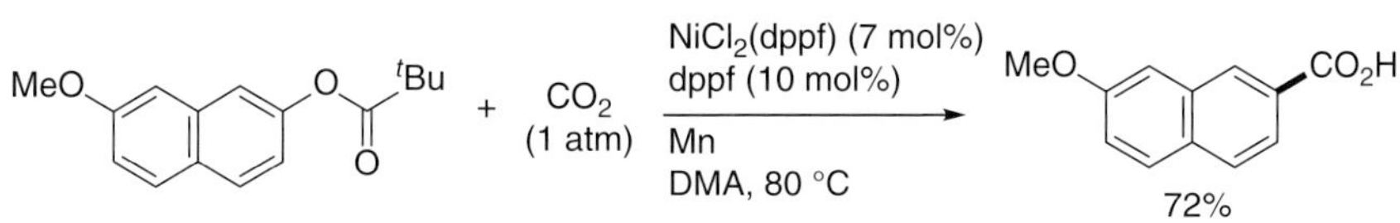

(7.3)

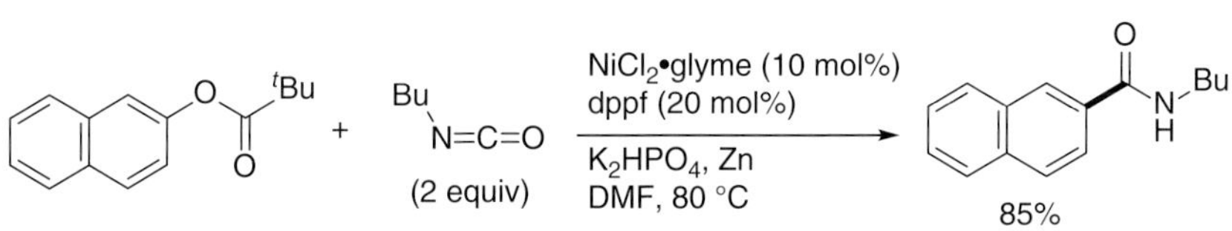

(7.4)

7.2.2 Aryl Ethers

C(aryl)—O bonds in aryl ethers are much less reactive than aryl esters and carbamates (Table 7.1), which makes the development of such reactions much more challenging. Nevertheless, the catalytic transformation of inert C(aryl)—O bonds was first accomplished by Wenkert et al. as early as in 1979 in the reaction of aryl ether substrates with Grignard reagents (Eq. (7.5)) [53].

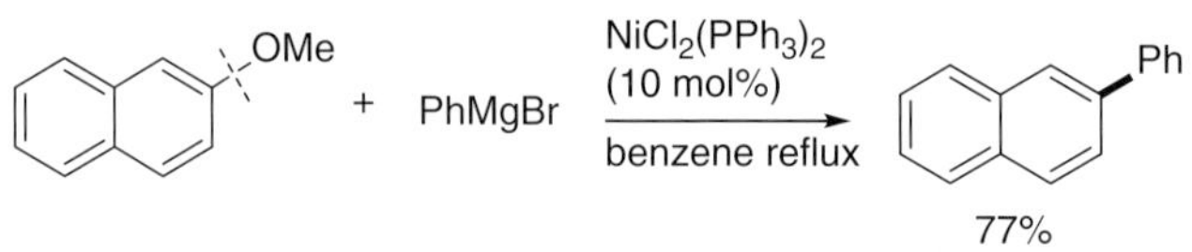

(7.5)

The synthetic utility of this method was revisited in 2004 by Dankwardt, who demonstrated that the nickel/PCy_3 system is an excellent catalyst for the cross-coupling of methoxyarenes with aryl Grignard reagents [54]. Since then, continuous advances have been made in the cross-coupling of methoxyarenes with organometallic reagents (Table 7.6). Grignard reagents are the most reliable coupling partners in terms of the scope of both methoxyarenes and Grignard reagents [55]. Alkynylation [56] and alkylation without β-hydrogens [57] are best performed with a nickel catalyst in conjunction with an ICy ligand (Entry 3). β-Hydrogen-containing alkyl Grignard reagents can be successfully cross-coupled with a range of anisole derivatives when a bidentate dcype ligand is used (Entry 4) [58]. In this alkylative cross-coupling, the addition of MgI_2 accelerates the reaction. The nickel-catalyzed Suzuki–Miyaura cross-coupling of aryl ethers was first accomplished by Tobisu and Chatani in 2008 using a Ni/PCy_3 system (Entry 5) [59]. However, this system was only applicable to π-extended substrates such as methoxynaphthalene. This limitation was later overcome by using Ni/ICy catalyst [60]. Rueping and coworkers reported that alkylative cross-coupling was also possible with *B*-alkyl-9-BBN reagents using a Ni/dcype catalyst (Entry 6) [61]. The Lewis acidic nature of alkylboron reagents is likely to play a key role in the activation of C(aryl)—O bond in methoxyarenes. Several other organometallic reagents, including organozinc (Entry 7) [62], lithium (Entries 8 and 9) [63], aluminum (Entries 10–12) [64], and scandium (Entry 13) [65], have also been reported to participate in the nickel-catalyzed cross-coupling of methoxyarenes (Table 7.6).

Regarding the substituent that is located on the ether moiety, a methoxy group is the most synthetically useful in terms of availability and atom economy. However, other alkoxy groups can also serve as leaving groups in nickel-catalyzed cross-coupling reactions. While longer chain alkoxy groups are less reactive than a methoxy group, a phenoxy group is more readily cleaved. The ring-opening cross-coupling of dibenzofurane is also possible, which is used in the synthesis of triphenylene derivatives (Scheme 7.2) [66].

Carbon-heteroatom bond-forming reactions for methoxyarene substrates are quite limited compared with aryl halides (Table 7.7). Tobisu and Chatani reported on the amination of methoxyarenes using a nickel/IPr catalyst (Entry 1) [67]. However, this amination method requires significantly harsher reaction conditions (20 mol% catalyst, 5 equiv of amine, 6 equiv of NaO^tBu, 120 °C,

Table 7.6 Nickel-catalyzed cross-coupling of methoxyarenes with organometallic nucleophiles.

$$\text{Ar–OMe} + \text{R'–M} \xrightarrow{\text{Ni catalyst}} \text{Ar–R'}$$

Entry	Nucleophile	Catalyst/ligand	Product	References
1	Ar′MgX	$NiCl_2(PPh_3)_2$ $NiCl_2(PCy_3)_2$ (allyl)NiCl(IPr)	Ar–Ar′	[53–55]
2	MeMgX	$NiCl_2(PCy_3)_2$	Ar–Me	[57a]
3	Me_3SiCH_2MgX $PhCH_2MgX$ $^iPr_3Si$–≡–MgX	$Ni(cod)_2$/ICy	Ar–CH_2SiMe_3 Ar–CH_2Ph Ar–≡–Si^iPr_3	[56, 57b]
4	Alkyl-Mgl	$Ni(cod)_2$/dcype	Ar–alkyl	[58]
5	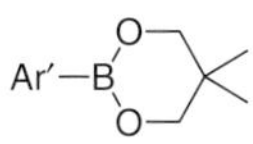	$Ni(cod)_2/PCy_3$ $Ni(cod)_2$/ICy	Ar–Ar′	[59, 60]
6	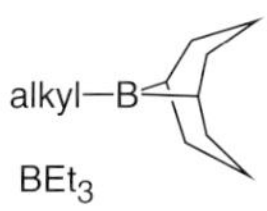 BEt_3	$Ni(cod)_2/PCy_3$	Ar–alkyl Ar–Et	[61]
7	$Ar'ZnMe_3Li_2$	$Ni(cod)_2/PCy_3$	Ar–Ar′	[62]
8	Me_3SiCH_2Li	$Ni(cod)_2$	Ar–CH_2SiMe_3	[63a]
9	Ar′Li	$NiCl_2(PPh_3)$(IPr)	Ar–Ar′	[63b]
10	Me_3Al	$Ni(cod)_2$/ICy	Ar–Me	[64a]
11	$(alkyl)_3Al$	$Ni(cod)_2$/dcype	Ar–alkyl	[14]
12	$Ar'Al^iBu_2$	$Ni(cod)_2$/ICy	Ar–Ar′	[64b]
13	$(Me_3SiCH_2)_3Sc(THF)_3$	$Ni(cod)_2/PCy_3$	Ar–CH_2SiMe_3	[65]

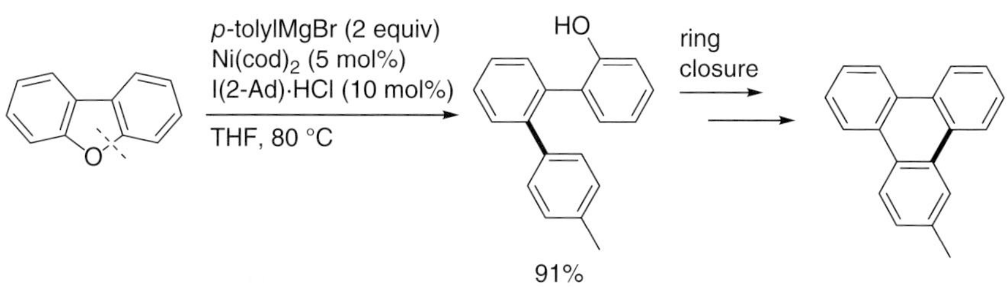

Scheme 7.2 Nickel-catalyzed ring-opening cross-coupling of dibenzofuran and its application to the synthesis of triphenylene derivatives.

Table 7.7 Nickel-catalyzed cross-coupling of methoxyarenes with heteroatom-based nucleophiles.

$$\text{Ar–OMe} + \text{Reagent} \xrightarrow{\text{Ni catalyst}} \text{Ar–E} \qquad \text{E = N, Si, B, P, etc.}$$

Entry	Reagent	Catalyst/ligand	Product	References
1	$HNR'R''$	$Ni(cod)_2$/IPr	$Ar–NR'R''$	[67]
2	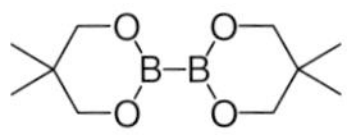	$Ni(cod)_2/PCy_3$	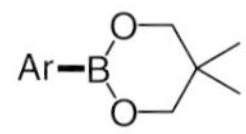	[68]
3	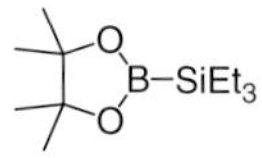	$Ni(cod)_2$	$Ar–SiEt_3$	[71, 72]

48 hours) than are needed for the amination of aryl esters. The borylation of methoxyarenes was reported to be catalyzed by the nickel/PCy_3 system (Entry 2) [68]. Interestingly, when the borylation of methoxyarenes is performed with a nickel/ICy catalyst using 0.8 equiv of a diboron reagent, the homocoupling product is formed (Eq. (7.6)) [69]. The reaction presumably proceeds through the borylation of a methoxy group, followed by the Suzuki–Miyaura type cross-coupling of the resulting borylated product with the starting methoxyarene. This protocol allows the rapid assembly of π-extended molecules from readily available methoxyarenes.

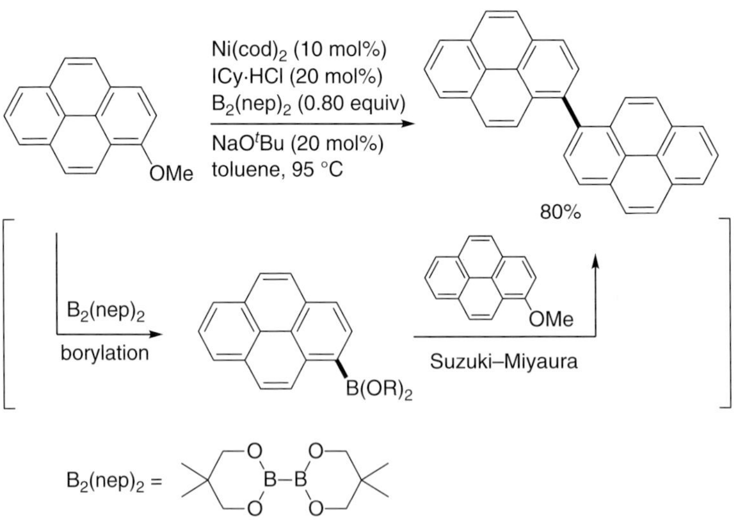

(7.6)

When the nickel-catalyzed borylation is applied to benzofuran substrates, the formal insertion of a boryl group into a C(alkenyl)—O bond occurs

(Eq. (7.7)) [70]. The resulting oxaborin derivatives serve as building blocks for cross-coupling as well as fluorescent materials.

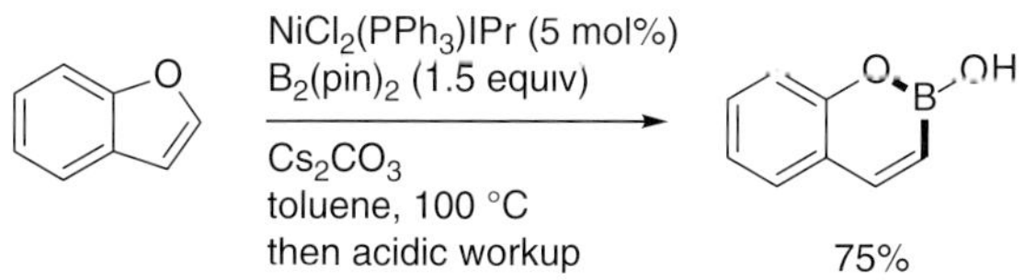

(7.7)

Martin and coworkers reported that the silylation of methoxyarenes takes place when a silylborane reagent is used (Entry 3) [71]. The striking feature of this silylation reaction is that the difficult C(aryl)—O bond cleavage occurs in the absence of any strong σ-donor ligands. It has been proposed that an anionic silylnickel(0) ate complex is responsible for this catalysis [72].

Montgomery and coworkers reported on a related silylation reaction in which hydrosilane reagents were used, but the applicable substrates are limited to aryl silyl ethers and a methoxy group cannot be silylated under the reaction conditions used (Eq. (7.8)) [73].

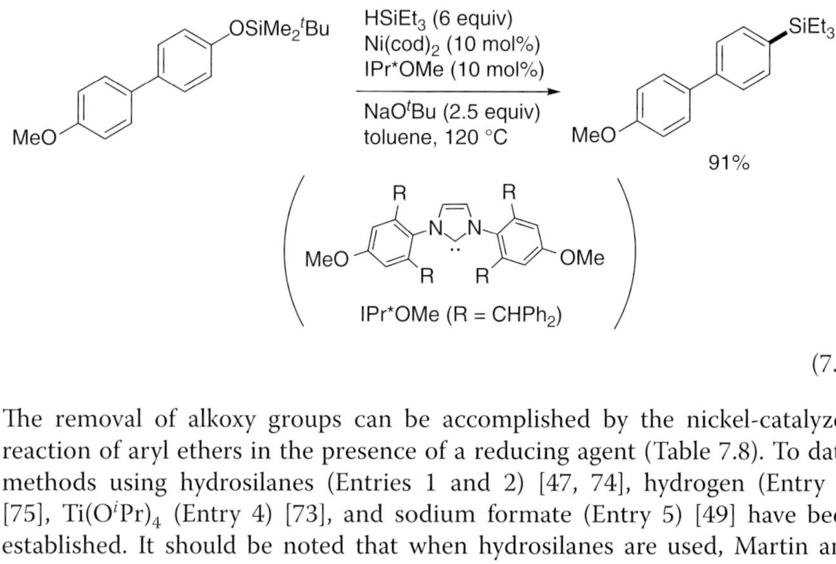

(7.8)

The removal of alkoxy groups can be accomplished by the nickel-catalyzed reaction of aryl ethers in the presence of a reducing agent (Table 7.8). To date, methods using hydrosilanes (Entries 1 and 2) [47, 74], hydrogen (Entry 3) [75], $Ti(O^iPr)_4$ (Entry 4) [73], and sodium formate (Entry 5) [49] have been established. It should be noted that when hydrosilanes are used, Martin and coworkers proposed a unique mechanism that involves a silyl-Ni(I)-mediated C(aryl)—O bond cleavage [76]. The reductive cleavage of aryl ethers can proceed even in the absence of an external reducing agent (Entry 6) [77]. A labeling study determined that the hydrogen delivered to the reduced product was primarily derived from the methoxy group of the substrate, indicating that the reaction proceeds through a β-hydrogen elimination from a nickel-alkoxide intermediate.

Reductive cross-coupling of two different aryl electrophiles has gained significant attention as alternative to the conventional cross-coupling of aryl halides with organometallic reagents, because it helps avoid the preparation of the organometallic reagents [78]. In this context, the formal reductive cross-coupling of methoxyarenes with aryl halides is reported to occur in the presence of metallic magnesium (Eq. (7.9)) [79]. It is proposed that the reaction

Table 7.8 Nickel-catalyzed reductive cleavage of aryl ethers.

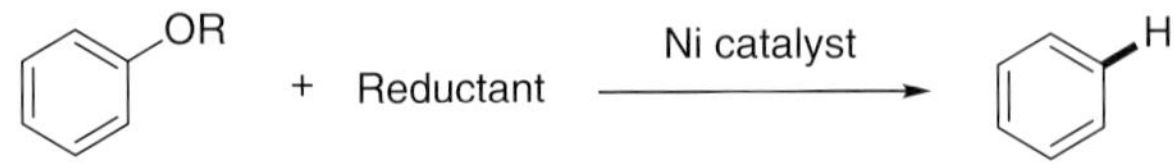

Entry	R	Reductant	Catalyst/ligand	References
1	Me	$Me_2Si(H)$–O–$SiMe_2(H)$	$NiCl_2(PCy_3)_2$	[74]
2	Alkyl	$HSiMe(OMe)_2$	$Ni(cod)_2/PCy_3$	[47]
4	Ar, alkyl	H_2	$Ni(cod)_2$/SIPr	[75]
5	SiR_3	$Ti(O^iPr)_4$	$Ni(acac)_2/IPr^{Me}$	[73]
6	Me	HCO_2Na	$Ni(cod)_2$/IPr	[76]
7	Alkyl	None	$Ni(cod)_2$/I(2-Ad)	[77]

is not a simple consequence of the generation of a Grignard reagent and a subsequent Kumada–Tamao–Corriu type cross-coupling, based on the fact that PCy_3, which is one of the most efficient ligands for cross-coupling using Grignard reagents, is unable to promote this process.

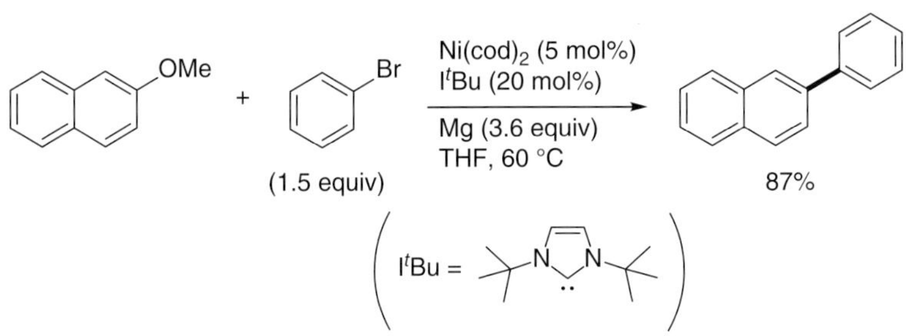

(7.9)

7.2.3 Arenols

There are only sporadic reports on the direct catalytic transformation of phenol derivatives without conversion into activated forms, such as triflates. Shi and coworkers reported on the cross-coupling of naphthol derivatives with Grignard reagents (Eq. (7.10)) [80]. In this reaction, magnesium naphthoate is initially generated, which then undergoes the C(aryl)—O bond cleavage by a nickel complex.

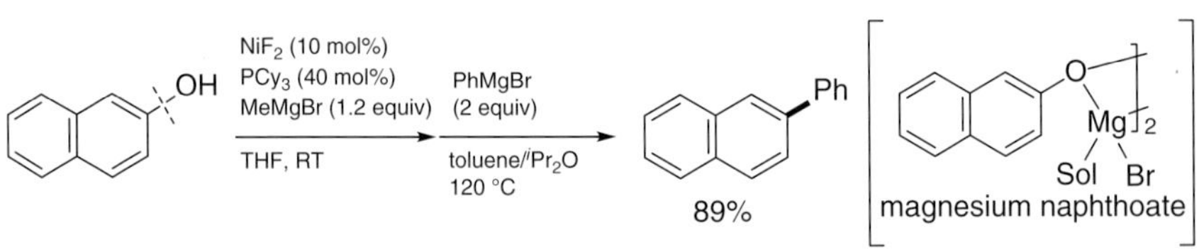

(7.10)

A related arylation of naphthol derivatives with organoboron reagents was also reported (Eq. (7.11)) [81]. The addition of BEt_3 is essential for an efficient reaction, although its role remains elusive at this stage.

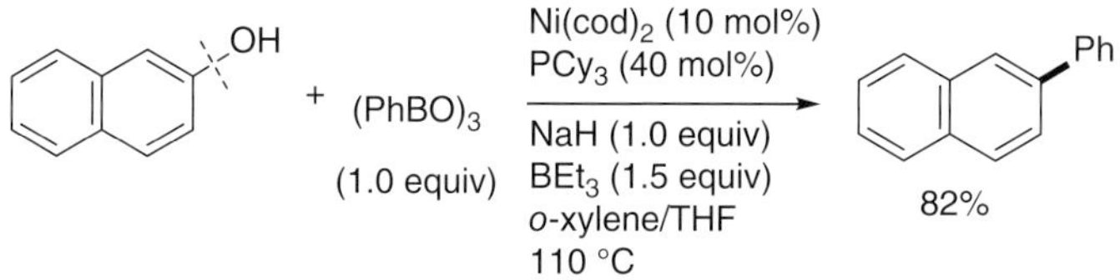

(7.11)

One isolated example, in which naphthol undergoes ipso borylation by nickel-catalyzed reaction using a diboron reagent, has been reported (Eq. (7.12)) [82].

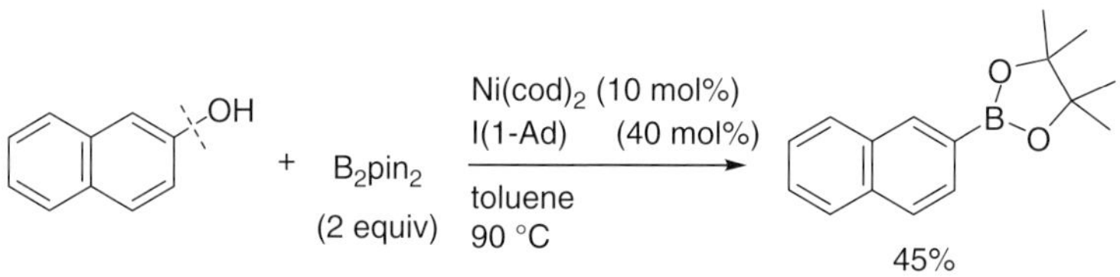

(7.12)

Interestingly, de-hydroxylation becomes a major pathway rather than borylation, when the ligand is switched from I(1-Ad) to PCy_3, even though both reactions involve the use of a diboron reagent (Eq. (7.13)) [83].

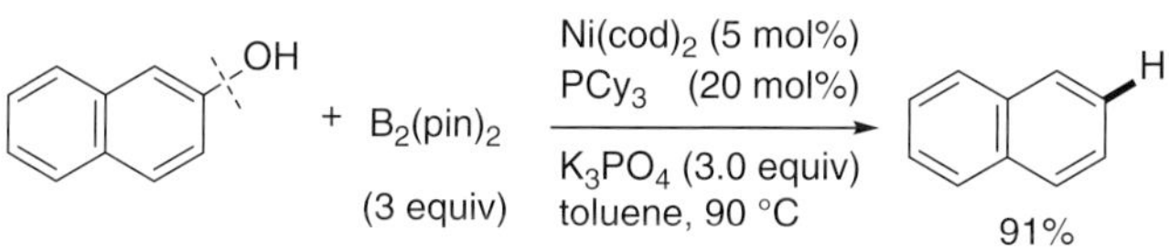

(7.13)

The nickel-catalyzed hydrogenolysis of phenols using $HSi(OMe)_2Me$ was reported by Nakao and coworker (Eq. (7.14)) [84]. This reaction proceeds through the silylation of phenol, followed by the reductive cleavage of the resulting silyl ether. A methoxy group is tolerated under these conditions, indicating that an $OSi(OMe)_2Me$ group is a better leaving group than a methoxy group in this reaction.

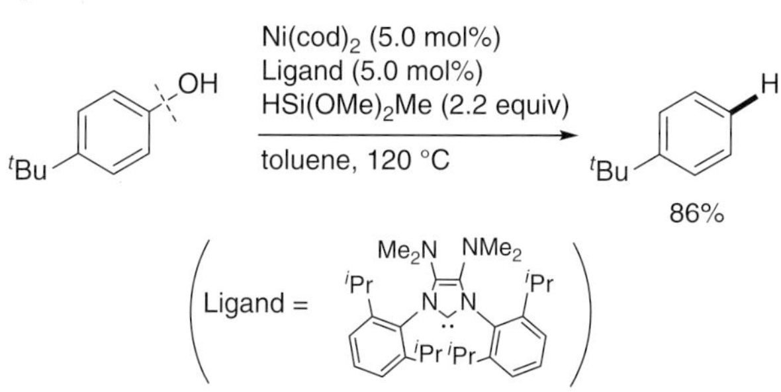

(7.14)

7.3 C(benzyl)—O Bond Cleavage

Nickel-catalyzed C(sp^3)—O bond activation reactions of esters and ethers are limited to allylic and benzylic substrates. Most of the catalytic reactions of aryl esters, carbamates, and ethers that are listed in Section 7.2 are also applicable to the corresponding benzylic substrates, which are not discussed in this section. One interesting feature of the use of benzylic substrates is the stereospecificity that is achieved when secondary benzylic esters and ethers are used. This aspect will be the focus of this section [85].

7.3.1 Benzyl Esters and Carbamates

In 2013, Jarvo and coworkers reported that secondary benzylic pivalates and carbamates undergo cross-coupling with organoboron reagents with a high

Table 7.9 Nickel-catalyzed stereospecific cross-coupling of benzyl esters and carbamates.

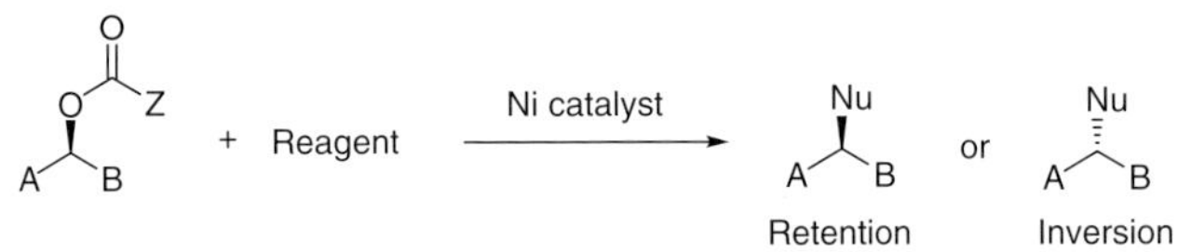

Entry	Substrate	Reagent	Catalyst/ligand	Product	References
1	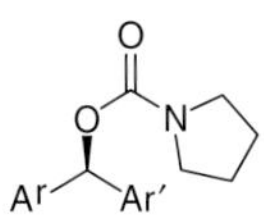	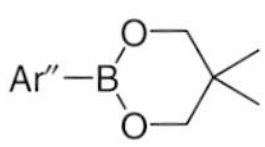	$Ni(cod)_2/PCy_3$	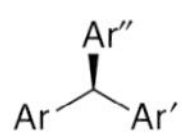	[86]
2	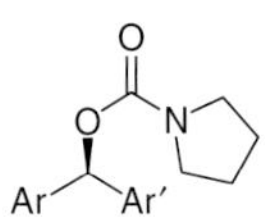	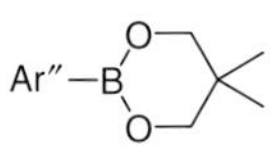	$Ni(cod)_2$/SIMes	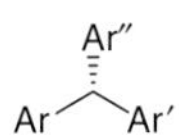	[86]
3	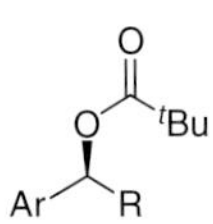 (R = alkyl, aryl)	$(Ar''BO)_3$	$Ni(cod)_2$	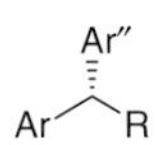	[87]
4	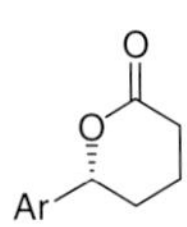	$ZnMe_2$	$Ni(acac)_2$/Xantphos	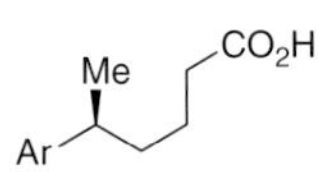	[88]
5	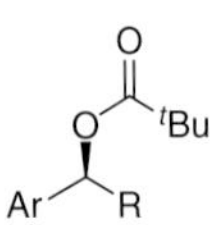 (R = alky)	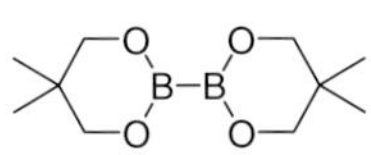	$Ni(cod)_2/PCy_3/CuF_2$	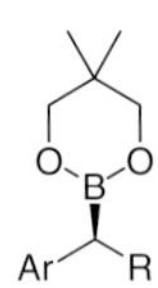	[89]

Xantphos = 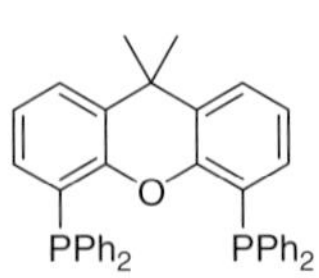

degree of stereospecificity [86]. Notably, the stereochemical outcome can be controlled by the ligand used: retention of configuration is observed in the case of a PCy_3 ligand (Table 7.9, Entry 1), while the SIMes system leads to inversion at the reaction center (Entry 2). Since the transmetalation and reductive elimination steps proceed with retention of configuration, the stereochemical course of the overall reaction is determined by the oxidative addition step. DFT studies indicate that the oxidative addition proceeds though a three-centered concerted pathway with the benzylic configuration being retained when PCy_3 is used [86b]. In contrast, oxidative addition proceeds via an S_N2-type mechanism, resulting in the inversion at the stereocenter. Watson and coworkers independently reported a similar reaction in which an external ligand was not used, which gives a cross-coupling product with inversion of the configuration (Entry 3) [87]. Jarvo and coworkers also reported on the nickel/Xantphos-catalyzed ring-opening cross-coupling of lactones using a Me_2Zn reagent, which gives the products with inversion (Entry 4) [88]. Stereoretentive borylation of secondary benzylic pivalates is also reported by Martin and coworkers using a nickel/copper dual catalyst (Entry 5) [89].

A stereospecific C(benzyl)—O bond cross-coupling was successfully combined with an intramolecular alkene insertion process (Eq. (7.15)) [90]. Thus, the chiral cyclic carbonate **A** undergoes oxidative addition with inversion at the benzylic stereocenter to form the benzylnickel intermediate **B**. A pendant alkene then inserts into the carbon–nickel bond in **B** to generate cyclopropylmethylnickel **C**, which reacts with a diboron reagent to provide the borylated product **D**. The use of arylboron reagents, instead of a diboron reagent, allows the introduction of an aryl group to the chiral cyclopropane skeleton.

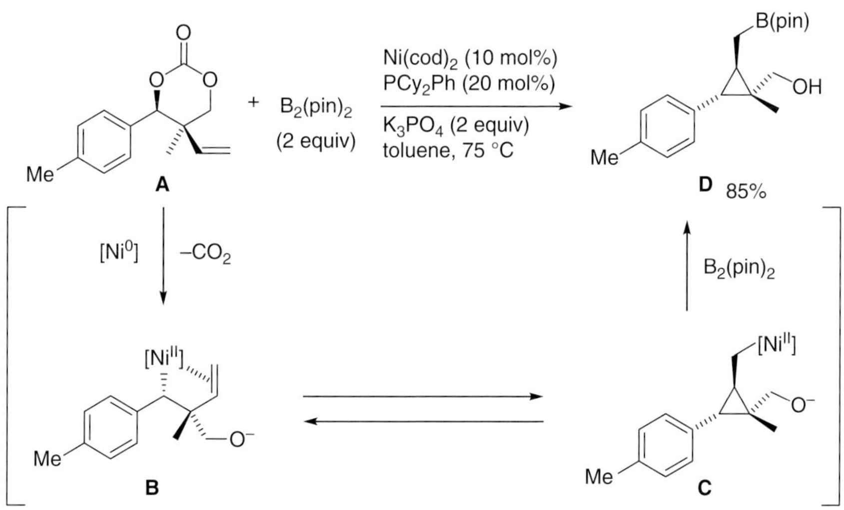

(7.15)

Intramolecular reductive cross-electrophile coupling between benzyl pivalate and aryl bromide is reported to occur by nickel catalysis using a stoichiometric amount of metallic zinc as a reducing agent (Eq. (7.16)) [91].

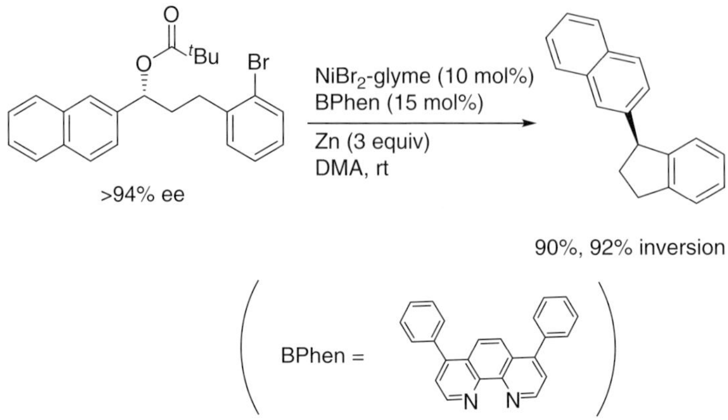

(7.16)

7.3.2 Benzyl Ethers

The stereospecific cross-coupling of benzylic ethers is also known. All reported reactions require Grignard reagents as coupling partners, presumably due to the inertness of the C—O bonds in ethers (Table 7.10) [88, 92, 93]. In these reactions, benzylic substitution occurs with inversion of the benzylic configuration. To overcome the low reactivity of simple benzylic ethers, the introduction of an additional heteroatom to an alkoxy group (2-methoxyethoxy, etc.) often leads to an improved reactivity because such bidentate alkoxy groups can coordinate to a metal cation derived from an organometallic coupling partner, thus increasing leaving group ability [94]. However, this section focuses on cross-coupling reactions of simple benzyl alkyl ethers.

Table 7.10 Nickel-catalyzed stereospecific cross-coupling of simple benzyl ethers.

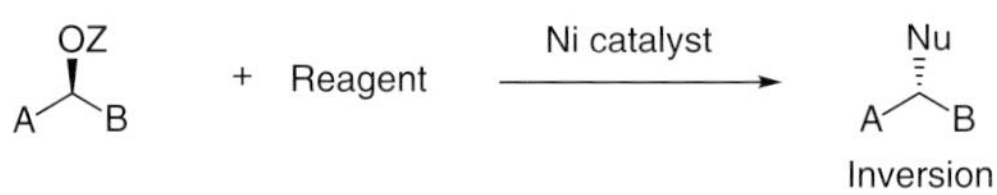

Entry	Substrate	Reagent	Catalyst/ligand	Product	References
1	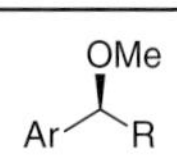 (R = alkyl, aryl)	MeMgX	Ni(cod)$_2$/*rac*-BINAP	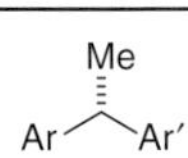	[92]
2	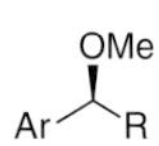 (R = alkyl, aryl)	Alkyl-MgX	$NiCl_2$(dppe)	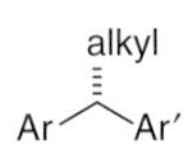	[93]
3	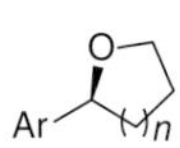 (*n* = 1,2)	R′MgX (R′ = alkyl, aryl)	Ni(cod)$_2$/*rac*-BINAP	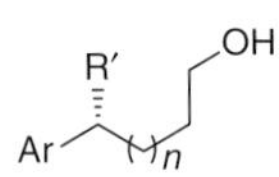	[88]

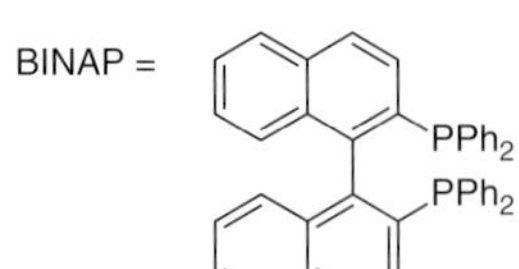

A stereospecific intramolecular Mizoroki–Heck reaction of chiral benzylic ethers was also developed by Jarvo (Eq. (7.17)) [95]. Although an external nucleophile is not incorporated in this reaction, the addition of 2 equiv of MeMgX was found to be essential for the reaction to proceed.

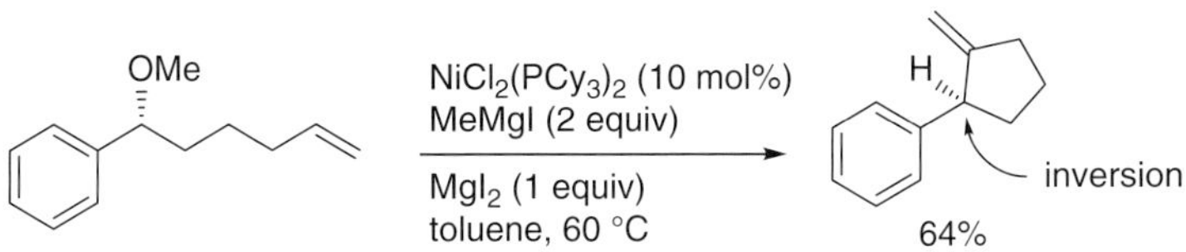

(7.17)

They also reported on the intramolecular reductive cross-electrophile coupling of cyclic benzylic ethers with alkyl chlorides, resulting in a ring contraction to form chiral cyclopropanes (Eq. (7.18)) [96]. The reaction proceeds in a stereospecific manner with respect to both stereocenters. The benzylic substitution occurs with net retention of configuration, whereas the reaction of the alkyl chloride moiety involves inversion.

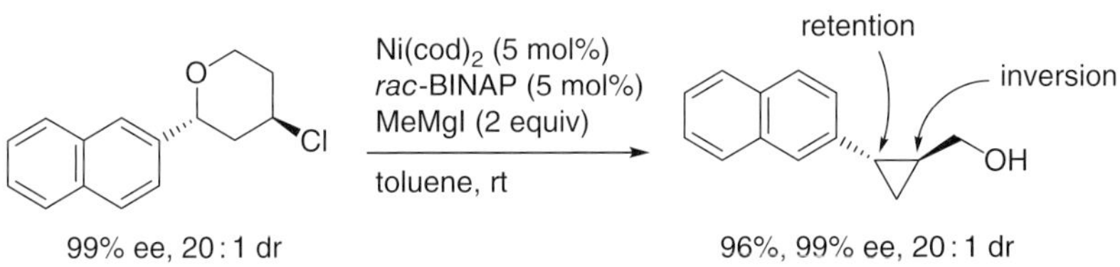

(7.18)

Shi and coworker reported that a nickel-catalyzed reaction of di(naphthylmethyl) ether in the presence of a diboron reagent and metallic Zn resulted in the formation of 1,2-dinaphthylethane (Eq. (7.19)) [97]. The reaction of an asymmetrically substituted ether provided not only the expected deoxygenation product, but also crossover products, indicating that the reaction occurs in an intermolecular manner.

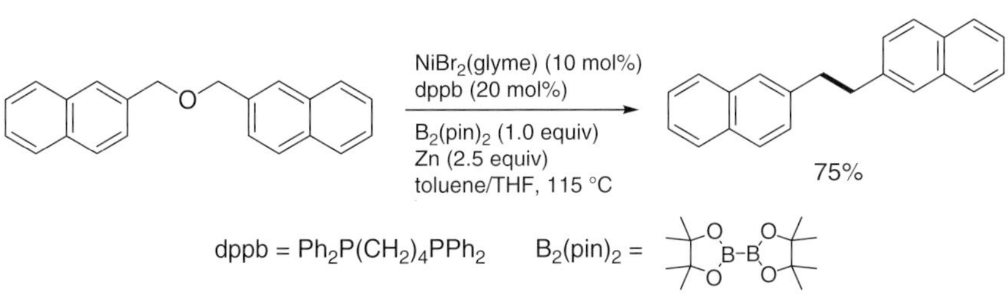

(7.19)

7.4 C(acyl)—O Bond Cleavage

Although several catalytic C(acyl)—O bond cleavage reactions of acid anhydrides [98], activated esters (i.e. *p*-nitrophenyl esters) [99], and esters having a directing group [100] have been reported to date, this section focuses on nickel-catalyzed C(acyl)—O bond transformations of simple ester substrates. The nickel-mediated activation of esters dates back to Yamamoto's seminal report in 1976 that phenyl propionate reacts with $Ni(cod)(PPh_3)_2$ to generate phenol, ethylene, and $Ni(CO)(PPh_3)_3$ (Eq. (7.20)) [101]. The formation of these compounds can be rationalized by the sequential oxidative addition of a C(acyl)—O bond to Ni(0), decarbonylation, β-hydrogen elimination, and reductive elimination.

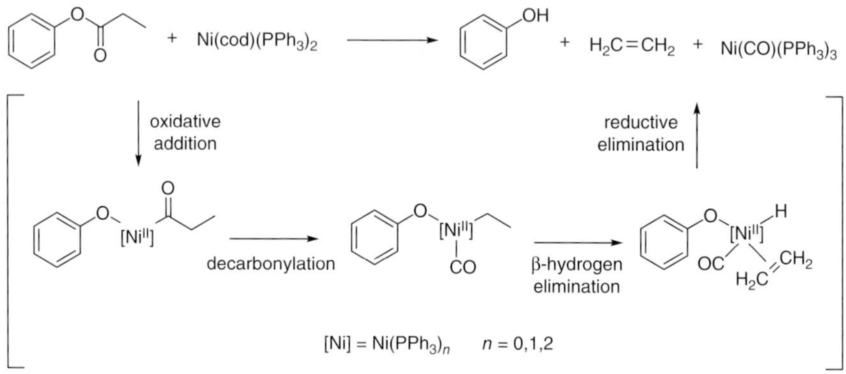

(7.20)

Catalytic applications based on Yamamoto's findings were stimulated by a seminal work by Itami and coworker, who developed the nickel-catalyzed decarbonylative cross-coupling of heteroaryl esters with azoles (Eq. (7.21)) [102]. In this reaction, a phenoxy carbonyl moiety serves as a leaving group in cross-coupling reactions.

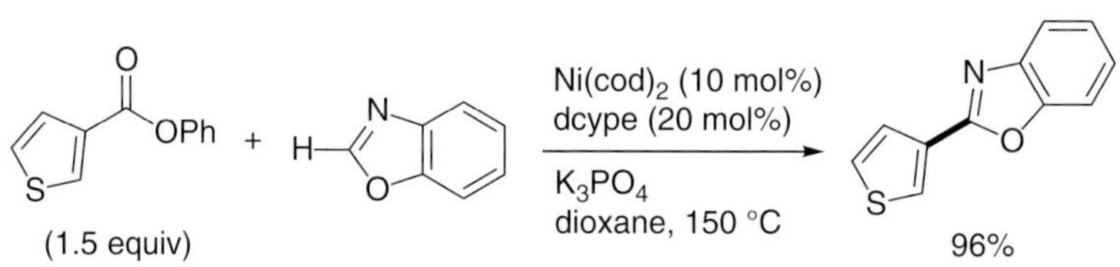

(7.21)

Since this report, a number of nickel-catalyzed decarbonylative cross-coupling reactions of aryl esters have appeared (Table 7.11). The Suzuki–Miyaura type reaction using phenyl esters can be catalyzed by a nickel/monophosphine system for coupling with aromatic boron reagents (Entries 1 and 2) [103] and by a nickel/bisphosphine system for reactions with alkylboron reagents (Entry 3) [104]. An alkylation method in which organozinc reagents are used was also reported (Entry 4) [13]. Carbon-heteroatom bond formation via the catalytic decarbonylative coupling of aryl esters was also disclosed. The conversion of carboxylic acids to amines is classically performed through a Schmidt-type rearrangement. Rueping and coworkers established an alternative approach, wherein a nickel-catalyzed decarbonylative amination of phenyl esters with benzophenone imine provides primary anilines upon hydrolysis (Entry 5) [105]. Itami and coworkers demonstrated that decarbonylative C—O bond formation is also possible by the nickel-catalyzed reaction of aryl esters derived from azine

Table 7.11 Nickel-catalyzed decarbonylative cross-coupling of aryl esters via C(acyl)—O bond cleavage.

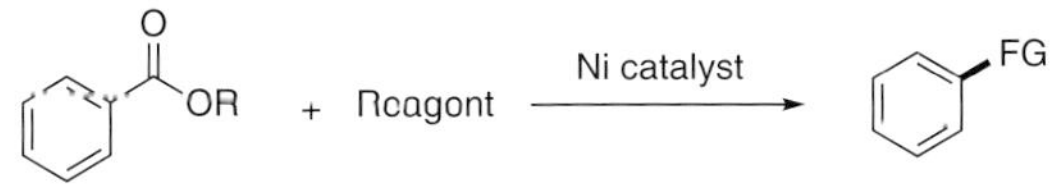

Entry	Substrate	Reagent	Catalyst/ligand	Product	References
1	Ar—CO_2Ph	Ar′B$(OH)_2$	Ni$(OAc)_2$/P^{n}Bu$_3$	Ar—Ar′	[103a]
2	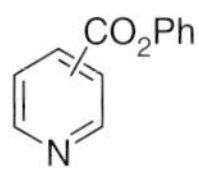	Ar′B$(OH)_2$	Ni$(cod)_2$/PCy$_3$	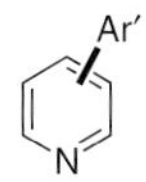	[103b]
3	Ar—CO_2Ph	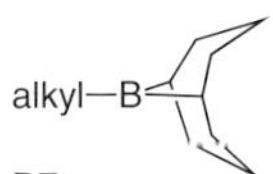 BEt$_3$	Ni$(cod)_2$/dcype	Ar—alkyl	[104]
4	Ar—CO_2Ph	Alkyl—ZnBr	Ni$(cod)_2$/dcype	Ar—alkyl	[13]
5	Het—CO_2Ph (Het = heteroaryl)	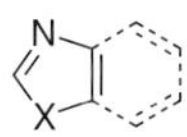 (X = O, S)	Ni$(cod)_2$/dcype or 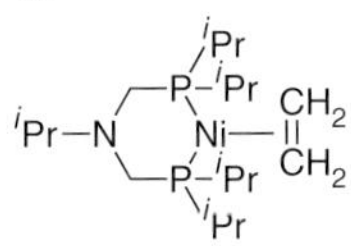	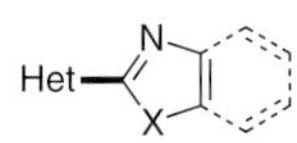	[102]
6	Ar—CO_2Ph	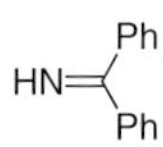	Ni$(cod)_2$/dcype	Ar—NH_2	[105]
7	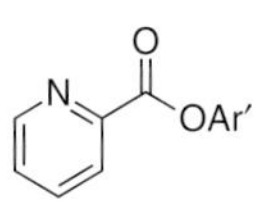	None	Ni$(cod)_2$/dcypt	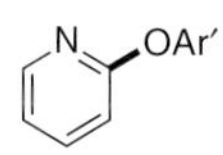	[106]
8	Ar—CO_2Ph	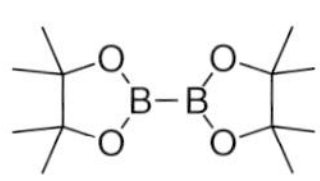	Ni$(cod)_2$/P^{n}Bu$_3$ or Ni$(cod)_2$/ICy	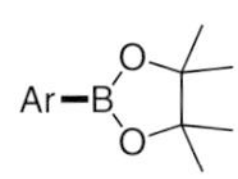	[107]
9	Ar—CO_2Ph	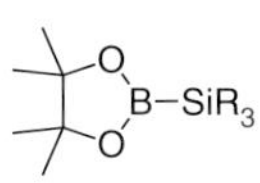	Ni$(cod)_2$/dcype/CuF$_2$ or Ni$(cod)_2$/P^{n}Bu$_3$/CuF$_2$	Ar—SiR$_3$	[107, 108]
10	Ar—CO_2R (R = alkyl, Ph)	Me$_3$Si—SnBu$_3$	Ni$(cod)_2$ or NiBr$_2$/dppp	Ar—SnBu$_3$	[109]
11	Ar—CO_2Ph	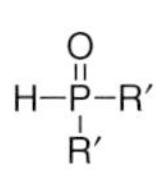	Ni$(OAc)_2$/dcype	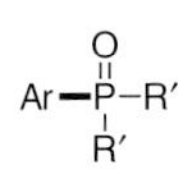	[110]
12	Ar—CO_2Ph	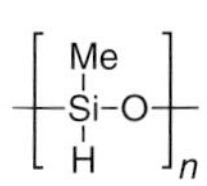	Ni$(OAc)_2$/dcype	Ar—H	[112]

carboxylic acids (Entry 6) [106]. In this reaction, C—O bond formation occurred intramolecularly by the decarbonylation of these esters without the need for adding an external alkoxide. Other heteroatoms, including boron (Entry 7) [107], silicon (Entry 8) [107, 108], stannane (Entry 9) [109], and phosphorus (Entry 10) [110], can also be introduced to aryl esters by nickel-catalyzed decarboxylative cross-coupling. Although the catalytic systems used in the reactions shown in Table 7.11 are similar to those used for the cross-coupling of aryl esters (Table 7.4), C(aryl)—O bond cleavage products are not formed. This is presumably because C(acyl)—O bond cleavage is basically preferred over the cleavage of C(aryl)—O bonds, particularly in the case of phenyl esters, unless steric bias is embedded in the substrates (i.e. pivalate esters) [8, 111]. The catalytic removal of a phenyl ester moiety is possible by decarbonylative reduction using a hydrosilane reagent (Entry 11) [112].

Nickel-catalyzed C(acy)—O bond activation reactions that do not involve decarbonylation, i.e. catalytic acylation, have also been reported, although the number of reports is limited so far, compared with decarbonylative reactions. Garg reported on the nickel-catalyzed amidation of methyl esters with aniline derivatives (Eq. (7.22)) [113]. This endergonic amidation process was made thermodynamically feasible by adding a stoichiometric amount of $Al(O^tBu)_3$, which stabilizes the amide product by complexation. A related reaction in combination with nitroarene reduction was also reported [114].

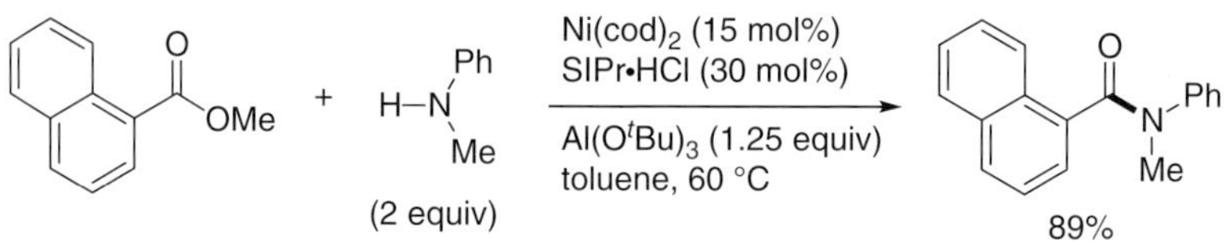

(7.22)

Another example of a non-decarbonylative reaction involves the nickel-catalyzed cross-coupling of phenyl esters with alkylboron reagents (Eq. (7.23)) [104]. This combination of reagents results in a decarbonylative cross-coupling when bidentate phosphine is used at 150 °C (Table 7.11, Entry 3). In contrast, the use of monophosphine at 80 °C leads to the formation of ketone products.

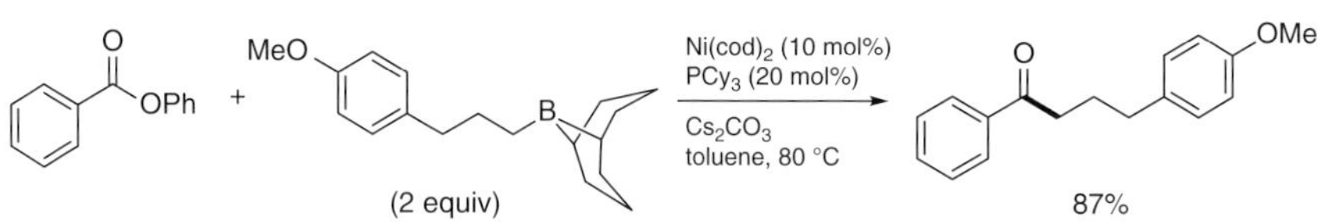

(7.23)

7.5 Summary and Outlook

Recent progress in the nickel-catalyzed transformation of inert C—O bonds has been overviewed. Since the pioneering discovery by Wenkert of the nickel-catalyzed cross-coupling of aryl ethers with Grignard reagents (Eq. (7.5)) in 1979, significant advances have been made in terms of the scope of both substrates and coupling partners. In particular, aryl esters and carbamates have now become viable substrates for use in a range of cross-coupling processes. More reliable catalysts need to be developed for more challenging methoxyarene and arenol substrates, especially when coupling partners other than Grignard reagents are used. One of the common problems associated with the nickel-catalyzed activation of C(aryl)—O bonds in inert phenol derivatives is the decreased reactivity of benzene derivatives compared with π-extended compounds, a subject that has not been fully addressed in many of the reactions reported to date. Activation of $C(sp^3)$—O bonds is currently possible only with allylic and benzylic substrates, and catalysts that can activate regular $C(sp^3)$—O bonds promise to be the subject of future studies. Although nickel complexes will continue to play a key role in developing new C—O bond activation reactions, merging with other metal catalysis and/or electron-transfer process is expected to play a major role in overcoming the remaining challenges in this field.

References

1 (a) Miyaura, N. (ed.) (2002). *Cross-Coupling Reactions: A Practical Guide.* Berlin: Springer-Verlag. (b) Diederich, F. and de Meijere, A. (eds.) (2004). *Metal-Catalyzed Cross-Coupling Reactions.* Weinheim: Wiley-VCH. (c) Suzuki, A. and Yamamoto, Y. (2011). *Chem. Lett.* 40: 894.

2 Rosen, B.M., Quasdorf, K.W., Wilson, D.A. et al. (2011). *Chem. Rev.* 111: 1346.

3 Li, B.-J., Yu, D.-G., Sun, C.-L., and Shi, Z.-J. (2011). *Chem. Eur. J.* 17: 1728.

4 Yamaguchi, J., Muto, K., and Itami, K. (2013). *Eur. J. Org. Chem.*: 19.

5 Cornella, J., Zarate, C., and Martin, R. (2014). *Chem. Soc. Rev.* 43: 8081.

6 Tobisu, M. and Chatani, N. (2015). *Acc. Chem. Res.* 48: 1717.

7 Tasker, S.Z., Standley, E.A., and Jamison, T.F. (2014). *Nature* 509: 299.

8 Li, Z., Zhang, S.-L., Fu, Y. et al. (2009). *J. Am. Chem. Soc.* 131: 8815.

9 (a) Shimasaki, T., Konno, Y., Tobisu, M., and Chatani, N. (2009). *Org. Lett.* 11: 4890. (b) Meng, L., Kamada, Y., Muto, K. et al. (2013). *Angew. Chem. Int. Ed.* 52: 10048.

10 Sengupta, S., Leite, M., Raslan, D.S. et al. (1992). *J. Org. Chem.* 57: 4066.

11 Yoshikai, N., Matsuda, H., and Nakamura, E. (2009). *J. Am. Chem. Soc.* 131: 9590.

12 Li, B.-J., Li, Y.-Z., Lu, X.-Y. et al. (2008). *Angew. Chem. Int. Ed.* 47: 10124.

13 Liu, X., Jia, J., and Rueping, M. (2017). *ACS Catal.* 7: 4491.

14 Ogawa, H., Yang, Z.-K., Minami, H. et al. (2017). *ACS Catal.* 7: 3988.

15 (a) Miyaura, N. and Suzuki, A. (1995). *Chem. Rev.* 95: 2457. (b) Suzuki, A. and Brown, H.C. (2003). *Organic Synthesis via Boranes.* Milwaukee, WI: Aldrich. (c) Hall, D.G. (2005). *Boronic Acids.* Weinheim: Wiley-VCH. (d) Miyaura, N. (2008). *Bull. Chem. Soc. Jpn.* 81: 1535.

16 (a) Quasdorf, K.W., Tian, X., and Garg, N.K. (2008). *J. Am. Chem. Soc.* 130: 14422. (b) Ramgren, S.D., Hie, L., Ye, Y., and Garg, N.K. (2013). *Org. Lett.* 15: 3950.

17 Guan, B.-T., Wang, Y., Li, B.-J. et al. (2008). *J. Am. Chem. Soc.* 130: 14468.

18 Xu, M., Li, X., Sun, Z., and Tu, T. (2013). *Chem. Commun.* 49: 11539.
19 Molander, G.A. and Beaumard, F. (2010). *Org. Lett.* 12: 4022.
20 Leowanawat, P., Zhang, N., and Percec, V. (2012). *J. Org. Chem.* 77: 1018.
21 (a) Quasdorf, K.W., Riener, M., Petrova, K.V., and Garg, N.K. (2009). *J. Am. Chem. Soc.* 131: 17748. (b) Baghbanzadeh, M., Pilger, C., and Kappe, C.O. (2011). *J. Org. Chem.* 76: 1507.
22 (a) Antoft-Finch, A., Blackburn, T., and Snieckus, V. (2009). *J. Am. Chem. Soc.* 131: 17750. (b) Quasdorf, K.W., Antoft-Finch, A., Liu, P. et al. (2011). *J. Am. Chem. Soc.* 133: 6352.
23 Xu, L., Li, B.-J., Wu, Z.-H. et al. (2010). *Org. Lett.* 12: 884.
24 Kuwano, R. and Shimizu, R. (2011). *Chem. Lett.* 40: 913.
25 Shi, W.-J., Zhao, H.-W., Wang, Y. et al. (2016). *Adv. Synth. Catal.* 358: 2410.
26 Wang, J., Ferguson, D.M., and Kalyani, D. (2013). *Tetrahedron* 69: 5780.
27 Song, W. and Ackermann, L. (2012). *Angew. Chem. Int. Ed.* 51: 8251.
28 Tobisu, M., Yasui, K., Aihara, Y., and Chatani, N. (2017). *Angew. Chem. Int. Ed.* 56: 1877.
29 (a) Muto, K., Yamaguchi, J., and Itami, K. (2012). *J. Am. Chem. Soc.* 134: 169. (b) Muto, K., Hatakeyama, T., Yamaguchi, J., and Itami, K. (2015). *Chem. Sci.* 6: 6792.
30 Muto, K., Yamaguchi, J., Lei, A., and Itami, K. (2013). *J. Am. Chem. Soc.* 135: 16384.
31 Iwai, T., Harada, T., Shimada, H. et al. (2017). *ACS Catal.* 7: 1681.
32 Wang, Y., Wu, S.-B., Shi, W.-J., and Shi, Z.-J. (2016). *Org. Lett.* 18: 2548.
33 Bellina, F. and Rossi, R. (2010). *Chem. Rev.* 110: 1082.
34 (a) Takise, R., Muto, K., Yamaguchi, J., and Itami, K. (2014). *Angew. Chem. Int. Ed.* 53: 6791. (b) Koch, E., Takise, R., Studer, A. et al. (2015). *Chem. Commun.* 51: 855.
35 Cornella, J., Jackson, E.P., and Martin, R. (2015). *Angew. Chem. Int. Ed.* 54: 4075.
36 Ehle, A.R., Zhou, Q., and Watson, M.P. (2012). *Org. Lett.* 14: 1202.
37 Anbarasan, P., Schareina, T., and Beller, M. (2011). *Chem. Soc. Rev.* 40: 5049.
38 Takise, R., Itami, K., and Yamaguchi, J. (2016). *Org. Lett.* 18: 4428.
39 (a) Hartwig, J.F. (2008). *Acc. Chem. Res.* 41: 1534. (b) Surry, D.S. and Buchwald, S.L. (2008). *Angew. Chem. Int. Ed.* 47: 6338. (c) Ruiz-Castillo, P. and Buchwald, S.L. (2016). *Chem. Rev.* 116: 12564.
40 Shimasaki, T., Tobisu, M., and Chatani, N. (2010). *Angew. Chem. Int. Ed.* 49: 2929.
41 (a) Mesganaw, T., Silberstein, A.L., Ramgren, S.D. et al. (2011). *Chem. Sci.* 2: 1766. (b) Hie, L., Ramgren, S.D., Mesganaw, T., and Garg, N.K. (2012). *Org. Lett.* 14: 4182.
42 Yue, H., Guo, L., Liu, X., and Rueping, M. (2017). *Org. Lett.* 19: 1788.
43 Huang, K., Yu, D.-G., Zheng, S.-F. et al. (2011). *Chem. Eur. J.* 17: 786.
44 Zarate, C. and Martin, R. (2014). *J. Am. Chem. Soc.* 136: 2236.
45 (a) Yang, J., Chen, T., and Han, L.-B. (2015). *J. Am. Chem. Soc.* 137: 1782. (b) Yang, J., Xiao, J., Chen, T., and Han, L.-B. (2016). *J. Org. Chem.* 81: 3911.
46 Gu, Y. and Martín, R. (2017). *Angew. Chem. Int. Ed.* 56: 3187.

47 Tobisu, M., Yamakawa, K., Shimasaki, T., and Chatani, N. (2011). *Chem. Commun.* 47: 2946.
48 Mesganaw, T., Fine Nathel, N.F., and Garg, N.K. (2012). *Org. Lett.* 14: 2918.
49 Xi, X., Chen, T., Zhang, J.-S., and Han, L.-B. (2018). *Chem. Commun.* 54: 1521.
50 Fujihara, T., Nogi, K., Xu, T. et al. (2012). *J. Am. Chem. Soc.* 134: 9106.
51 Correa, A., León, T., and Martin, R. (2014). *J. Am. Chem. Soc.* 136: 1062.
52 Correa, A. and Martin, R. (2014). *J. Am. Chem. Soc.* 136: 7253.
53 (a) Wenkert, E., Michelotti, E.L., and Swindell, C.S. (1979). *J. Am. Chem. Soc.* 101: 2246. (b) Wenkert, E., Michelotti, E.L., Swindell, C.S., and Tingoli, M. (1984). *J. Org. Chem.* 49: 4894.
54 Dankwardt, J.W. (2004). *Angew. Chem. Int. Ed.* 43: 2428.
55 (a) Xie, L.-G. and Wang, Z.-X. (2011). *Chem. Eur. J.* 17: 4972. (b) Zhao, F., Yu, D.-G., Zhu, R.-Y. et al. (2011). *J. Chem. Lett.* 40: 1001. (c) Iglesias, M.J., Prieto, A., and Nicasio, M.C. (2012). *Org. Lett.* 14: 4318. d) Zhang, J., Xu, J., Xu, Y. et al. (2015). *Organometallics* 34: 5792.
56 (a) Tobisu, M., Takahira, T., Ohtsuki, A., and Chatani, N. (2015). *Org. Lett.* 17: 680. (b) Chen, X.-C., Nishinaga, S., Okuda, Y. et al. (2015). *Org. Chem. Front.* 2: 536.
57 (a) Guan, B.-T., Xiang, S.-K., Wu, T. et al. (2008). *Chem. Commun.* 44: 1437. (b) Tobisu, M., Takahira, T., and Chatani, N. (2015). *Org. Lett.* 17: 4352.
58 Tobisu, M., Takahira, T., Morioka, T., and Chatani, N. (2016). *J. Am. Chem. Soc.* 138: 6711.
59 Tobisu, M., Shimasaki, T., and Chatani, N. (2008). *Angew. Chem. Int. Ed.* 47: 4866.
60 Tobisu, M., Yasutome, A., Kinuta, H. et al. (2014). *Org. Lett.* 16: 5572.
61 Guo, L., Liu, X., Baumann, C., and Rueping, M. (2016). *Angew. Chem. Int. Ed.* 55: 15415.
62 Wang, C., Ozaki, T., Takita, R., and Uchiyama, M. (2012). *Chem. Eur. J.* 18: 3482.
63 (a) Leiendecker, M., Hsiao, C.-C., Guo, L. et al. (2014). *Angew. Chem. Int. Ed.* 53: 12912. (b) Heijnen, D., Gualtierotti, J.-B., Hornillos, V., and Feringa, B.L. (2016). *Chem. Eur. J.* 22: 3991.
64 (a) Morioka, T., Nishizawa, A., Nakamura, K. et al. (2015). *Chem. Lett.* 44: 1729. (b) Liu, X., Hsiao, C.-C., Kalvet, I. et al. (2016). *Angew. Chem. Int. Ed.* 55: 6093.
65 Yan, X., Yang, F., Cai, G. et al. (2018). *Org. Lett.* 20: 624.
66 Kurata, Y., Otsuka, S., Fukui, N. et al. (2017). *Org. Lett.* 19: 1274.
67 (a) Tobisu, M., Shimasaki, T., and Chatani, N. (2009). *Chem. Lett.* 38: 710. (b) Tobisu, M., Yasutome, A., Yamakawa, K. et al. (2012). *Tetrahedron* 68: 5157.
68 Zarate, C., Manzano, R., and Martin, R. (2015). *J. Am. Chem. Soc.* 137: 6754.
69 Nakamura, K., Tobisu, M., and Chatani, N. (2015). *Org. Lett.* 17: 6142.
70 Saito, H., Otsuka, S., Nogi, K., and Yorimitsu, H. (2016). *J. Am. Chem. Soc.* 138: 15315.
71 Zarate, C., Nakajima, M., and Martin, R. (2017). *J. Am. Chem. Soc.* 139: 1191.

72 (a) Wang, B., Zhang, Q., Jiang, J. et al. (2017). *Chem. Eur. J.* 23: 17249. (b) Jain, P., Pal, S., and Avasare, V. (2018). *Organometallics* 37: 1141.
73 Wiensch, E.M., Todd, D.P., and Montgomery, J. (2017). *ACS Catal.* 7: 5568.
74 Álvarez-Bercedo, P. and Martin, R. (2010). *J. Am. Chem. Soc.* 132: 17352.
75 Sergeev, A.G. and Hartwig, J.F. (2011). *Science* 332: 439.
76 Cornella, J., Gómez-Bengoa, E., and Martin, R. (2013). *J. Am. Chem. Soc.* 135: 1997.
77 Tobisu, M., Morioka, T., Ohtsuki, A., and Chatani, N. (2015). *Chem. Sci.* 6: 3410.
78 (a) Everson, D.A. and Weix, D.J. (2014). *J. Org. Chem.* 79: 4793. (b) Weix, D.J. (2015). *Acc. Chem. Res.* 48: 1767.
79 Cao, Z.-C., Luo, Q.-Y., and Shi, Z.-J. (2016). *Org. Lett.* 18: 5978.
80 (a) Yu, D.-G., Li, B.-J., Zheng, S.-F. et al. (2010). *Angew. Chem. Int. Ed.* 49: 4566. (b) Shi, W.J. and Shi, Z. (2018). *J. Chin. J. Chem.* 36: 183.
81 Yu, D.-G. and Shi, Z.-J. (2011). *Angew. Chem. Int. Ed.* 50: 7097.
82 Cao, Z.-C., Luo, F.-X., Shi, W.-J., and Shi, Z.-J. (2015). *Org. Chem. Front.* 2: 1505.
83 Shi, W.-J., Li, X.-L., Li, Z.-W., and Shi, Z.-J. (2016). *Org. Chem. Front.* 3: 375.
84 Ohgi, A. and Nakao, Y. (2016). *Chem. Lett.* 45: 45.
85 Tollefson, E.J., Hanna, L.E., and Jarvo, E.R. (2015). *Acc. Chem. Res.* 48: 2344.
86 (a) Harris, M.R., Hanna, L.E., Greene, M.A. et al. (2013). *J. Am. Chem. Soc.* 135: 3303. (b) Zhang, S.-Q., Taylor, B.L.H., Ji, C.-L. et al. (2017). *J. Am. Chem. Soc.* 139: 12994.
87 Zhou, Q., Srinivas, H.D., Dasgupta, S., and Watson, M.P. (2013). *J. Am. Chem. Soc.* 135: 3307.
88 Tollefson, E.J., Dawson, D.D., Osborne, C.A., and Jarvo, E.R. (2014). *J. Am. Chem. Soc.* 136: 14951.
89 Martin-Montero, R., Krolikowski, T., Zarate, C. et al. (2017). *Synlett* 28: 2604.
90 Guo, Y.-A., Liang, T., Kim, S.W. et al. (2017). *J. Am. Chem. Soc.* 139: 6847.
91 Konev, M.O., Hanna, L.E., and Jarvo, E.R. (2016). *Angew. Chem. Int. Ed.* 55: 6730.
92 Taylor, B.L.H., Swift, E.C., Waetzig, J.D., and Jarvo, E.R. (2011). *J. Am. Chem. Soc.* 133: 389.
93 Yonova, I.M., Johnson, A.G., Osborne, C.A. et al. (2014). *Angew. Chem. Int. Ed.* 53: 2422.
94 (a) Greene, M.A., Yonova, I.M., Williams, F.J., and Jarvo, E.R. (2012). *Org. Lett.* 14: 4293. (b) Taylor, B.L.H., Harris, M.R., and Jarvo, E.R. (2012). *Angew. Chem. Int. Ed.* 51: 7790.
95 Harris, M.R., Konev, M.O., and Jarvo, E.R. (2014). *J. Am. Chem. Soc.* 136: 7825.
96 Tollefson, E.J., Erickson, L.W., and Jarvo, E.R. (2015). *J. Am. Chem. Soc.* 137: 9760.
97 Cao, Z.-C. and Shi, Z.-J. (2017). *J. Am. Chem. Soc.* 139: 6546.
98 Kajita, Y., Kurahashi, T., and Matsubara, S. (2008). *J. Am. Chem. Soc.* 130: 17226.
99 Gooßen, L.J. and Paetzold, J. (2002). *Angew. Chem. Int. Ed.* 41: 1237.

100 (a) Chatani, N., Tatamidani, H., Ie, Y. et al. (2001). *J. Am. Chem. Soc.* 123: 4849. (b) Tatamidani, H., Kakiuchi, F., and Chatani, N. (2004). *Org. Lett.* 6: 3597. (c) Tatamidani, H., Yokota, K., Kakiuchi, F., and Chatani, N. (2004). *J. Org. Chem.* 69: 5615.

101 (a) Ishizu, J., Yamamoto, T., and Yamamoto, A. (1976). *Chem. Lett.* 5: 1091. (b) Yamamoto, T., Ishizu, J., Kohara, T. et al. (1980). *J. Am. Chem. Soc.* 102: 3758.

102 (a) Amaike, K., Muto, K., Yamaguchi, J., and Itami, K. (2012). *J. Am. Chem. Soc.* 134: 13573. (b) Kruckenberg, A., Wadepohl, H., and Gade, L.H. (2013). *Organometallics* 32: 5153.

103 (a) Muto, K., Yamaguchi, J., Musaev, D.G., and Itami, K. (2015). *Nat. Commun.* 6: 7508. (b) LaBerge, N.A. and Love, J.A. (2015). *Eur. J. Org. Chem.* 2015: 5546.

104 Chatupheeraphat, A., Liao, H.-H., Srimontree, W. et al. (2018). *J. Am. Chem. Soc.* 140: 3724.

105 Yue, H., Guo, L., Liao, H.H. et al. (2017). *Angew. Chem. Int. Ed.* 56: 4282.

106 Takise, R., Isshiki, R., Muto, K. et al. (2017). *J. Am. Chem. Soc.* 139: 3340.

107 (a) Guo, L. and Rueping, M. (2016). *Chem. Eur. J.* 22: 16787. (b) Pu, X., Hu, J., Zhao, Y., and Shi, Z. (2016). *ACS Catal.* 6: 6692.

108 Guo, L., Chatupheeraphat, A., and Rueping, M. (2016). *Angew. Chem. Int. Ed.* 55: 11810.

109 Yue, H., Zhu, C., and Rueping, M. (2018). *Org. Lett.* 20: 385.

110 Isshiki, R., Muto, K., and Yamaguchi, J. (2018). *Org. Lett.* 20: 1150.

111 (a) Hong, X., Liang, Y., and Houk, K.N. (2014). *J. Am. Chem. Soc.* 136: 2017. (b) Lu, Q., Yu, H., and Fu, Y. (2014). *J. Am. Chem. Soc.* 136: 8252. (c) Xu, H., Muto, K., Yamaguchi, J. et al. (2014). *J. Am. Chem. Soc.* 136: 14834.

112 Yue, H., Guo, L., Lee, S.C. et al. (2017). *Angew. Chem. Int. Ed.* 56: 3972.

113 Hie, L., Nathel, N.F.F., Hong, X. et al. (2016). *Angew. Chem. Int. Ed.* 55: 2810.

114 Cheung, C.W., Ploeger, M.L., and Hu, X. (2017). *Nat. Commun.* 8: 14878.

Part III

Coupling Reactions via Ni(I) and/or Ni(III)

8

Photo-Assisted Nickel-Catalyzed Cross-Coupling Processes

Christophe Lévêque, Cyril Ollivier, and Louis Fensterbank

Sorbonne Université, Institut Parisien de Chimie Moléculaire (IPCM), UMR CNRS 8232, 4, place Jussieu, 75252 Paris Cedex 5, France

8.1 Introduction

Transition-metal-catalyzed reactions have become essential tools for the formation of new carbon–carbon or carbon–heteroatom bonds in the elaboration of molecular building blocks. Despite its presumed hazards for health, nickel has found applications in efficient catalytic industrial processes (oil, pharmaceutical, food) and has recently received significant development in cross-coupling reactions. Located just above palladium in the periodic table, it has similar reactivity and can undergo oxidative addition or reductive elimination, the mechanisms of these steps relying on two-electron transfer steps. Although 3d transition metals of the fourth period are known to perform oxidative addition in carbon–halogen or carbon–hydrogen bonds, they can also promote single-electron transfer (SET) [1] and generate radicals. The latter can be further trapped by the metal and involved in organometallic processes. Nickel has also proved to be an exquisite catalyst involved in various transformations thanks to the large number of accessible oxidation states, from +I to +IV [2]. This ability of nickel allows different kinds of reactivities and radical pathways.

Photoredox catalysis has shown efficiency in triggering radical reactions [3]. Recently, it has proved to be compatible with transition metal catalysis such as nickel one in order to perform some dual catalytic cross-coupling reactions [4]. Traditional cross-coupling reactions have so far mainly relied on two strategies. The most established one consists in the transition-metal-catalyzed reaction of an organometallic nucleophile (R–M) with an electrophile (R–X). More recently, the alternative coupling between two electrophiles, (R–X) and (R′–X), in the presence of reducing agents has been worked out [5]. Both approaches however suffer from severe flaws. In the first case, you need a tamed nucleophile, compatible with functionality; in the second one, you have to combine electrophiles with different reactivity profiles and reducing agent in excess of the stoichiometric amount is needed. The recent third strategy, also referred as single-electron transmetalation [6–8], which consists in merging the photocatalytic formation of radical entities with a catalyzed cross-coupling, alleviates these issues and therefore alters the

Nickel Catalysis in Organic Synthesis: Methods and Reactions,
First Edition. Edited by Sensuke Ogoshi.

conventional paradigm for running coupling reactions. In this system, the photocatalyst is necessary to generate a radical by oxidation and the organometallic catalyst performs the cross-coupling steps between an electrophile and the generated radical. Therefore, metal complexes able to trap radical entities are the catalysts of choice for such transformations. In many cases, the excited photocatalyst enables a SET with a radical precursor providing a radical. Recent reports have also shown that the excited photocatalyst can also promote energy transfer to a transient nickel complex during the catalysis and facilitate a reductive elimination step.

Based on these features, new coupling reactions involving photo-assisted steps and Ni(I) to Ni(IV) intermediates for the formation of the carbon–carbon or carbon–heteroatom bonds have been envisaged and will be presented in this chapter.

8.2 Development of Visible-Light Photoredox/Nickel Dual Catalysis

8.2.1 For the Formation of Carbon–Carbon Bonds

Thanks to the dynamic initiated by the work of Sanford and coworkers [7] on CH functionalization of aromatics by visible-light photoredox/palladium dual catalysis and the work of Weix on nickel-catalyzed cross-electrophile coupling [5], two distinct reports from MacMillan and coworkers [8] and from Molander and coworkers [6] published back to back in *Science* proposed to merge photoredox and nickel catalysis.

8.2.1.1 Starting from Organotrifluoroborates

Csp^3–Csp^2 Cross-couplings with (Hetero)aryl/Vinyl Halides The methodology developed by Molander [8] benefits from preliminary reports by Akita and coworkers [9] and Chen and coworkers [10] and his own work [11] on the generation of stabilized alkyl radicals by photooxidation of alkyltrifluoroborates in the presence of $[Ir(dF(CF_3)ppy)_2bpy](PF_6)$ as photocatalyst under blue LED irradiation. The aim was to engage the radical formed in a nickel catalytic process to perform cross-coupling reactions with aryl halide electrophiles. Among the possible alkyltrifluoroborates, the benzyl precursor was first selected due to its lowest oxidation potential ($E_{1/2}^{ox}$ = +1.13 V vs. SCE in MeCN). In the presence of 2 mol% of the iridium photocatalyst, 3 mol% of $Ni(COD)_2$, and 4,4′-di-*tert*-butyl-2,2′-bipyridine (as the ancillary ligand) they managed to couple the benzyl radical with (hetero)aryl bromide derivatives in moderate to excellent yields (Scheme 8.1). The reaction could be performed with benzyltrifluoroborates bearing electron-donating or electron-withdrawing groups with high yields. Various aryl bromides could be coupled with excellent yields. Moreover, bromo-pyridine, bromo-pyrimidine, bromo-indole, bromo-quinoline, and bromo-thiophene have proved to be also suitable substrates.

In terms of mechanism for this reaction, the authors proposed the direct oxidative addition of aryl bromide on the nickel(0) complex (Scheme 8.2). For

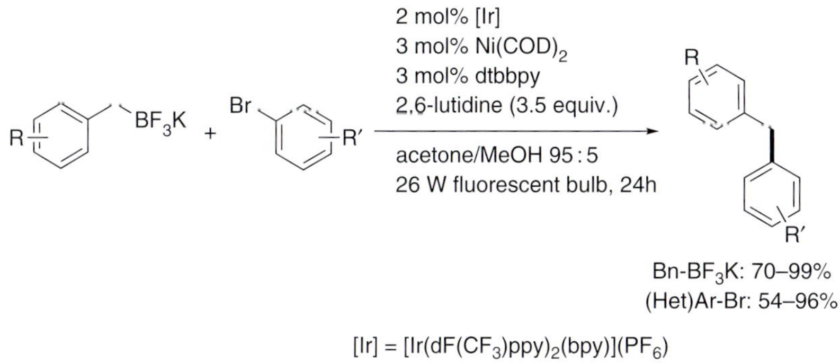

Scheme 8.1 Photoredox cross-coupling of benzyltrifluoroborates with aryl bromides.

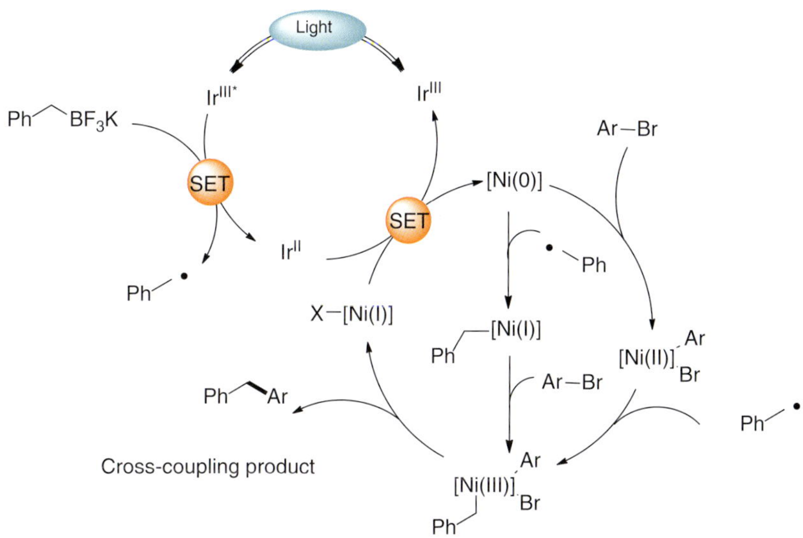

Scheme 8.2 Proposed single-electron transmetalation in photoredox/nickel cross-coupling.

its part, the photoexcited iridium complex ($E_{1/2}^{red}(Ir^{III*}/Ir^{II})$ = +1.21 V vs. SCE in MeCN) oxidizes the benzyltrifluoroborate by SET to give the benzyl radical and an Ir(II) intermediate. As envisaged by Weix and coworker [12], the radical may react with the complex $ArCH_2$[Ni(II)]X to give a Ni(III) intermediate, which after reductive elimination affords the cross-coupling product. Both catalysts would be regenerated from the reaction of Ni(I) with Ir(II) intermediate complexes by SET and completing both photoredox and organometallic catalytic cycles as the Ni(I) complex is well known to be unreactive toward aryl halides ArX [13]. DFT investigations suggested also a second mechanism where the alkyl radical adds to the nickel(0) complex generating a transient alkylnickel(I) complex, which evolves toward a nickel(III) complex [14]. What still remains unclear is the presence of oxygen, which may contribute to the success of this transformation [15].

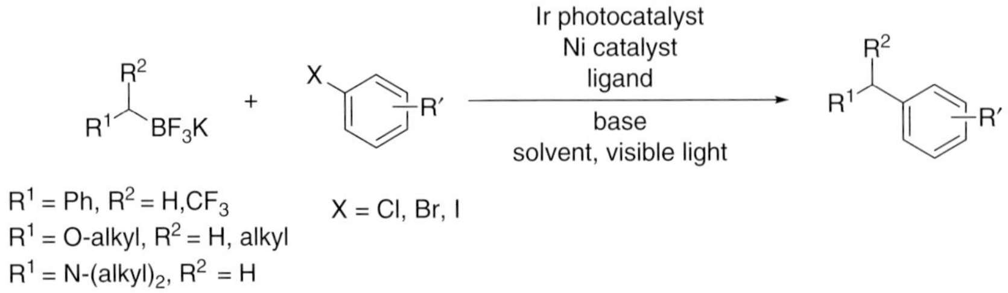

Scheme 8.3 Organotrifluoroborates in photoredox/nickel dual catalysis.

This efficient $C_{sp}{}^3–C_{sp}{}^2$ cross-coupling process was further extended to the reactivity of other activated alkyltrifluoroborates (Scheme 8.3) such as α-hydroxyalkyltrifluoroborates [16], α-acyloxyalkyltrifluoroborates [17], α-alkoxyalkyltrifluoroborates [17, 18] including 2-trifluoroboratochromanones [19], α-aminomethyltrifluoroborates [20], and α-trifluoromethylated benzyltrifluoroborates [21] with various types of bromo (hetero)aryl partners for the synthesis of benzylic alcohol derivatives, flavones, and amino acids respectively. For their part, unactivated secondary [22] and tertiary [23] alkyltrifluoroborates, particularly alkyl β-trifluoroboratoketones and β-esters [24], provide access to the corresponding alkyl cross-couplings, but specifically allow the construction of tertiary quaternary centers and the formation of secondary β-(hetero)arylated ketones and esters respectively. Interestingly, an approach to orthogonal cross-coupling between alkyltrifluoroborates and borylated aryl bromides was also devised [25]. Cross-coupling experiments between aryl bromides [26] or 2- and 3-bromo-7-azaindoles [27] and alkyltrifluoroborates could be performed in continuous flow under mild conditions.

Csp³–Csp² Cross-couplings with Carboxylic Acid Derivatives Examples of cross-coupling acylation reactions of alkyltrifluoroborates have been developed by using acyl chlorides [28], *N*-acylpyrrolidine-2,5-diones [29], anhydrides [30], or directly *in situ* activated carboxylic acids [31] as an acylation agent. These transformations represent efficient approaches to the synthesis of aliphatic ketones (Scheme 8.4).

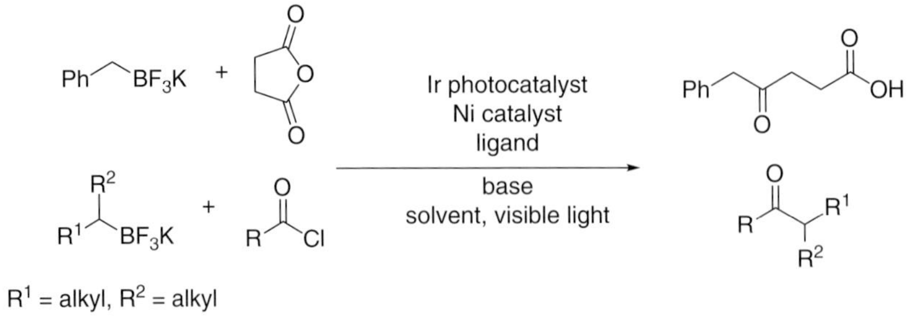

Scheme 8.4 Photoredox/nickel catalytic cross-coupling of activated carboxylic acids and alkyltrifluoroborates.

8.2.1.2 Starting from Carboxylates or Keto Acids or from Methylanilines

Csp^3–Csp^2 Cross-couplings with (Hetero)aryl/Vinyl Halides MacMillan and Doyle reported at the same time a similar approach with the participation of α-amino radicals [7]. Their idea was then to engage the radical formed in a nickel catalytic process to perform cross-coupling reactions. These radicals were easily obtained from α-amino carboxylic acids in the presence of cesium carbonate [32] as well as from dimethylaniline by photooxidation with [Ir(dF(CF_3)ppy)$_2$(dtbbpy)](PF$_6$) as photocatalyst. Both radicals were coupled with aryl halides through nickel catalysis. The reactions were realized with $NiCl_2$.dme as precatalyst and 4,4′-di-*tert*-butyl-2,2′-bipyridine as ligand. Chloro, bromo, and iodo arenes, and heteroaryl bromides could be engaged in the cross-coupling process with cyclic α-amino carboxylates in good yields (Scheme 8.5). As efficiently, non-cyclic α-amino radicals were also engaged with (hetero)aryl bromides [33]. A continuous flow processing approach was developed for cross-coupling reactions between α-amino (or α-oxy) acids with (hetero)aryl bromides [34].

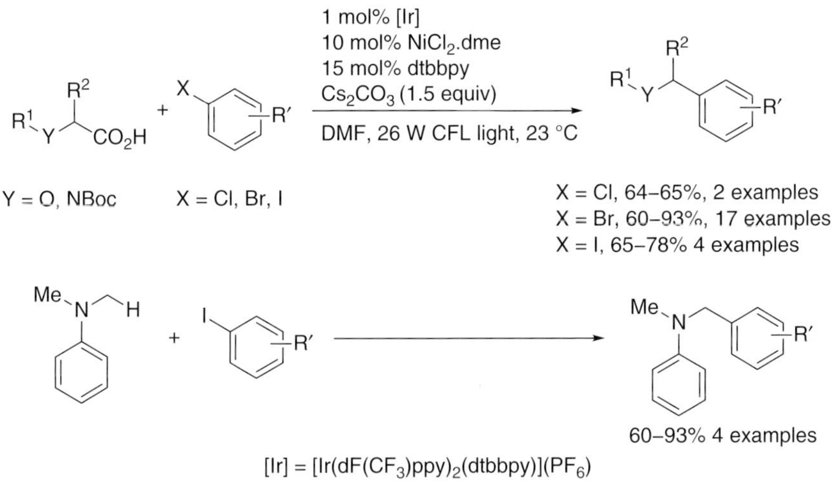

Scheme 8.5 Photoredox cross-coupling of α-amino radicals with aryl halides.

The authors proposed the same type of mechanism as proposed for trifluoroborates (Scheme 8.6). They envisaged the reduction of the [Ni(II)] complex by two SET events to give a [Ni(0)] active catalyst. This complex may undergo oxidative addition with the aryl halide to form the Ar[Ni(II)]X. Oxidation of the carboxylate ($E_{1/2}$(R-CO_2^-/R-$CO_2^\cdot$) = +0.95 to +1.16 V vs. SCE) [32a, 35] or dimethyl aniline ($E_{1/2}$(PhNMe$_2$/PhN$^+$.Me$_2^\cdot$) = +0.71 vs. SCE) by the photoexcited Ir(III) ($E_{1/2}^{red}$(Ir^{III*}/Ir^{II}) = +1.21 V vs. SCE in MeCN) leads to the corresponding α-amino radical, which adds on the nickel to form the Ar[Ni(III)]RX intermediate. After reductive elimination, benzylamine is obtained and a [Ni(I)]X complex is released. A final SET with the Ir(II) regenerates both catalysts.

This photoredox/nickel dual catalysis was then extended to the coupling of α-oxy carboxylic acids with vinyl iodide as electrophile. The α-oxy carboxylates

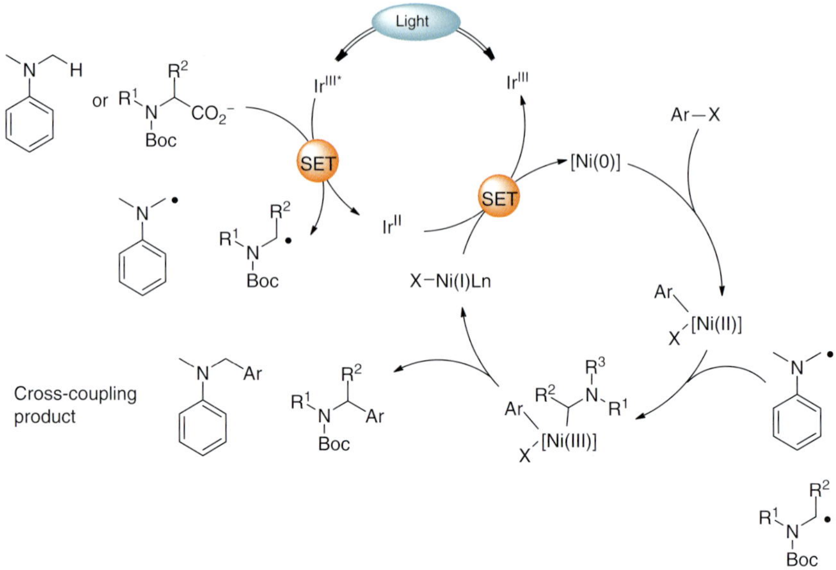

Scheme 8.6 Proposed mechanism by photoredox/nickel dual catalysis α-amino arylation.

are as easy as the α-amino carboxylate ($E_{1/2}^{ox}$ = +1.08 V vs. SCE in MeCN) to be photooxidized and so are able to provide α-oxy radicals after decarboxylation, which can be engaged in dual catalysis. Following the same mechanism, allyl ethers were obtained in good yields with only 1 mol% of iridium photocatalyst and 2 mol% of nickel catalyst. It is interesting to mention that the process proved to be stereospecific. No erosion of the *E* : *Z* ratio was observed during the process. Moreover, α-amino and alkyl carboxylic acids could be engaged in similar dual catalysis (Scheme 8.7) [36].

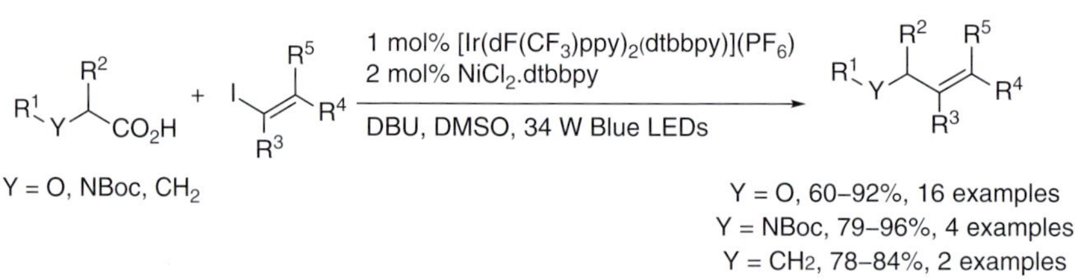

Scheme 8.7 Decarboxylative cross-coupling of carboxylic acids with vinyl iodides.

Csp³–Csp³ Cross-couplings with Alkylhalides In 2016, the group of MacMillan reported new reaction conditions for cross-coupling reactions between alkyl carboxylates and alkyl bromides [37]. The reported methodology involves the use of $[Ir(dF(CF_3)ppy)_2(dtbbpy)](PF_6)$ as oxidizing photocatalyst, $NiCl_2$.dme as nickel source, and 4,4′-dimethoxy-2,2′-bipyridine (4,4′-dOMe-bpy) as ancillary ligand. The potassium carboxylates reacted with various alkyl bromides under these conditions. Decarboxylation of α-amino or α-oxo carboxylates gave

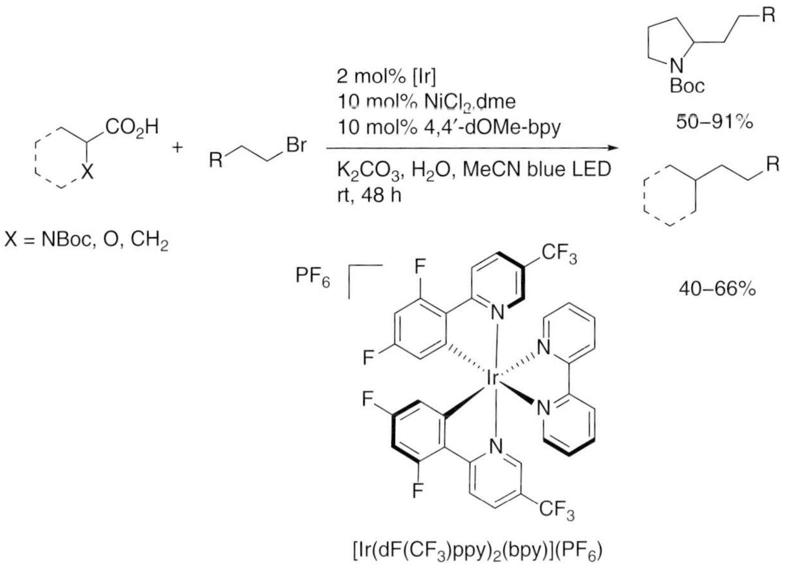

Scheme 8.8 Dual photoredox/nickel-catalyzed sp^3–sp^3 coupling reaction between alkyl carboxylates and alkyl bromides.

stabilized alkyl radicals that could be coupled with primary or secondary alkyl bromides in good yields (Scheme 8.8). The reaction conditions were tolerated by electrophiles bearing a free alcohol, carbonyl, or oxirane function. In addition, unactivated primary and secondary carboxylic acids were engaged in the process, giving the coupling products in moderate yields.

The authors proposed the following mechanism for this transformation (Scheme 8.9). The excited iridium photocatalyst oxidizes the carboxylate. After CO_2 extrusion, the generated radical adds onto the Ni(0) catalyst to form the Ni(I) species. Oxidative addition with the alkyl bromide leads to a Ni(III) intermediate. A reductive elimination step would give the desired product and liberate a Ni(I) complex. A final SET from Ir(II) to Ni(I) regenerates both catalysts to complete a single turnover.

Cross-couplings of Anhydrides, Keto Acids, and Oxalates MacMillan and coworker demonstrated all the potential of his methodology based on the generation of alkyl radicals by photooxidation of carboxylic acids followed by decarboxylation. Among the substrates susceptible to undergo CO_2 extrusion, anhydrides [38], keto acids [39], and oxalates [40] revealed also to be excellent radical precursors.

After in situ formations of anhydride, from a carboxylic acid and an acyl chloride, a Ni(0) complex can undergo an insertion into a C—O bond. The resulting intermediate RCO_2[Ni(II)]C(O)R′ is oxidized by a photoexcited iridium photocatalyst to a Ni(III) complex. Subsequent to the CO_2 extrusion, the new Ni(III) complex undergoes reductive elimination to give a ketone. Both catalytic species were restored by SET between the releasing Ni(I) complex and the Ir(II) photocatalyst (Scheme 8.10). Among the carboxylic acids encountered, cyclic and linear

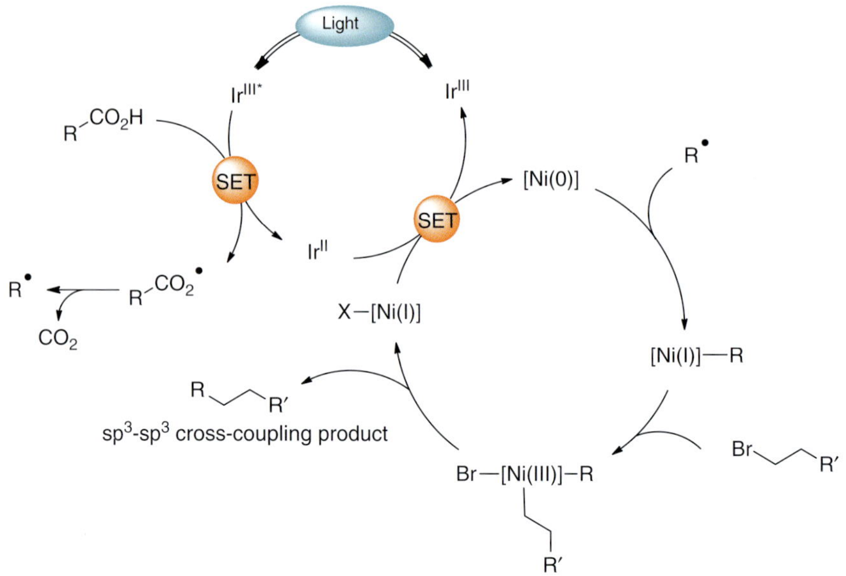

Scheme 8.9 Proposed mechanism for the cross-coupling of carboxylates with alkyl bromides.

α-amino acids were tolerated in the reported conditions, as well as acyl chloride bearing ester, nitrile, or ether moieties.

In the same line, they developed a methodology for the cross-coupling of aryl bromides/iodides with keto acids to provide aryl ketones. Aryl electrophiles bearing electron-withdrawing groups or electron-donating groups afforded the expected products in good yields. Aliphatic and aryl keto acids were also valuable coupling partners. Interestingly, vinyl bromides and secondary alkyl bromides could be engaged in the process (Scheme 8.11). The authors proposed a photooxidation of the keto carboxylate, which after loss of CO_2 provides the acyl radical. The postulated mechanism is then similar to those reported above except the nature of the generated radical.

Another kind of decarboxylative process was performed from oxalates. Indeed, secondary alkyl radicals could be obtained from photooxidative decarboxylation of oxalates, preformed by the reaction between secondary alcohols and oxalyl chloride (Scheme 8.12). The radicals were coupled with aryl bromides in the presence of a nickel(II) catalyst following the methodology already exposed. Various electrophiles bearing electron-donating or electron-withdrawing groups could be engaged in moderate yields, as well as heteroaryl bromides. Interestingly, benzyl alcohols gave good yields but these dramatically dropped when non-activated primary alcohols were tested.

8.2.1.3 Starting from Alkylsilicates

Csp^3–Csp^2 Cross-couplings with (Hetero)aryl/Vinyl Halides or Tosylates The photooxidation of potassium or ammonium bis-catecholato alkylsilicates provides alkyl radicals efficiently, including non-stabilized primary alkyl radicals [41].

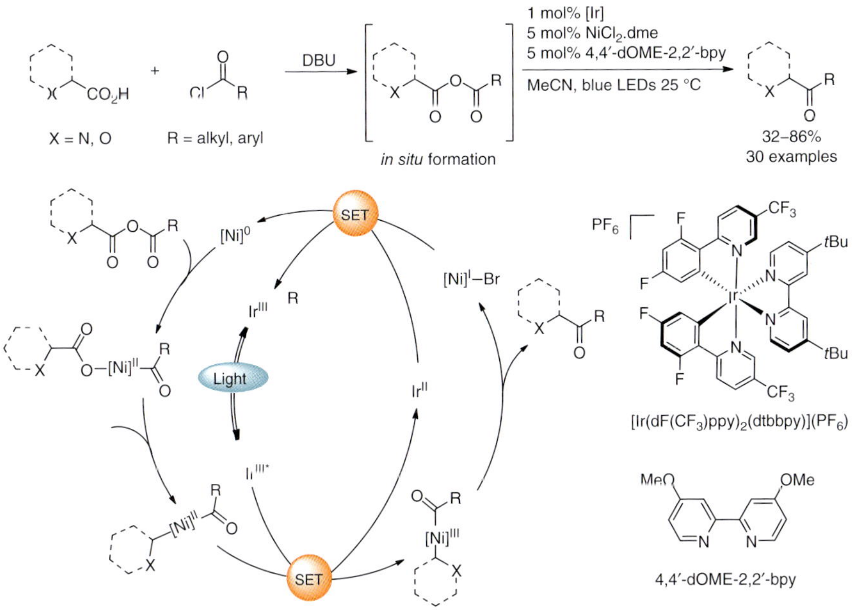

Scheme 8.10 Nickel-catalyzed ketone formation via decarboxylation of anhydrides.

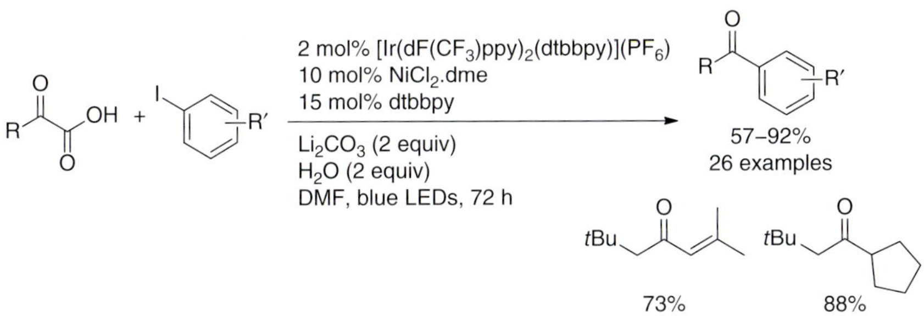

Scheme 8.11 Photoredox/nickel-catalyzed decarboxylation of keto-acids.

Exploiting the features of the nickel catalysis, efficient cross-coupling reactions involving ammonium/potassium bis-catecholato alkylsilicates and (hetero)aryl or vinyl electrophiles have been developed. So far, two methodologies have been developed involving either [18-C-6] potassium alkylsilicates, by Goddard, Ollivier, and Fensterbank, or the ammonium analogs by Molander, with various types of electrophiles.

Initially, it was shown that [18-C-6] potassium alkylsilicates can be coupled with *p*-bromobenzonitrile in the presence of $Ir[(dF(CF_3)ppy)_2(bpy)](PF_6)$ as photocatalyst, $Ni(COD)_2$ as nickel source, and dtb-bpy as ligand. Mainly, primary alkylsilicates were engaged with various aryl bromides and heteroaryl bromides (Scheme 8.13) to give coupling adducts in good yields [42]. The efficiency of the system depends on the electronic density on the aromatic ring

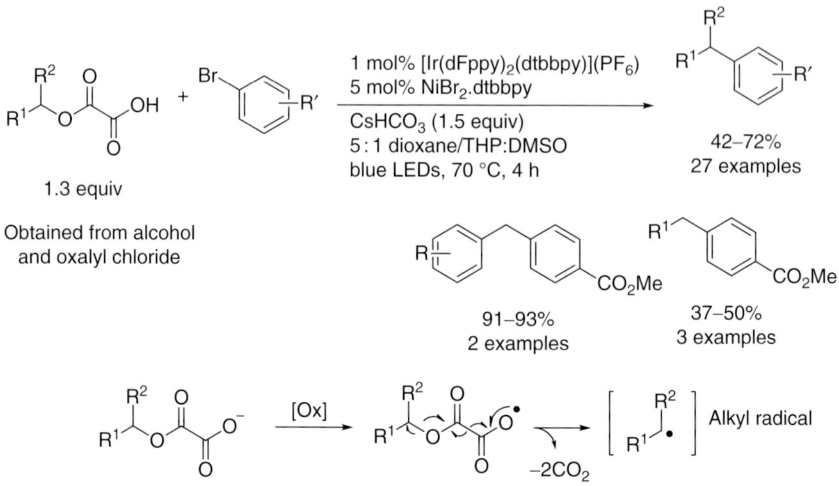

Scheme 8.12 Photoredox/nickel-catalyzed cross-coupling reactions from oxalates.

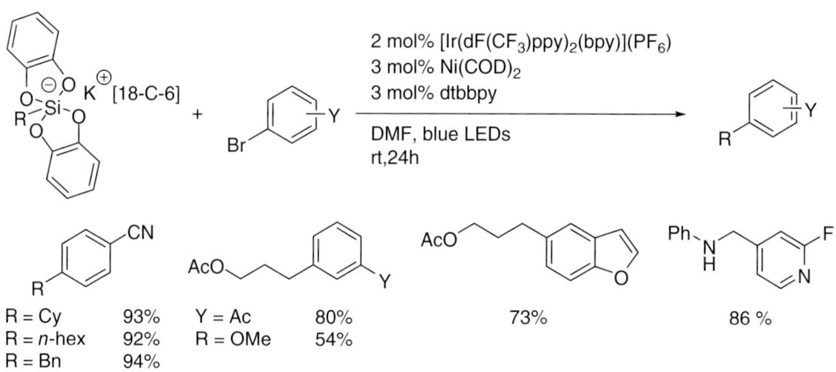

Scheme 8.13 Photoredox/nickel cross-coupling of (hetero)aryl bromides and [18-C-6] potassium alkylsilicates.

since electron-poor electrophiles gave better yields than electron-rich arene substrates.

As an alternative, ammonium silicates can be involved in a similar process [43]. The catalytic system reported requires $Ru(bpy)_3(PF_6)_2$ as photocatalyst, $NiCl_2$.dme as nickel catalyst, and dtb-bpy as ligand (Scheme 8.14). The methodology is equally well tolerant to primary and secondary silicates and compatible with aryl and heteroarylbromide derivatives, leading to good yields of cross-coupling products.

The dual-catalytic systems previously developed have been mainly reported using metal-based photocatalysts. Zhang and coworkers developed a dual-catalytic system for the coupling of organocarboxylates or organotrifluoroborates with arylhalides, involving 1,2,3,5-tetrakis-(carbazole-yl)-4-dicyanobenzene (4CzIPN) as photocatalyst with a long excited state lifetime

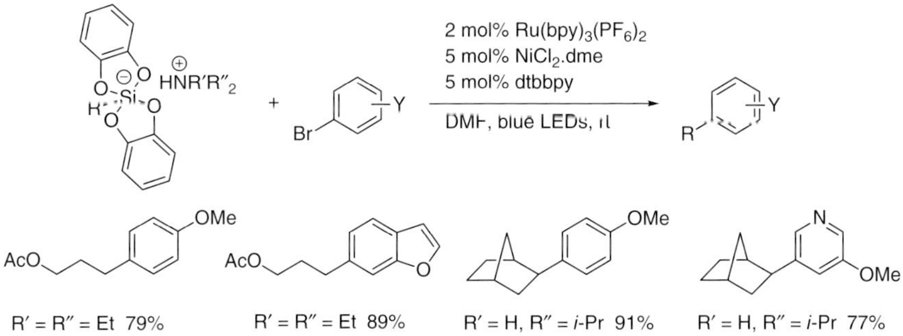

Scheme 8.14 Photoredox/nickel cross-coupling of (hetero)aryl bromides and ammonium alkylsilicates.

(5.1 μs) [44]. Regarding electrochemical properties after light absorption ($E_{1/2}$ (4CzIPN*/[4CzIPN]$^-$) = +1.59 V vs. SCE and $E_{1/2}$ (4CzIPN/[4CzIPN]$^-$) = −1.21 V vs. SCE) [45], this organic dye is able to oxidize the alkylsilicates. Fensterbank and coworkers [46] showed that a catalytic condition involving 4CzIPN, $NiCl_2$.dme, and dtb-bpy provides a dual catalytic system for the cross-coupling reactions of (hetero)aryl bromides and potassium alkylsilicates (Scheme 8.15). Interestingly, the efficiency of the reaction is similar to those involving Ir- and Ru-based photocatalysts. Moreover, ammonium silicates can be used as well without loss of reactivity [47].

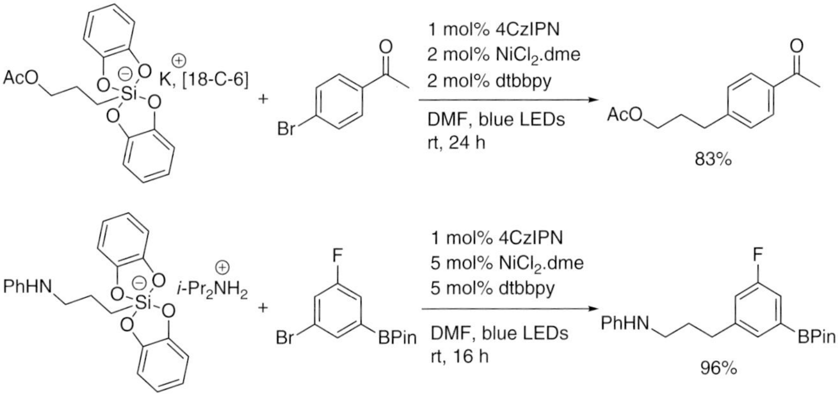

Scheme 8.15 Organic-photoredox/nickel cross-coupling of (hetero)aryl bromides and alkylsilicates.

Among the electrophiles, (hetero)aryl bromides proved to be efficient in the process. Furthermore, other cross-coupling partners showed their compatibility in the dual-catalytic system. In a similar catalytic system, alkenyl halides can be coupled with ammonium alkylsilicates efficiently and potassium alkylsilicates. The reactions are tolerant to *gem-*, *cis-*, and *trans-*substituted alkenyl halides (Scheme 8.16). These reactions can be performed with the two types of reported silicates in the presence of $Ru(bpy)_3(PF_6)_2$ or 4CzIPN as photocatalysts.

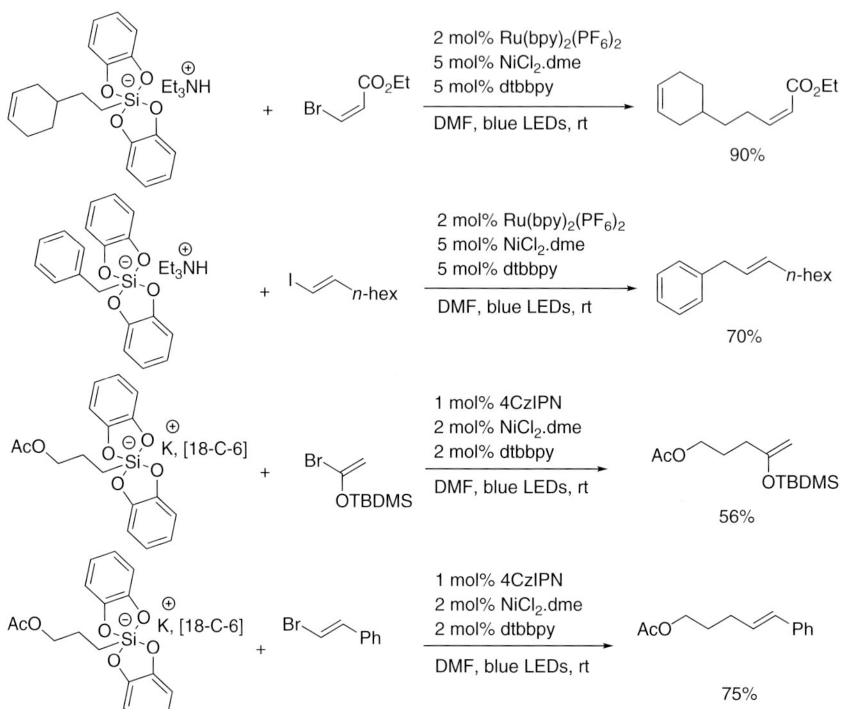

Scheme 8.16 Photoredox/nickel cross-coupling of alkenyl halides and alkylsilicates.

Moreover, full diastereospecificity was shown for this reaction with complete retention of configuration of the C=C bond, even for activated alkenes.

Phenol derivatives could also be engaged in such processes as electrophiles with alkylsilicates as radical precursors [48]. Indeed, arylsulfonates revealed their suitability in a catalytic system made with $Ru(bpy)_3(PF_6)_2$ (2 mol%), $NiCl_2$.dme (10 mol%), and dtb-bpy (10 mol%). Triflate, tosylates, and mesylates were engaged with ammonium alkylsilicates to provide the corresponding coupling adducts. However, triflates seemed to be more reactive than their tosylate analogs and even better than the mesylate ones (Scheme 8.17).

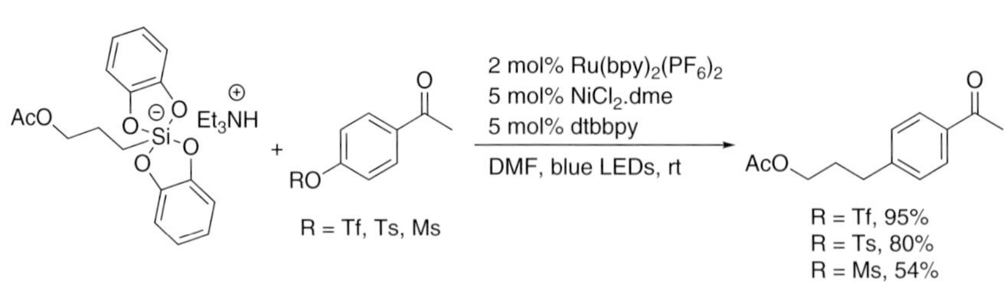

Scheme 8.17 Photoredox/nickel cross-coupling of phenolate derivatives and ammonium alkylsilicates.

Csp³–Csp² Cross-couplings with Isocyanates The formation of amides from isocyanates usually requires the use of highly reactive organometallic nucleophiles [49]. Nickel-catalyzed cross-electrophilic coupling reactions have been also developed but must be performed in the presence of additional reductant [50]. Thanks to the photoredox catalysis, an amidation reaction from isocyanates, in mild conditions without external additives, has been developed [51]. Indeed, a catalytic system made of $Ru(bpy)_3(PF_6)_2$ and $NiCl_2$(phen) enables the cross-coupling reaction between trimethyl or triethylammonium alkylsilicates and isocyanates to provide amides (Scheme 8.18). Contrary to previous methods, no additives are needed and various functionalized isocyanates or silicates are tolerated. The methodology allows the formation of amides bearing various functions (alcohol, esters, etc.) and heterocycles (pyridine, pyrrole, thiophene).

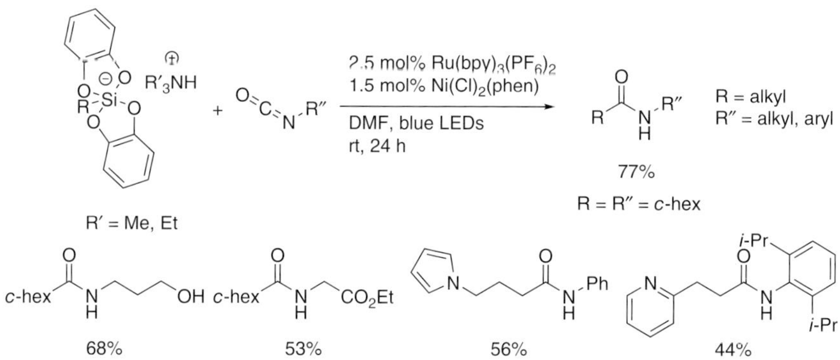

Scheme 8.18 Photoredox/nickel cross-coupling of isocyanates and ammonium potassium alkylsilicates.

Csp³–Csp³ Cross-couplings with Alkylhalides The ability of bis-catecholato alkylsilicates to provide non-stabilized primary alkyl radicals was exploited by the group of Ollivier and Fensterbank to develop a Csp^3–Csp^3 cross-coupling dual catalytic system [52]. The reaction conditions led to the formation of cross-coupling adducts between [18-C-6] potassium bis-catecholato alkylsilicates and alkyl bromides. However, the catalytic system provides also a side product coming from the homocoupling of the electrophile, which is known to be obtained by nickel-catalyzed electophilic cross-coupling [53]. The efforts of the authors to minimize or avoid the formation of the homocoupling product focused on the screening of various bidentate ligands and electrophiles. Modified bipyridyl ligand, bioxazoles, bis(oxazoline)pyridine, and terpyridine were tested without succeeding in removing the side product. As electrophilic partner, while bromide derivatives gave moderate yields of products, chloro, iodo, tosylate, and triflate substrates proved to be less effective/noneffective. Even if a high conversion of the alkyl bromide can be obtained, the formation of the homocoupling product was significant as shown in Table 8.1.

Various primary alkylsilicates bearing functional groups are tolerated in this process despite the formation of side product. Secondary alkylsilicates showed

Table 8.1 Optimization of photoredox/nickel cross-coupling reaction of alkylhalides and [18-C-6] potassium alkylsilicates.

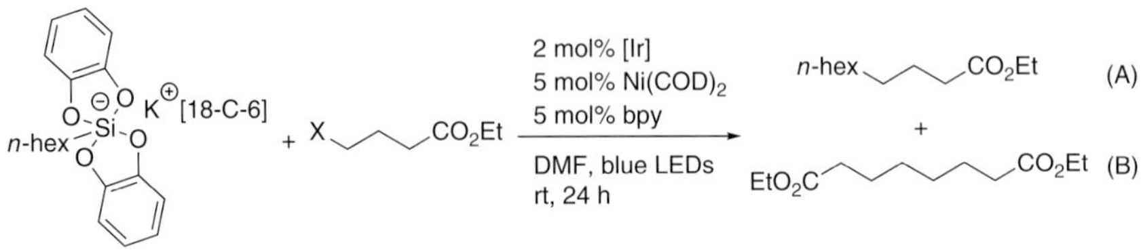

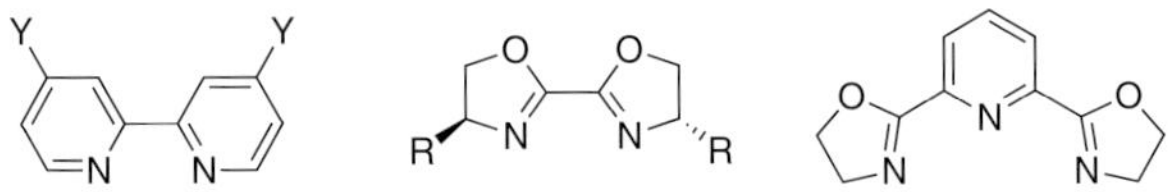

Y = H, bpy
Y = *t*-Bu, dtb-bpy
Y = OMe, dMeO-bpy

L1 R=H
L2 R=Bn

L3

Electrophile (X)	Ligand	Conversion (%)	Yield (%) A/B
Br	bpy	92	42/50
Br	dtb-bpy	79	41/38
Br	dMeO-bpy	0	0/0
Br	L1	15	9/6
Br	L2	14	8/6
Br	L3	51	29/22
Br	Terpyridine	92	42/50
Cl	bpy	0	0/0
I	bpy	25	11/14
OTs	bpy	0	0/0
OTf	bpy	0	0/0

to be reactive as well but the reaction conditions seemed to be sensitive to the nature of the silicate salt (Scheme 8.19). Indeed, ammonium silicates gave worse results than their potassium analogs.

Among the electrophiles, the 5-hexenyl bromide gave a mixture of adducts including a 5-*exo-trig* cross-coupling product presumably coming from a 5-hexenyl radical subsequently trapped by the nickel catalyst.

8.2.1.4 Starting from 1,4-Dihydropyridines

In the continuous quest for reliable alkyl radical precursors, 1,4-dihydropyridines hold a special position since they involve the aldehyde feedstock. Indeed, by condensing ethylacetoacetate to an aldehyde in the presence of an ammonia source (Hantzsch di-hydropyridines synthesis), 1,4-dihydropyridines are readily obtained. With oxidation potentials just above 1 V vs. SCE, the latter are readily oxidized by a series of photocatalysts to give an alkyl radical and the corresponding pyridine by-product. It should nevertheless be emphasized that

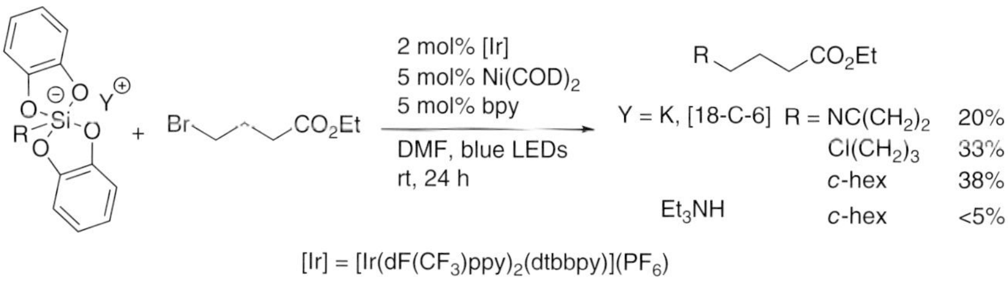

Scheme 8.19 Photoredox/nickel cross-coupling of alkyl halides and [18-C-6] potassium alkylsilicates.

the vast majority of successful cases has involved the generation of somehow stabilized radicals: from benzylic, heteroatom stabilized to secondary alkyl radicals. Three groups [54] initially investigated this chemistry and in two of them reaction conditions were developed in order to merge the radical generation with nickel-catalyzed cross-coupling. In Nishibayashi's study, only benzylic radicals were efficiently coupled to iodoarenes. This study also shed some light on the mechanism and supports the formation of a Ni(I) intermediate resulting from radical addition followed by oxidative addition to give the Ni(III) intermediate, which undergoes reductive elimination. Molander's study shows a larger scope of alkyl dihydropyridine derivatives that are coupled to aryl bromide derivatives, and in this case Adachi's carbazole based 4-CzIPN photocatalyst is used (Scheme 8.20).

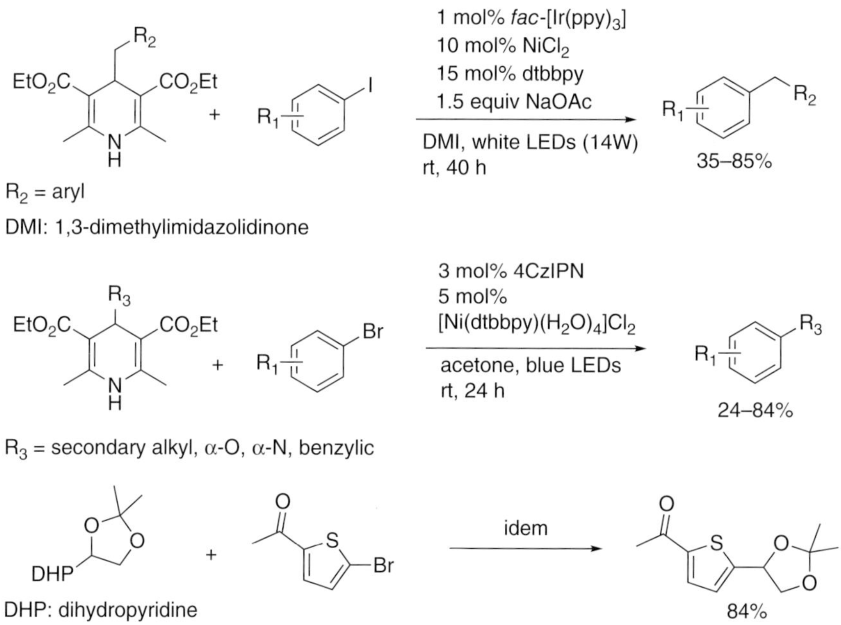

Scheme 8.20 Photoredox/nickel cross-coupling of (hetero)aryl halides and 1,4-dihydropyridines as alkyl radical precursors.

8.2.1.5 Starting from Alkylsulfinates

The cleavage of the C—S bond was investigated in the context of the photoredox/Ni dual catalytic desulfinative cross-coupling of alkylsulfinates and aryl halides (Scheme 8.21) [55]. Alkylsulfinates have low oxidation potentials, well below 1.0 V (vs. SCE), which renders them valuable substrates to test in dual catalysis transformations and the less oxidative $Ru(bpy)_3Cl_2$ photocatalyst could be used. Non-stabilized secondary sulfinates as well as primary ones can also react, but the latter gave cross coupling products in ca 30% yields.

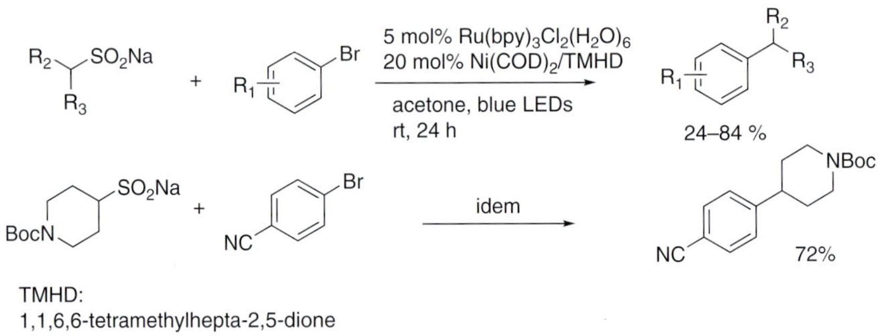

Scheme 8.21 Photoredox/nickel cross-coupling of (hetero)aryl halides and alkylsulfinates.

8.2.1.6 Starting from Alkyl Bromides

Following oxidative addition of RX to give an R[Ni(II)]X intermediate, a radical R′· adds upon it to furnish the transient RR′[Ni(III)]X complex. The way to generate the carbon-centered radical is therefore an essential point. The direct oxidation of radical precursors proved to be effective toward these dual/catalytic processes. Recently, the formation of radicals by atom abstraction further involved in this kind of synthetic approach has been reported. MacMillan and coworkers accomplished the cross-coupling reaction of alkyl bromides with aryl bromides [56]. This type of reaction previously reported by Weix could be performed at room temperature by using a radical mediator. The authors proposed the formation of a bromine radical in the process, which can abstract a hydrogen atom from $(TMS)_3SiH$. The generated silyl radical abstracts the bromine of the alkyl bromide and liberates the reactive alkyl radical (Scheme 8.22).

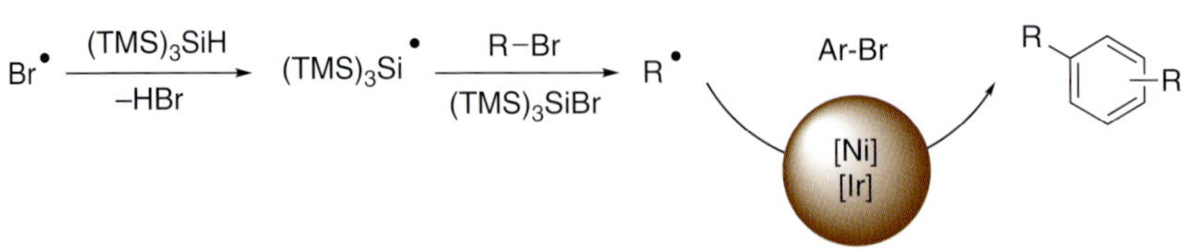

Scheme 8.22 Photoredox/nickel dual catalysis with the generation of radicals by atom abstraction.

8.2.1.7 Starting from Xanthates

A deoxygenation reaction has also been envisaged as a source of radicals for Ni/photoredox dual catalysis. O-benzyl xanthate esters are derived from highly encountered benzyl alcohols. To promote the C—O bond cleavage, a secondary trifluoroborate substrate was used as radical precursor, which is oxidized through the photocatalytic cycle. Various methylene derivatives flanked with two distinct aryl or heteroaryl moieties could be obtained [57].

8.2.1.8 Starting from Sp^3 CH Bonds

Unarguably, the most opportune bond to activate toward homolytic scission is the C–H one. This is not an easy task but it can be accomplished by playing with the bond dissociation energy (BDE) of the targeted sites. But to ensure high level of regioselectivity, a good lever is to rely on polar effects. An electrophilic radical will abstract a hydrogen with hydridic character. Conversely, an electron-rich nucleophilic radical will abstract a hydrogen with a proton character. Based on this concept of polarity match, a wide array of C—H bonds have been activated and engaged in cross-coupling reactions.

The first item to solve consisted in finding the right radical entity for abstraction. Halide radicals have been engaged. For instance, bromine radical can be generated presumably by energy transfer from excited Ir[dF(CF_3)ppy]$_2$(bpy)PF_6 to a Ni(II)–Br intermediate, which promotes the Ni–Br homolysis [58]. The resulting bromine radical abstracts the hydrogen in alpha position to the oxygen of THF and related substrates to generate a C-centered radical, which can then be arylated by an arylbromide electrophile after addition onto an intermediate Ni(I) center and subsequent reductive elimination. In that case, no photoredox cycle would be involved but the mechanism still remains to be unambiguously determined (Scheme 8.23). A related transformation has been reported by Doyle and coworker using arylchlorides and dioxolane in similar reaction conditions [59]. After acidic workup, high yields of aromatic aldehydes are obtained. In that case, the homolysis of a Ni—Cl bond on a Ni(III) intermediate resulting from the classical photoredox/cross-coupling catalysis dual cycles is proposed. The generated chlorine radical then abstracts the activated hydrogen to produce the radical entering in the cross-coupling cycle (Scheme 8.23).

As shown by MacMillan and coworkers, the judicious use of the quinuclidinium radical cation as an electrophilic H abstractor allows the highly selective generation of radical entities at the electron richest C—H bond and not at the weakest ones [60]. Thus, the polarity matching, although possibly contra-thermodynamic, promotes the efficient generation of radical intermediates in alpha position to heteroatoms such as oxygen, nitrogen, and sulfur. The quinuclidinium radical cation is generated by reductive quenching of the excited Ir[dF(CF_3)ppy]$_2$(dtbbpy)PF_6 photocatalyst and a second classical photoredox serves as the central gear of three intertwined catalytic cycles. More precisely, the iridium photocatalyst allows the oxidation of quinuclidine into the quinuclidinium radical cation and the reduction of a nickel(I) species into the nickel(0) catalytic species of the third organometallic cross-coupling catalytic cycle. The nickel(0) species is trapped by the alkyl radical intermediate to generate a new nickel(I) complex, which engages in oxidative addition with different

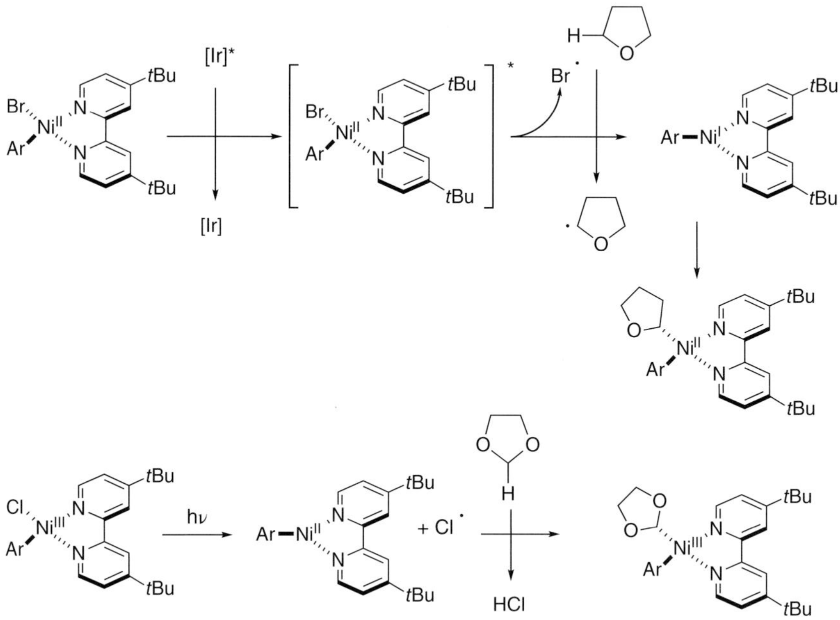

Scheme 8.23 Bromine (or chlorine) radical generation through Ni-Br (or Ni-Cl) homolysis for C(sp^3)—H bond abstraction.

electrophiles. The resulting Ni(III) intermediate undergoes reductive elimination to liberate the coupling product and restore the Ni(I) species. This chemistry has recently culminated in Csp^3–Csp^3 cross-coupling [61]. This selective sp^3 C–H alkylation reaction is particularly powerful for the functionalization in alpha position to amines, ethers, and thioethers, which are ubiquitous functions in bioactive compounds (Scheme 8.24).

Based on the principle of triple catalysis, acyl radicals from aldehydes could also be generated and engaged to give various ketones [62], and primary alkoxides of Zn-based Lewis acids could be selectively arylated at the position alpha to the oxygen [63].

8.2.2 For the Formation of Carbon–Heteroatom Bonds

8.2.2.1 Formation of C—O Bond

Nickel also proved to be a valuable candidate to photoredox/transition metal dual catalysis. MacMillan developed an efficient intermolecular cross-coupling reaction between arylbromides and alcohols involving photoredox/nickel dual catalysis. Studies on nickel catalysis revealed that C—O bond formation using this metal is tricky. Reductive elimination of the C–Ni^{II}–O is endothermic [64] and does not favor the C—O bond formation except for nickel metallacycles [65]. However, reductive elimination from a nickel(III) intermediate was suspected to be much easier. In order to reach this oxidation state, the idea of this group was to use a photocatalyst to oxidize the [C–Ni^{II}–O] intermediate. Using $NiCl_2$.dme, a bipyridine type ligand, and $[Ir(dF(CF_3)ppy)_2(dtbbpy)]PF_6$ as

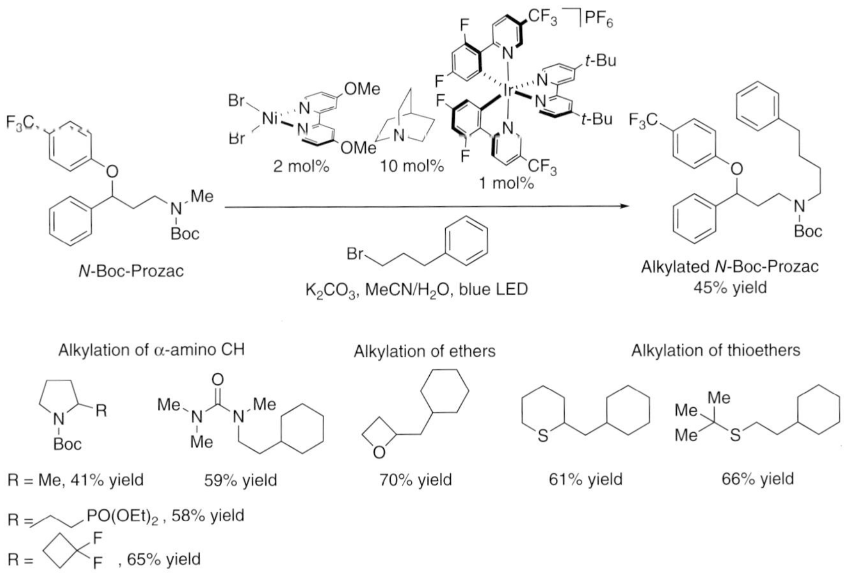

Scheme 8.24 Cross-coupling of α-hetero-substituted alkanes with alkyl bromides via photoredox, nickel, and hydrogen-atom triple catalysis.

photocatalyst, they obtained the expected aryl ether products in excellent yields (Scheme 8.13). Regarding the mechanism, they proposed an oxidative addition of the aryl bromide on the Ni(0)(dtbbpy) active catalyst. Ligand exchange with the alcohol would then generate the Ni(II) aryl alkoxide. At this stage, the crucial oxidation to Ni(III) was carried out by the photoexcited iridium complex ($E_{1/2}(Ni^{III}/Ni^{II+})$ = +0.71 V vs. Ag/AgCl [66] and $E_{1/2}(Ir^{III*}/Ir^{II})$ = +1.21 V vs. SCE). The Ni(III) aryl alkyloxide complex then underwent reductive elimination to give the aryl ether and a Ni(I) complex reduced by the Ir(II) intermediate to regenerate both catalysts (Scheme 8.25).

8.2.2.2 Formation of C—P Bond

Radicals are not only carbon-centered. The formation of C—X bonds by photoredox/nickel dual catalysis has been also envisioned. In 2015, the group of Xiao reported an efficient method for C—P bond formation under mild conditions. Thanks to a ruthenium photocatalyst and an *in situ* formed nickel catalyst, they worked out the cross-coupling reaction between (hetero)aryl iodides and phosphine oxides in high yields. The authors proposed the same catalytic cycles as above except that a phosphinyl radical is at stake. This radical is obtained by a base-assisted photooxidation of the phosphinous acid (Scheme 8.26). More recently, another group engaged phosphonates as radical precursors and aryl/alkenyl tosylate, mesylate, or sulfamate as electrophiles.

8.2.2.3 Formation of C—S Bond

Thiyl radicals showed to be reactive intermediates in radical chain reactions [67]. Johannes and coworkers demonstrated that these radicals can be involved

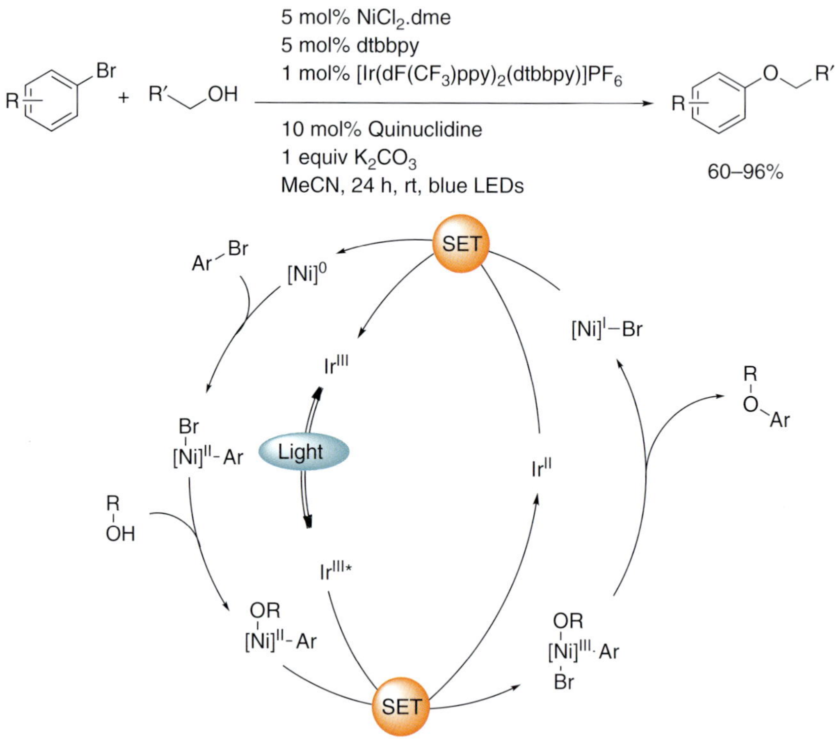

Scheme 8.25 Photo-assisted nickel-catalyzed C—O bond formation.

in photoredox-mediated and nickel-catalyzed cross-coupling reactions with (hetero)aryl iodides. Thiophenol, alkyl, and benzyl thiols could be coupled with electrophiles bearing electron-donating and electron-withdrawing groups in good to excellent yields (Scheme 8.27). In their case, a $NiCl_2$.dme was used as precatalyst with 4,4′-di-*tert*-butyl-2,2′-bipyridine as ligand. In addition, $NiCl_2$.dppe as catalyst gave also the expected products. However, no reaction was observed with a Ni(0) precatalyst under an inert atmosphere. The authors considered that the mechanism of the reaction would be different than those proposed before. The photoexcited iridium catalyst oxidizes the thiol, which after deprotonation provides the thiyl radical. Redox potentials for $NiCl_2$.dppe/NiCl.dppe and NiCl.dppe/Ni(0)dppe are respectively −0,88 V vs. SCE and −1.41 V vs. SCE in DMF/THF (4 : 1). A reduction step of the $NiCl_2$.dppe by Ir(II) ($E(Ir^{III}/Ir^{II}) = -1.37$ V vs. SCE in MeCN) was therefore considered as an initiation step. According to the authors, the same step would be possible with $NiCl_2$.dtbbpy. The resulting active complex [Ni(I)]Cl would react with the thiyl radical to generate the intermediate RS[Ni(II)]Cl. Another reduction step from Ir(II) may give the complex RS[Ni(I)], which would undergo oxidative addition with an aryl iodide. After reductive elimination, the RS[Ni(III)]X was converted to the active nickel complex, and the thioether was obtained.

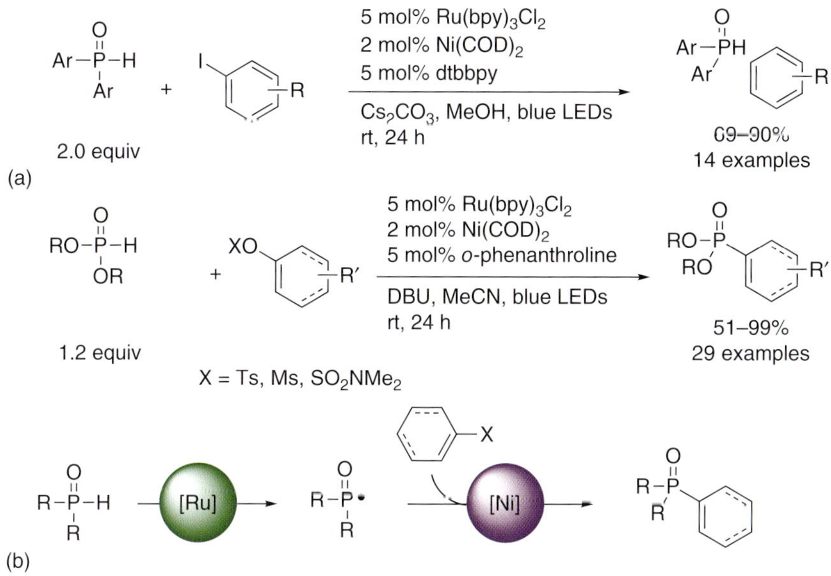

Scheme 8.26 Photoredox/nickel dual-catalyzed C—P bond formation.

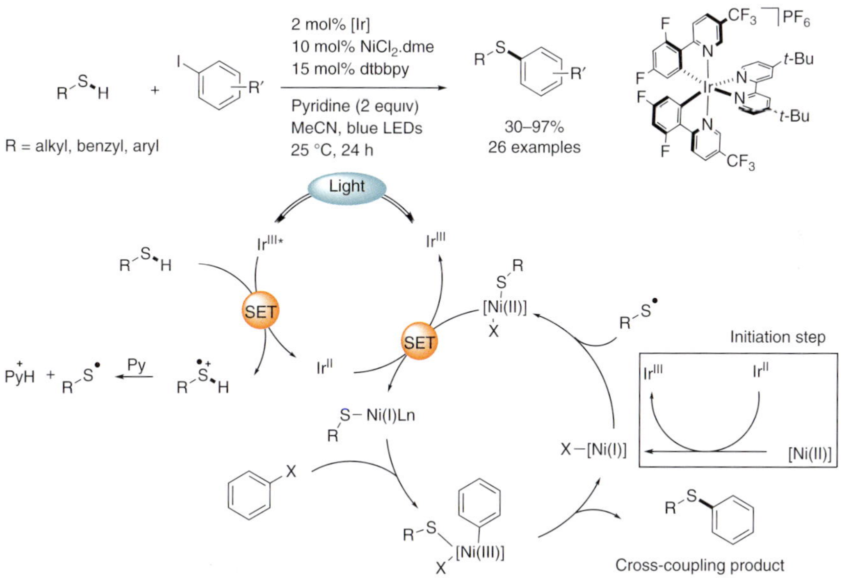

Scheme 8.27 Photoredox/nickel dual-catalyzed C—S bond formation.

8.3 Energy-Transfer-Mediated Nickel Catalysis

While triplet sensitization by energy transfer has been known for a long time as a powerful activation mode in classical organic photochemistry, its utilization in dual catalysis events was reported only recently by McCusker and coworkers

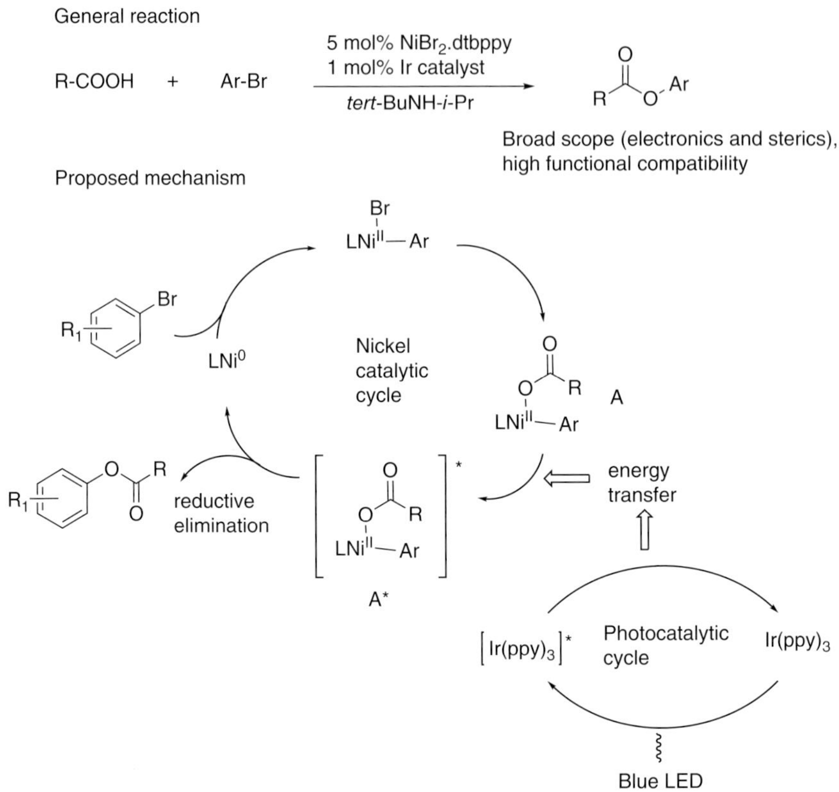

Scheme 8.28 C—O Bond formation of aryl and heteroaryl halides via energy-transfer photosensitized nickel catalysis.

(Scheme 8.28) [68]. Connected to the general objective of excited-state transition metal catalysis, the underlying idea of this work was to introduce a discrete photosensitizer in order to distinguish the light harvesting species from the key cross-coupling intermediate and optimize the overall process. Thus, it was found that blue LED excitation of an iridium photocatalyst induces the triplet sensitization of a Ni(II)–aryl carboxylate intermediate complex through an energy-transfer mechanism. The resulting excited Ni(II) complex of type **A***undergoes reductive elimination and provides the cross-coupling adduct from an aryl halide and a carboxylic acid. While this so far elusive reaction proved here to be fairly general with a series of carboxylic acid and aryl bromide coupling partners of various electronic and steric demands, mechanistic studies provided a lot of precious information.

Preliminary studies showed that the prototypical intermediate **A** was thermally stable to reductive elimination. Under blue LED irradiation C—O bond formation can take place but quite inefficiently. Still, it demonstrates that direct visible light excitation of the metal species occurs. When an iridium photocatalyst is present, the process is much faster. The pending question was whether the excited photocatalyst acts as a redox trigger or as an energy transfer shuttle.

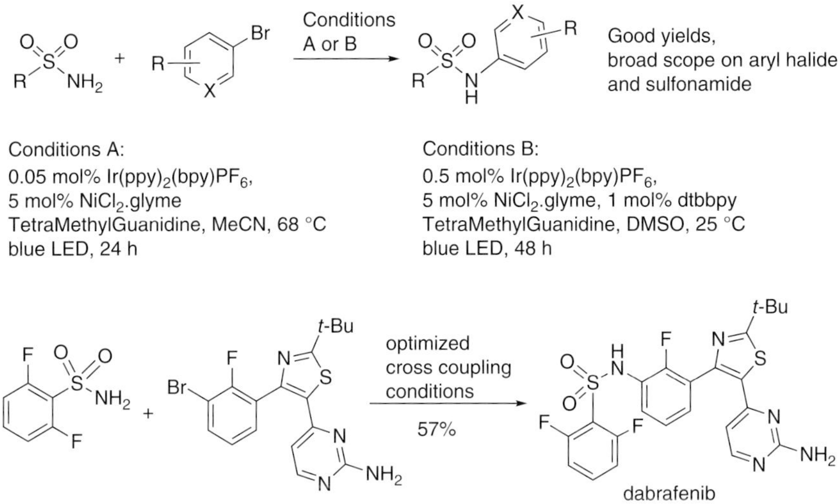

Scheme 8.29 Sulfonamidation of aryl and heteroaryl halides via energy-transfer photosensitized nickel catalysis.

However, several key facts supported the energy transfer mechanism, generating **A*** versus an electron-transfer pathway. First, higher reaction yields were observed when [Ir(ppy$_2$)ligand]PF$_6$ photocatalysts with higher triplet energies and lower oxidation potentials were used; second, it was shown by cyclic voltammetry that redox pathways between the excited Ir(ppy)$_3$ (ppy: 2-phenylpyridine) and Ni(II)–aryl carboxylate complex are thermodynamically unfavorable. While at present, it is still difficult to predict whether a given transformation will proceed via an electron-transfer or energy-transfer pathway, this pioneering report opens a novel and likely general strategy for applying photocatalysis to the field of transition metal catalysis.

Very recently, the MacMillan group could apply this principle for the sulfonamidation of aryl and heteroaryl halides (Scheme 8.29) [69]. The working hypothesis was that energy exchange between a photoexcited iridium photocatalyst and a nickel(II) intermediate would promote reductive elimination corresponding to C—N bond formation. Two types of reaction conditions proved fecund. Conditions B that rely on the adjunction of a bipyridyl ligand and (dtbbpy: di-*tert*-butyl-bipyridine) and the use of DMSO allowed to work at milder temperature (25 °C) than ligand-free conditions A in MeCN (68 °C). An interesting application provided with the synthesis of the B-raf kinase inhibitor dabrafenib in 57% yield.

8.4 Conclusion

Since the seminal works by MacMillan and Doyle and Molander on the merging of photoredox catalysis and nickel catalysis for the cross-coupling of carboxylates and trifluoroborates with aryl bromides, a series of other related dual catalysis based studies have ensued. Extension of the nature of the radical precursors to dimethyl anilines and related amines, silicates, sulfinates, dihydropyridines, and others – all acting as oxidative quench of the photocatalyst – and the nature of electrophiles (aryl, vinyl and alkyl halides) provides valuable cross-coupling transformations. Interestingly, this methodology has proved to be competent for the forging of C–heteroatom bonds (C—O, C—S, C—P), which greatly expands the synthetic opportunities. Recently, it has also been shown that the light activation that is mediated by the photocatalyst can trigger energy transfer. This energy transfer is necessary for some elementary steps of the catalytic cycle to take place. All these recent developments are highly suggestive that a new way of envisaging C—C and C—X bond formation is available, which should enjoy broad utilizations in various contexts.

References

1 Kochi, J.K. (1974). Electron-transfer mechanisms for organometallic intermediates in catalytic reactions. *Acc. Chem. Res.* 7: 351–360.
2 Tasker, S.Z., Standley, E.A., and Jamison, T.F. (2014). Recent advances in homogeneous nickel catalysis. *Nature* 509: 299–309.
3 For recent books, see: (a)König, B. (ed.) (2013). *Chemical Photocatalysis.* DeGruyter Berlin. (b) Albini, A. and Fagnoni, M. (eds.) (2013). *Photochemically Generated Intermediates in Synthesis.* Hoboken, NJ: Wiley. (c) Stephenson, C.R.J., Yoon, T.P., and MacMillan, D.W.C. (eds.) (2018). *Visible Light Photocatalysis in Organic Chemistry.* Wiley-VCH.
4 Twilton, J., Le, C., Zhang, P. et al. (2017). The merger of transition metal and photocatalysis. *Nat. Rev. Chem.* 1: 0052.
5 Weix, D. (2015). Methods and mechanisms for cross-electrophile coupling of Csp^2 halides with alkyl electrophiles. *Acc. Chem. Res.* 48: 1767–1775.
6 Tellis, J.C., Primer, D.N., and Molander, G.A. (2014). Dual catalysis. Single-electron transmetalation in organoboron cross-coupling by photoredox/nickel dual catalysis. *Science* 345: 433–436.
7 Kalyani, D., McMurtrey, K.B., Neufeldt, S.R., and Sanford, M.S. (2011). Room-temperature C—H arylation: merger of Pd-catalyzed C—H functionalization and visible-light photocatalysis. *J. Am. Chem. Soc.* 133: 18566–18569.
8 Zuo, Z., Ahneman, D.T., Chu, L. et al. (2014). Dual catalysis. Merging photoredox with nickel catalysis: coupling of α-carboxyl sp^3-carbons with aryl halides. *Science* 345: 437–440.
9 (a) Yasu, Y., Koike, T., and Akita, M. (2012). Visible light-induced selective generation of radicals from organoborates by photoredox catalysis. *Adv. Synth. Catal.* 354: 3414–3420. (b) Li, Y., Miyazawa, K., Koike, T., and Akita, M. (2015). Alkyl- and aryl-thioalkylation of olefins with organotrifluoroborates by photoredox catalysis. *Org. Chem. Front.* 2: 319–323. (c) Koike, T. and Akita, M. (2016). Combination of organotrifluoroborates with photoredox catalysis marking a new phase in organic radical chemistry. *Org. Biomol. Chem.* 14: 6886–6890.

10 Huang, H., Zhang, G., Gong, L. et al. (2014). Visible-light-induced chemoselective deboronative alkynylation under biomolecule-compatible conditions. *J. Am. Chem. Soc.* 136: 2280–2283.

11 For a review, see: Molander, G.A. (2015). Organotrifluoroborates: another branch of the mighty oak. *J. Org. Chem.* 80: 7837–7848.

12 Biswas, S. and Weix, D.J. (2013). Mechanism and selectivity in nickel-catalyzed cross-electrophile coupling of aryl halides with alkyl halides. *J. Am. Chem. Soc.* 135: 16192–16197.

13 Zhang, K., Conda-Sheridan, M., Cooke, S.R., and Louie, J. (2011). N-heterocyclic carbene bound nickel(I) complexes and their roles in catalysis. *Organometallics* 30: 2546–2552.

14 Gutierrez, O., Tellis, J.C., Primer, D.N. et al. (2015). Nickel-catalyzed cross-coupling of photoredox-generated radicals: uncovering a general manifold for stereoconvergence in nickel-catalyzed cross-couplings. *J. Am. Chem. Soc.* 137: 4896–4899.

15 Oderinde, M.S., Varela-Alvarez, A., Aquila, B. et al. (2015). Effects of molecular oxygen, solvent, and light on iridium-photoredox/nickel dual-catalyzed cross-coupling reactions. *J. Org. Chem.* 80: 7642–7651.

16 Alam, R. and Molander, G.A. (2017). Direct synthesis of secondary benzylic alcohols enabled by photoredox/Ni dual-catalyzed cross-coupling. *J. Org. Chem.* 82: 13728–13734.

17 Karimi-Nami, R., Tellis, J.C., and Molander, G.A. (2016). Single-electron transmetalation: protecting-group-independent synthesis of secondary benzylic alcohol derivatives via photoredox/nickel dual catalysis. *Org. Lett.* 18: 2572–2575.

18 Karakaya, I., Primer, D.N., and Molander, G.A. (2015). Photoredox cross-coupling: Ir/Ni dual catalysis for the synthesis of benzylic ethers. *Org. Lett.* 17: 3294–3297.

19 Matsui, J.K. and Molander, G.A. (2017). Direct α-arylation/heteroarylation of 2-trifluoroboratochromanones via photoredox/nickel dual catalysis. *Org. Lett.* 19: 436–439.

20 El Khatib, M., Serafim, R.A.M., and Molander, G.A. (2016). α-arylation/heteroarylation of chiral α-aminomethyltrifluoroborates by synergistic iridium photoredox/nickel cross-coupling catalysis. *Angew. Chem. Int. Ed.* 55: 254–258.

21 Ryu, D., Primer, D.N., Tellis, J.C., and Molander, G.A. (2016). Single-electron transmetalation: synthesis of 1,1-diaryl-2,2,2-trifluoroethanes by photoredox/nickel dual catalytic cross-coupling. *Chem. Eur. J.* 22: 120–123.

22 Primer, D.N., Karakaya, I., Tellis, J.C., and Molander, G.A. (2015). Single-electron transmetalation: an enabling technology for secondary alkylboron cross-coupling. *J. Am. Chem. Soc.* 137: 2195–2198.

23 Primer, D.N. and Molander, G.A. (2017). Enabling the cross-coupling of tertiary organoboron nucleophiles through radical-mediated alkyl transfer. *J. Am. Chem. Soc.* 139: 9847–9850.

24 Tellis, J.C., Amani, J., and Molander, G.A. (2016). Single-electron transmetalation: photoredox/nickel dual catalytic cross-coupling of secondary

alkyl β-trifluoroboratoketones and –esters with aryl bromides. *Org. Lett.* 18: 2994–2997.

25 Yamashita, Y., Tellis, J.C., and Molander, G.A. (2015). Protecting group-free, selective cross-coupling of alkyltrifluoroborates with borylated aryl bromides via photoredox/nickel dual catalysis. *Proc. Natl. Acad. Sci. U.S.A.* 112: 12026–12029.

26 DeLano, T.J., Bandarage, U.K., Palaychuk, N. et al. (2016). Application of the photoredox coupling of trifluoroborates and aryl bromides to analog generation using continuous flow. *J. Org. Chem.* 81: 12525–12531.

27 Palaychuk, N., DeLano, T.J., Boyd, M.J. et al. (2016). Synthesis of cycloalkyl substituted 7-azaindoles via photoredox nickel dual catalytic cross-coupling in batch and continuous flow. *Org. Lett.* 18: 6180–6183.

28 (a) Amani, J. and Molander, G.A. (2017). Synergistic photoredox/nickel coupling of acyl chlorides with secondary alkyltrifluoroborates: dialkyl ketone synthesis. *J. Org. Chem.* 82: 1856–1863. (b) Amani, J., Sodagar, E., and Molander, G.A. (2016). Visible light photoredox cross-coupling of acyl chloride with potassium alkoxymethyltrifluoroborates: synthesis of α-alkoxyketones. *Org. Lett.* 18: 732–735.

29 Amani, J., Alam, R., Badir, S., and Molander, G.A. (2017). Synergistic visible-light photoredox/nickel-catalyzed synthesis of aliphatic ketones via N—C cleavage of imines. *Org. Lett.* 19: 2426–2429.

30 Stache, E.E., Rovis, T., and Doyle, A.G. (2017). Dual nickel- and photoredox-catalyzed enantioselective desymmetrization of cyclic meso-anhydrides. *Angew. Chem. Int. Ed.* 56: 3679–3683.

31 Amani, J. and Molander, G.A. (2017). Direct conversion of carboxylic acids to alkyl ketones. *Org. Lett.* 19: 3612–3615.

32 (a) Zuo, Z. and MacMillan, D.W.C. (2014). Decarboxylative arylation of α-amino acids via photoredox catalysis: a one-step conversion of biomass to drug pharmacophore. *J. Am. Chem. Soc.* 136: 5257–5260. (b) Noble, A. and MacMillan, D.W.C. (2014). Photoredox α-vinylation of α-amino acids and *N*-aryl amines. *J. Am. Chem. Soc.* 136: 11602–11605.

33 Zuo, Z., Cong, H., Li, W., and Choi, J. (2016). Enantioselective decarboxylative arylation of α-amino acids via the merger of photoredox and nickel catalysis. *J. Am. Chem. Soc.* 138: 1832–1835.

34 Abdiaj, I. and Alcázar, J. (2017). Improving the throughput of batch photochemical reactions using flow: dual photoredox and nickel catalysis in flow for $C(sp^2)$-$C(sp^3)$ cross-coupling. *Bioorg. Med. Chem.* 25: 6190–6196.

35 McNally, A., Prier, C.K., and MacMillan, D.W.C. (2011). Discovery of an α-amino C—H arylation reaction using the strategy of accelerated serendipity. *Science* 334: 1114–1117.

36 Noble, A., McCarver, S.J., and MacMillan, D.W.C. (2015). Merging photoredox and nickel catalysis: decarboxylative cross-coupling of carboxylic acids with vinyl halides. *J. Am. Chem. Soc.* 137: 624–627.

37 Johnston, C.P., Smith, R.T., Allmendinger, S., and MacMillan, D.W.C. (2016). Metallaphotoredox-catalysed sp^3–sp^3 cross-coupling of carboxylic acids with alkyl halides. *Nature* 536: 322–325.

38 Le, C.C. and MacMillan, D.W.C. (2015). Fragment couplings via CO_2 extrusion–recombination: expansion of a classic bond-forming strategy via metallaphotoredox. *J. Am. Chem. Soc.* 137: 11938–11941.
39 Chu, L., Lipshultz, J.M., and MacMillan, D.W.C. (2015). Merging photoredox and nickel catalysis: the direct synthesis of ketones by the decarboxylative arylation of α-oxo acids. *Angew. Chem. Int. Ed.* 54: 7929–7933.
40 Zhang, X. and MacMillan, D.W.C. (2016). Alcohols as latent coupling fragments for metallaphotoredox catalysis: sp^3–sp^2 cross-coupling of oxalates with aryl halides. *J. Am. Chem. Soc.* 138: 13862–13865.
41 Corcé, V., Chamoreau, L.-M., Derat, E. et al. (2015). Silicates as latent alkyl radical precursors: visible-light photocatalytic oxidation of hypervalent bis-catecholato silicon compounds. *Angew. Chem. Int. Ed.* 54: 11414–11418.
42 Lévêque, C., Chenneberg, L., Corcé, V. et al. (2016). Primary alkyl bis-catecholato silicates in dual photoredox/nickel catalysis: aryl- and heteroaryl-alkyl cross coupling reactions. *Org. Chem. Front.* 3: 462–465.
43 Jouffroy, M., Primer, D.N., and Molander, G.A. (2016). Base-free photoredox/nickel dual-catalytic cross-coupling of ammonium alkylsilicates. *J. Am. Chem. Soc.* 138: 475–478.
44 Uoyama, H., Goushi, K., Shizu, K. et al. (2012). Highly efficient organic light-emitting diodes from delayed fluorescence. *Nature* 492: 234–238.
45 (a) Luo, J. and Zhang, J. (2016). Donor–acceptor fluorophores for visible-light-promoted organic synthesis: photoredox/Ni dual catalytic $C(sp^3)$–$C(sp^2)$ cross-coupling. *ACS Catal.* 6: 873–877. (b) Vaillant, F.L., Garreau, M., Nicolai, S. et al. (2018). Fine-tuned organic photoredox catalysts for fragmentation-alkynylation cascades of cyclic oximes ethers. *Chem. Sci.* 9: 5883–5889.
46 Lévêque, C., Chenneberg, L., Corcé, V. et al. (2016). Organic photoredox catalysis for the oxidation of silicates: applications in radical synthesis and dual catalysis. *Chem. Commun.* 52: 9877–9880.
47 Vara, B.A., Jouffroy, M., and Molander, G.A. (2017). $C(sp^3)$–$C(sp^2)$ cross-coupling of alkylsilicates with borylated aryl bromides – an iterative platform to alkylated aryl- and heteroaryl boronates. *Chem. Sci.* 8: 530–535.
48 Patel, N.R. and Molander, G.A. (2016). Phenol derivatives as coupling partners with alkylsilicates in photoredox/nickel dual catalysis. *J. Org. Chem.* 81: 7271–7275.
49 (a) Pace, V., Castoldia, L., and Holzera, W. (2013). Addition of lithium carbenoids to isocyanates: a direct access to synthetically useful N-substituted 2-haloacetamides. *Chem. Commun.* 49: 8383–8385. (b) Coldham, I., Dufour, S., Haxell, T.F.N. et al. (2006). Dynamic thermodynamic and dynamic kinetic resolution of 2-lithiopyrrolidines. *J. Am. Chem. Soc.* 128: 10943–10951. (c) Schäfer, G., Matthey, C., and Bode, J.W. (2012). Eine einfache synthese von sterisch gehinderten und elektronenarmen, sekundären Amiden aus Isocyanaten. *Angew. Chem.* 124: 9307–9310.
50 (a) Hsieha, J.-C. and Cheng, C.-H. (2005). Nickel-catalyzed coupling of isocyanates with 1,3-iodoesters and halobenzenes: a novel method for the synthesis of imide and amide derivatives. *Chem. Commun.*: 4554–4556. (b) Correa, A. and Martin, R. (2014). Ni-catalyzed direct reductive amidation via

C—O bond cleavage. *J. Am. Chem. Soc.* 136: 7253–7256. (c) Serrano, E. and Martin, R. (2016). Nickel-catalyzed reductive amidation of unactivated alkyl bromides. *Angew. Chem. Int. Ed.* 55: 11207–112011.

51 Patel, N.R., Kelly, C.B., Siegenfeld, A.P., and Molander, G.A. (2017). Mild, redox-neutral alkylation of imines enabled by an organic photocatalyst. *ACS Catal.* 7: 1766–1770.

52 Lévêque, C., Corcé, V., Chenneberg, L. et al. (2017). Photoredox/nickel dual catalysis for the $C(sp^3)$–$C(sp^3)$ cross-coupling of alkylsilicates with alkyl halides. *Eur. J. Org. Chem.* 15: 2118–2121.

53 Prinsell, M.R., Everson, D.A., and Weix, D.J. (2010). Nickel-catalyzed, sodium iodide-promoted reductive dimerization of alkyl halides, alkyl pseudohalides, and allylic acetates. *Chem. Commun.* 46: 5743–5745.

54 (a) Chen, W., Liu, Z., Tian, J. et al. (2016). Building congested ketone: substituted hantzsch ester and nitrile as alkylation reagents in photoredox catalysis. *J. Am. Chem. Soc.* 138: 12312–12315. (b) Nakajima, K., Nojima, S., and Nishibayashi, Y. (2016). Nickel- and photoredox-catalyzed cross-coupling reactions of aryl halides with 4-alkyl-1,4-dihydropyridines as formal nucleophilic alkylation reagents. *Angew. Chem. Int. Ed.* 55: 14106–14110.

55 Knauber, T., Chandrasekaran, R., Tucker, J.W. et al. (2017). Ru/Ni dual catalytic desulfinative photoredox Csp^2–Csp^3 cross-coupling of alkyl sulfinate salts and aryl halides. *Org. Lett.* 19: 6566–6569.

56 Zhang, P., Le, C.C., and MacMillan, D.W.C. (2016). Silyl radical activation of alkyl halides in metallaphotoredox catalysis: a unique pathway for cross-electrophile coupling. *J. Am. Chem. Soc.* 138: 8084–8087.

57 Vara, B.A., Patel, N.R., and Molander, G.A. (2017). O-benzyl xanthate esters under Ni/photoredox dual catalysis: selective radical generation and Csp^3–Csp^2 cross-coupling. *ACS Catal.* 7: 3955–3959.

58 Heitz, D.R., Tellis, J.C., and Molander, G.A. (2016). Photochemical nickel-catalyzed C—H arylation: synthetic scope and mechanistic investigations. *J. Am. Chem. Soc.* 138: 12715–12718.

59 Shields, B.J. and Doyle, A.G. (2016). Direct $C(sp^3)$—H cross coupling enabled by catalytic generation of chlorine radicals. *J. Am. Chem. Soc.* 138: 12719–12722.

60 Shaw, M.H., Shurtleff, V.W., Terrett, J.A. et al. (2016). Native functionality in triple catalytic cross-coupling: sp^3 C—H bonds as latent nucleophiles. *Science* 352: 1304–1308.

61 Le, C., Liang, Y., Evans, R.W. et al. (2017). Selective sp^3 C—H alkylation via polarity-match-based cross-coupling. *Nature* 547: 79–83.

62 Zhang, X. and MacMillan, D.W.C. (2017). Direct aldehyde C—H arylation and alkylation via the combination of nickel, hydrogen atom transfer, and photoredox catalysis. *J. Am. Chem. Soc.* 139: 11353–11356.

63 Twilton, J., Christensen, M., DiRocco, D.A. et al. (2018). Selective hydrogen atom abstraction through induced bond polarization: direct α-arylation of alcohols through photoredox, HAT, and nickel catalysis. *Angew. Chem. Int. Ed.* 57: 5369–5373.

64 Macgregor, S.A., Neave, G.W., and Smith, C. (2003). Theoretical studies on C-heteroatom bond formation via reductive elimination from group 10

$M(PH_3)_2(CH_3)(X)$ species ($X = CH_3$, NH_2, OH, SH) and the determination of metal-X bond strengths using density functional theory. *Faraday Discuss.* 124: 111–127.

65 (a) Matsunaga, P.T., Hillhouse, G.L., and Rheingold, A.L. (1993). Oxygen-atom transfer from nitrous oxide to a nickel metallacycle. Synthesis, structure, and reactions of [cyclic] (2,2′-bipyridine)$Ni(OCH_2CH_2CH_2CH_2)$. *J. Am. Chem. Soc.* 115: 2075–2077. (b) Matsunaga, P.T., Mavropoulos, J.C., and Hillhouse, G.L. (1995). Oxygen-atom transfer from nitrous oxide (NNO) to nickel alkyls. Syntheses and reactions of nickel(II) alkoxides. *Polyhedron* 14: 175–185. (c) Han, R. and Hillhouse, G.L. (1997). Carbon–oxygen reductive-elimination from nickel(II) oxametallacycles and factors that control formation of ether, aldehyde, alcohol, or ester products. *J. Am. Chem. Soc.* 119: 8135–8136.

66 Klein, A., Kaiser, A., Wielandt, W. et al. (2008). Halide ligands—more than just σ-donors? A structural and spectroscopic study of homologous organonickel complexes. *Inorg. Chem.* 47: 11324–11333.

67 (a) Feldman, K.S., Romanelli, A.L., Ruckle, R.E., and Miller, R.F. (1988). Cyclopentane synthesis via free radical mediated addition of functionalized alkenes to substituted vinyl cyclopropanes. *J. Am. Chem. Soc.* 110: 3300–3302. (b) Feldman, K.S. and Schildknegt, K. (1993). Thioamide and thioester cyclopentane synthesis via trimethyltin radical catalyzed alkenylation of substituted (thiocarbonyl)cyclopropanes. *J. Org. Chem.* 59: 1129–1134. (c) Feldman, K.S. and Vong, A.K.K. (1990). *Tetrahedron Lett.* 31: 823–826. (d) Miura, K., Fugami, K., Oshima, K., and Utimoto, K. (1998). Synthesis of vinylcyclopentanes from vinylcyclopropanes and alkenes promoted by benzenethiyl radical. *Tetrahedron Lett.* 29: 5135–5138. (e) Kim, S. and Lee, S. (1991). Free radical mediated ring expansion of vinyl epoxides. *Tetrahedron Lett.* 32: 6575–6578.

68 Welin, E.R., Le, C., Arias-Rotondo, D.M. et al. (2017). Photosensitized, energy transfer-mediated organometallic catalysis through electronically excited nickel(II). *Science* 355: 380–385.

69 Kim, T., McCarver, S.J., Lee, C., and MacMillan, D.W.C. (2018). Sulfonamidation of aryl and heteroaryl halides through photosensitized nickel catalysis. *Angew. Chem. Int. Ed.* 57: 3488–3492.

9

Cross-Electrophile Coupling: Principles and New Reactions

Matthew M. Goldfogel, Liangbin Huang, and Daniel J. Weix

University of Wisconsin, Department of Chemistry, 1101 University Avenue, Madison, WI, 53706, USA

9.1 Introduction

The development of modern cross-coupling methods for carbon–carbon bond formation in the latter part of the twentieth century led to a major change in academic and pharmaceutical synthesis based upon the modular construction of small molecules through pairing of aryl electrophiles with aryl nucleophiles [1–3]. The employed nucleophiles are principally arylboronic acid derivatives, but arylzinc, aryl magnesium, aryltin, and arylsilicon are also utilized, whereas the electrophilic coupling partners are generally aryl halides (iodides, bromides, and, more recently, chlorides) or sulfonate esters of phenols [4]. The success of these methods has been such that, by several accounts, these biaryl-forming reactions are 11.5% of all C—C bond-forming reactions used in medicinal chemistry [5, 6]. Cross-coupling initially had the largest impact in discovery efforts, but has since become common in API production and even in the larger scale production of pesticides [7]. The resulting proliferation of biaryls has proved controversial and has led to a push to develop similarly robust couplings that form $C(sp^3)$—$C(sp^2)$ and $C(sp^3)$—$C(sp^3)$ bonds [8, 9]. As cross-coupling matured, the limiting factor in many cases was the synthesis of the carbon nucleophile component because relatively few carbon nucleophiles are commercially available compared to carbon electrophiles (Figure 9.1).

The cross-coupling of two electrophiles is a logical solution to this problem because the largest pools of substrates, outside of C—H bonds, are carbon electrophiles. We, and others, have defined cross-electrophile coupling as *a catalytic method that couples two different carbon electrophiles under reductive conditions* (Figure 9.1) [10–14]. Cross-electrophile coupling has several advantages that have led to its rapid uptake in the past decade: (i) both coupling partners are bench stable, (ii) the substrate pools are large, (iii) the reducing conditions are compatible with many sensitive functional groups, and (iv) the formation of $C(sp^3)$—$C(sp^x)$ bonds is relatively easy by this method. The central challenge of cross-electrophile coupling is to differentiate the substrates in a way that leads to robust cross-selectivity (Figure 9.2).

Nickel Catalysis in Organic Synthesis: Methods and Reactions,
First Edition. Edited by Sensuke Ogoshi.

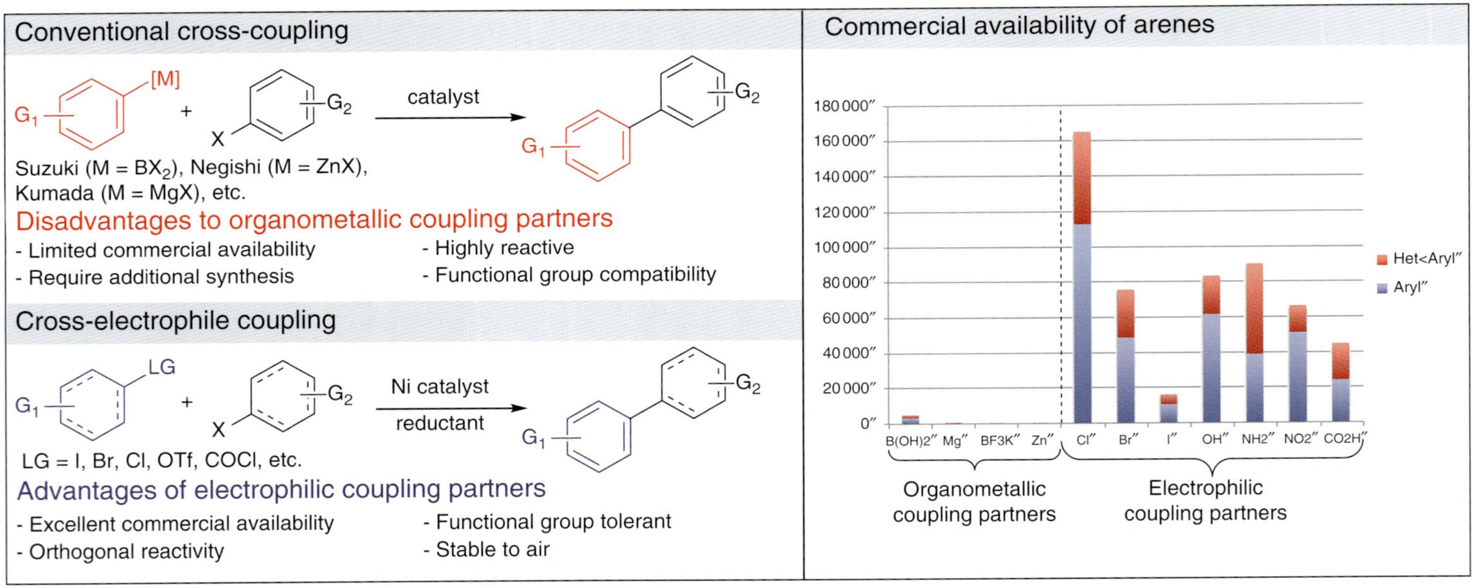

Figure 9.1 A comparison of cross-coupling and cross-electrophile coupling.

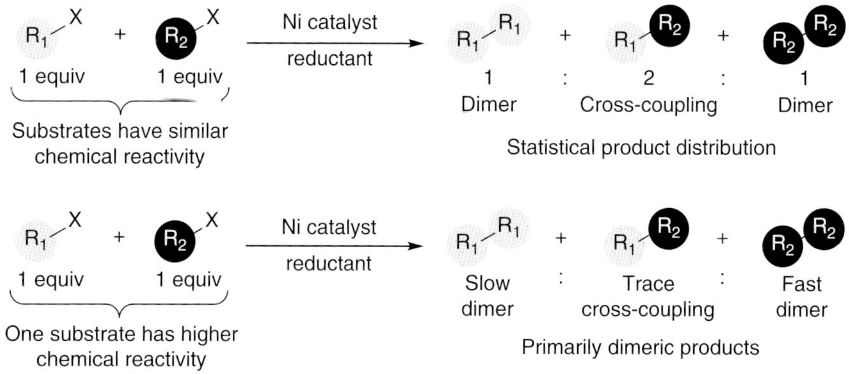

Figure 9.2 Selectivity challenges for cross-electrophile coupling.

While cross-electrophile coupling has only recently become a useful tool in organic synthesis, its roots go back over 100 years to the Wurtz–Fittig coupling of aryl halides with alkyl halides mediated by sodium metal [15, 16]. This, and the related homocoupling reactions using stoichiometric metals as both reductant and mediator [17–20], were limited by high temperatures and poor functional-group compatibility, yet still found a place in synthesis. Decades later, work on nickel-mediated Ullman homocoupling of aryl halides led to milder conditions and the first use of zinc to reduce the metal mediator [21–23]. This work, and subsequent studies with chemical and especially electrochemical methods [24, 25], led to some of the first catalytic, cross-selective methods. This separation of the catalyst from the reductant also set the stage for catalyst control of selectivity by ancillary ligands, which was crucial for the development of robust, cross-selective and even enantioselective methods. In fact, there are already cases where cross-electrophile coupling succeeded where cross-coupling with carbon nucleophiles failed [26, 27].

This chapter is intended as a brief review of the field of nickel-catalyzed cross-electrophile coupling for the formation of C—C bonds up to mid-2018. This review is limited to couplings between organic halides and related pseudohalides with some discussion of additions to π-electrophiles, such as olefins or carbonyls. After discussing the challenges and mechanistic considerations of cross-electrophile coupling, catalytic methods will be separated by the type of C—C bond being formed and further subdivided by the types of electrophiles employed. Only nickel-catalyzed methods will be discussed in this chapter, but it should be noted that important, useful work with copper [28], iron [29, 30], palladium [31–33], and cobalt [34–39] has been reported [11].

9.2 Mechanistic Discussion of Cross-Electrophile Coupling

While the utility of cross-electrophile coupling reactions has rapidly progressed in recent years, in most cases detailed mechanisms are still unknown. Several different mechanisms are possible and the relevant pathway for any reaction depends strongly on the substrates, the catalyst, and the conditions employed. We will discuss the challenges and mechanistic possibilities of cross-electrophile coupling in general and refer the reader to specific papers for further details. The central challenge of cross-electrophile coupling is the differentiation of the substrates by the catalyst. Compared to the coupling of a nucleophile with an electrophile, where only one substrate can undergo oxidative addition, cross-electrophile coupling pairs substrates that are both prone to oxidative addition. As such, the origins of selectivity must be derived from slight reactivity differences between the two electrophiles, such as electronics (leaving group, activating groups), steric effects, or the relative stability of radicals (Figure 9.3).

How catalysts leverage chemical differences to provide selectivity is not intuitively obvious, but typically relies on different mechanisms of oxidative addition [40]. An analysis of the literature shows substrate differentiation based upon (i) relative propensity for radical formation vs. oxidative addition [C(sp^2)-X vs. C(sp^3)-X] [41, 42], (ii) differences in reactivity due to leaving group or activating group [43, 44], (iii) the polarity of the oxidative addition transition state [C(sp^2)-X vs. C(sp^2)-OTf] [45], (iv) the steric environment of the electrophile [C(sp^3)–C(sp^3), C(sp^2)–C(sp^3)] [46], and (v) S_N2 vs. radical oxidative pathways [C(sp^2)–C(sp^3) and C(sp^3)–C(sp^3)]. The different oxidation states that the nickel catalyst reaches during the catalytic cycle change its selectivity to allow for a single catalyst to react selectively with both substrates [47]. Alternatively, a co-catalyst system can be used to independently activate each electrophile [48].

In addition to these chemical origins of selectivity, cross-coupling can be enhanced by adding an excess of a less reactive coupling partner [46, 49]. This provides a probabilistic improvement in yield, but is inherently inefficient; adding unequal ratios of substrate increases the formation of by-products (homocoupling and reduction products) and can complicate product isolation. Despite these drawbacks, many cross-electrophile coupling methods use unequal substrate ratios to bias selectivity [12].

Mechanistic hypotheses are integral to rationalizing the origins of selectivity in cross-electrophile coupling, but the mechanistic differences between reactions

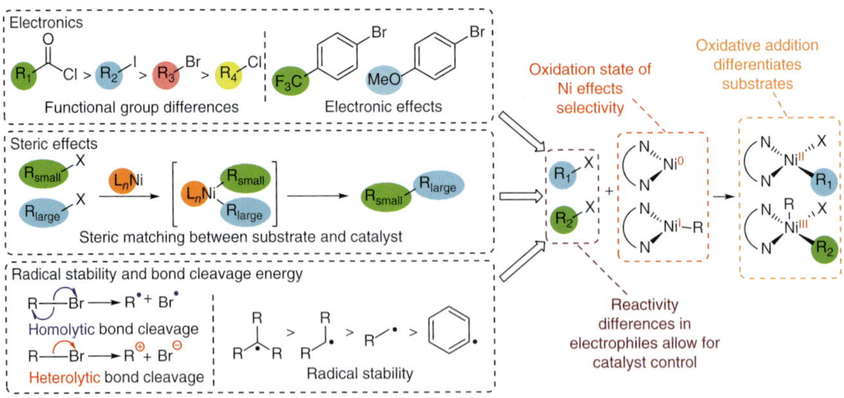

Figure 9.3 Origins of selectivity in cross-electrophile coupling.

are often subtle and difficult to elucidate experimentally. The mechanisms cited for C—C bond formation via nickel-catalyzed cross-electrophile coupling can be broadly characterized as either occurring through transmetalation between organometallic species or by sequential oxidative activations at a single metal center (Figure 9.4) [50].

Reactions occur by a transmetalation mechanism when the metal reductant (i.e. Zn, Mn, Mg) interacts directly with the electrophile to generate an organometallic nucleophile (Figure 9.4a) [51, 52]. Activation of the electrophile can occur stoichiometrically – referred to as ***in situ* organometallic formation** [29, 33, 53, 54], or through a catalyst that is regenerated by a reductant – referred to as **multimetallic transmetalation** [45, 55, 56]. The resulting organometallic reagent then participates in the oxidative addition, transmetalation, and reductive elimination pathway common to cross-coupling reactions [4]. Cross-electrophile coupling differs mechanistically from cross-coupling in that the organometallic intermediate is maintained at a low steady-state concentration and consumed rapidly upon generation [29]. This prevents buildup of a reactive carbon nucleophile, which can improve functional group tolerance to sensitive electrophiles and/or acidic functional groups.

Other cross-electrophile coupling mechanisms are characterized by sequential oxidations of the nickel catalyst in order to activate both electrophiles at the same metal center. Mechanisms that operate via sequential oxidative activations can be further subdivided into (i) **double oxidative additions with intermediate reduction** [23, 57–60] or (ii) **radical chain reactions** [25, 61, 62]. Double oxidative addition (shown in Figure 9.4b) is defined by sequential oxidative additions of each electrophilic component to a single metal catalyst. These two-electron oxidative additions are separated by an intermediate reduction before subsequent reductive elimination from **B4** generates the C—C bond. This mechanism was first proposed for electrochemical processes [63, 64] and later invoked to describe reactions with metal reductants where radical traps and/or the absence of radical rearrangements suggested the absence of radical intermediates [65–67]. Conversely, radical chain reactions are defined by single-electron transfer between

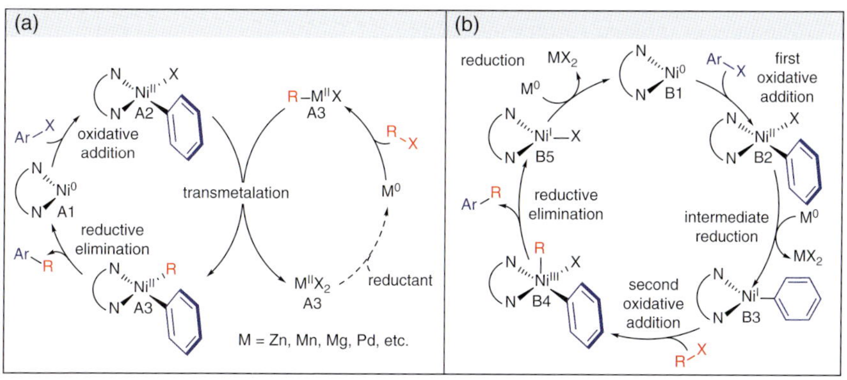

Figure 9.4 Two mechanisms for selective cross-electrophile coupling. (a) Transmetalation and (b) sequential oxidative activation.

a nickel complex and an organic radical – this results in the one-electron oxidation of nickel with concomitant formation of the Ni–carbon bond. The first oxidative addition is a two-electron process that occurs prior to radical capture, whereas the second oxidative activation of the electrophile is a single-electron process that generates the alkyl radical. Further subsets of radical capture mechanisms are defined by whether the alkyl radical is formed via atom transfer with Ni [50] or from a separate co-catalytic cycle [48, 68, 69]. In all cases, the nickel catalyst is turned over by reduction.

The proliferation of cross-electrophile coupling reactions can be traced to the discovery of new ways to provide cross-selectivity through subtle differences between electrophilic coupling partners. The proliferation of cross-electrophile coupling methods continues to provide variations on the general mechanisms discussed here. We anticipate that this field will continue to grow as academic and industrial chemists seek more efficient approaches to synthetic targets while avoiding the challenges associated with reactive carbon nucleophiles [70–73].

9.3 C(sp^2)—C(sp^3) Bond Formation

The majority of cross-electrophile coupling reactions developed for C—C bond formation have focused on the formation of C(sp^2)—C(sp^3) bonds for two reasons [11–14]: C(sp^2)–C(sp^3) bond formation with palladium cross-coupling methods is not as versatile as biaryl synthesis [74, 75] and the inherent differences of C(sp^2) and C(sp^3) electrophiles have enabled robust cross-selectivity. This section will review the field of cross-electrophile coupling for the formation of C(sp^2)—C(sp^3) bonds and is further subdivided based on the coupling partners used in each transformation to make it easier for interested readers to find a specific bond disconnection that may be relevant to a synthetic application.

9.3.1 Cross-Electrophile Coupling of Aryl-X and Alkyl-X

Aryl–alkyl linkages are widely prevalent in bioactive molecules and drugs and are of significant interest to synthetic chemists. Cross-electrophile coupling has flourished in this field because it provides a direct and efficient strategy for making these bonds from commercially available and synthetically convenient electrophilic reagents. Cross-electrophile coupling benefits from the pairing of aryl and alkyl coupling partners, as the differences in hybridization can be translated to discrete reactivity with metal catalysts based on differences in the mechanisms of oxidative addition (*vide supra*) [10]. This chemistry is also the most used of the cross-electrophile coupling reactions developed to date, both academically and industrially [70–73].

Early work in forming aryl–alkyl linkages relied on electrochemical reduction with sacrificial anodes and was limited to activated alkyl halides, defined as substrates with a resonance contributing functional group (i.e. ester, cyanide) alpha to the halide, which stabilize intermediate radicals and rapidly form organometallic reagents (primarily enolates) [24]. Slow addition of the activated alkyl halide was essential to achieving good cross-selectivity and avoiding alkyl dimerization [76]. Vinyl halides and allylic halides could be used in place of aryl halides and alpha-chloro carbonyls [77] and stereochemistry could be set via chiral auxiliaries [78]. Building on the nickel-catalyzed Ullman literature [22, 23], Gosmini and coworkers established that these electrochemical successes could be translated to conditions with chemical reductants, such as Mn powder, setting the stage for later developments [52].

The first general nickel-catalyzed cross-electrophile coupling of aryl halides with unactivated alkyl halides was reported by the Weix group in 2010 [41] (Figure 9.5). This coupling replaced unfamiliar electrochemical apparatuses with a stoichiometric reductant and demonstrated that unactivated alkyl electrophiles were viable substrates. The key advance was that modification of the ligands on nickel could improve selectivity for the cross-coupled product over dimeric products (Figure 9.6). In the initial study, an unusual synergistic effect was noted, where the combination of (o-$(Ph_3P)_2C_6H_4$)NiI_2 and (dtbbpy)NiI_2 (dtbbpy = 4,4′-di-*tert*-butyl-2,2′-bipyridine) provided the best selectivity for the product over the aryl dimer. Complementarity to cross-coupling methods was shown by coupling protic substrates and substrates containing C—B bonds that would be reactive in conventional cross-coupling. This publication spurred a renewed interest in cross-electrophile coupling with stoichiometric metal reductants, whereas previous studies were largely focused on electrochemical methods

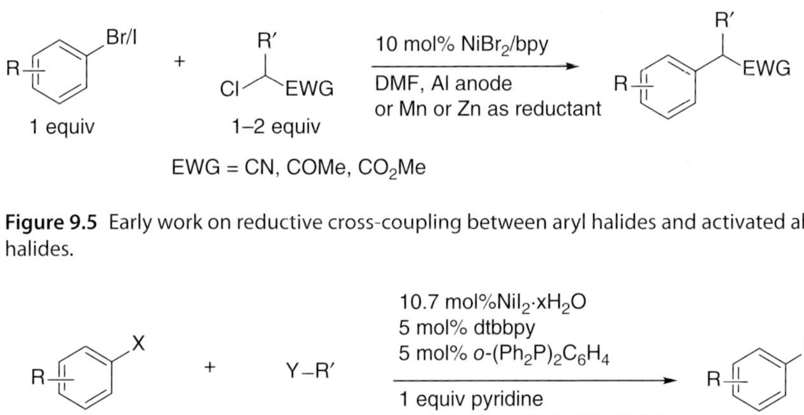

Figure 9.5 Early work on reductive cross-coupling between aryl halides and activated alkyl halides.

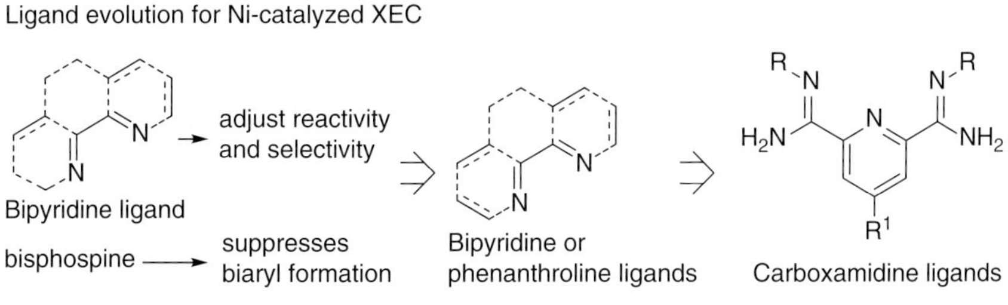

Figure 9.6 Ligand evolution for Ni-catalyzed cross-electrophile coupling.

[25, 52, 77, 78], and started a search for improved selectivity by ligand design (Figure 9.6). Subsequent studies found that a 5–10 mol% loading of a single nickel bipyridine or phenanthroline catalyst could be used in place of the two-ligand system and that the inclusion of a sodium iodide (NaI) additive enabled the coupling of aryl bromides, vinyl bromides, and activated aryl chlorides with alkyl bromides [42]. Later methods by other groups demonstrated the value of salt additives (i.e. $MgCl_2$) [79] and π-basic ligands (i.e. ethyl crotonate) [80]. A recent collaboration between Pfizer and the Weix group discovered that pyridine carboxamidines (PyCam) and pyridine bis(carboxamidines) (PyBCam) allow for the coupling of a range of difficult substrates, such as arenes with *ortho*-coordinating groups, Lewis basic nitrogen heterocycles, aryl chlorides, and aryl triflates [70, 71]. These newly discovered ligand types were later employed in the electrochemical cross-electrophile coupling of aryl and alkyl bromides [72].

The preceding reactions were the focus of a detailed mechanistic study by our laboratory that suggested that these reactions occur through a radical chain mechanism [50]. This mechanism was originally proposed by Hegedus and Miller for the stoichiometric reactivity of allylnickel(II) reagents [61], but later abandoned in favor of a mechanism that avoided carbon-centered radicals [55] more similar to the proposals of Kochi [63, 81, 82]. Durandetti et al. proposed the original Hegedus mechanism as one of two possibilities for nickel cross-electrophile coupling, but were unable to rule either one out [25, 62]. Our later studies built upon this work by providing the experimental basis for radical chain mechanisms in many cross-electrophile coupling reactions. Key experiments included (i) using an organic reductant to show that organometallic reagents were not being formed *in situ*, (ii) using stoichiometrically reacting Ni(II)-aryl and Ni(II)-alkyl complexes to show that multimetallic transmetalation was not operative, and (iii) measuring the amount of radical rearrangement observed with varying catalyst concentration to demonstrate that multiple metal centers must be involved in product formation. This data suggests that selectivity for cross-coupling is derived from the preference for each substrate to undergo oxidative addition or radical formation, respectively.

The synthetic value of cross-electrophile coupling for aryl–alkyl bond formation is increasing rapidly because new methods addressing challenging substrates are now appearing weekly from a large number of research groups. Modifications to the reaction conditions, ligands, and additives have enabled coupling reactions with both highly reactive alkyl halides, such as methyl iodide [83] and methyl tosylate [84], and hindered quaternary alkyl halides [85]. These methods demonstrate how specific reactivity challenges can often be overcome through judicious ligand selection and the addition of stoichiometric additives. These trends in reaction optimization were further demonstrated in the medicinally relevant [86–88] coupling of fluorinated alkyl halides to (hetero)aryl iodides [89], where α-fluorine elimination was prevented through the use of a stoichiometric $FeBr_2$ additive. Although the role of this additive is unknown, it shows how selectivity of the Ni catalysts can be tuned to accommodate a variety of substrates.

Although significant progress has been made in nickel-catalyzed cross-electrophile coupling between aryl halides and alkyl halides, specific challenges still exist when unreactive aryl chlorides are used as coupling partners. In 2018, Zhang and coworkers reported the first general cross-electrophile coupling of (hetero)aryl chlorides with chlorodifluoromethane [90]. The electron-rich bipyridine ligand facilitated the oxidative addition step and promoted product formation. The reliable practicability and scalability of those transformations were demonstrated by several 10-g scale reactions without loss of reaction efficiency (Figure 9.7).

Researchers have also developed methods for expanding cross-electrophile coupling beyond halide leaving groups, as oxygen electrophiles (tosylates, triflates, etc.) have been used in the place of halides. Currently, the most general method for forming aryl–alkyl bonds with $C(sp^3)$-oxygen electrophiles is from Komeyama et al. [91]. Their work built upon a previous study from the Weix group that used cobalt phthalocyanine to activate benzylic mesylates by nucleophilic displacement and reductive homolysis of the resulting C—Co bond [68]. The use of cobalamine improved upon the previous substrate scope by allowing for coupling reactions with non-benzylic alcohols. Molander et al. had also demonstrated that the coupling of secondary alkyl tosylates with aryl electrophiles can be achieved in high yield at higher temperature (80 °C) through the addition of KI to promote nucleophilic substitution of the oxygen leaving group to form an alkyl iodide [92]. Coupling then occurs under reductive conditions with a nickel catalyst without the need for cobalt (Figure 9.8).

A desire to access non-halide electrophiles as coupling partners has frequently inspired new approaches to cross-electrophile coupling, as evidenced by the cobalt-catalyzed activation of alcohol derivatives described above. This pattern repeats in the development of coupling reactions with alkyl carboxylic acids. Interest in these substrates prompted our group to employ redox active esters as alkyl radical sources [93]. Synthetic methods with redox active esters were initially developed by Okada et al. and Overman and coworker as a method for converting carboxylic acids to reactive alkyl radicals [94, 95]. Success with other radical precursors prompted the exploration of these reagents in cross-electrophile coupling and culminated in the coupling of methyl, primary,

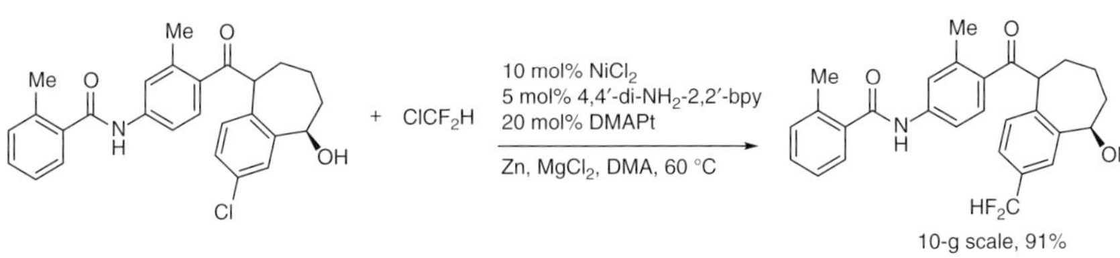

Figure 9.7 Ni-catalyzed reductive coupling between aryl chloride and chlorodifluoromethane.

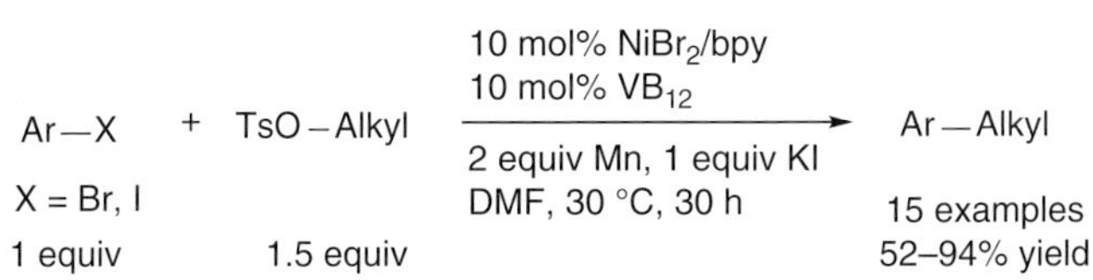

Figure 9.8 Ni-catalyzed reductive coupling between aryl halides and alkyl-OTs.

and secondary alkyl *N*-hydroxyphthalimide esters (NHP esters) with aryl iodides in good yields (77% average yield). One coupling between an acid chloride (3 equiv) and NHP ester was also presented, but additional attempts to apply this method to less reactive aryl electrophiles (e.g. Ar-Br, Ar-Cl) were unsuccessful and resulted in consumption of the NHP ester but not the aryl halide. Stoichiometric reactions of (dtbbpy)Ni(2-tolyl)I with an NHP ester show that an arylnickel(II) complex could directly react with an NHP ester to form an alkylated arene, suggesting that the reaction mechanism is similar to previous cross-electrophile coupling methods [50] (Figure 9.9).

While diastereoselective methods had previously been reported [78], catalytic enantioselective cross-electrophile coupling remains a significant challenge. Recently, Reisman and coworker developed an asymmetric cross-electrophile coupling of heteroaryl iodides and α-chloronitriles to furnish enantioenriched α,α-disubstituted nitriles in good yield and high enantioselectivity [96]. Optimal reaction conditions required the use of electron-rich phosphine-oxazoline ligands, which suppressed hydrodehalogenation of the α-chloronitrile reagents, over more common chiral bisoxazoline ligands. Notably, the reaction tolerates many heterocycles, but enantioselective reactions with unactivated alkyl halides are still unknown (Figure 9.10).

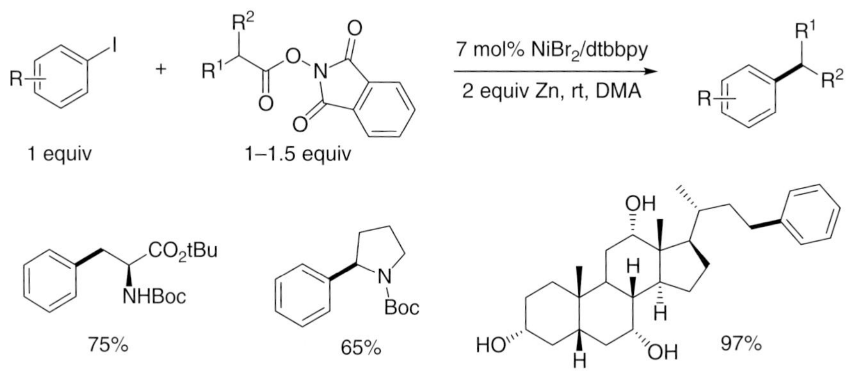

Figure 9.9 Ni-catalyzed reductive coupling of aryl iodides with *N*-hydroxyphthalimide (NHP) esters.

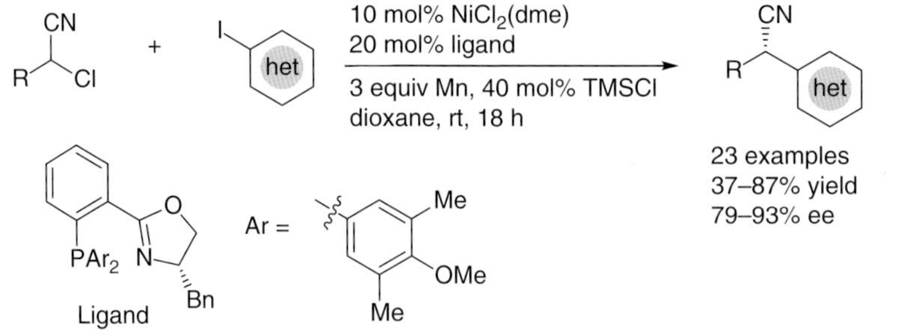

Figure 9.10 Ni-catalyzed enantioselective coupling of aryl iodides with α-chloronitriles.

Epoxide and aziridine electrophiles provide an opportunity for catalyst-controlled regioselectivity and allow for the installation of multiple carbon units while maintaining a functional group handle. Cross-electrophile coupling with epoxides has been explored using co-catalysts to activate the epoxide and provide regio-divergent reactivity. Initial studies in our group showed that couplings between epoxides and aryl halides could be accomplished using an iodide co-catalyst to open the epoxide at the less hindered position [48]. The resulting iodohydrin intermediate could then be coupled with an aryl iodide. An analogous iodine ring opening was used by Yamamoto and coworker in the stereoretentive coupling of aryl iodides with 3,4-epoxy alcohols to furnish enantioenriched 4,4-diaryl alkanes [97]. Our studies also showed that the regiochemistry of the epoxide opening could be switched by using a titanocene co-catalyst to enable ring opening at the more hindered position of the epoxide [48]. Regioselectivity was presumably due to the preferential formation of a Ti(III)-mediated radical on the more substituted carbon. Later, the Ti(III)-mediated radical formation was extended to the enantioselective synthesis of *trans*-β-arylcycloalkanols via a parallel kinetic resolution [98] through the use of a chiral titanocene catalyst [99]. The success of these transformations demonstrates how a variety of radical sources can be efficiently interfaced with nickel catalysts (Figure 9.11).

The reductive ring opening of aziridines has also been developed and mirrors epoxides in that a co-catalyst is necessary for activation of the electrophile. Doyle and coworkers reported the Ni-catalyzed cross-electrophile coupling of styrenyl aziridines with aryl iodides in the presence of 2 equiv of NaI [100]. The reaction proceeds by a steroconvergent mechanism and is amenable to asymmetric catalysis using a chiral bisoxazoline ligand. The process allows facile access to highly enantioenriched 2-arylphenethylamines from racemic aziridines (Figure 9.12).

The reductive methods described thus far have primarily relied on stoichiometric metals or electrochemical reduction using metal electrodes. The most common metal reductants are zinc and manganese, which benefit from their abundance and air stability. Despite these advantages, the use of these

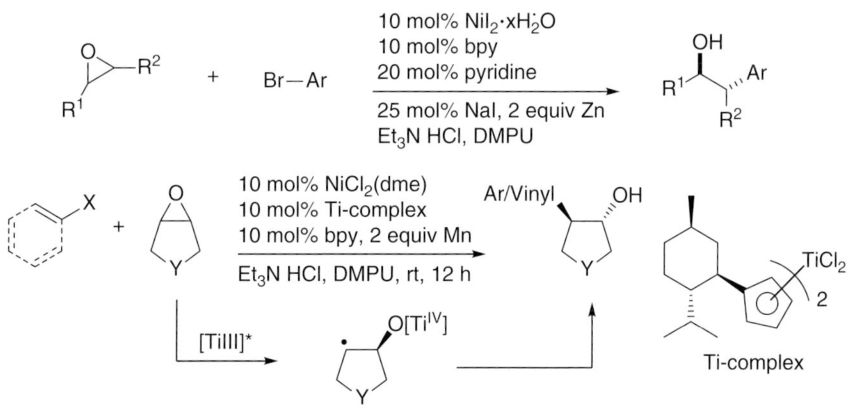

Figure 9.11 Ni/Ti-catalyzed cross-coupling of aryl bromides with epoxide.

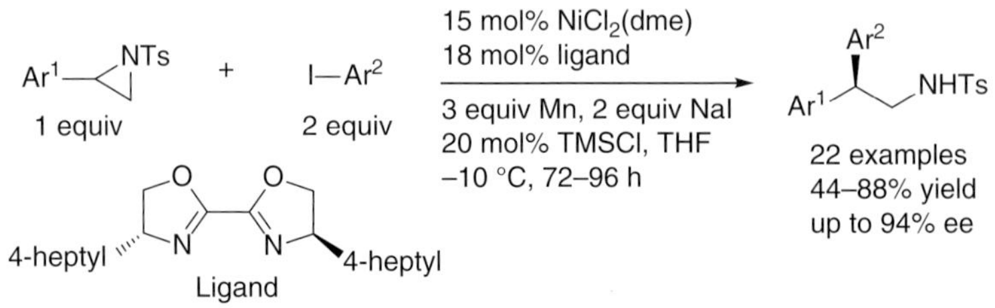

Figure 9.12 Ni-catalyzed enantioselective coupling of aryl iodides with aziridines.

reductants generates stoichiometric metal salts, creates heterogeneous reaction mixtures, and can complicate waste disposal. To combat these limitations, several groups have experimented with organic amine reductants, such as tetrakis(dimethylamino) ethylene (TDAE), as stoichiometric reductants. These reagents have been particularly important as mechanistic probes to rule out *in situ* organometallic formation by proving that product formation can still occur in the absence of a metal reductant (Figure 9.4). Furthermore, organic reductants were used to overcome synthetic limitations by allowing for coupling in non-amide solvents such as acetonitrile or propylene oxide, and by preventing the reduction of sensitive functional groups [101]. Cross-electrophile coupling reactions are rarely published in non-amide solvents, but the Molander group has reported some additional examples of aryl–alkyl coupling reactions in methanol, cyclopentyl methyl ether, and toluene [102, 103]. Other strategies for eliminating stoichiometric metal reductants include merging photoredox and nickel catalysis. This was explored by the MacMillan group through the photochemical generation of silyl radical species that could abstract a halogen atom from an alkyl halide to generate an alkyl radical [69]. This radical could then be captured by a nickel catalyst and coupled with aryl electrophiles. Lei, Vannucci, and later Jamison also demonstrated that photoredox and nickel catalysis could be combined to directly use Et_3N or triethanolamine as terminal reductants [73, 104, 105] (Figure 9.13).

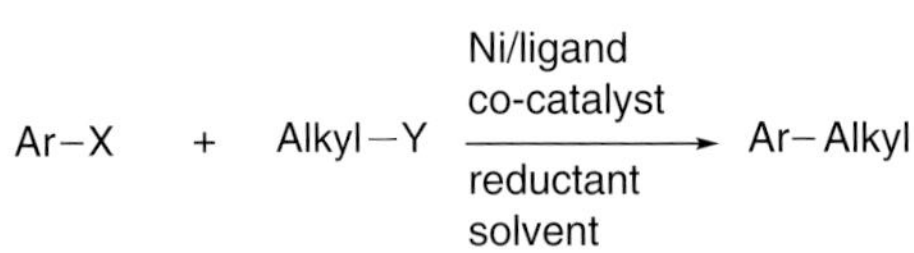

Entry	Key additive	Co-catalyst	Reductant	Solvent	References
1	—	—	TDAE	MeCN	[101]
2	$MgCl_2$	Ir-photocatalyst	Et_3N	DMF	[104]
3	—	Ir-photocatalyst	$N(CH_2CH_2OH)_3$	MeCN	[105]
4	Na_2CO_3 (base)	Ir-photocatalyst	$(TMS)_3SiH$	DME	[69]
3	NaI (electrolyte)	—	Anode (Zn)	DMA	[72]
4	Bu_4NPF_6 (electrolyte)	—	Et_3N	DMA	[73]

Figure 9.13 Alternative reduction systems for Ni-catalyzed reductive coupling between aryl and alkyl electrophiles.

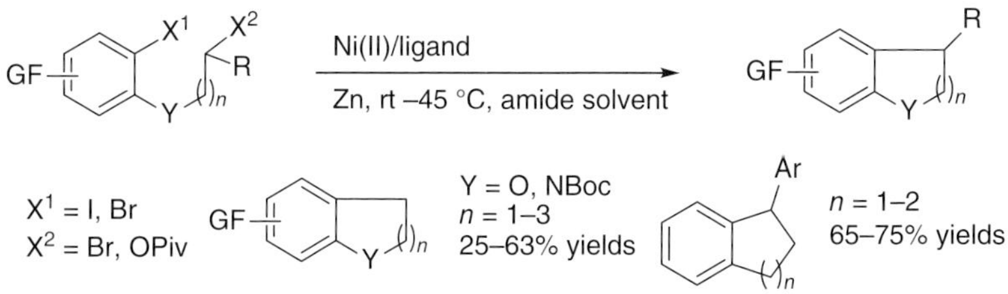

Figure 9.14 Ni-catalyzed intramolecular cross-electrophile coupling.

Compared to intermolecular cross-electrophile coupling, intramolecular variants are far less studied. This is particularly apparent for C(sp^2)—C(sp^3) bond formations where there are only two reported examples to date. Peng and coworkers reported a cross-electrophile coupling mediated by stoichiometric nickel that can construct five to seven member rings from the intramolecular coupling of an aryl iodide and alkyl bromide [80], and a related intramolecular cyclization of an aryl bromide and benzylic pivlate to form indanes and tetralines was accomplished with catalytic $NiBr_2$-dmethoxyethane ($NiBr_2$-dme) and a bathophenanthroline ligand (BPhen) [106] (Figure 9.14).

9.3.2 Cross-Electrophile Coupling of ArX and Bn-X

Simple, unstabilized alkyl electrophiles account for much of the cross-electrophile coupling literature, but the reactivity of a C(sp^3) electrophile can be enhanced by incorporating a benzyl or allyl group. These functional groups stabilize transient intermediates via resonance contributions from π-systems and have been used in cross-electrophile coupling to perturb the relative reactivity of C(sp^2) and C(sp^3) electrophiles. Matching the relative reactivity of each electrophile can be vital for ensuring that cross-coupling outcompetes homocoupling. The following sections will examine how benzylic and allylic electrophiles have been used in cross-electrophile coupling reactions, as these are the most studied classes of stabilized alkyl electrophiles and the reactivity trends translate to other resonance-stabilized electrophiles.

Our discussions of coupling reactions with alkyl bromide and iodide electrophiles emphasize the absence of chloride substrates. Many cross-electrophile coupling methods struggle to overcome the increased bond homolysis energy of the C—Cl bond and its decreased reactivity toward oxidative addition. The inclusion of stabilizing benzylic groups can overcome these limitations, but this increased activity does create an additional selectivity challenge – benzylic electrophiles often dimerize more readily than the corresponding alkyl electrophile. This selectivity challenge was overcome by forming the organometallic reagent *in situ* for the synthesis of diarylmethanes from benzyl chlorides and aryl chlorides/fluorides [51]. The Ni/Mg catalyst pairing was also applied to the coupling of O-based benzylic electrophiles with aryl bromides [107] and mechanistic studies support that these reactions proceed through the *in situ* organometallic formation of a benzylic Grignard reagent [51] (Figure 9.15).

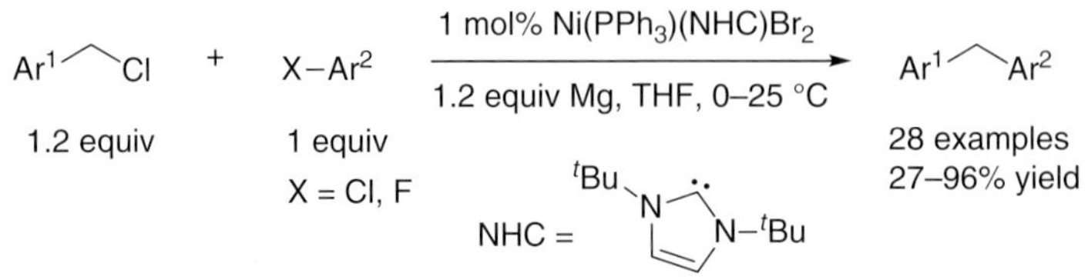

Figure 9.15 Ni-catalyzed reductive coupling of aryl halides with benzylic chlorides.

The challenge of benzylic dimerization can also be solved by using sterics to dissuade homocoupling. Gong and coworkers demonstrated that the increased sterics of secondary benzylic chlorides slows dimerization and enables selective couplings with aryl halides [108]. The use of secondary benzylic electrophiles also provides an opportunity for enantioselective cross-electrophile coupling, and an asymmetric Ni-catalyzed reductive cross-coupling of (hetero)aryl iodides and benzylic chlorides was reported by the Reisman group for the preparation of enantioenriched 1,1-diarylalkanes [109] (Figure 9.16).

Although we will not be discussing cross-electrophile couplings with π-systems, aldehydes are still an attractive class of arene electrophiles. The Doyle laboratory was able to convert aldehydes to benzylic electrophiles by transforming them into benzylic acetals, which could be reduced to form alpha-oxy radicals in the presence of TMSCl and Zn [110]. The resulting radicals were then coupled with aryl iodides. The coupling of benzylic acetals was followed by three-component coupling to generate secondary benzhydryl amines where catalysis proceeded through the *in situ* formation of an imminium ion via condensation between an *N*-trimethylsilyl amine and aryl aldehyde [111]. The generated benzylic imminium could then be reduced by zinc to form a benzylic radical stabilized by an α-amine (Figure 9.17).

9.3.3 Cross-Electrophile Coupling of ArX and Allyl-X

Allyl groups have similar stabilizing effects as benzylic groups. This increases their electrophilic reactivity relative to simple alkyl electrophiles and can result in undesired dimerization due to the correspondingly faster rates of oxidative addition. Additionally, allylic substrates can react to form linear or branched products, although cross-electrophile coupling reactions developed to date favor linear regioselectivities. Our own group developed the coupling of allyl acetates and aryl halides as a method for the formation of allylarenes with high linear selectivity [112] (Figure 9.18). A tridentate ligand was essential to preventing undesired

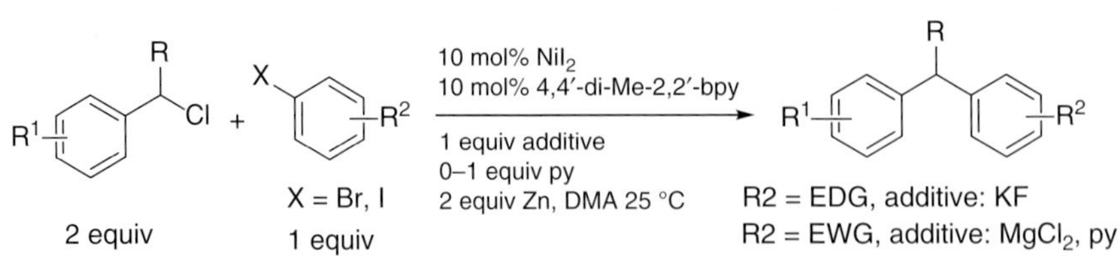

Figure 9.16 Ni-catalyzed coupling of aryl halides with secondary benzylic chlorides.

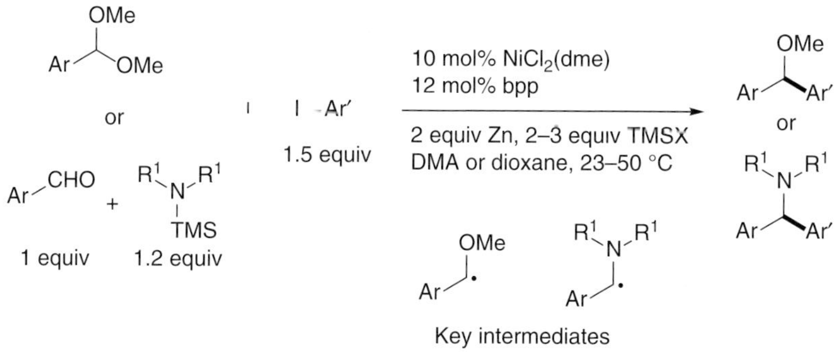

Figure 9.17 Ni-catalyzed coupling of aryl halides with acetals and imines.

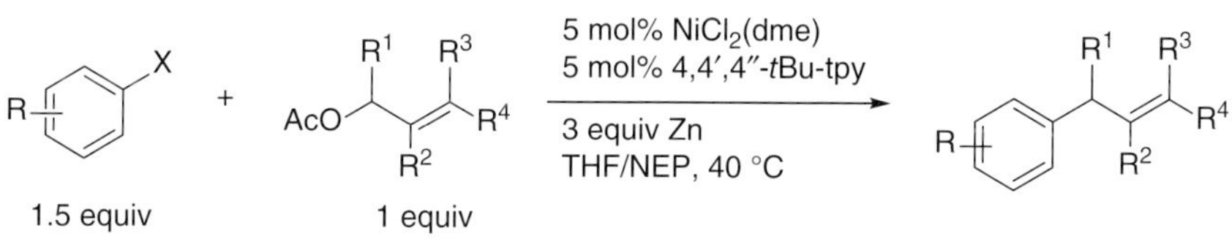

Figure 9.18 Ni-catalyzed reductive coupling of aryl halides with allyl acetates.

homocoupling in this reaction and related coupling reactions have extended allylarenes formation to electron-rich aryl bromides with various allylic acetates by adding $MgCl_2$ and NBu_4Br additives [79, 113]. Further research by the Shu group allowed for the direct coupling of allylic alcohols without functionalization to the allylic ester [114]. This was accomplished through the use of a zirconium co-catalyst that activated the alcohol by binding to the oxygen lone pair as a Lewis acid.

Another interesting example of allylic cross-electrophile coupling reaction originates not from an allyl alcohol derivative but from an enone. Allylic nickel intermediates were generated from the reaction of $Ni(cod)_2$ with an enone and a chlorosilane in a reversed-polarity conjugate allylation mechanism that was first described by Mackenzie and coworkers [115]. These allylnickel intermediates were then used as electrophilic coupling partners in the cross-electrophile coupling with aryl electrophiles to form γ-aryl substituted silyl enol ethers [116]. Selectivity is attributed to the rapid and selective reaction of Ni(0) with chlorotriethylsilane and enone in the presence of other organic electrophiles, and is assisted by the minimization of enone dimer through ligand steric effects (Figure 9.19).

9.3.4 Vinyl-X with R-X

Cross-electrophile coupling methods for vinylation are closely related to arylation, as both are $C(sp^2)$ electrophiles. However, reactivity trends in cross-electrophile coupling show that vinyl electrophiles are generally more active than arenes due to their decreased sterics [42]. As seen for benzylic and allylic electrophiles, perturbing the balance in substrate reactivity often results in

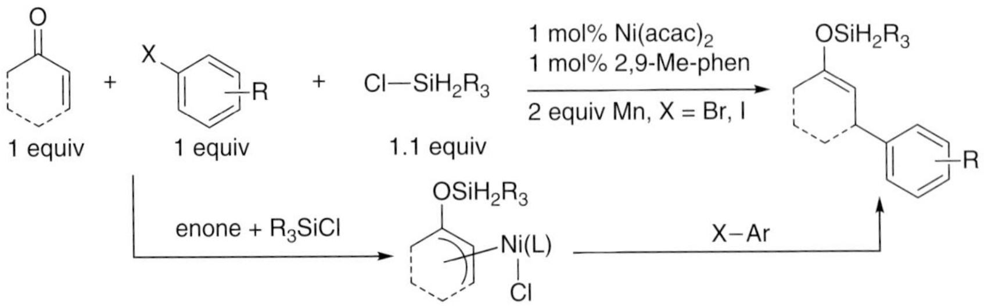

Figure 9.19 Ni-catalyzed reductive coupling of aryl halides with chlorotriethylsilane and enones.

competitive homocoupling [117]. Two principle strategies have been applied to overcome these reactivity differences: (i) lowering catalyst loadings to minimize dimerization of vinyl electrophiles and (ii) pairing vinylic reagents with more reactive electrophiles to encourage cross-reactivity. Both of these strategies were used in the cross-electrophile coupling of vinyl bromides and triflates with alkyl halides – conditions for arene electrophiles [42] were modified to suit vinyl substrates by decreasing the catalyst loading and converting alkyl bromides to more reactive alkyl iodides *in situ* [117] (Figure 9.20). This method was rapidly adopted in the total syntheses of (−)-6,7-dideoxysqualestatin [27] and (−)-Rasfonin [118]. Switching the ligand to a hindered phenanthroline allowed for the coupling of vinyl bromides with allyl electrophiles [112]. Successful reactions were achieved by the Condon group by matching the increased reactivity of vinyl electrophiles with equally reactive α-stabilized electrophiles. Vinyl halides were paired with α-chloro esters, α-chloro ketones, and α-chloro cyanides using electrochemical reduction at an aluminum anode [119]. Similarly, Gong and coworkers used Mn in the place of electrochemical methods for the coupling of α-halo carbonyl compounds with vinyl bromides [120].

The Reisman group has contributed significantly to cross-electrophile coupling by publishing several asymmetric methods with vinyl electrophiles [9]. The first enantioselective cross-electrophile coupling with vinyl electrophiles was demonstrated with benzyl halides using a Mn reductant, and allowed for the synthesis of tertiary allylic stereocenters in 85–97% ee [67]. The scope of this reaction was later extended to couplings with benzylic α-chloro silanes and vinyl triflates [121] (Figure 9.21). The steric contributions of the bulky silyl group slowed the rate of oxidative addition into the benzyl electrophile and prompted the use of a cobalt phthalocyanine [Co(Pc)] co-catalyst as an alkyl radical initiator. Parallel to the

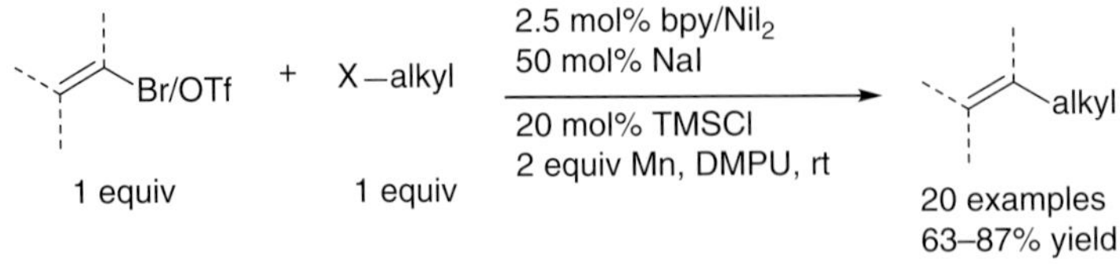

Figure 9.20 Ni-catalyzed reductive coupling of vinyl electrophiles with unactivated alkyl halides.

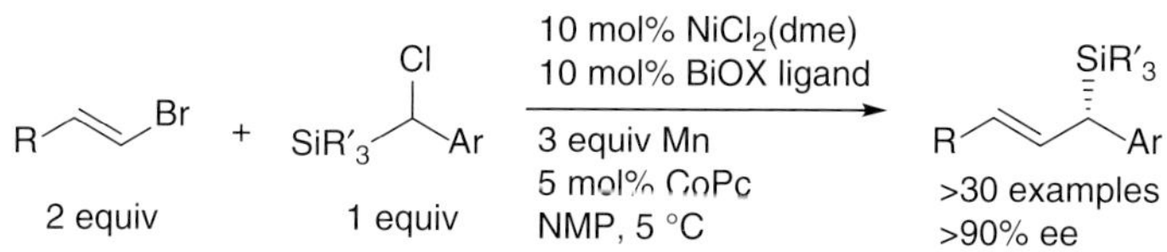

Figure 9.21 Ni-catalyzed enantioconvergent reductive coupling of vinyl bromides with benzylic chlorides.

development of these methods, the enantioselective coupling between benzylic redox active esters and vinyl bromides was also discovered [122]. Redox active esters were previously demonstrated in achiral cross-electrophile coupling methods [93], but their use in stereoselective catalysis marked a significant advance. All of the enantioselective methods developed in the Resiman group use bisoxazoline ligated nickel complexes to engender enantioselectivity, and the success of these ligands encourages further exploration of ligand types beyond bipyridine and phenanthroline.

9.3.5 Acyl-X with Alkyl-X

The reaction of carboxylic acid derivatives with organometallic nucleophiles provides an efficient and convergent strategy for the synthesis of unsymmetrical ketones [123]. Replacing organometallics with more abundant carbon electrophiles can further improve the scope of this strategy (Figure 9.22).

The seminal work in this field was the reductive cross-coupling between carboxylic derivatives and primary alkyl iodides reported by Mukaiyama and coworkers in 1981 [124] (Figure 9.23). Unsymmetrical ketones were obtained in

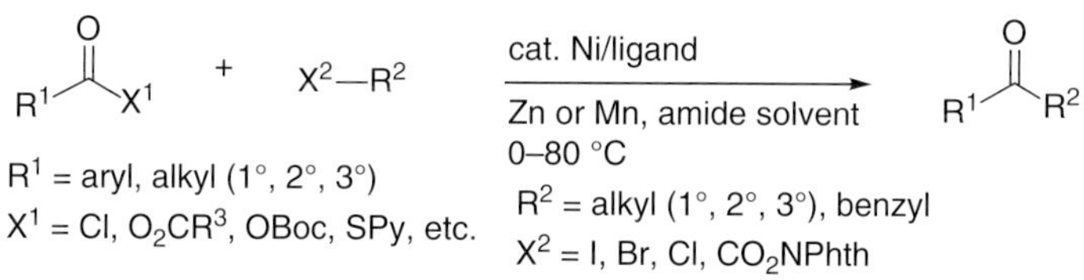

Figure 9.22 Ni-catalyzed reductive coupling of carboxylic acid derivatives with alkyl electrophiles.

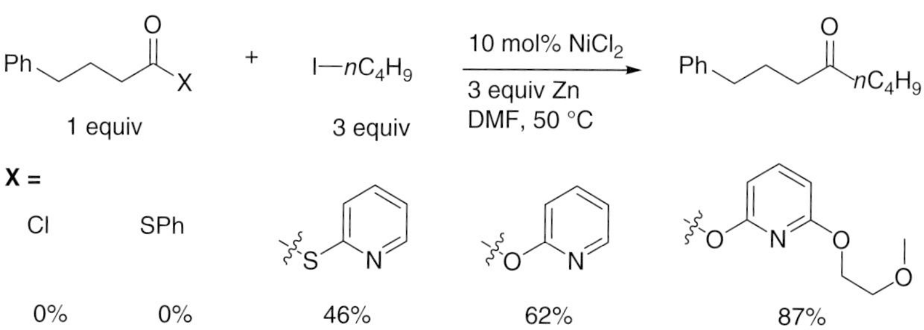

Figure 9.23 Ni-catalyzed reductive coupling of various carboxylic acid derivatives with primary alkyl iodides.

good yield using a nickel catalyst and zinc reductant. Amatore et al. examined the mechanism of a closely related cross-coupling between phenylacetyl chloride (BnCOCl) and benzyl bromide (BnBr) that used electrochemical reduction rather than stoichiometric zinc [47]. These mechanistic studies have been highly influential to the field of cross-electrophile coupling and provide much of the early evidence for the intermediacy of a double oxidative addition mechanism (Figure 9.4b).

Improved selectivity obtained with bipyridine ligands prompted researchers to revisit this coupling and our own group found that limitations to the acid coupling partner could be overcome by using dtbbpy as a ligand [44, 125]. These studies were followed by a series of publications from the Gong group that found that alkyl iodides and bromides could be coupled with aryl acid chlorides to form a variety of unsymmetrical alkyl aryl ketones [126]. Competition studies between alkyl zinc and alkyl iodide suggested that the reaction did not occur through an alkyl zinc. This reaction was also translated to gram-scale synthesis and activated substrates, such as benzyl bromide [127]. Gong and coworkers utilized a similar system for the methylation of acid chlorides with methyl *p*-tosylate [128]. The Kishi group observed that electrophiles containing beta heteroatoms often underwent undesired elimination or epimerization reactions [129] and were able to overcome this limitation in a stereoretentive formation of unsymmetrical ketones using stoichiometric Cp_2ZrCl_2 (Figure 9.24).

One limitation to acyl chlorides is that they require an additional synthetic step from the substrate pool of carboxylic acids. The Gong group addressed this using di-*tert*-butyl dicarbonate (Boc_2O) for the *in situ* functionalization of carboxylic acids to anhydrides [130]. The resulting anhydride was directly coupled with alkyl iodides and bromides in a one-pot cross-electrophile coupling [130]. This method was improved for the coupling of primary, secondary, and tertiary alkyl bromides [131]. This *in situ* activation of carboxylic acids was extended from alkyl to aryl carboxylic acids [132], and to chloroformate electrophiles [133] (Figure 9.25).

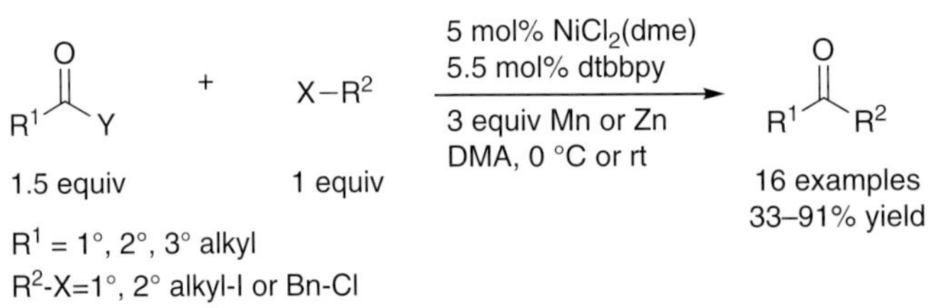

Figure 9.24 Ni-catalyzed reductive coupling of carboxylic acid derivatives with alkyl electrophiles.

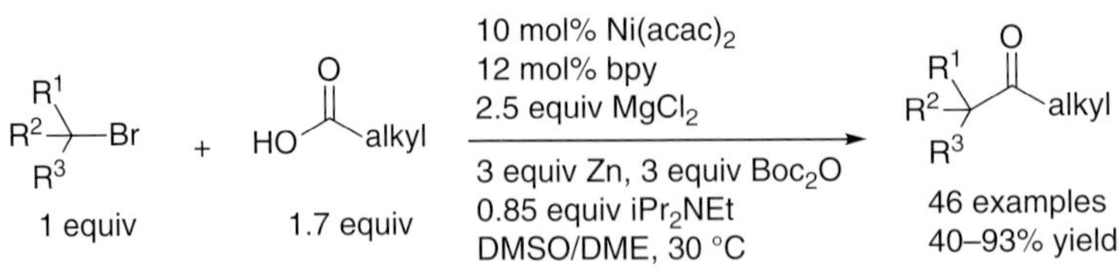

Figure 9.25 Ni-catalyzed reductive coupling of mixed anhydrides with tertiary alkyl bromides.

The Reisman group was able to demonstrate the first enantioselective method for cross-electrophile coupling with carboxylic acid derivatives. They reported the synthesis of enantioenriched acyclic α,α'-disubstituted ketones from the coupling of acid chlorides and racemic secondary benzyl chlorides using a Ni(II)/bis(oxazoline) catalyst system with Mn as a reductant [66].

9.4 C(sp^2)–C(sp^2) Coupling

9.4.1 Aryl-X/Vinyl-X + Aryl-X/Vinyl-X

In comparison with cross-electrophile coupling to form C(sp)^2–C(sp)^3 bonds, the corresponding C(sp)^2–C(sp)^2 bond formation is more challenging due to the similar reactivity of two C(sp^2)-electrophiles. Differences in bond homolysis energy are less pronounced for carbon centers with the same hybridization, which restricts one of the three chemical origins of selectivity in cross-electrophile coupling (Figure 9.3). In many cases, differences in the sterics and electronics of the C(sp^2)-electrophiles are sufficient for differentiation by a nickel catalyst, but the use of an excess of one coupling partner to bias reactions toward cross-selectivity is prevalent with these substrates.

Nickel-catalyzed cross-electrophile coupling for the formation of C(sp)^2—C(sp)^2 bonds has built upon electrochemical studies with activated alkyl electrophiles that identified nickel–bipyridine complexes as uniquely selective catalysts [25]. Durandetti et al. demonstrated that this catalyst system could use a zinc anode for electrochemical reduction to couple aryl halides with 2-chloro- and 2-bromopyridine [59]. Further studies by Gosmini et al. found that replacing the zinc anode with a milder iron electrode enabled couplings with more challenging heterocycles such as 2-chloropyrimidines or 2-chloropyrazines [134, 135]. Gosmini et al. also acknowledged that electrochemical methods could be operationally challenging and difficult to translate to a larger scale. To improve synthetic utility, a version of the cross-electrophile coupling was developed that used stoichiometric Mn as a reductant [136], and a recent study used Zn to extend the scope to 2-(*p*-toluenesulfonyl)-pyrimidines [65]. Other researchers have also built upon these landmark studies by introducing electrochemical methods for coupling amine heterocycles [137–140]. More electronically similar heteroarenes are correspondingly more difficult to couple selectively and Duan and coworkers developed the heterocoupling of 2-bromopyridines using slight variations in electronics and a 2.5 : 1 ratio of the electron-rich and electron-poor arene coupling partner [141]. The Gong group employed a similar strategy to generate hetero-biphenyl products [142] and couple heteroaryl iodides with styrenyl bromides [60] (Figure 9.26).

Our own group published a fundamentally different approach to selective hetero-biphenyl formation in 2015 [45]. We hypothesized that arene electrophiles could be differentiated by their electrophilic functional groups (i.e. halide leaving groups vs. oxygen leaving groups) rather than by their electronic properties. This led to a method for the selective cross-electrophile coupling of aryl bromides with aryl triflates using a Pd and Ni multimetallic catalyst

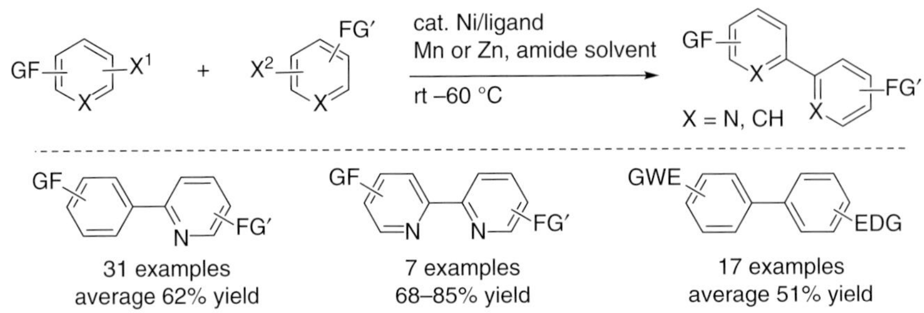

Figure 9.26 Ni-catalyzed reductive cross-coupling between two aryl electrophiles.

system in which palladium selectively inserts into the aryl triflate substrate while nickel prefers to undergo oxidative addition into the aryl bromide. The proposed multimetallic mechanism allowed for selective biarene formation with electronically matched or mismatched arenes using a 1 : 1 ratio of starting materials. An expansion of this multimetallic strategy expanded the reaction scope to include the cross-coupling vinyl bromides with vinyl triflates to form complex 1,3-diene motifs [143]. Mechanistic studies strongly suggested that a vinyl-zinc intermediate was involved in catalysis and could have an important role in mediating the transmetalation of the vinyl reagent between the Ni and Pd catalysts. A similar strategy of differentiating arenes by their leaving groups was employed by Shi et al. in the coupling of 2-napthol ethers and bromoarenes, although the reaction was proposed to occur through the *in situ* formation of an organometallic reagent (Figure 9.27).

9.4.2 Aryl-X + Acyl-X

Compared to the cross-electrophile coupling of acyl halides with alkyl electrophiles, acylation with aryl halides is far less explored. Still, aryl ketones are

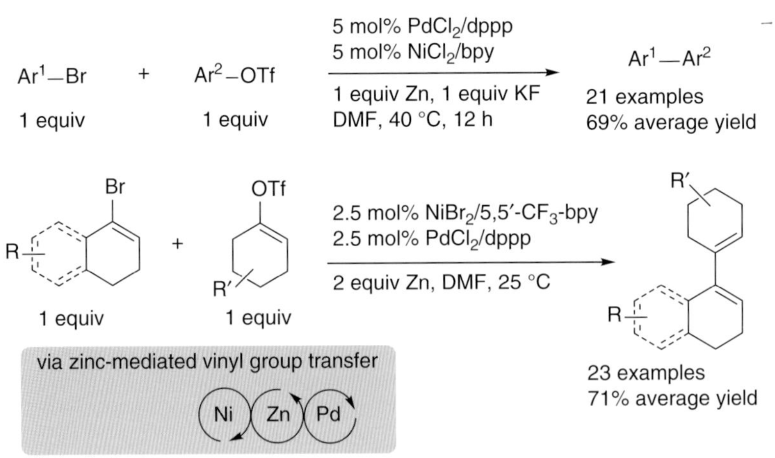

Figure 9.27 Pd/Ni-catalyzed reductive cross-coupling of Csp^2-Br substrates with Csp^2-OTf substrates.

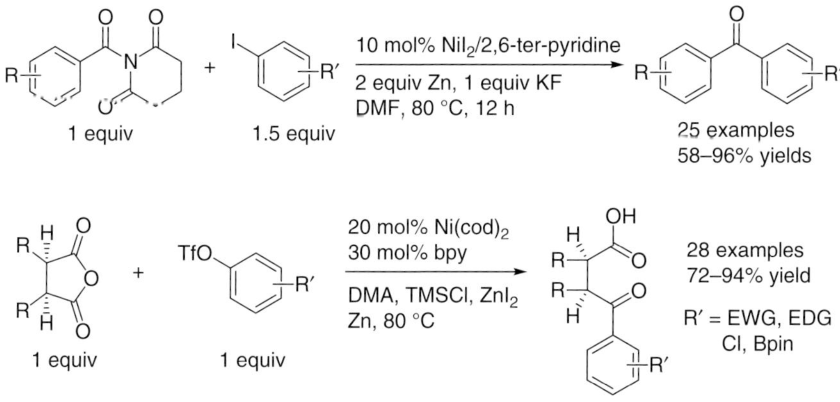

Figure 9.28 Ni-catalyzed reductive coupling of aryl electrophiles with carboxylic acid derivatives.

valuable synthetic targets and some efforts have shown that they can be synthesized via cross-electrophile coupling. *N*-Acyl-glutarimides have emerged as stable acyl reagents in catalytic cross-coupling reactions [144], which prompted Han and coworkers to use aryl versions of these acyl equivalents as coupling partners with aryl iodides [145]. Since then, an impressive nickel-catalyzed desymmetrization reaction was shown to occur between aryl triflates and meso-anhydrides [43]. Aryl triflates were used in place of the more commonly employed aryl halides in order to minimize the formation of the aryl dimer. Effective catalysis required both a catalytic amount of TMS-Cl and a stoichiometric ZnI_2 additive, presumably to assist in activating the meso-anhydride. Meso-anhydrides were previously unexplored as coupling partners in cross-electrophile coupling and their clever use allowed Walsh and coworkers to synthesize a diverse array of valuable γ-keto acid building blocks with excellent functional group tolerance and stereochemical fidelity (Figure 9.28).

9.5 C(sp^3)–C(sp^3) Coupling

The formation of C(sp^3)—C(sp^3) bonds is extremely important in organic synthesis and is regarded as the most difficult bond disconnection for conventional cross-coupling [8]. This leaves a notable gap in the literature that can be filled by cross-electrophile coupling. However, C(sp^3)—C(sp^3) bond formation via cross-electrophile coupling faces similar selectivity challenges to C(sp^2)—C(sp^2) bonds, as the identical hybridization of the carbon centers removes one chemical origin of substrate differentiation (Figure 9.3). This has made cross-selectivity particularly challenging in the formation of alkyl–alkyl bonds.

The Gong group pioneered this sub-field and reported the cross-electrophile coupling of allylic carbonates with primary and secondary alkyl bromide in 2012 [146]. Screening studies found that additives were important for suppressing undesired dimerizations and that either CuI or $MgCl_2$ improved selectivity. Likewise, a tridentate amine ligand proved vital to obtaining high selectivity.

Our group discovered a similar reaction between cinnamyl acetate derivatives and alkyl bromides [112], while continued efforts by Gong and coworkers later expanded the reaction scope to include tertiary alkyl bromide coupling partners [147] (Figure 9.29). The Shu group also contributed to methods for coupling C–O electrophiles by coupling of benzyl oxalates with alkyl bromides [148].

Nickel-allyl intermediates can also be generated through the activation of an α,β-unsaturated ketone with silyl-chloride in the presence of a low-valent nickel complex [115]. This was employed in cross-electrophile coupling to form γ-substituted enol silanes from the combination of an enone, silyl chloride, and an alkyl bromide [149]. In contrast to the related coupling with aryl electrophiles (*vide supra*) [114], the utilization of bidentate amine ligands resulted in significant dimerization of the silyl enol ether, whereas the use of a tridentate amine ligand allowed for selective cross-coupling. The initial method was only successful with secondary and tertiary alkyl bromides, but later improvements, specifically changing the solvent from DMF to THF and using 6,6′-dibromo-terpyridine as a ligand, allowed for the selective coupling of primary alkyl bromides with minimal dimerization of the silyl enol ether [150] (Figure 9.30).

The preceding examples of alkyl–alkyl bond formation through cross-electrophile coupling derive selectivity from the pairing of an activated allylic electrophile with an alkyl halide. The next logical progression is the coupling of two unactivated alkyl halides, but this is exceptionally challenging due to rapid homocoupling [151, 152]. Initial efforts to overcome this selectivity challenge by the Gong group relied on using a stoichiometric excess of one coupling partner to bias product formation [49]. Selectivity was dependent on the statistical distribution of crossed and dimeric products, although sterics and the identity of the halides did have some influence. Chemical selectivity was improved by using a bulky B_2pin_2 reductant to enhance the sensitivity of the nickel catalyst toward the steric contributions of the alkyl substrates (Figure 9.31). The cross-electrophile coupling of primary alkyl bromides with secondary alkyl halides was reported using only a 1.5 equiv excess of the primary

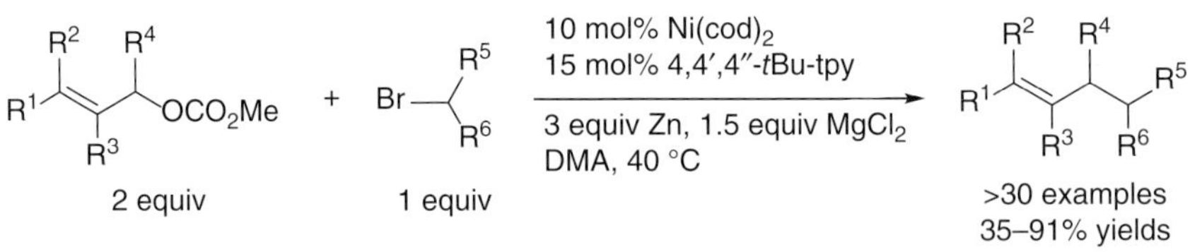

Figure 9.29 Ni-catalyzed reductive coupling of allylic alcohol derivatives with alkyl bromides.

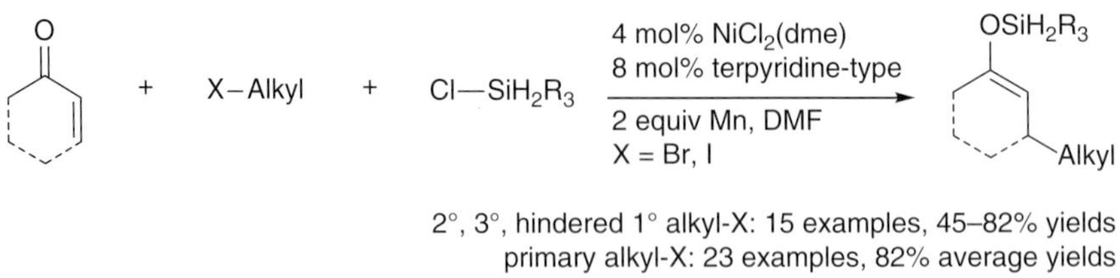

Figure 9.30 Ni-catalyzed reductive coupling of alkyl halides with enones and silyl chlorides.

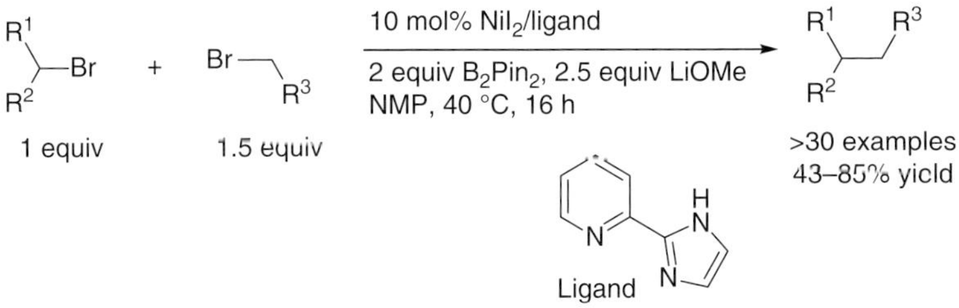

Figure 9.31 Ni-catalyzed reductive coupling between primary and secondary alkyl halides.

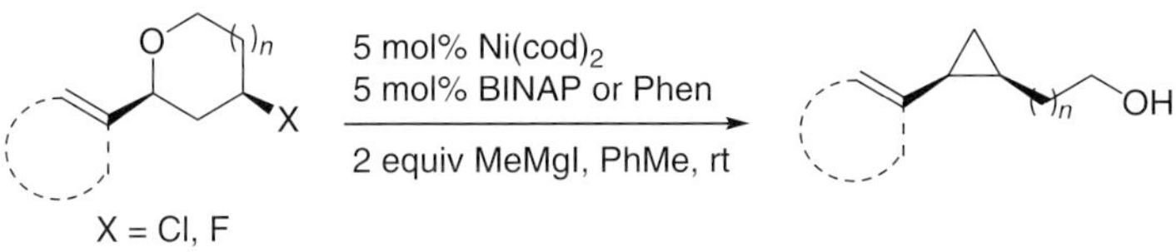

Figure 9.32 Stereospecific Ni-catalyzed intramolecular reductive coupling to form cyclopropanes.

substrate in modest yields [46]. The same catalytic system was also used for the methylation of unactivated alkyl halides [128].

An alternative strategy for obtaining selectivity is to favor the desired bond formation entropically via an intramolecular cyclization. This was demonstrated by Gong and coworkers in the cyclization of dihaloalkanes to form five- to seven-membered rings [153]. A creative application of intramolecular cross-electrophile coupling was demonstrated by the Jarvo group in a stereospecific intramolecular ring contraction to form cyclopropanes [154, 155]. It had previously been shown that nickel catalysts could oxidatively insert into benzylic and allylic ether linkages and they reasoned that the resulting organonickel intermediate could participate in cross-electrophile coupling with a second intramolecular electrophile. This insight resulted in the publication of the stereospecific intramolecular cross-electrophile coupling of 2-aryl-4-chlorotetrahydropyrans to afford disubstituted cyclopropane rings [154]. Further studies found that benzylic ethers could be replaced with styrenyl ethers and that the secondary alkyl chloride could be substituted with an array of other leaving groups including fluoride and tosylate [155] (Figure 9.32).

9.6 C(sp)–C(sp^3) Coupling

The success of the Sonogashira coupling as a method for directly converting terminal alkynes to carbon nucleophiles *in situ* has led to cross-electrophile coupling being largely overlooked in the literature as a method for forming C(sp)—C(sp^3) bonds [156]. However, the functional group compatibility of cross-electrophile coupling can still provide advantages over known methods. Our own group reported a decarboxylative cross-electrophile coupling between alkyl NHP esters and bromoalkynes, which builds upon prior work with redox

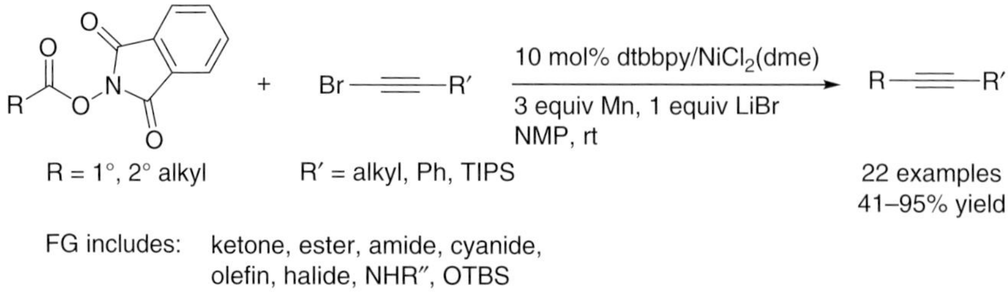

Figure 9.33 Ni-catalyzed decarboxylative reductive coupling of alkyl N-hydroxyphthalimide esters with bromoalkynes to form internal alkynes.

active esters [93] that access the large pool of carboxylic acid substrates as alkyl coupling partners [157]. Internal alkynes were obtained in moderate to excellent yields in the presence of various functional groups including carbamates, halogens, nitriles, olefins, ketones, and esters (Figure 9.33).

9.7 Multicomponent Reactions

Tandem reactions have emerged as a popular strategy for efficiently forming multiple bonds in a single synthetic step and have been reviewed extensively in methodology [158, 159] and synthesis [160, 161]. Cross-electrophile coupling reactions are particularly amenable to the development of tandem processes because many reaction mechanisms involve radical intermediates that can be intercepted by π-systems, such as alkenes. Interception of the radical converts the π-bonds of the alkene to additional σ-bonds before the radical is recaptured by nickel prior to bond formation with the electrophilic coupling partner. This constrains cross-electrophile coupling tandem processes to couplings where at least one reaction partner is a C(sp^3)-electrophile that can generate an alkyl radical intermediate. In a related development, sequential insertion/elimination with alkene intermediates (chain-walking [162]) has been a fruitful research avenue for coupling reactions with alkyl halides where the bond formed is remote from the C—X bond in the starting material [163, 164].

The majority of tandem cross-electrophile coupling reactions use intramolecularity to offset the entropic penalty of three-component coupling and avoid intermolecular dimerization, and fully intramolecular cross-electrophile couplings have been reported by Peng et al. to form *trans*-tetrahydronaphtho[2,3-*b*]furans [165] and tetrahydro-2*H*-indeno[1,2-*b*]furans [166]. However, substrates can also be designed with partial intramolecularity by only linking the alkyl radical source to the alkene. Connecting the alkyl electrophile to the alkene ensures that the radical will be rapidly trapped before competing homocoupling can occur. Peng and coworkers contributed to this class of tandem reactions by diastereoselectively forming linear fused perhydrofuro[2,3-*b*]furan(pyran) moieties [80] along with [5,5]- and [5,6]-spiroketals [167]. Diao and coworkers demonstrated that aryl bromides could be coupled in the place of aryl iodides [168]. Tandem cross-electrophile coupling was expanded to fully intermolecular variants by the Nevado group in a three-component reductive dicarbofunctionalization of alkenes with aryl iodides and tertiary alkyl iodides [169]. The sterically hindered tertiary radical, generated from tertiary alkyl iodide, does not couple efficiently with the nickel catalyst employed; [85] however, reacting with the alkene generates a secondary radical that can be productively captured by arylnickel(II). Subsequent reductive elimination provides the product (Figure 9.34).

Ketones are ubiquitous in organic synthesis and reductive cross-carbonylation would be an attractive strategy for accessing ketones through a convergent, tandem process. This proves to be challenging because the concentration of CO is difficult to control in a reductive heterogeneous system [170]; deficiency of CO results in direct R^1–R^2 coupling, whereas excess of CO poisons the catalyst. This difficulty can be addressed using a weighable metal carbonyl complex, such as $Fe(CO)_5$ or $Co_2(CO)_8$, and the Troupel group employed these reagents in the first reductive cross-carbonylation [171]. Our own group conducted a mechanistic investigation of such tandem reductive carbonylation and found that the reaction proceeds through initial formation of an acyl nickel(II) complex [172]. This discovery was translated to the catalytic cross-carbonylation of bromoarenes and primary bromo alkanes (Figure 9.35).

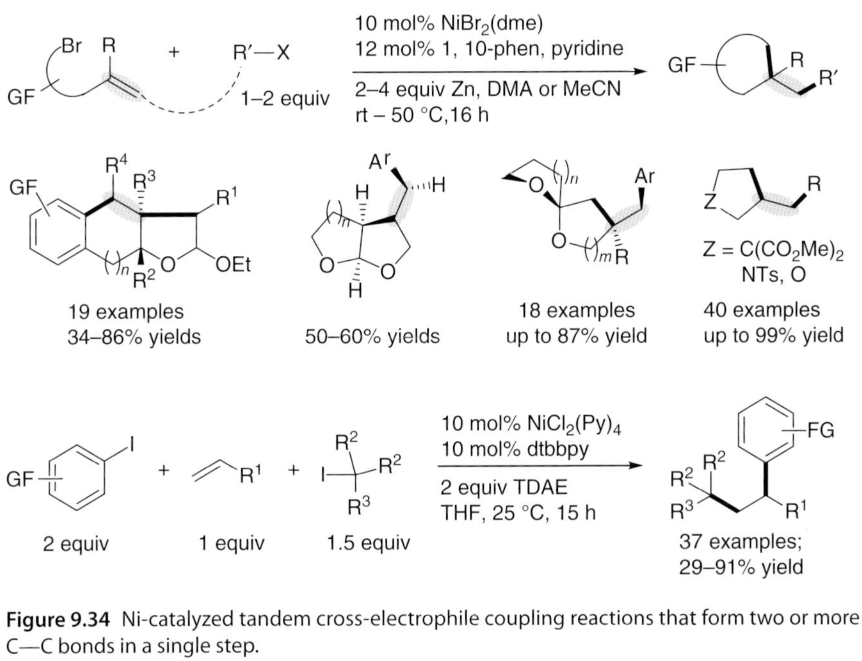

Figure 9.34 Ni-catalyzed tandem cross-electrophile coupling reactions that form two or more C—C bonds in a single step.

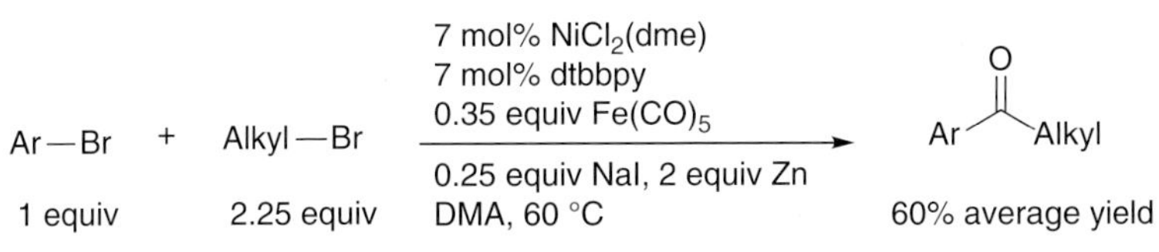

Figure 9.35 Ni-catalyzed carbonylative cross-electrophile coupling to form ketones.

9.8 Future of the Field

Cross-electrophile coupling uses abundant, stable, and commercially available carbon electrophiles and avoids the organometallic reagents necessary in conventional cross-coupling. This skips the extra synthetic steps necessary for the formation of organometallic reagents, provides increased functional group tolerance, and improves the stability of the process toward moisture and air. The principal difficulty for cross-electrophile coupling is overcoming the selectivity challenges inherent to the slight differences in reactivity between electrophilic starting materials. This hurdle slowed the development of cross-electrophile coupling, but a renaissance in the field occurred with the discovery that selectivity can be tuned through judicious ligand choice. Mechanistic work in this field has empowered the development of many cross-electrophile coupling methods and continues to expand our understanding. The improvements in this field have not gone unnoticed by synthetic and industrial chemists, and these methods are already being applied to laboratory and industrial syntheses. This progress has encouraged further work in the field and we anticipate significant improvements to the following aspects in the future:

(1) Now that the synthetic value of cross-electrophile coupling has been established, we anticipate that many groups will look to developing a better mechanistic understanding in order to improve known methods. General mechanisms are established in the literature and new methods often cite a proposed mechanism; however, detailed studies that confidently support these claims are rare. This is particularly important in cross-electrophile coupling, as mechanisms are the key to explaining the origins of cross-selectivity. Work in the isolation and characterization of catalytically relevant Ni(III) complexes is ongoing [173–175], and further studies on the reactivity of organonickel compounds toward different electrophiles would significantly benefit this field.

(2) The strength of cross-electrophile coupling is its exceptional functional group tolerance – reactions do not require harsh bases and generally occur at temperatures under 60 °C. Despite this, there are still substrate classes that prove challenging; coupling reactions with heterocycles are underexplored and access to nitrogen electrophiles (i.e. nitro groups, amines, cyanides) is still limited. New electrophiles, such as *N*-hydroxyphthalimide esters, have improved the synthetic diversity of cross-electrophile coupling reactions and we anticipate that future developments will access underexplored classes of electrophiles.

(3) The need for a stoichiometric reductant is a significant limitation to cross-electrophile coupling. Early studies in electrochemical reduction benefited from their atom economy, but were operationally obtuse and acted as a barrier to entry for many synthetic applications. On the other hand, metal reductants benefit from their abundance and stability, but result in heterogeneous reactions that complicate scale-up. Researchers have begun to explore other methods for regenerating low-valent nickel intermediates including (i) organic reductants, which are homogeneous but rarely serve as a direct

replacement for metal reductants; (ii) photoredox catalysts, which allow for the use of homogenous amines as electron donors; and (iii) electrochemical methods, which are being revived to provide application-specific methods. We anticipate that new reducing strategies will continue to improve the efficiency and scalability of cross-electrophile coupling reactions.

(4) The frontier of reaction development is stereoselective bond formation. Thus far, the Riesman group has been a major contributor to enantioselective cross-electrophile coupling, but the importance of stereoselective synthesis ensures that other researchers will explore this problem. The challenge of enantioselectivity in cross-electrophile coupling is magnified by the competing cross-selectivity issue, which has been primarily addressed with pyridine-based ligands – unfortunately, chiral versions of pyridine ligands are underdeveloped. We anticipate that this will remain a productive research field in the foreseeable future as new ligands and enantioselective methods are developed.

(5) Cyclic molecules are common in drugs and bioactive compounds, but intramolecular and tandem cross-electrophile coupling methods have only just begun to explore the limits of reductive coupling. Ring formation through cross-electrophile coupling is uncommon and has been limited to the formation of five- to six-membered rings. Further advances in reductive cyclization would be attractive to the pharmaceutical industry and to laboratory-scale preparation of macrocycles.

(6) Lastly, the discovery of new types of ligands and air-stable nickel pre-catalysts would improve the selectivity, catalyst loading, and operational simplicity of cross-electrophile coupling. New ligand development has been disincentivized by the success of bipyridine ligands, but further reactivity challenges will necessitate research into new ligand classes and drive our understanding of the connection between ligand and catalyst activity/selectivity. The development of stable precatalysts and preformed metal–ligand complexes would also benefit the field by increasing the utility and reliability of cross-electrophile coupling reactions. A major benefit of cross-electrophile coupling is that reductive conditions allow for the use of air-stable Ni(II) salts in place of unstable low-valent Ni catalysts. Conversely, nearly all cross-electrophile coupling catalysts are formed *in situ* and the use of preformed complexes is rare outside of mechanistic studies. We anticipate that inorganic syntheses of isolable metal complexes will become increasingly important as researchers tackle challenging substrates and probe reaction mechanisms.

References

1 de Meijere, A., Brase, S., and Oestreich, M. (2014). *Metal Catalyzed Cross-Coupling Reactions and More*. Weinheim: Wiley-VCH.

2 Nicolaou, K.C., Bulger, P.G., and Sarlah, D. (2005). Palladium-catalyzed cross-coupling reactions in total synthesis. *Angew. Chem. Int. Ed.* 44 (29): 4442–4489. https://doi.org/10.1002/anie.200500368.

3 Biajoli, A.F.P., Schwalm, C.S., Limberger, J. et al. (2014). Recent progress in the use of Pd-catalyzed C—C cross-coupling reactions in the synthesis of pharmaceutical compounds. *J. Braz. Chem. Soc.* https://doi.org/10.5935/0103-5053.20140255.

4 Xue, L. and Lin, Z. (2010). Theoretical aspects of palladium -catalysed carbon–carbon cross-coupling reactions. *Chem. Soc. Rev.* 39 (5): 1692–1705. https://doi.org/10.1039/B814973A.

5 Carey, J.S., Laffan, D., Thomson, C., and Williams, M.T. (2006). Analysis of the reactions used for the preparation of drug candidate molecules. *Org. Biomol. Chem.* 4 (12): 2337–2347. https://doi.org/10.1039/B602413K.

6 Roughley, S.D. and Jordan, A.M. (2011). The medicinal chemist's toolbox: an analysis of reactions used in the pursuit of drug candidates. *J. Med. Chem.* 54 (10): 3451–3479. https://doi.org/10.1021/jm200187y.

7 Magano, J. and Dunetz, J.R. (2011). Large-scale applications of transition metal-catalyzed couplings for the synthesis of pharmaceuticals. *Chem. Rev.* 111 (3): 2177–2250. https://doi.org/10.1021/cr100346g.

8 Choi, J. and Fu, G.C. (2017). Transition metal–catalyzed alkyl-alkyl bond formation: another dimension in cross-coupling chemistry. *Science* 356 (6334): eaaf7230. https://doi.org/10.1126/science.aaf7230.

9 Cherney, A.H., Kadunce, N.T., and Reisman, S.E. (2015). Enantioselective and enantiospecific transition-metal-catalyzed cross-coupling reactions of organometallic reagents to construct C—C bonds. *Chem. Rev.* 115 (17): 9587–9652. https://doi.org/10.1021/acs.chemrev.5b00162.

10 Everson, D.A. and Weix, D.J. (2014). Cross-electrophile coupling: principles of reactivity and selectivity. *J. Org. Chem.* 79 (11): 4793–4798. https://doi.org/10.1021/jo500507s.

11 Knappke, C.E.I., Grupe, S., Gärtner, D. et al. (2014). Reductive cross-coupling reactions between two electrophiles. *Chem. Eur. J.* 20 (23): 6828–6842. https://doi.org/10.1002/chem.201402302.

12 Gu, J., Wang, X., Xue, W., and Gong, H. (2015). Nickel-catalyzed reductive coupling of alkyl halides with other electrophiles: concept and mechanistic considerations. *Org. Chem. Front.* 2 (10): 1411–1421. https://doi.org/10.1039/C5QO00224A.

13 Wang, X., Dai, Y., and Gong, H. (2016). Nickel-catalyzed reductive couplings. *Top. Curr. Chem.* 374 (4): 43. https://doi.org/10.1007/s41061-016-0042-2.

14 Lucas, E.L. and Jarvo, E.R. (2017). Stereospecific and stereoconvergent cross-couplings between alkyl electrophiles. *Nat. Rev. Chem.* 1 (9): 0065. https://doi.org/10.1038/s41570-017-0065.

15 Wurtz, A. (1855). Ueber eine neue Klasse organischer Radicale. *Ann. Chem. Pharm.* 96: 364–375.

16 Tollens, B. and Fittig, R. (1864). Ueber Die Synthese Der Kohlenwasserstoffe Der Benzolreihe. *Justus Liebigs Ann. Chem.* 131 (3): 303–323. https://doi.org/10.1002/jlac.18641310307.

17 Fanta, P.E. (1946). The ullmann synthesis of biaryls. *Chem. Rev.* 38 (1): 139–196. https://doi.org/10.1021/cr60119a004.

18 Fanta, P.E. (1964). The ullmann synthesis of biaryls, 1945–1963. *Chem. Rev.* 64 (6): 613–632. https://doi.org/10.1021/cr60232a002.

19 Fanta, P.E. (1974). The ullmann synthesis of biaryls. *Synthesis* 1974 (01): 9–21. https://doi.org/10.1055/s-1974-23219.

20 Hassan, J., Sévignon, M., Gozzi, C. et al. (2002). Aryl–aryl bond formation one century after the discovery of the ullmann reaction. *Chem. Rev.* 102 (5): 1359–1470. https://doi.org/10.1021/cr000664r.

21 Kende, A.S., Liebeskind, L.S., and Braitsch, D.M. (1975). In situ generation of a solvated zerovalent nickel reagent. Biaryl formation. *Tetrahedron Lett.* 16 (39): 3375–3378. https://doi.org/10.1016/S0040-4039(00)91402-3.

22 Zembayashi, M., Tamao, K., Yoshida, J., and Kumada, M. (1977). Nickel-phosphine complex-catalyzed homo coupling of aryl halides in the presence of zinc powder. *Tetrahedron Lett.* 18 (47): 4089–4091. https://doi.org/10.1016/S0040-4039(01)83434-1.

23 Colon, I. and Kelsey, D.R. (1986). Coupling of aryl chlorides by nickel and reducing metals. *J. Org. Chem.* 51 (14): 2627–2637. https://doi.org/10.1021/jo00364a002.

24 Chaussard, J., Folest, J.-C., Nedelec, J.-Y. et al. (1990). Use of sacrificial anodes in electrochemical functionalization of organic halides. *Synthesis* 1990 (05): 369–381. https://doi.org/10.1055/s-1990-26880.

25 Durandetti, M., Nédélec, J.-Y., and Périchon, J. (1996). Nickel-catalyzed direct electrochemical cross-coupling between aryl halides and activated alkyl halides. *J. Org. Chem.* 61 (5): 1748–1755. https://doi.org/10.1021/jo9518314.

26 Matsui, J.K., Lang, S.B., Heitz, D.R., and Molander, G.A. (2017). Photoredox-mediated routes to radicals: the value of catalytic radical generation in synthetic methods development. *ACS Catal.* 7 (4): 2563–2575. https://doi.org/10.1021/acscatal.7b00094.

27 Fegheh-Hassanpour, Y., Arif, T., Sintim, H.O. et al. (2017). Synthesis of (−)-6,7-dideoxysqualestatin H5 by carbonyl ylide cycloaddition–rearrangement and cross-electrophile coupling. *Org. Lett.* 19: 3540–3543. https://doi.org/10.1021/acs.orglett.7b01513.

28 Liu, J.-H., Yang, C.-T., Lu, X.-Y. et al. (2014). Copper-catalyzed reductive cross-coupling of nonactivated alkyl tosylates and mesylates with alkyl and aryl bromides. *Chem. Eur. J.* 20 (47): 15334–15338. https://doi.org/10.1002/chem.201405223.

29 Czaplik, W.M., Mayer, M., and Jacobi von Wangelin, A. (2008). Domino iron catalysis: direct aryl–alkyl cross-coupling. *Angew. Chem. Int. Ed.* 48 (3): 607–610. https://doi.org/10.1002/anie.200804434.

30 Czaplik, W.M., Mayer, M., and Jacobi von Wangelin, A. (2011). Iron-catalyzed reductive aryl–alkenyl cross-coupling reactions. *ChemCatChem* 3 (1): 135–138. https://doi.org/10.1002/cctc.201000276.

31 Hassan, J., Hathroubi, C., Gozzi, C., and Lemaire, M. (2001). Preparation of unsymmetrical biaryls via palladium-catalyzed coupling reaction of aryl halides. *Tetrahedron* 57 (37): 7845–7855. https://doi.org/10.1016/S0040-4020(01)00752-9.

32 Wang, L., Zhang, Y., Liu, L., and Wang, Y. (2006). Palladium-catalyzed homocoupling and cross-coupling reactions of aryl halides in poly(ethylene glycol). *J. Org. Chem.* 71 (3): 1284–1287. https://doi.org/10.1021/jo052300a.

33 Krasovskiy, A., Duplais, C., and Lipshutz, B.H. (2009). Zn-mediated, Pd-catalyzed cross-couplings in water at room temperature without prior formation of organozinc reagents. *J. Am. Chem. Soc.* 131 (43): 15592–15593. https://doi.org/10.1021/ja906803t.
34 Gomes, P., Fillon, H., Gosmini, C. et al. (2002). Synthesis of unsymmetrical biaryls by electroreductive cobalt-catalyzed cross-coupling of aryl halides. *Tetrahedron* 58 (42): 8417–8424. https://doi.org/10.1016/S0040-4020(02)01030-X.
35 Gomes, P., Gosmini, C., and Périchon, J. (2003). Cobalt-catalyzed electrochemical vinylation of aryl halides using vinylic acetates. *Tetrahedron* 59 (17): 2999–3002. https://doi.org/10.1016/S0040-4020(03)00404-6.
36 Amatore, M., Gosmini, C., and Périchon, J. (2005). Cobalt-catalyzed vinylation of functionalized aryl halides with vinyl acetates. *Eur. J. Org. Chem.* 2005 (6): 989–992. https://doi.org/10.1002/ejoc.200400897.
37 Amatore, M. and Gosmini, C. (2008). Efficient cobalt-catalyzed formation of unsymmetrical biaryl compounds and its application in the synthesis of a sartan intermediate. *Angew. Chem. Int. Ed.* 47 (11): 2089–2092. https://doi.org/10.1002/anie.200704402.
38 Moncomble, A., Floch, P.L., Lledos, A., and Gosmini, C. (2012). Cobalt-catalyzed vinylation of aromatic halides using β-halostyrene: experimental and DFT studies. *J. Org. Chem.* 77 (11): 5056–5062. https://doi.org/10.1021/jo3005149.
39 Qian, X., Auffrant, A., Felouat, A., and Gosmini, C. (2011). Cobalt-catalyzed reductive allylation of alkyl halides with allylic acetates or carbonates. *Angew. Chem. Int. Ed.* 50 (44): 10402–10405. https://doi.org/10.1002/anie.201104390.
40 Hartwig, J. (2010). *Organotransition Metal Chemistry: From Bonding to Catalysis*. University Science Books.
41 Everson, D.A., Shrestha, R., and Weix, D.J. (2010). Nickel-catalyzed reductive cross-coupling of aryl halides with alkyl halides. *J. Am. Chem. Soc.* 132: 920–921. https://doi.org/10.1021/ja9093956.
42 Everson, D.A., Jones, B.A., and Weix, D.J. (2012). Replacing conventional carbon nucleophiles with electrophiles: nickel-catalyzed reductive alkylation of aryl bromides and chlorides. *J. Am. Chem. Soc.* 134: 6146–6159. https://doi.org/10.1021/ja301769r.
43 Lin, T., Mi, J., Song, L. et al. (2018). Nickel-catalyzed desymmetrizing cross-electrophile coupling of cyclic meso-anhydrides. *Org. Lett.* 20 (4): 1191–1194. https://doi.org/10.1021/acs.orglett.8b00114.
44 Wotal, A.C. and Weix, D.J. (2012). Synthesis of functionalized dialkyl ketones from carboxylic acid derivatives and alkyl halides. *Org. Lett.* 14: 1476–1479. https://doi.org/10.1021/ol300217x.
45 Ackerman, L.K.G., Lovell, M.M., and Weix, D.J. (2015). Multimetallic catalysis enabled cross-coupling of aryl bromides with aryl triflates. *Nature* 524: 454–457. https://doi.org/10.1038/nature14676.
46 Xu, H., Zhao, C., Qian, Q. et al. (2013). Nickel-catalyzed cross-coupling of unactivated alkyl halides using bis(pinacolato)diboron as reductant. *Chem. Sci.* 4: 4022–4029. https://doi.org/10.1039/C3SC51098K.

47 Amatore, C., Jutand, A., Périchon, J., and Rollin, Y. (2000). Mechanism of the nickel-catalyzed electrosynthesis of ketones by heterocoupling of acyl and benzyl halides. *Monatsh. Chem.* 131: 1293–1304. https://doi.org/10.1007/s007060070008.

48 Zhao, Y. and Weix, D.J. (2014). Nickel-catalyzed regiodivergent opening of epoxides with aryl halides: co-catalysis controls regioselectivity. *J. Am. Chem. Soc.* 136: 48–51. https://doi.org/10.1021/ja410704d.

49 Yu, X., Yang, T., Wang, S. et al. (2011). Nickel-catalyzed reductive cross-coupling of unactivated alkyl halides. *Org. Lett.* 13 (8): 2138–2141. https://doi.org/10.1021/ol200617f.

50 Biswas, S. and Weix, D.J. (2013). Mechanism and selectivity in nickel-catalyzed cross-electrophile coupling of aryl halides with alkyl halides. *J. Am. Chem. Soc.* 135: 16192–16197. https://doi.org/10.1021/ja407589e.

51 Zhang, J., Lu, G., Xu, J. et al. (2016). Nickel-catalyzed reductive cross-coupling of benzyl chlorides with aryl chlorides/fluorides: a one-pot synthesis of diarylmethanes. *Org. Lett.* 18: 2860–2863. https://doi.org/10.1021/acs.orglett.6b01134.

52 Durandetti, M., Gosmini, C., and Périchon, J. (2007). Ni-catalyzed activation of α-chloroesters: a simple method for the synthesis of α-arylesters and β-hydroxyesters. *Tetrahedron* 63: 1146–1153. https://doi.org/10.1016/j.tet.2006.11.055.

53 Walters, I.A.S. (2006). A convenient one-pot Negishi coupling of amino-heteroaryl chlorides and alkyl bromides. *Tetrahedron Lett.* 47 (3): 341–344. https://doi.org/10.1016/j.tetlet.2005.11.013.

54 Amatore, M. and Gosmini, C. (2008). Synthesis of functionalised diarylmethanes via a cobalt-catalysed cross-coupling of arylzinc species with benzyl chlorides. *Chem. Commun.* (40): 5019–5021. https://doi.org/10.1039/B809626K.

55 Hegedus, L.S. and Thompson, D.H.P. (1985). The reactions of organic halides with (.Pi.-Allyl)nickel halide complexes: a mechanistic study. *J. Am. Chem. Soc.* 107 (20): 5663–5669. https://doi.org/10.1021/ja00306a012.

56 Yamamoto, T., Wakabayashi, S., and Osakada, K. (1992). Mechanism of C—C coupling reactions of aromatic halides, promoted by $Ni(COD)_2$ in the presence of 2,2′-bipyridine and PPh_3, to give biaryls. *J. Organomet. Chem.* 428 (1): 223–237. https://doi.org/10.1016/0022-328X(92)83232-7.

57 Amatore, C. and Jutand, A. (1988). Rates and mechanism of biphenyl synthesis catalyzed by electrogenerated coordinatively unsaturated nickel complexes. *Organometallics* 7 (10): 2203–2214. https://doi.org/10.1021/om00100a019.

58 Yakhvarov, D.G., Budnikova, Y.G., and Sinyashin, O.G. (2003). Organonickel σ-complexes—key intermediates of electrocatalytic cycles. *Russ. J. Electrochem.* 39 (11): 1261–1270. https://doi.org/10.1023/B:RUEL.0000003456.78545.0d.

59 Gosmini, C., Lasry, S., Nedelec, J.-Y., and Perichon, J. (1998). Electrochemical cross-coupling between 2-halopyridines and aryl or heteroaryl halides catalysed by nickel-2,2′-bipyridine complexes. *Tetrahedron* 54 (7): 1289–1298. https://doi.org/10.1016/S0040-4020(97)10225-3.

60 Liu, J., Ren, Q., Zhang, X., and Gong, H. (2016). Preparation of vinyl arenes by nickel-catalyzed reductive coupling of aryl halides with vinyl bromides. *Angew. Chem. Int. Ed.* 55: 15544–15548. https://doi.org/10.1002/anie.201607959.

61 Hegedus, L.S. and Miller, L.L. (1975). Reaction of .Pi.-allylnickel bromide complexes with organic halides. Stereochemistry and mechanism. *J. Am. Chem. Soc.* 97 (2): 459–460. https://doi.org/10.1021/ja00835a061.

62 Durandetti, M., Devaud, M., and Périchon, J. (1996). Investigation of the reductive coupling of aryl halides and/or ethyl chloroacetate electrocatalyzed by the precursor NiX_2(Bpy) with X- = Cl-, Br- or $MeSO_3$- and Bpy = 2,2'-dipyridyl. *New J. Chem.* 20: 659.

63 Kochi, J.K. (1974). Electron-transfer mechanisms for organometallic intermediates in catalytic reactions. *Acc. Chem. Res.* 7 (10): 351–360. https://doi.org/10.1021/ar50082a006.

64 Schiavon, G., Bontempelli, G., and Corain, B. (1981). Coupling of organic halides electrocatalyzed by the NiII/NiI/Ni0–PPh_3 system. A mechanistic study based on an electroanalytical approach. *J. Chem. Soc., Dalton Trans.* (5): 1074–1081. https://doi.org/10.1039/DT9810001074.

65 Gong, C., Huo, C., Wang, X., and Quan, Z. (2017). Nickel-catalyzed cross-electrophile coupling of aryl bromides with pyrimidin-2-Yl tosylates. *Chin. J. Chem.* 35: 1366–1370. https://doi.org/10.1002/cjoc.201700071.

66 Cherney, A.H., Kadunce, N.T., and Reisman, S.E. (2013). Catalytic asymmetric reductive acyl cross-coupling: synthesis of enantioenriched acyclic α,α-disubstituted ketones. *J. Am. Chem. Soc.* 135: 7442–7445. https://doi.org/10.1021/ja402922w.

67 Cherney, A.H. and Reisman, S.E. (2014). Nickel-catalyzed asymmetric reductive cross-coupling between vinyl and benzyl electrophiles. *J. Am. Chem. Soc.* 136: 14365–14368. https://doi.org/10.1021/ja508067c.

68 Ackerman, L.K.G., Anka-Lufford, L.L., Naodovic, M., and Weix, D.J. (2015). Cobalt co-catalysis for cross-electrophile coupling: diarylmethanes from benzyl mesylates and aryl halides. *Chem. Sci.* 6 (2): 1115–1119. https://doi.org/10.1039/C4SC03106G.

69 Zhang, P., Le, C.C., and MacMillan, D.W.C. (2016). Silyl radical activation of alkyl halides in metallaphotoredox catalysis: a unique pathway for cross-electrophile coupling. *J. Am. Chem. Soc.* 138: 8084–8087. https://doi.org/10.1021/jacs.6b04818.

70 Hansen, E.C., Pedro, D.J., Wotal, A.C. et al. (2016). New ligands for nickel catalysis from diverse pharmaceutical heterocycle libraries. *Nat. Chem.* https://doi.org/10.1038/nchem.2587.

71 Hansen, E.C., Li, C., Yang, S. et al. (2017). Coupling of challenging heteroaryl halides with alkyl halides via nickel-catalyzed cross-electrophile coupling. *J. Org. Chem.* 82 (14): 7085–7092. https://doi.org/10.1021/acs.joc.7b01334.

72 Perkins, R.J., Pedro, D.J., and Hansen, E.C. (2017). Electrochemical nickel catalysis for Sp^2-Sp^3 cross-electrophile coupling reactions of unactivated alkyl halides. *Org. Lett.* 19: 3755–3758. https://doi.org/10.1021/acs.orglett.7b01598.

73 Li, H., Breen, C.P., Seo, H. et al. (2018). Ni-catalyzed electrochemical decarboxylative C—C couplings in batch and continuous flow. *Org. Lett.* 20 (5): 1338–1341. https://doi.org/10.1021/acs.orglett.8b00070.

74 Frisch, A.C. and Beller, M. (2005). Catalysts for cross-coupling reactions with non-activated alkyl halides. *Angew. Chem. Int. Ed.* 44 (5): 674–688. https://doi.org/10.1002/anie.200461432.

75 Kambe, N., Iwasaki, T., and Terao, J. (2011). Pd-catalyzed cross-coupling reactions of alkyl halides. *Chem. Soc. Rev.* 40 (10): 4937–4947. https://doi.org/10.1039/C1CS15129K.

76 Durandetti, M., Sibille, S., Nédélec, J.-Y., and Périchon, J. (1994). A novel method of arylation of α-chloroketones. *Synth. Commun.* 24 (2): 145–151. https://doi.org/10.1080/00397919408013812.

77 Durandetti, M. and Périchon, J. (2004). Nickel-catalyzed electrochemical coupling of aryl, heteroaryl or vinyl halides with activated alkyl chlorides: synthetic and stereochemical aspects. *Synthesis* 2004 (18): 3079–3083. https://doi.org/10.1055/s-2004-834896.

78 Durandetti, M., Périchon, J., and Nédélec, J.-Y. (1997). Asymmetric induction in the electrochemical cross-coupling of aryl halides with α-chloropropionic acid derivatives catalyzed by nickel complexes. *J. Org. Chem.* 62: 7914–7915. https://doi.org/10.1021/jo971279d.

79 Wang, S., Qian, Q., and Gong, H. (2012). Nickel-catalyzed reductive coupling of aryl halides with secondary alkyl bromides and allylic acetate. *Org. Lett.* 14: 3352–3355. https://doi.org/10.1021/ol3013342.

80 Yan, C.-S., Peng, Y., Xu, X.-B., and Wang, Y.-W. (2012). Nickel-mediated inter- and intramolecular reductive cross-coupling of unactivated alkyl bromides and aryl iodides at room temperature. *Chem. Eur. J.* 18 (19): 6039–6048. https://doi.org/10.1002/chem.201200190.

81 Tsou, T.T. and Kochi, J.K. (1979). Mechanism of biaryl synthesis with nickel complexes. *J. Am. Chem. Soc.* 101 (25): 7547–7560. https://doi.org/10.1021/ja00519a015.

82 Kochi, J.K. (1980). The role of electron transfer and charge transfer in organometallic chemistry. *Pure Appl. Chem.* 52 (3): 571–605. https://doi.org/10.1351/pac198052030571.

83 Hu, L., Liu, X., and Liao, X. (2016). Nickel-catalyzed methylation of aryl halides with deuterated methyl iodide. *Angew. Chem. Int. Ed.* 55: 9743–9747. https://doi.org/10.1002/anie.201604406.

84 Wang, J., Zhao, J., and Gong, H. (2017). Nickel-catalyzed methylation of aryl halides/tosylates with methyl tosylate. *Chem. Commun.* 53: 10180–10183. https://doi.org/10.1039/C7CC06106D.

85 Wang, X., Wang, S., Xue, W., and Gong, H. (2015). Nickel-catalyzed reductive coupling of aryl bromides with tertiary alkyl halides. *J. Am. Chem. Soc.* 137: 11562–11565. https://doi.org/10.1021/jacs.5b06255.

86 Kirk, K.L. (2008). Fluorination in medicinal chemistry: methods, strategies, and recent developments. *Org. Process Res. Dev.* 12 (2): 305–321. https://doi.org/10.1021/op700134j.

87 Hagmann, W.K. (2008). The many roles for fluorine in medicinal chemistry. *J. Med. Chem.* 51 (15): 4359–4369. https://doi.org/10.1021/jm800219f.

88 Gillis, E.P., Eastman, K.J., Hill, M.D. et al. (2015). Applications of fluorine in medicinal chemistry. *J. Med. Chem.* 58 (21): 8315–8359. https://doi.org/10.1021/acs.jmedchem.5b00258.

89 Li, X., Feng, Z., Jiang, Z.-X., and Zhang, X. (2015). Nickel-catalyzed reductive cross-coupling of (hetero)aryl iodides with fluorinated secondary alkyl bromides. *Org. Lett.* 17: 5570–5573. https://doi.org/10.1021/acs.orglett.5b02716.

90 Xu, C., Guo, W.-H., He, X. et al. (2018). Difluoromethylation of (hetero)aryl chlorides with chlorodifluoromethane catalyzed by nickel. *Nat. Commun.* 9 (1): 1170. https://doi.org/10.1038/s41467-018-03532-1.

91 Komeyama, K., Ohata, R., Kiguchi, S., and Osaka, I. (2017). Highly nucleophilic vitamin B12-assisted nickel-catalysed reductive coupling of aryl halides and non-activated alkyl tosylates. *Chem. Commun.* 53: 6401–6404. https://doi.org/10.1039/C7CC01932G.

92 Molander, G.A., Traister, K.M., and O'Neill, B.T. (2015). Engaging nonaromatic, heterocyclic tosylates in reductive cross-coupling with aryl and heteroaryl bromides. *J. Org. Chem.* 80: 2907–2911. https://doi.org/10.1021/acs.joc.5b00135.

93 Huihui, K.M.M., Caputo, J.A., Melchor, Z. et al. (2016). Decarboxylative cross-electrophile coupling of *N*-hydroxyphthalimide esters with aryl iodides. *J. Am. Chem. Soc.* 138: 5016–5019. https://doi.org/10.1021/jacs.6b01533.

94 Okada, K., Okamoto, K., and Oda, M. (1988). A new and practical method of decarboxylation: photosensitized decarboxylation of *N*-acyloxyphthalimides via electron-transfer mechanism. *J. Am. Chem. Soc.* 110 (26): 8736–8738. https://doi.org/10.1021/ja00234a047.

95 Schnermann, M.J. and Overman, L.E. (2012). A concise synthesis of (−)-aplyviolene facilitated by a strategic tertiary radical conjugate addition. *Angew. Chem. Int. Ed.* 51 (38): 9576–9580. https://doi.org/10.1002/anie.201204977.

96 Kadunce, N.T. and Reisman, S.E. (2015). Nickel-catalyzed asymmetric reductive cross-coupling between heteroaryl iodides and α-chloronitriles. *J. Am. Chem. Soc.* 137: 10480–10483. https://doi.org/10.1021/jacs.5b06466.

97 Banerjee, A. and Yamamoto, H. (2017). Nickel catalyzed regio-, diastereo-, and enantioselective cross-coupling of 3,4-epoxyalcohol with aryl iodides. *Org. Lett.* 19: 4363–4366. https://doi.org/10.1021/acs.orglett.7b02076.

98 Vedejs, E. and Chen, X. (1997). Parallel kinetic resolution. *J. Am. Chem. Soc.* 119 (10): 2584–2585. https://doi.org/10.1021/ja963666v.

99 Zhao, Y. and Weix, D.J. (2015). Enantioselective cross-coupling of meso-epoxides with aryl halides. *J. Am. Chem. Soc.* 137: 3237–3240. https://doi.org/10.1021/jacs.5b01909.

100 Woods, B.P., Orlandi, M., Huang, C.-Y. et al. (2017). Nickel-catalyzed enantioselective reductive cross-coupling of styrenyl aziridines. *J. Am. Chem. Soc.* 139: 5688–5691. https://doi.org/10.1021/jacs.7b03448.

101 Anka-Lufford, L.L., Huihui, K.M.M., Gower, N.J. et al. (2016). Nickel-catalyzed cross-electrophile coupling with organic reductants in non-amide solvents. *Chem. Eur. J.* 22 (33): 11564–11567. https://doi.org/10.1002/chem.201602668.

102 Molander, G.A., Traister, K.M., and O'Neill, B.T. (2014). Reductive cross-coupling of nonaromatic, heterocyclic bromides with aryl and heteroaryl bromides. *J. Org. Chem.* 79 (12): 5771–5780. https://doi.org/10.1021/jo500905m.

103 Molander, G.A., Wisniewski, S.R., and Traister, K.M. (2014). Reductive cross-coupling of 3-bromo-2,1-borazaronaphthalenes with alkyl iodides. *Org. Lett.* 16 (14): 3692–3695. https://doi.org/10.1021/ol501495d.

104 Duan, Z., Li, W., and Lei, A. (2016). Nickel-catalyzed reductive cross-coupling of aryl bromides with alkyl bromides: Et_3N as the terminal reductant. *Org. Lett.* 18: 4012–4015. https://doi.org/10.1021/acs.orglett.6b01837.

105 Paul, A., Smith, M.D., and Vannucci, A.K. (2017). Photoredox-assisted reductive cross-coupling: mechanistic insight into catalytic aryl–alkyl cross-couplings. *J. Org. Chem.* 82: 1996–2003. https://doi.org/10.1021/acs.joc.6b02830.

106 Konev, M.O., Hanna, L.E., and Jarvo, E.R. (2016). Intra- and intermolecular nickel-catalyzed reductive cross-electrophile coupling reactions of benzylic esters with aryl halides. *Angew. Chem. Int. Ed.* 55 (23): 6730–6733. https://doi.org/10.1002/anie.201601206.

107 Cao, Z.-C., Luo, Q.-Y., and Shi, Z.-J. (2016). Practical cross-coupling between O-based electrophiles and aryl bromides via Ni catalysis. *Org. Lett.* 18 (23): 5978–5981. https://doi.org/10.1021/acs.orglett.6b02656.

108 Zhang, Q., Wang, X., Qian, Q., and Gong, H. (2016). Nickel-catalyzed reductive cross-coupling of benzyl halides with aryl halides. *Synthesis* 48 (17): 2829–2836. https://doi.org/10.1055/s-0035-1562442.

109 Poremba, K.E., Kadunce, N.T., Suzuki, N. et al. (2017). Nickel-catalyzed asymmetric reductive cross-coupling to access 1,1-diarylalkanes. *J. Am. Chem. Soc.* 139: 5684–5687. https://doi.org/10.1021/jacs.7b01705.

110 Arendt, K.M. and Doyle, A.G. (2015). Dialkyl ether formation by nickel-catalyzed cross-coupling of acetals and aryl iodides. *Angew. Chem. Int. Ed.* 54: 9876–9880. https://doi.org/10.1002/anie.201503936.

111 Heinz, C., Lutz, J.P., Simmons, E.M. et al. (2018). Ni-catalyzed carbon–carbon bond-forming reductive amination. *J. Am. Chem. Soc.* 140 (6): 2292–2300. https://doi.org/10.1021/jacs.7b12212.

112 Anka-Lufford, L.L., Prinsell, M.R., and Weix, D.J. (2012). Selective cross-coupling of organic halides with allylic acetates. *J. Org. Chem.* 77: 9989–10000. https://doi.org/10.1021/jo302086g.

113 Cui, X., Wang, S., Zhang, Y. et al. (2013). Nickel-catalyzed reductive allylation of aryl bromides with allylic acetates. *Org. Biomol. Chem.* 11: 3094–3097. https://doi.org/10.1039/C3OB40232K.

114 Jia, X.-G., Guo, P., Duan, J., and Shu, X.-Z. (2018). Dual nickel and lewis acid catalysis for cross-electrophile coupling: the allylation of aryl halides with allylic alcohols. *Chem. Sci.* https://doi.org/10.1039/C7SC03140H.

115 Johnson, J.R., Tully, P.S., Mackenzie, P.B., and Sabat, M. (1991). A practical reversed-polarity alternative to organocuprate conjugate addition chemistry. Halocarbon coupling reactions of enal- and enone-derived allylnickel

reagents. *J. Am. Chem. Soc.* 113 (16): 6172–6177. https://doi.org/10.1021/ja00016a037.

116 Shrestha, R., Dorn, S.C.M., and Weix, D.J. (2013). Nickel-catalyzed reductive conjugate addition to enones via allylnickel intermediates. *J. Am. Chem. Soc.* 135 (2): 751–762. https://doi.org/10.1021/ja309176h.

117 Johnson, K.A., Biswas, S., and Weix, D.J. (2016). Cross-electrophile coupling of vinyl halides with alkyl halides. *Chem. Eur. J.* 22: 7399–7402. https://doi.org/10.1002/chem.201601320.

118 Boeckman, R.K., Niziol, J.M., and Biegasiewicz, K.F. (2018). Scalable synthesis of (−)-rasfonin enabled by a convergent enantioselective α-hydroxymethylation strategy. *Org. Lett.* 20 (16): 5062–5065. https://doi.org/10.1021/acs.orglett.8b02222.

119 Cannes, C., Condon, S., Durandetti, M. et al. (2000). Nickel-catalyzed electrochemical couplings of vinyl halides: synthetic and stereochemical aspects. *J. Org. Chem.* 65 (15): 4575–4583. https://doi.org/10.1021/jo000182f.

120 Qiu, C., Yao, K., Zhang, X., and Gong, H. (2016). Ni-catalyzed reductive coupling of α-halocarbonyl derivatives with vinyl bromides. *Org. Biomol. Chem.* 14 (48): 11332–11335. https://doi.org/10.1039/C6OB02269C.

121 Hofstra, J.L., Cherney, A.H., Ordner, C.M., and Reisman, S.E. (2018). Synthesis of enantioenriched allylic silanes via nickel-catalyzed reductive cross-coupling. *J. Am. Chem. Soc.* 140 (1): 139–142. https://doi.org/10.1021/jacs.7b11707.

122 Suzuki, N., Hofstra, J.L., Poremba, K.E., and Reisman, S.E. (2017). Nickel-catalyzed enantioselective cross-coupling of *N*-hydroxyphthalimide esters with vinyl bromides. *Org. Lett.* 19: 2150–2153. https://doi.org/10.1021/acs.orglett.7b00793.

123 Gooßen, L.J., Rodríguez, N., and Gooßen, K. (2008). Carboxylic acids as substrates in homogeneous catalysis. *Angew. Chem. Int. Ed.* 47 (17): 3100–3120. https://doi.org/10.1002/anie.200704782.

124 Onaka, M., Matsuoka, Y., and Mukaiyama, T. (1981). A convenient method for the direct preparation of ketones from 2-(6-(2-methoxyethyl)pyridyl)carboxylates and alkyl iodides by use of zinc dust and a catalytic amount of nickel dichloride. *Chem. Lett.* 10 (4): 531–534. https://doi.org/10.1246/cl.1981.531.

125 Wotal, A. (2016). Nickel-catalyzed synthesis of ketones from alkyl halides and acid chlorides: preparation of ethyl 4-oxododecanoate. *Org. Synth.* 93: 50–62. https://doi.org/10.15227/orgsyn.093.0050.

126 Wu, F., Lu, W., Qian, Q. et al. (2012). Ketone formation via mild nickel-catalyzed reductive coupling of alkyl halides with aryl acid chlorides. *Org. Lett.* 14: 3044–3047. https://doi.org/10.1021/ol3011198.

127 Lu, W., Liang, Z., Zhang, Y. et al. (2013). Gram-scale ketone synthesis by direct reductive coupling of alkyl iodides with acid chlorides. *Synthesis* 45: 2234–2240. https://doi.org/10.1055/s-0033-1339297.

128 Liang, Z., Xue, W., Lin, K., and Gong, H. (2014). Nickel-catalyzed reductive methylation of alkyl halides and acid chlorides with methyl *p*-tosylate. *Org. Lett.* 16: 5620–5623. https://doi.org/10.1021/ol502682q.

129 Ai, Y., Ye, N., Wang, Q. et al. (2017). Zirconium/nickel-mediated one-pot ketone synthesis. *Angew. Chem. Int. Ed.* 56 (36): 10791–10795. https://doi.org/10.1002/anie.201705520.
130 Yin, H., Zhao, C., You, H. et al. (2012). Mild ketone formation via Ni-catalyzed reductive coupling of unactivated alkyl halides with acid anhydrides. *Chem. Commun.* 48 (56): 7034–7036. https://doi.org/10.1039/C2CC33232A.
131 Zhao, C., Jia, X., Wang, X., and Gong, H. (2014). Ni-catalyzed reductive coupling of alkyl acids with unactivated tertiary alkyl and glycosyl halides. *J. Am. Chem. Soc.* 136: 17645–17651. https://doi.org/10.1021/ja510653n.
132 Jia, X., Zhang, X., Qian, Q., and Gong, H. (2015). Alkyl-aryl ketone synthesis via nickel-catalyzed reductive coupling of alkyl halides with aryl acids and anhydrides. *Chem. Commun.* 51: 10302–10305. https://doi.org/10.1039/C5CC03113C.
133 Zheng, M., Xue, W., Xue, T., and Gong, H. (2016). Ester formation via nickel-catalyzed reductive coupling of alkyl halides with chloroformates. *Org. Lett.* 18: 6152–6155. https://doi.org/10.1021/acs.orglett.6b03158.
134 Gosmini, C., Nédélec, J.Y., and Périchon, J. (2000). Electrochemical cross-coupling between functionalized aryl halides and 2-chloropyrimidine or 2-chloropyrazine catalyzed by nickel 2,2′-bipyridine complex. *Tetrahedron Lett.* 41 (2): 201–203. https://doi.org/10.1016/S0040-4039(99)02037-7.
135 Gosmini, C., Nédélec, J.Y., and Périchon, J. (2000). Electrosynthesis of functionalized 2-arylpyridines from functionalized aryl and pyridine halides catalyzed by nickel bromide 2,2′-bipyridine complex. *Tetrahedron Lett.* 41 (26): 5039–5042. https://doi.org/10.1016/S0040-4039(00)00760-7.
136 Gosmini, C., Bassene-Ernst, C., and Durandetti, M. (2009). Synthesis of functionalized 2-arylpyridines from 2-halopyridines and various aryl halides via a nickel catalysis. *Tetrahedron* 65: 6141–6146. https://doi.org/10.1016/j.tet.2009.05.044.
137 Sengmany, S., Léonel, E., Polissaint, F. et al. (2007). Preparation of functionalized aryl- and heteroarylpyridazines by nickel-catalyzed electrochemical cross-coupling reactions. *J. Org. Chem.* 72 (15): 5631–5636. https://doi.org/10.1021/jo070429+.
138 Sengmany, S., Vitu-Thiebaud, A., Le Gall, E. et al. (2013). An electrochemical nickel-catalyzed arylation of 3-amino-6-chloropyridazines. *J. Org. Chem.* 78: 370–379. https://doi.org/10.1021/jo3022428.
139 Sengmany, S., Le Gall, E., and Léonel, E. (2011). An electrochemical synthesis of functionalized arylpyrimidines from 4-amino-6-chloropyrimidines and aryl halides. *Molecules* 16: 5550–5560.
140 Sengmany, S., Vasseur, S., Lajnef, A. et al. (2016). Beneficial effects of electrochemistry in cross-coupling reactions: electroreductive synthesis of 4-aryl- or 4-heteroaryl-6-pyrrolylpyrimidines. *Eur. J. Org. Chem.* 2016 (28): 4865–4871. https://doi.org/10.1002/ejoc.201600790.
141 Liao, L.-Y., Kong, X.-R., and Duan, X.-F. (2014). Reductive couplings of 2-halopyridines without external ligand: phosphine-free nickel-catalyzed synthesis of symmetrical and unsymmetrical 2,2′-bipyridines. *J. Org. Chem.* 79: 777–782. https://doi.org/10.1021/jo402084m.

142 Qian, Q., Zang, Z., Wang, S. et al. (2013). Nickel-catalyzed reductive cross-coupling of aryl halides. *Synlett* 24 (05): 619–624. https://doi.org/10.1055/s-0032-1318237.

143 Olivares, A.M. and Weix, D.J. (2018). Multimetallic Ni- and Pd-catalyzed cross-electrophile coupling to form highly substituted 1,3-dienes. *J. Am. Chem. Soc.* 140 (7): 2446–2449. https://doi.org/10.1021/jacs.7b13601.

144 Szostak, R. and Szostak, M. (2018). *N*-acyl-glutarimides: resonance and proton affinities of rotationally-inverted twisted amides relevant to N—C(O) cross-coupling. *Org. Lett.* 20 (5): 1342–1345. https://doi.org/10.1021/acs.orglett.8b00086.

145 Ni, S., Zhang, W., Mei, H. et al. (2017). Ni-catalyzed reductive cross-coupling of amides with aryl iodide electrophiles via C—N bond activation. *Org. Lett.* 19 (10): 2536–2539. https://doi.org/10.1021/acs.orglett.7b00831.

146 Dai, Y., Wu, F., Zang, Z. et al. (2012). Ni-catalyzed reductive allylation of unactivated alkyl halides with allylic carbonates. *Chem. Eur. J.* 18 (3): 808–812. https://doi.org/10.1002/chem.201102984.

147 Chen, H., Jia, X., Yu, Y. et al. (2017). Nickel-catalyzed reductive allylation of tertiary alkyl halides with allylic carbonates. *Angew. Chem. Int. Ed.* 56 (42): 13103–13106. https://doi.org/10.1002/anie.201705521.

148 Yan, X.-B., Li, C.-L., Jin, W.-J. et al. (2018). Reductive coupling of benzyl oxalates with highly functionalized alkyl bromides by nickel catalysis. *Chem. Sci.* 9 (19): 4529–4534. https://doi.org/10.1039/C8SC00609A.

149 Shrestha, R. and Weix, D.J. (2011). Reductive conjugate addition of haloalkanes to enones to form silyl enol ethers. *Org. Lett.* 13 (10): 2766–2769. https://doi.org/10.1021/ol200881v.

150 Huihui, K.M.M., Shrestha, R., and Weix, D.J. (2017). Nickel-catalyzed reductive conjugate addition of primary alkyl bromides to enones to form silyl enol ethers. *Org. Lett.* 19 (2): 340–343. https://doi.org/10.1021/acs.orglett.6b03509.

151 Prinsell, M.R., Everson, D.A., and Weix, D.J. (2010). Nickel-catalyzed, sodium iodide-promoted reductive dimerization of alkyl halides, alkyl pseudohalides, and allylic acetates. *Chem. Commun.* 46 (31): 5743–5745. https://doi.org/10.1039/C0CC01716G.

152 Goldup, S.M., Leigh, D.A., McBurney, R.T. et al. (2010). Ligand-assisted nickel-catalysed Sp^3–Sp^3 homocoupling of unactivated alkyl bromides and its application to the active template synthesis of rotaxanes. *Chem. Sci.* 1 (3): 383–386. https://doi.org/10.1039/C0SC00279H.

153 Xue, W., Xu, H., Liang, Z. et al. (2014). Nickel-catalyzed reductive cyclization of alkyl dihalides. *Org. Lett.* 16: 4984–4987. https://doi.org/10.1021/ol502207z.

154 Tollefson, E.J., Erickson, L.W., and Jarvo, E.R. (2015). Stereospecific intramolecular reductive cross-electrophile coupling reactions for cyclopropane synthesis. *J. Am. Chem. Soc.* 137: 9760–9763. https://doi.org/10.1021/jacs.5b03870.

155 Erickson, L.W., Lucas, E.L., Tollefson, E.J., and Jarvo, E.R. (2016). Nickel-catalyzed cross-electrophile coupling of alkyl fluorides: stereospecific

synthesis of vinylcyclopropanes. *J. Am. Chem. Soc.* 138 (42): 14006–14011. https://doi.org/10.1021/jacs.6b07567.
156 Chinchilla, R. and Nájera, C. (2011). Recent advances in Sonogashira reactions. *Chem. Soc. Rev.* 40 (10): 5084–5121. https://doi.org/10.1039/C1CS15071E.
157 Huang, L., Olivares, A.M., and Weix, D.J. (2017). Reductive decarboxylative alkynylation of *N*-hydroxyphthalimide esters with bromoalkynes. *Angew. Chem. Int. Ed.* 56 (39): 11901–11905. https://doi.org/10.1002/anie.201706781.
158 Kim, J.H., Ko, Y.O., Bouffard, J., and Lee, S. (2015). Advances in tandem reactions with organozinc reagents. *Chem. Soc. Rev.* 44 (8): 2489–2507. https://doi.org/10.1039/C4CS00430B.
159 Tietze, L.F. (1996). Domino reactions in organic synthesis. *Chem. Rev.* 96 (1): 115–136. https://doi.org/10.1021/cr950027e.
160 Parsons, P.J., Penkett, C.S., and Shell, A.J. (1996). Tandem reactions in organic synthesis: novel strategies for natural product elaboration and the development of new synthetic methodology. *Chem. Rev.* 96 (1): 195–206. https://doi.org/10.1021/cr950023+.
161 Nicolaou, K.C., Edmonds, D.J., and Bulger, P.G. (2006). Cascade reactions in total synthesis. *Angew. Chem. Int. Ed.* 45 (43): 7134–7186. https://doi.org/10.1002/anie.200601872.
162 Sommer, H., Juliá-Hernández, F., Martin, R., and Marek, I. (2018). Walking metals for remote functionalization. *ACS Cent. Sci.* 4 (2): 153–165. https://doi.org/10.1021/acscentsci.8b00005.
163 Chen, F., Chen, K., Zhang, Y. et al. (2017). Remote migratory cross-electrophile coupling and olefin hydroarylation reactions enabled by in situ generation of NiH. *J. Am. Chem. Soc.* 139: 13929–13935. https://doi.org/10.1021/jacs.7b08064.
164 Juliá-Hernández, F., Moragas, T., Cornella, J., and Martin, R. (2017). Remote carboxylation of halogenated aliphatic hydrocarbons with carbon dioxide. *Nature* 545 (7652): 84–88. https://doi.org/10.1038/nature22316.
165 Peng, Y., Xiao, J., Xu, X.-B. et al. (2016). Stereospecific synthesis of tetrahydronaphtho[2,3-*b*]furans enabled by a nickel-promoted tandem reductive cyclization. *Org. Lett.* 18: 5170–5173. https://doi.org/10.1021/acs.orglett.6b02665.
166 Xiao, J., Wang, Y.-W., and Peng, Y. (2017). Nickel-promoted reductive cyclization cascade: a short synthesis of a new aromatic strigolactone analogue. *Synthesis* 49: 3576–3581. https://doi.org/10.1055/s-0036-1588804.
167 Peng, Y., Xu, X.-B., Xiao, J., and Wang, Y.-W. (2014). Nickel-mediated stereocontrolled synthesis of spiroketals via tandem cyclization-coupling of [small beta]-bromo ketals and aryl iodides. *Chem. Commun.* 50: 472–474. https://doi.org/10.1039/C3CC47780K.
168 Kuang, Y., Wang, X., Anthony, D., and Diao, T. (2018). Ni-catalyzed two-component reductive dicarbofunctionalization of alkenes via radical cyclization. *Chem. Commun.* https://doi.org/10.1039/C8CC00358K.
169 García-Domínguez, A., Li, Z., and Nevado, C. (2017). Nickel-catalyzed reductive dicarbofunctionalization of alkenes. *J. Am. Chem. Soc.* 139: 6835–6838. https://doi.org/10.1021/jacs.7b03195.

170 Oçafrain, M., Devaud, M., Troupel, M., and Périchon, J. (1995). New electrochemical synthesis of ketones from organic halides and carbon monoxide. *J. Chem. Soc., Chem. Commun.* (22): 2331–2332. https://doi.org/10.1039/C39950002331.

171 Dolhem, E., Barhdadi, R., Folest, J. et al. (2001). Nickel catalysed electrosynthesis of ketones from organic halides and iron pentacarbonyl. Part 2: Unsymmetrical ketones. *Tetrahedron* 57: 525–529. https://doi.org/10.1016/S0040-4020(00)01017-6.

172 Wotal, A.C., Ribson, R.D., and Weix, D.J. (2014). Stoichiometric reactions of acylnickel(II) complexes with electrophiles and the catalytic synthesis of ketones. *Organometallics* 33: 5874–5881. https://doi.org/10.1021/om5004682.

173 Zheng, B., Tang, F., Luo, J. et al. (2014). Organometallic nickel(III) complexes relevant to cross-coupling and carbon–heteroatom bond formation reactions. *J. Am. Chem. Soc.* 136 (17): 6499–6504. https://doi.org/10.1021/ja5024749.

174 Schultz, J.W., Fuchigami, K., Zheng, B. et al. (2016). Isolated organometallic nickel(III) and nickel(IV) complexes relevant to carbon–carbon bond formation reactions. *J. Am. Chem. Soc.* 138 (39): 12928–12934. https://doi.org/10.1021/jacs.6b06862.

175 Watson, M.B., Rath, N.P., and Mirica, L.M. (2016). Oxidative C—C bond formation reactivity of organometallic Ni(II), Ni(III), and Ni(IV) complexes. *J. Am. Chem. Soc.* https://doi.org/10.1021/jacs.6b10303.

10

Organometallic Chemistry of High-Valent Ni(III) and Ni(IV) Complexes

Liviu M. Mirica[1,2], *Sofia M. Smith*[2,*], *and Leonel Griego*[1,*]

[1] *University of Illinois at Urbana-Champaign, Department of Chemistry, 600 S. Mathews Avenue, Urbana, IL 61801, USA*
[2] *Washington University in St. Louis, Department of Chemistry, One Brookings Drive, St. Louis, MO 63130, USA*

10.1 Introduction

Nickel (Ni) is commonly used as a catalyst for various cross-coupling reactions, including Negishi, Kumada, and Suzuki couplings. Although the 0, I, and II oxidation states of Ni are commonly involved in these catalytic transformations, recent studies show that the III and IV oxidation states of Ni also play a key role in organic transformations such as C–C and C–heteroatom bond formation reactions.

The first organometallic Ni^{III} and Ni^{IV} complexes were reported almost 40 years ago, when they were observed by Kochi in 1978 and Smart in 1982, respectively. However, it is only in the past decade that many advances have been made in the synthesis of high-valent organometallic Ni complexes as also the detailed study of their reactivity and involvement in various organic transformations. The once elusive Ni^{III} and Ni^{IV} species are now promoting a wide range of organic transformations in a complementary manner to the reactions involving Ni^0 and Ni^{II} intermediates.

In this chapter, we will present accomplishments made in the past 40 years involving organometallic Ni^{III} and Ni^{IV} complexes, with a focus on the more recent systems that exhibit rich reactivity profiles. It is important to note that there are other high-valent, non-organometallic Ni complexes that provide key insight into the role of various Ni intermediates in organic transformations, and such species are also presented herein.

10.2 Organometallic Ni(III) Complexes

Kochi and coworker published in 1978 the first organometallic Ni^{III} species, **1** (Figure 10.1), that was observed by EPR and UV–Vis spectroscopy. Oxidation of the Ni^{II} precursor with sodium hexachloroiridate at −50 °C yielded an EPR-active

* These authors contributed equally.

Nickel Catalysis in Organic Synthesis: Methods and Reactions,
First Edition. Edited by Sensuke Ogoshi.

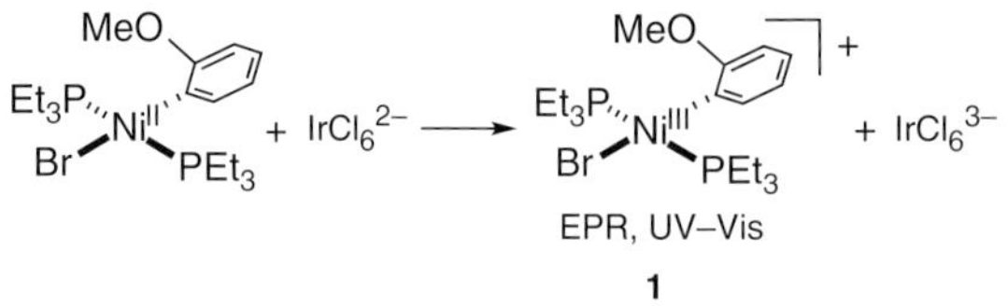

Figure 10.1 The first observed organometallic Ni^{III} species by Kochi and coworkers in 1978.

species with an absorption maximum at 410 nm, suggestive of a Ni^{III} intermediate. Warming up of the solution led to the disappearance of EPR signal, which was associated with a reductive elimination from the Ni^{III} species [1].

It was not until 1983 that van Koten and coworkers isolated the first high-valent nickel organometallic complex using the pincer ligand 1,1′-(1,3-phenylene) bis(*N*,*N*-dimethylmethanamine) (NCN, Figure 10.2). The binding of the tridentate NCN pincer ligand and two anionic ligands yielded stable trigonal bipyramidal Ni^{III} complexes **2**. The auxiliary ligands in the *cis* coordination sites could easily be exchanged with nitrite or nitrate to generate **3**, or with isothiocyanate and pyridine to yield **4** and **5**. All these complexes are unusually stable and have only been proposed as intermediates in the Kharasch radical addition of CCl_4 to alkenes [2–6].

In 2018 Zargarian and Cloutier revisited the reactivity of complexes **2** and were able to show that they undergo C—X bond formation at the *ipso* C atom of the

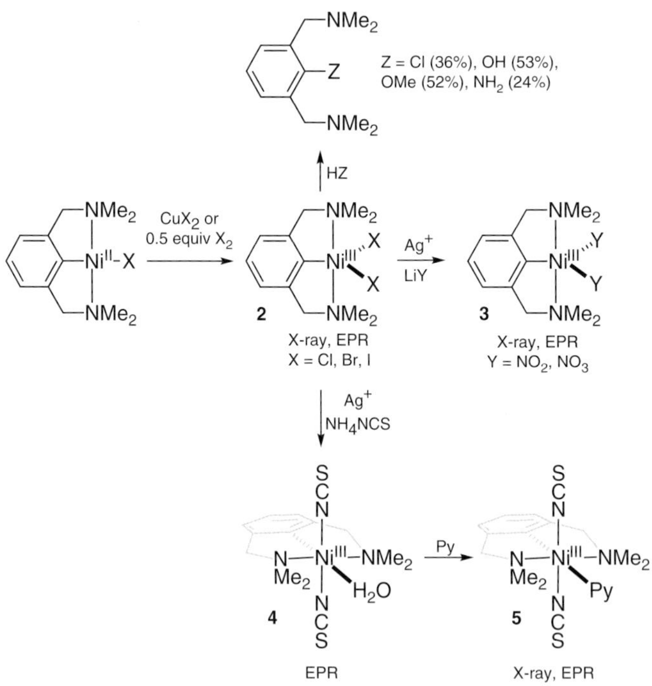

Figure 10.2 The first isolated organometallic Ni^{III} complexes stabilized by an NCN pincer ligand reported by van Koten and coworkers, and their subsequent reactivity reported recently by Zargarian and coworkers.

phenyl group with water, alcohols, amines, and strong acids under mild conditions [7].

In 1995, van Koten and coworkers introduced a chiral pincer ligand containing amino-acid-derived arms. In contrast to their original NCN ligand, this histidine-type NCN pincer ligand could coordinate to the Ni center either in a tridentate or a pentadentate manner. In this case, the Ni^{III} complex **6** adopts an octahedral geometry and can be isolated in 72% yield (Figure 10.3) [8].

In 1995, Hillhouse and coworkers suggested the formation a Ni^{III} intermediate after exposing the Ni^{II} precursor to either O_2 or I_2 at room temperature (Figure 10.4). The reactions resulted in clean formation of *N*-*p*-tolylpyrrolidine, while the one-electron oxidant acetylferrocenium tetrafluoroborate also cleanly converts the Ni^{II} to **7**, supporting the formation of the transient Ni^{III} species [9, 10].

In addition, the same authors reported the reactivity of the dimeric seven-membered nickel(II) oxametallacycle, which in presence of O_2 leads to the formation of a new C—O bond (Figure 10.5). An oxidatively induced mechanism was proposed that involves the formation of a transient Ni^{III} intermediate, **8**, followed by reductive elimination to give 4,4-dimethylchroman [10, 11].

Figure 10.3 The first isolated chiral Ni^{III} organometallic complex stabilized by an amino acid pincer ligand.

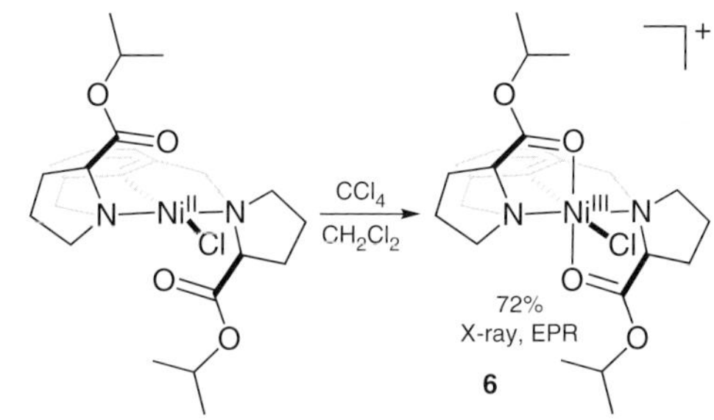

Figure 10.4 Oxidatively induced C—N bond formation from a Ni^{II} complex through a transient Ni^{III} species.

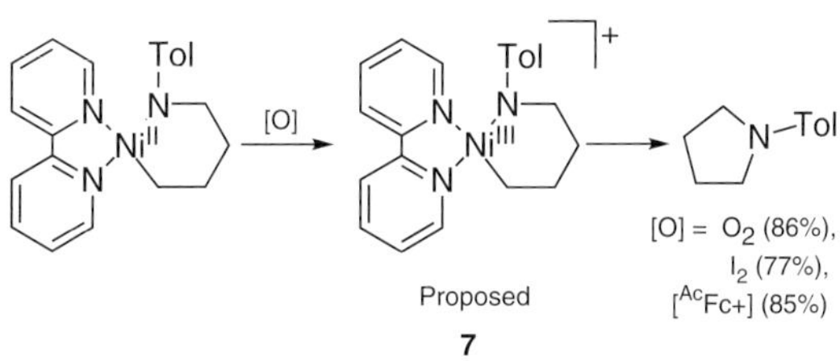

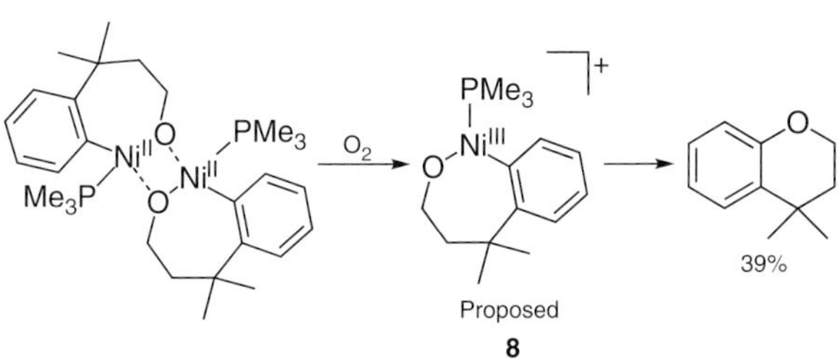

Figure 10.5 Oxidatively induced reactivity of a dimeric seven-membered N^{II} oxametallacycle.

In 2002, Hillhouse and coworkers described the stereochemical synthesis of aziridine by an oxidatively induced reductive elimination (Figure 10.6). Reaction of *N*-tosylaziridine with (bpy)Ni^0(COD) gives only one regioisomer of the azametallacyclobutane complex, the oxidative addition occurring exclusively at the least hindered C—N bond. Addition of O_2 generates a proposed Ni^{III} intermediate **9**, which undergoes reductive elimination with inversion of stereochemistry to reform the aziridine with 92% inversion, as observed by ^1H NMR [12].

In 2005, Warren and coworkers isolated a trigonal planar Ni^{III} complex, **10**, stabilized by a bidentate β-diketiminate ligand and an imidoadamantane group (Figure 10.7). The reactivity of the terminal imido group with CO, ^{t}BuNC, and PMe_3 result in the formation of OC=N, NC=N, and P=N bonds, respectively [13].

Starting in 2007, Zargarian and coworkers have employed the 1,3-bis((diisopropylphosphaneyl)oxy)propane (PCP) and 1-(3-((diisopropylphosphaneyl)oxy) phenyl)-*N*,*N*-diN^Rmethanamine (PCN) phosphonite pincer ligands to stabilize Ni^{III} complexes **11** and **12**, respectively (Figure 10.8). Similar to the (NCN)Ni complexes reported by van Koten and coworkers, the PCP/PCN-supported Ni^{III} species could promote radical Kharasch addition of CCl_4 to alkenes in good yields [14–16].

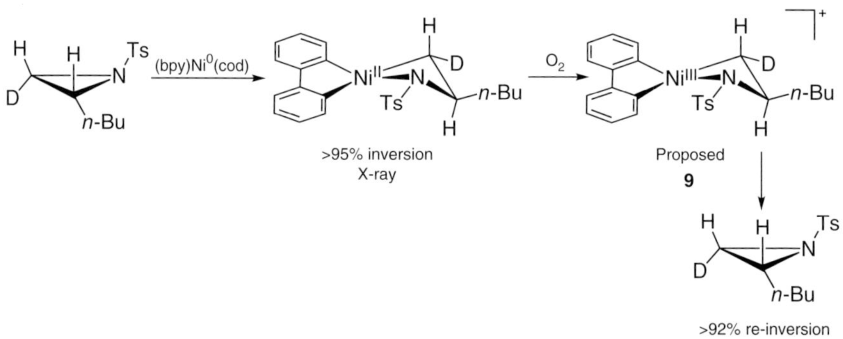

Figure 10.6 The formation of aziridine by an oxidatively induced reductive elimination from Ni^{III}.

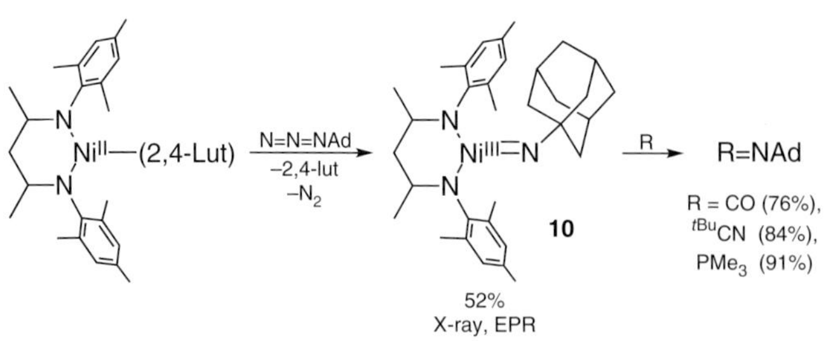

Figure 10.7 Reactivity of a trigonal planar Ni^{III}-imido complex to give N—X bond formation products.

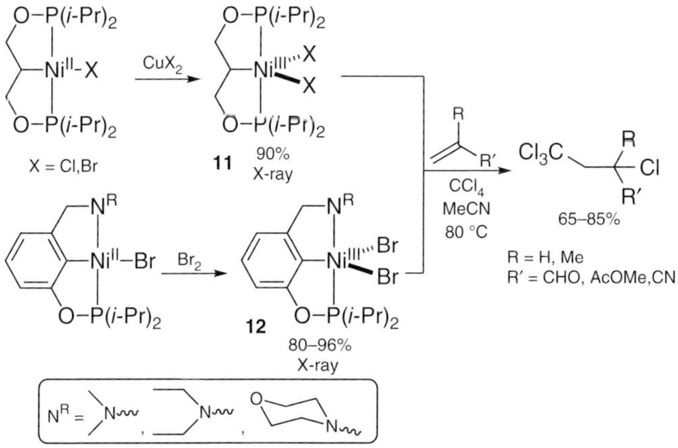

Figure 10.8 Ni^{III} complexes supported by phosphonite pincer ligands isolated by Zargarian and coworkers, and their radical Kharasch addition reactivity.

In 2008, Zargarian and coworkers used the PCP pincer ligand 1,5-bis(diisopropylphosphaneyl) pentyl to stabilize a Ni^{III} complex **13** with a trigonal bipyramidal geometry (Figure 10.9). In contrast to the complexes supported by the PCP phosphonite ligands, **13** does not exhibit significant reactivity at the Ni^{III} center [17].

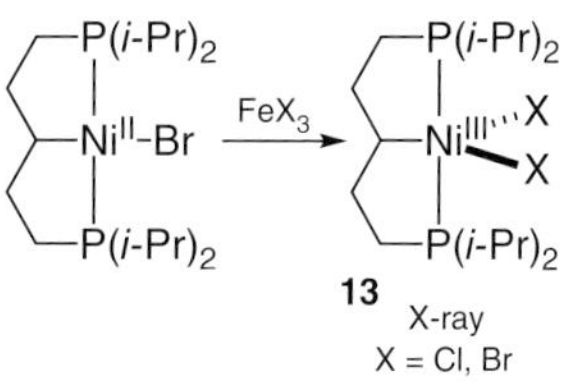

Figure 10.9 Aliphatic PCP-Ni^{III} complexes isolated by Zargarian and coworkers in 2008.

Hillhouse and coworkers synthesized in 2011 two Ni^{III}-imido complexes supported by 1,2-bis (di-*tert*-buylphosphino)ethane (Figure 10.10). The oxidation of the Ni^{II} complexes led to the formation

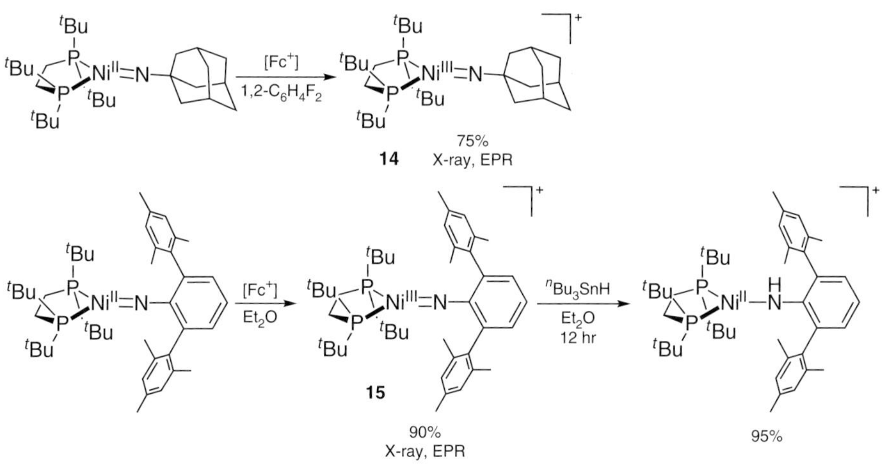

Figure 10.10 Isolated Ni^{III} alkyl/aryl-amido complexes reported by Hillhouse and coworkers in 2011.

of two Ni^{III} aryl/alkyl-imido complexes. Both Ni^{III} species exhibit analogous EPR spectra at low temperature, yet different variable-temperature magnetic properties and a low-spin/high-spin equilibrium was proposed to take place for the alkylimido Ni^{III} complex **14** to account for the difference. The Ni^{III}-arylimido complex **15** can undergo hydrogen atom abstraction when reacted with $^{n}Bu_3SnH$ to generate a Ni^{II}-amide complex [18].

Vicic and coworkers reported in 2012 the one-electron oxidation of Ni^{II} complexes supported by the BOXAM ligand (BOXAM = bis((4-isopropyl-4,5-dihydrooxazol-2-yl)phenyl)amine (Figure 10.11). The resulting Ni^{III} complexes **16** were characterized by EPR and UV–Vis spectroscopy, and it was shown that their redox behavior strongly depends on the co-ligand (CH_3, CF_3, or Cl) [19, 20].

In addition, in 2013 Vicic and coworkers reported a five-coordinate Ni^{III} complex supported by a terpyridine ligand and two fluoroalkyl ligands (Figure 10.12). Ferrocenium hexafluorophosphate was used to oxidize the Ni^{II} complex to a transient Ni^{III} species **17**, which undergoes elimination of one CF_3 group, likely through a reductive homolysis of a trifluoromethyl radical [21].

In 2013, Tilley and coworkers reported an isolable Ni^{III}-alkyl complex, **18**, which has been prepared through a two-electron oxidative addition of MeI to a Ni^{I} complex (Figure 10.13). Complex **18** is thermally unstable and decomposes in one day in benzene to give ethane and a Ni^{II} complex [22, 23]. Interestingly, the bis-amido ligand framework is capable of supporting Ni complexes in three different oxidation states.

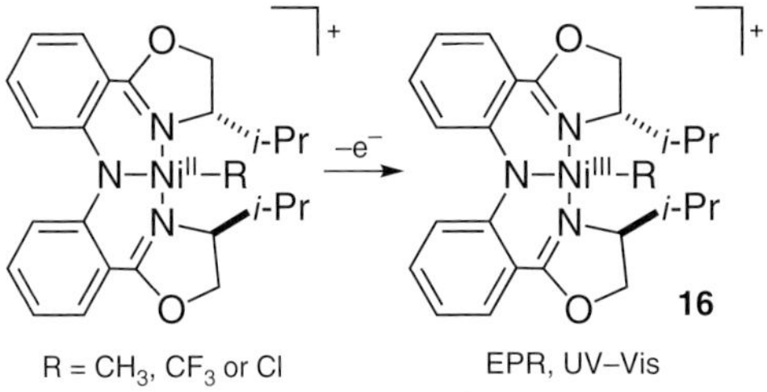

Figure 10.11 Formation of Ni^{II} complexes supported by a BOXAM ligand, reported by Vicic and coworkers.

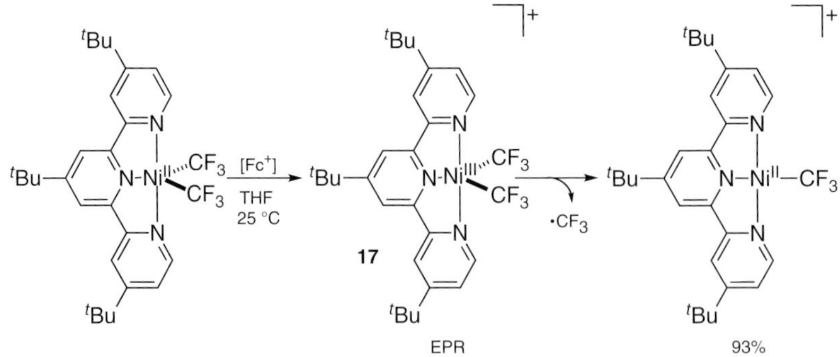

Figure 10.12 Detected five-coordinate Ni^{III} complex stabilized by two CF_3 ligands and a terpyridine ligand.

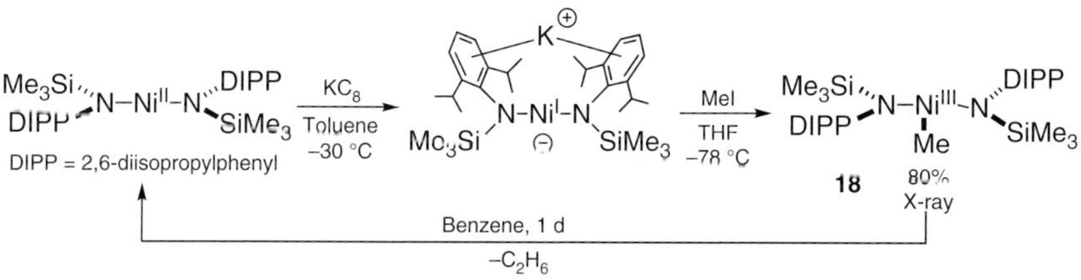

Figure 10.13 Synthesis of a Ni^{III}-alkyl species formed upon oxidative addition of MeI to a Ni^{I} precursor.

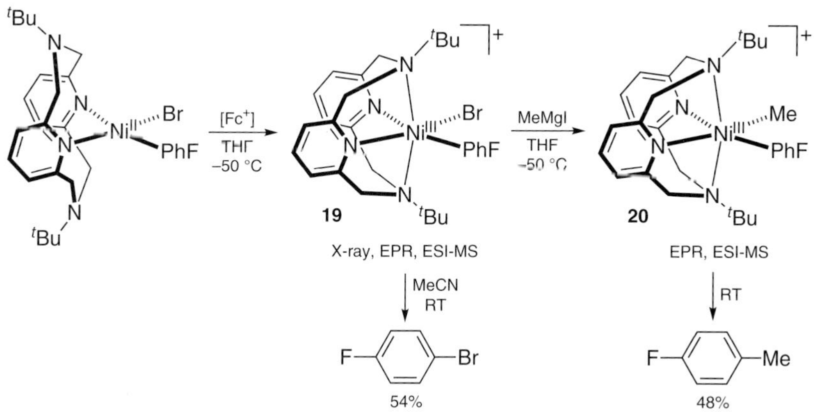

Figure 10.14 Synthesis and reactivity of Ni^{III}(aryl)halide and Ni^{III}(aryl)alkyl complexes reported by Mirica and coworkers in 2014.

Mirica and coworkers published in 2014 a series of Ni^{III}(aryl)halide complexes stabilized by the tetradentate ligand *N,N′*-di-*tert*-butyl-2,11-diaza[3,3](2,6) pyridophane (Figure 10.14). The Ni^{III}(aryl)halide complex **19** is stable at low temperature, yet undergoes rapid C–halide bond formation at room temperature. The Ni^{III}(aryl)halide complex also undergoes a rapid transmetalation reaction with Grignard reagents to yield a detectable Ni^{III}(aryl)alkyl complex **20**, followed by rapid C—C bond formation at RT [24].

In addition, Mirica and coworkers published in 2015 the synthesis of stable organometallic Ni^{III} complexes that contain two trifluoromethyl ligands and are supported by tetradentate N-donor ligands ^{R}N4 (R = Me or *t*Bu, Figure 10.15). Interestingly, the corresponding Ni^{II} precursors undergo facile oxidation,

Figure 10.15 Synthesis of stable bis(trifluoromethyl)nickel^{III} complexes supported by tetradentate pyridinophane ligands.

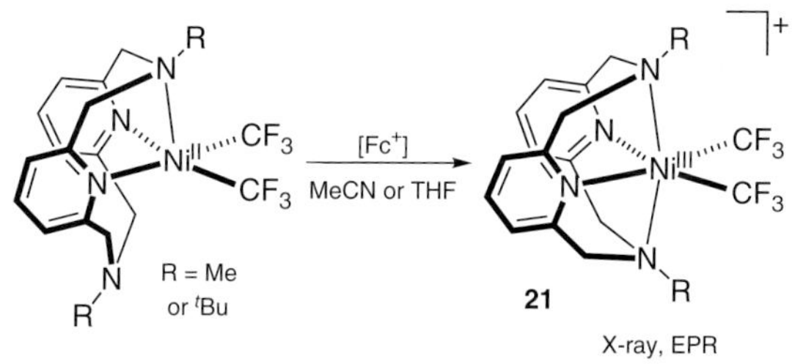

including aerobic oxidation to generate the Ni^{III} complexes **21**. Unlike most other organometallic Ni^{III} complexes, complexes **21** are indefinitely stable at RT under N_2, while heating at 80 °C for 24 hours or photolysis with visible light at RT produces only trace amounts of CF_3H and C_2F_6 [25].

Mirica and coworkers reported in 2015 the synthesis of several Ni^{III} complexes supported by a modified tetradentate pyridinophane ligand containing one phenyl group (Figure 10.16). The Ni^{III}(MeCN)Br complex **22** was generated by one-electron oxidation of the Ni^{II} precursor, while halogen abstraction with silver gives a Ni^{III}-disolvento complex **23**, which is an RT-stable dicationic complex. Interestingly, the Ni^{III}-dimethyoxide complex **24** undergoes aryl methoxylation that is favored by addition of the $PhI(PyOMe)_2OTf_2$ oxidant-suggesting the formation of a Ni^{IV} intermediate, followed by rapid reductive elimination to generate the C—O bond formation product [26].

In 2015, Vicic and coworkers reported the isolation of a $[(tpy)Ni^{III}(C_4F_8)]^+$ complex **25** (tpy = terpyridine) stabilized by the cycloperfluorobutyl $-C_4F_8-$ ligand (Figure 10.17). The Ni center in **25** adopts a distorted octahedral geometry,

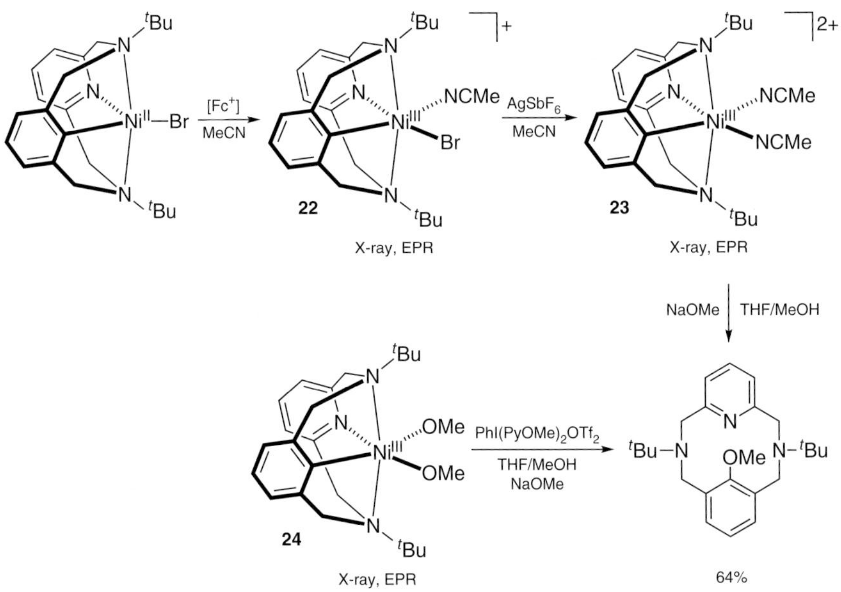

Figure 10.16 Synthesis of Ni^{III} complexes supported by a modified pyridinophane ligand containing one phenyl group.

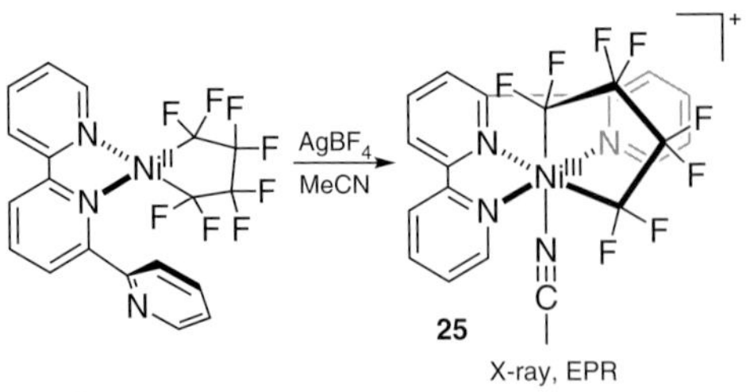

Figure 10.17 Formation of a stable $(tpy)Ni^{III}(C_4F_8)$ complex reported by Vicic and coworkers in 2015.

where Ni—N bonds *trans* to the fluoroalkyl ligand were shorter by about 0.2 Å than the Ni—N bonds that are *cis* to the fluoroalkyl groups [27].

In 2016, Diao and coworkers proposed a square pyramidal Ni^{III}-alkyl complex **26** stabilized by the bidentate anionic ligand 3,5-dimethyl-2-(2-pyridyl)pyrrole (Figure 10.18). This unstable intermediate is thought to form an iodide-bridged Ni^{III} dinuclear complex **27**, which undergoes rapid reductive elimination to form ethane in 43% yield [28].

In addition, a mixed-valent Ni^{II}/Ni^{III} dinuclear complex **28** stabilized by four triazabicyclodecene (TBD) ligands was reported by the same group (Figure 10.19). Attempts to oxidize **28** with 1.5 equiv $PhICl_2$ did not lead to the isolation of the di-Ni^{III} complex **29**; instead, the N–N coupling of two TBD ligands was observed [29].

In 2016, Mirica and coworkers reported a series of Ni^{III} complexes **30** and **31** that undergo aromatic cycanoalkylation (Figure 10.20). Addition of base to the $Ni^{III}(MeCN)_2$ complex **31** gives the cyanomethylated product in quantitative yield. Regioselective α-cycanoalkylation was also observed with various nitrile substrates to generate secondary and tertiary nitriles [30].

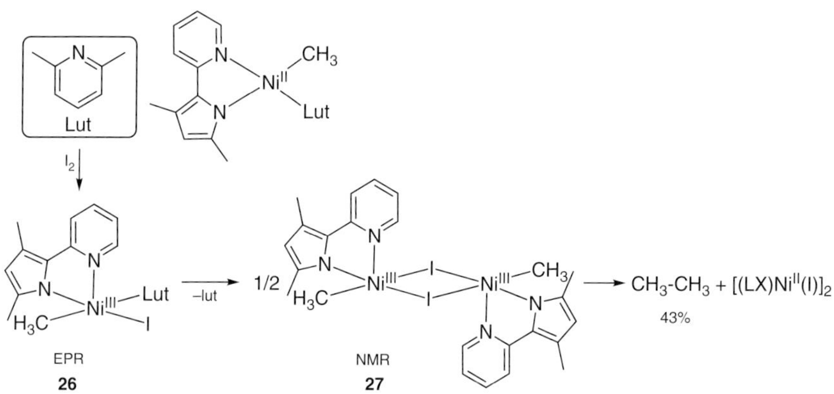

Figure 10.18 Synthesis of mono- and dinuclear square pyramidal Ni^{III}-alkyl complexes.

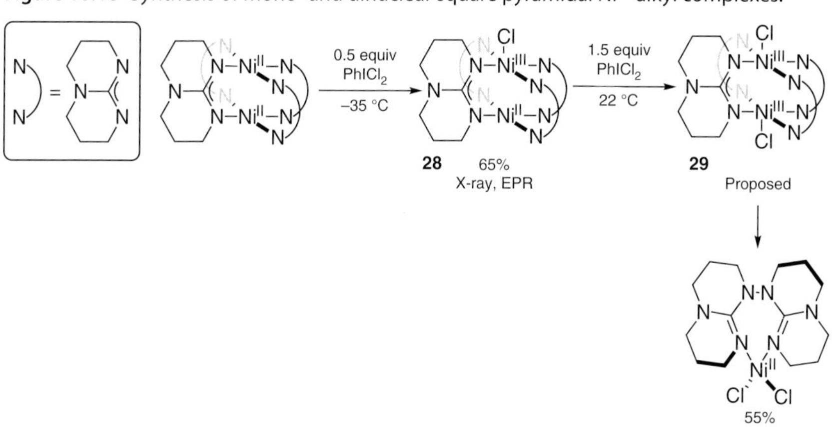

Figure 10.19 N—N bond formation from a Ni^{III} dimer published by Diao and coworkers in 2016.

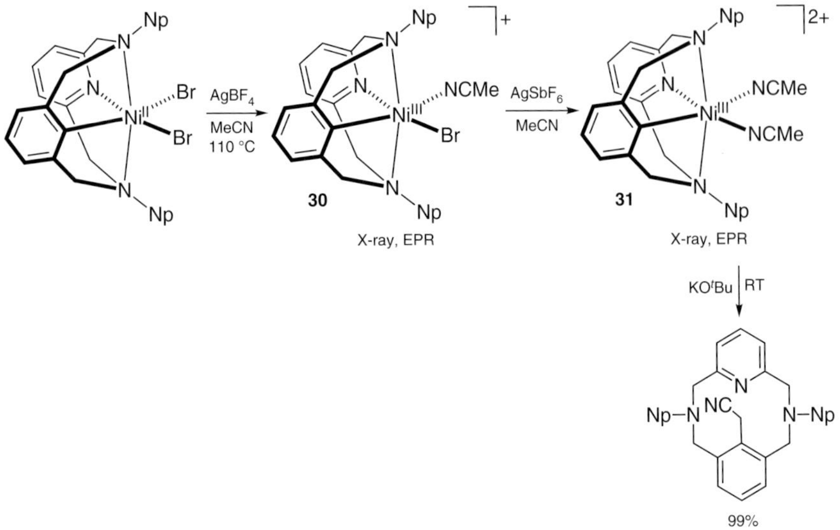

Figure 10.20 Synthesis of Ni^{III} complexes and their aromatic cyanoalkylation reactivity.

In 2016, Mirica and coworkers synthesized the first isolated Ni^{III}-dialkyl complex, stabilized by the *N,N′*-dimethyl-2,11-diaz[3.3](2,6)pyridinophane ligand (Figure 10.21). The Ni^{III} complex **32** was shown to generate ethane in 54% yield over 24 hours at RT. However, the addition of 1 equiv of oxidant to the Ni^{III} complex **32** led to clean formation of ethane in 84% yield, while addition of 2 equiv of oxidant to the Ni^{II} precursor generated an almost quantitative amount of ethane within 30 minutes at RT. These results suggest that access to a Ni^{IV} intermediate should lead to fast reductive elimination and C—C bond formation [31].

In 2016, Company and coworkers observed a Ni^{III}-OCl intermediate stabilized by a tetradentate macrocyclic ligand (Figure 10.22). Oxidation of the Ni^{II} complex with 3 equiv $Ca(OCl)_2$ leads to the formation of a square pyramidal complex

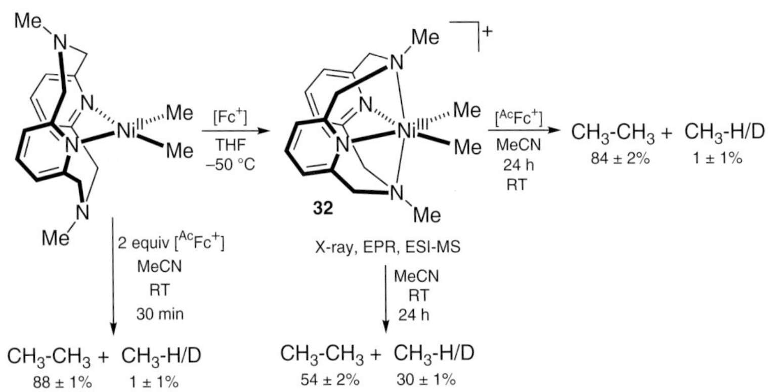

Figure 10.21 Synthesis and reactivity of the Ni^{III}-dialkyl complex reported by Mirica and coworkers in 2016.

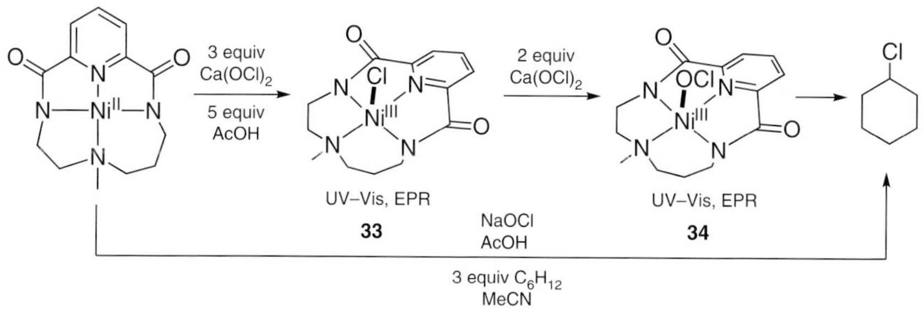

Figure 10.22 Catalytic oxidation of cyclohexane through a Ni^{III}-OCl intermediate.

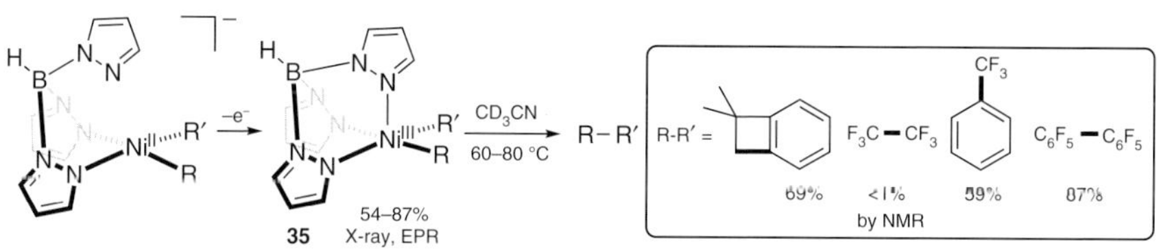

Figure 10.23 C—C bond formation from isolated organometallic Ni^{III} complexes stabilized by a trispyrazolylborate ligand.

33. However, addition of 2 equiv more of $Ca(OCl)_2$ leads to the formation of the Ni^{III}-perchlorate complex **34**, which is thought to be the active intermediate in the catalytic oxidation of cyclohexane [32].

In 2016, Sanford and coworkers isolated a series of Ni^{III} complexes **35**, stabilized by an anionic trispyrazolylborate ligand and different C(sp^2), C(sp^3), and CF_3 ligands (Figure 10.23). Upon heating at mild temperatures, these complexes undergo reductive elimination of the organic ligands to generate the corresponding C—C bond formation products [33].

In 2017, Diao and coworkers observed the formation of dinuclear Ni^{III} complexes **36** and **37** stabilized by benzoquinoline and carboxylate ligands (Figure 10.24). While complex **36** undergoes reductive elimination to generate

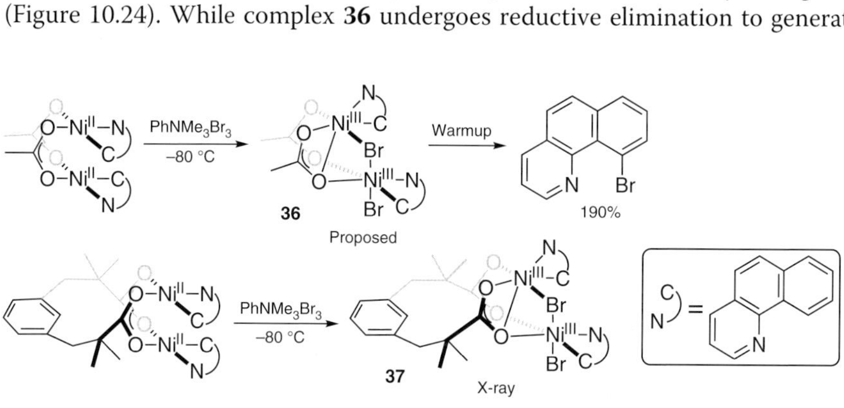

Figure 10.24 C—Br bond formation from a dinuclear Ni^{III} complex supported by benzoquinoline and carboxylate ligands.

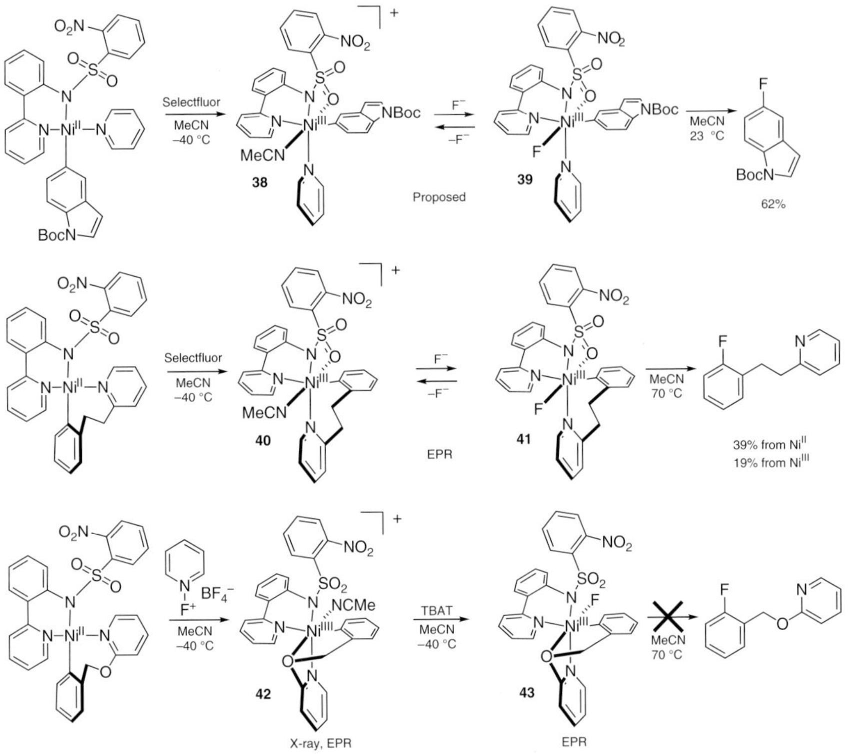

Figure 10.25 C—F bond formation from well-defined Ni^{III}(aryl)fluoride complexes reported by Ritter and coworkers in 2017.

10-bromobenzoquinoline, the structurally characterized complex **37** is stable and does not lead to C—Br bond formation [34].

In 2017, Ritter and coworkers reported C—F bond formation from well-defined Ni^{III}(aryl)fluorides complexes stabilized by the bidentate ligand 2-(2-pyridinyl)phenyl-2-nitrobenzenesulfonamide (Figure 10.25). These Ni^{III} species were generated by addition of Selectfluor to the Ni^{II} precursors. While complexes **38** and **39** could not be isolated since they rapidly generate the C—F bond formation product, complexes **40** and **41** could be detected by EPR prior to the reductive elimination step. By contrast, an isolable Ni^{III} complex **42** could be generated when an aryl ligand containing an O atom was employed. Although the Ni^{III}-F complex **43** was also detected by EPR, no C—F bond formation was detected in this case [35].

10.3 Organometallic Ni(IV) Complexes

In 2003, Dimitrov and Linden reported a quasi-tetrahedral Ni^{IV} complex **44** stabilized by four anionic ligands (three 1-norbornyl ligands and a halide ion) generated upon oxidation of a nickelate precursor with O_2 at −60 °C

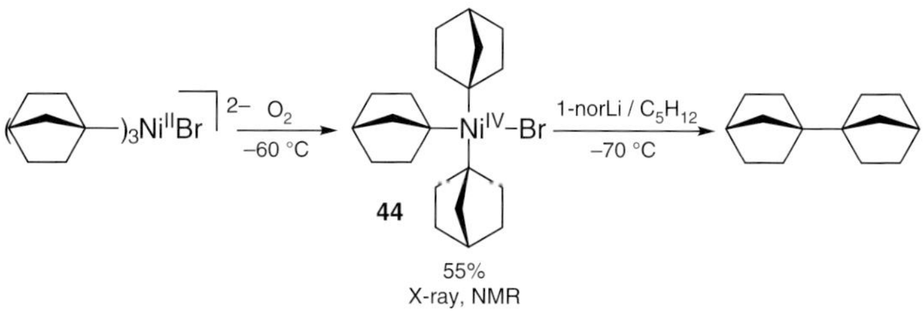

Figure 10.26 Formation of a quasi-tetrahedral Ni^{IV} complex and its C—C bond formation reactivity published by Dimitrov and coworkers in 2003.

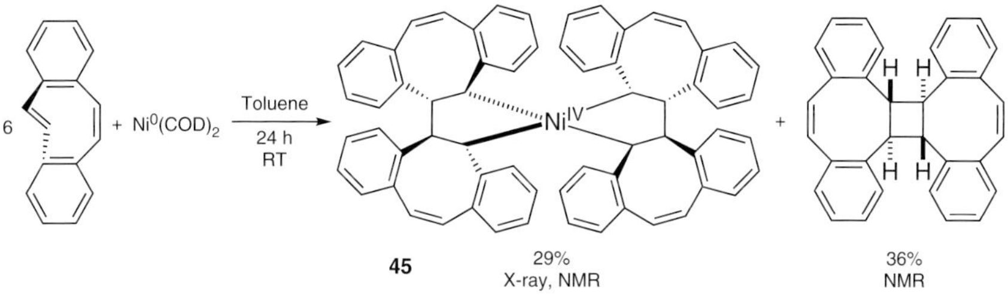

Figure 10.27 Synthesis of a stable tetraalkyl Ni^{IV} complex published by Nuckols and coworkers.

(Figure 10.26). Addition of 1 equiv of 1-norbornyl lithium to **44** in an attempt to remove the halide ligand and isolate a homoleptic Ni^{IV} complex was unsuccessful; instead, the C—C bond formation product di-norbornene was observed [36].

Nuckols and coworkers reported in 2009 a stable tetraalkyl Ni^{IV} complex **45** stabilized by two (5*Z*,11*E*)-dibenzo[*a*,*e*]cyclooctatetraene ligands (Figure 10.27). Interestingly, this Ni^{IV} complex is air-stable and can be heated to above 290 °C before any decomposition is observed, although during its synthesis a C—C bond formation product is obtained [37].

It is important to mention that starting in 2013, Chatani and coworkers reported Ni-catalyzed alkylation and arylation reactions involving C—H bond activation of benzamides and acrylamides. The proposed mechanisms involve a C—H bond activation via a concerted metalation–deprotonation step at a Ni^{II} species **46** to generate **47** (Figure 10.28), followed by oxidative addition of butyl bromide to generate a Ni^{IV} intermediate **48**. Subsequent reductive elimination to form **49** followed by protonation yields the alkylation product and regenerates the Ni^{II} precursor [38, 39].

In 2015, Sanford and Camasso isolated an octahedral Ni^{IV} complex **50** stabilized by a neutral scorpionate tris(2-pyridyl)methane ligand, a cyclic alky/aryl C-donor ligand (–CH_2CMe_2-*o*-C_6H_4–), and a trifluoromethyl group (Figure 10.29). Upon heating **50** at 95 °C for seven hours, the C—C bond formation benzocyclobutane product was obtained in quantitative yield, while addition of acetate generates the $C(sp^3)$—O bond formation product at RT [40].

Similarly, the anionic trispyrazolylborate ligand stabilizes a Ni^{IV} complex **51**, which does not undergo reductive elimination of benzocyclobutane. Addition

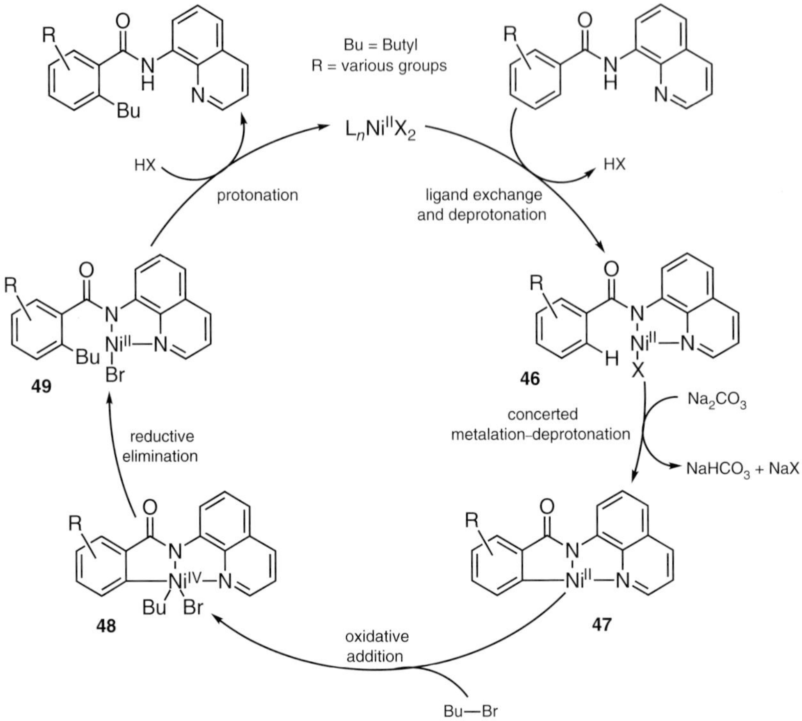

Figure 10.28 Proposed nickel-catalyzed alkylation mechanism involving Ni^{IV} intermediates by Chatani and coworkers.

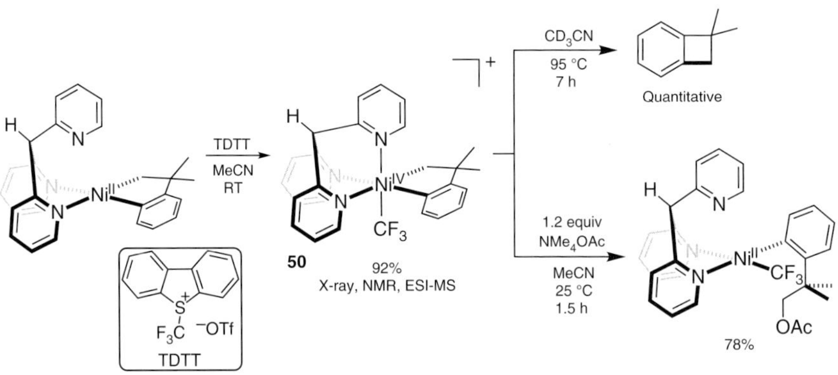

Figure 10.29 An isolated [tris(2-pyridyl)methane]Ni^{IV}(cycloneophyl)(CF_3) complex and its C(sp^2)—C(sp^3) and C(sp^3)—OAc bond formation reactivity.

of different nucleophiles results in quantitative formation of the corresponding C(sp^3)—X bond formation product (Figure 10.30), while addition of azide results in the formation of indoline in quantitative yields [40]. In addition, other Ni^{III} and Ni^{IV} complexes supported by the trispyrazolylborate ligand and various axial ligands were reported in 2017 [41].

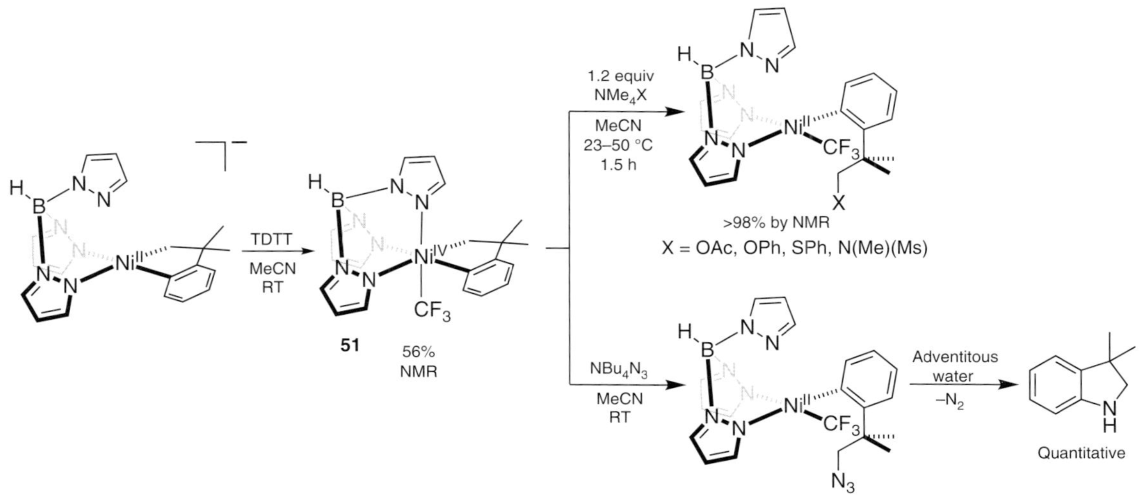

Figure 10.30 An isolated (trispyrazolylborate)Ni^{IV}(cycloneophyl)(CF_3) complex and its C(sp^3)—X bond formation reactivity.

In 2015, Sanford and coworkers reported the synthesis of various (trispyrazolylborate)Ni^{IV}(aryl)$(CF_3)_2$ complexes **52** using either a trifluoromethylating agent or aryl electrophiles (Figure 10.31). Upon heating, these complexes undergo trifluoromethylation of the Ni-aryl group, the electron-donating aryl group undergoing reductive elimination more rapidly [42].

In 2016, Fout and coworkers isolated an octahedral Ni^{IV} complex **53** supported by an anionic, tridentate bis(diisopropylphenyl-benzimidazol-2-ylidene)phenyl ligand and three bromide ligands (Figure 10.32). Complex **53** acts as a bromide transfer reagent for the bromination of different organic substrates including styrene, cyclohexene, mesitylmagnesium bromide, and lithium hexamethyldisilazide [43].

In 2017, Sanford and coworkers probed the intramolecular C—H bond activation at a Ni^{IV} center by using the bis(2-pyridyl)(2-^{R}phenyl)-fluoromethane ligand as a model system. Oxidation of the Ni^{II} complex under mild conditions results in C_{ipso}–H activation of the aryl ring on the ligand (Figure 10.33), and a range of (bis(2-pyridyl)(2-^{R}phenyl)-fluoromethane)$Ni^{IV}F(CF_3)_2$ octahedral complexes **54** were isolated in good yields [44].

Mirica and coworkers reported in 2017 the synthesis of organometallic Ni^{III} and Ni^{IV} complexes **55** and **56** supported by a triazacyclononane ligand (Me_3tacn) and a cyclic C-donor ligand (–CH_2CMe_2-*o*-C_6H_4–, Figure 10.34). Interestingly, the five-coordinate Ni^{III} complex **55** is quite stable, and only

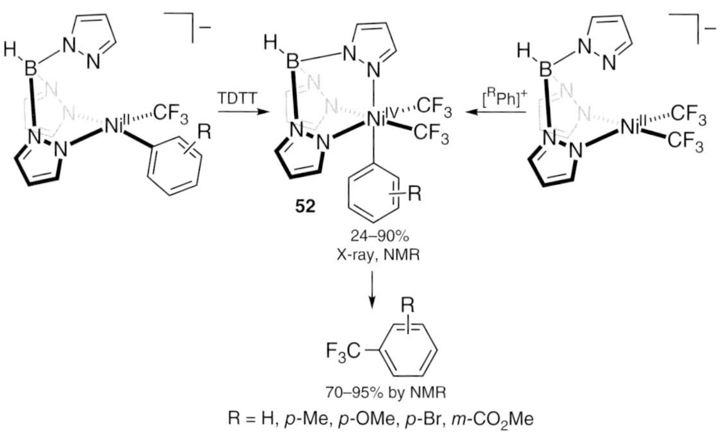

Figure 10.31 Isolated (trispyrazolylborate)Ni^{IV}(aryl)$(CF_3)_2$ complexes and their trifluormethylation reactivity.

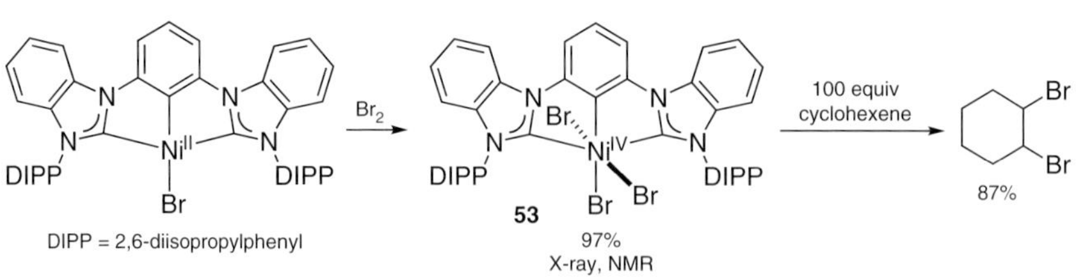

Figure 10.32 Bromination of alkenes using an isolated Ni^{IV}-tribromide complex.

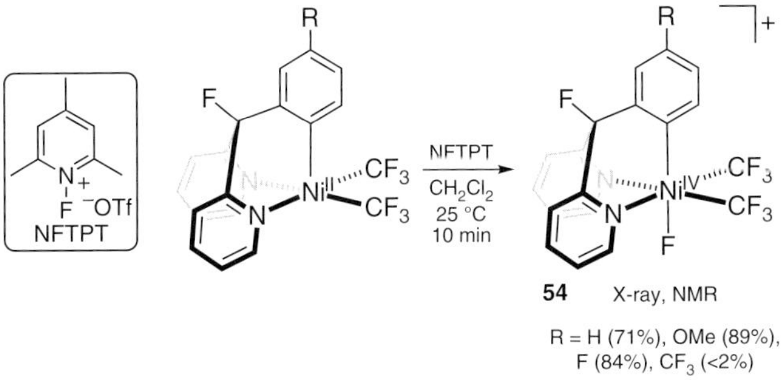

Figure 10.33 Oxidatively induced C–H activation of bis(2-pyridyl)(2-^{R}phenyl)-fluoromethane Ni–aryl$(CF_3)_2$ complexes.

gives the C—C bond formation product in moderate yields upon heating. By contrast, exposure of the Ni^{IV} complex **56** to blue LED light leads to rapid and quantitative C—C bond formation. When heated in the presence of chloride, the Ni^{IV} complex **56** generates both C(sp^2)—X and C(sp^3)—X bond formation products [45].

In 2017, Nebra and coworkers reported the isolation of a stable octahedral Ni^{IV} complex **57** supported by the monodentate ligands pyridine, trifluoromethyl, and fluoride, along with mononuclear and dinuclear Ni^{III} complexes **58** and **59** (Figure 10.35). These complexes were generated by the addition of different amounts of XeF_2 to the same $(py)_2Ni^{II}(CF_3)_2$ precursor. Interestingly, the Ni^{IV} complex **57** promotes the trifluoromethylation of 1,2-dichlorobenzene in quantitative yields [46].

10.4 Other High-Valent Ni Complexes

10.4.1 Additional Ni^{III} Complexes

Included here are several other Ni^{III} complexes that have been isolated, yet they exhibit limited reactivity (Figure 10.36). In 1982, Smart and coworkers isolated the Ni^{III} complex **60** upon the one-electron oxidation of decamethylnickelocene [47, 48]. Starting in 2006, Liaw and coworkers isolated a Ni^{III} species **6**, supported by a trianionic, tetradentate triphenylphosphine-type ligand with thiolate arms [49, 50]. Using the same ligand framework, in 2010 Lee and coworkers isolated the first Ni^{III}-alkyl complexes **61** [51]. One year later, Kinoshita and coworkers obtained a halide-bridged Ni^{III} dinuclear complex **62** supported by the tetradentate ligand tris(2-pyridylthio)methanide [52]. Complex **62** slowly converts into the mononuclear Ni^{III} complex **63**, during which the ligand isomerizes to bis(2-pyridylthio)(2-thiopyridinium)methyl [52]. In 2012, Ray and coworkers observed a trigonal bipyramidal high-valent Ni^{III}-oxo complex **64** stabilized by the tetradentate tris[2-(*N*-tetramethylguanidyl)ethyl]amine ligand [53]. In the same year, Patra and coworkers observed a high-valent Ni^{III} species **65** supported by two tridentate *N*-2-methylthiophenyl-2′-pyridinecarboxamide ligands [54].

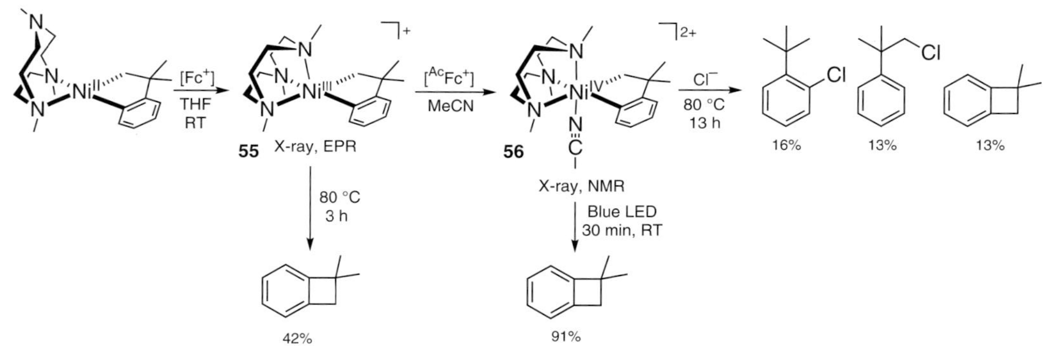

Figure 10.34 Synthesis of $(Me_3tacn)Ni^{III/IV}(CH_2CMe_2\text{-}o\text{-}C_6H_4)$ complexes and their C—C and C—Cl bond formation reactivity.

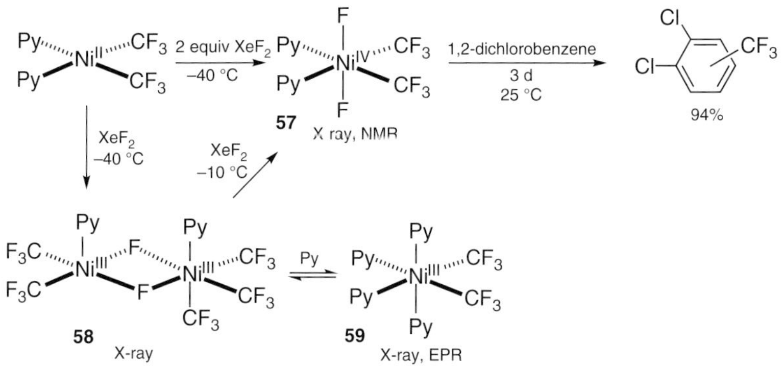

Figure 10.35 Trifluoromethylation of 1,2-dichlorobenzene using an isolated Ni^{IV} complex supported by pyridine.

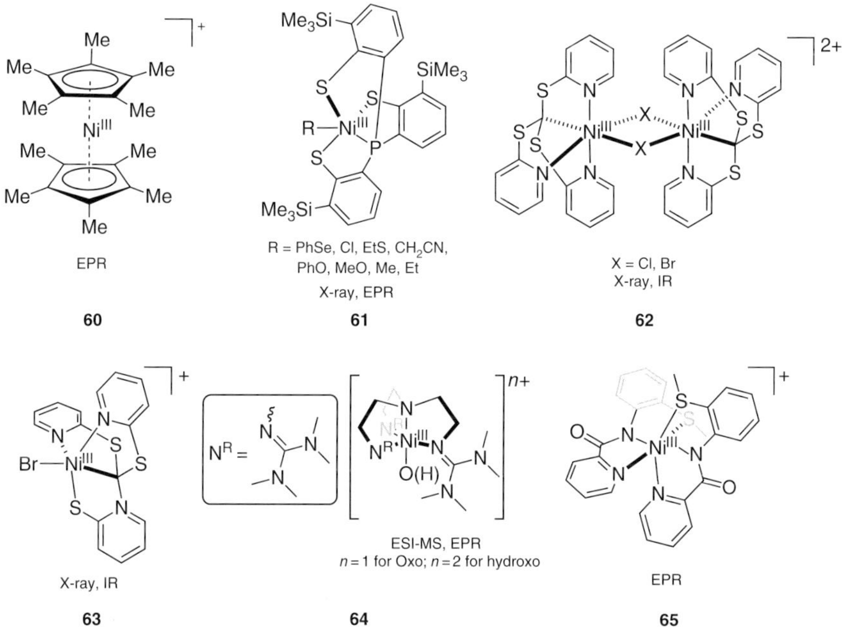

Figure 10.36 Additional isolated Ni^{III} complexes.

10.4.2 Additional Ni^{IV} Complexes

Included here are several other Ni^{IV} complexes that have been isolated, yet they exhibit limited reactivity (Figure 10.37). In 1982, the Ni^{IV} complex **66** was observed by Smart and coworkers upon oxidation by two electrons of the decamethylnickelocene Ni^{II} precursor [47, 48]. In 1994, Klein and coworkers published the first organometallic Ni^{IV} complex, **67**, which is stabilized by a chelating acylphenolato dianion, two trimethylphosphine ligands, and an iodide and methyl group [55, 56]. Another similar Ni^{IV} complex, **68**, which also adopts an octahedral geometry was reported a few years later [57]. In 1995, Gould and

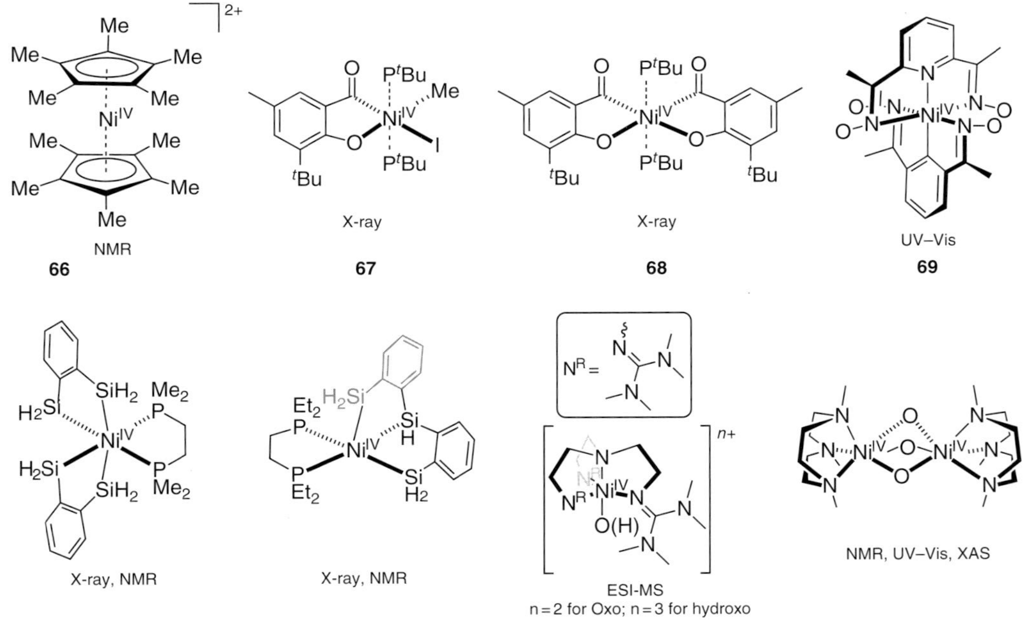

Figure 10.37 Additional isolated Ni^{IV} complexes.

coworkers isolated an octahedral Ni^{IV} complex **69** stabilized by two tridentate 2,6-diacetylpyridine dioxime ligands [58, 59]. A bis(disilyl)-Ni^{IV} complex **70** was isolated in 1999 by Tanka and coworkers [60], while in 2004 Saigo and coworkers reported a (trisilyl)Ni^{IV} complex **71** [61]. Ray and coworkers reported in 2012 a trigonal bipyramidal Ni^{IV}-oxo species **72** supported by the tetradentate ligand (tris[2-(*N*-tetramethylguanidyl)ethyl] amine) and that was observed by ESI-MS [53]. Interestingly, the first dinuclear Ni^{IV} complex **73** was recently reported by Browne and coworkers, in which the Ni^{IV} centers are stabilized by two 1,4,7-trimethyl-1,4,7-triazacyclononane ligands and bridged by three μ-oxido groups [62].

10.5 Conclusions and Outlook

This chapter provides an overview of the organometallic chemistry of isolated Ni^{III} and Ni^{IV} complexes and the proposed involvement of Ni^{III} and Ni^{IV} intermediates in various organometallic transformations. Although most C–C and C–heteroatom coupling reactions that are catalyzed by Ni are proposed to involve mostly $Ni^{0/II}$ oxidation states, in the past decade there has been increasing evidence for the participation of Ni^{III} and Ni^{IV} intermediates in various organometallic transformations.

The first mononuclear organometallic Ni^{III} complex was observed in 1978, but it was not until 1983 when an organometallic Ni^{III} complex was actually isolated. Starting in 1995 there have been several reports proposing the involvement of Ni^{III} species during oxidatively induced C–C and C–heteroatom bond formation reactions from Ni^{III} precursors. However, it is only during the past decade that several isolated organometallic Ni^{III} complexes have been reported and their C–C and C–heteroatom bond formation reactivity studied in detail.

The first isolated organometallic Ni^{IV} complex was reported in 1982, but it was not until 2003 when the first C—C bond formation reactivity from Ni^{IV} organometallic complexes was observed. During the past five years there have been several reports showing the direct involvement of Ni^{IV} species in C–C and C–heteroatom bond formation reactions. In addition, dinuclear Ni^{III} complexes that are not stabilized by a Ni—Ni bond and their role in catalytic processes have also been reported. Most importantly, the interconversion between Ni^{III} and Ni^{IV} complexes has been demonstrated, suggesting that it is not always possible to unambiguously confirm the oxidation state responsible for the observed reactivity.

Finally, we envision that in the near future the chemistry of these less common oxidation states of Ni will lead to novel organometallic reactions that involve one- and two-electron oxidation processes. This includes higher yielding C–C and C–heteroatom bond formation reactions and a better understanding of the catalytic cycles involving high-valent Ni species. Specifically, we anticipate that Ni^{IV} complexes can be tailored to promote selective $C(sp^3)$- and $C(sp^2)$–heteroatom coupling reactions. A key challenge for achieving this objective will be to define a class of oxidants and supporting ligands that selectively support the Ni^{IV} oxidation state rather than a Ni^{III} oxidation state. This should enable the design of catalytic sequences in which a Ni—C bond forming step is coupled with oxidation and C–heteroatom coupling via a $Ni^{II/III/IV}$ catalytic cycle.

References

1 Tsou, T.T. and Kochi, J.K. (1978). Reductive coupling of organometals induced by oxidation - detection of metastable paramagnetic intermediates. *J. Am. Chem. Soc.* 100 (5): 1634–1635.

2 Grove, D.M., Van Koten, G., Zoet, R. et al. (1983). Unique stable organometallic nickel(III) complexes; syntheses and the molecular structure of $Ni[C_6H_3(CH_2NMe_2)_2$-*o,o'*$]I_2$. *J. Am. Chem. Soc.* 105 (5): 1379–1380.

3 Grove, D.M., Van Koten, G., Mul, P. et al. (1986). Arylnickel(III) species containing nitrogen trioxide, nitrogen dioxide, and thiocyanate ligands. ESR data and the X-ray crystal structure of hexacoordinate (pyridine)bis(thiocyanato)[*o,o'*-bis{(dimethylamino)methyl}phenyl]nickel(III). *Organometallics* 5 (2): 322–326.

4 Grove, D.M., Van Koten, G., Mul, P. et al. (1988). Syntheses and characterization of unique organometallic nickel(III) aryl species. ESR and electrochemical studies and the X-ray molecular study of square-pyramidal $[Ni\{C_6H_3(CH_2NMe_2)_2$-*o,o'*$\}I_2]$. *Inorg. Chem.* 27 (14): 2466–2473.

5 Grove, D.M., Vankoten, G., Ubbels, H.J.C. et al. (1984). Organonickel(Ii) complexes of the terdentate monoanionic ligand-*Omicron,Omicron'*-bis[(dimethylamino)methyl]phenyl (N-C-N) - syntheses and the X-ray crystal-structure of the stable nickel(Ii) formate $[Ni(N\text{-}C\text{-}N)O_2CH]$. *Organometallics* 3 (7): 1003–1009.

6 vandeKuil, L.A., Grove, D.M., Gossage, R.A. et al. (1997). Mechanistic aspects of the Kharasch addition reaction catalyzed by organonickel(II) complexes containing the monoanionic terdentate arlydiamine ligand system $[C_6H_2(CH_2NMe_2)_2$-2,6-R-4]. *Organometallics* 16 (23): 4985–4994.

7 Cloutier, J.P. and Zargarian, D. (2018). Functionalization of the aryl moiety in the pincer complex $(NCN)Ni^{III}Br_2$: insights on Ni^{III}-promoted carbon–heteroatom coupling. *Organometallics* 37 (9): 1446–1455.

8 van de Kuil, L.A., Veldhuizen, Y.S.J., Grove, D.M. et al. (1995). A novel enantiopure proline-based organonickel(III) halide monocation with a pentadentate C, N_2, O_2-bonded bis(*ortho*-chelating) aryldiamine ligand. *J. Organomet. Chem.* 488 (1–2): 191–197.

9 Koo, K. and Hillhouse, G.L. (1995). Carbon–nitrogen bond formation by reductive elimination from nickel(II) amido alkyl complexes. *Organometallics* 14 (9): 4421–4423.

10 Han, R.Y. and Hillhouse, G.L. (1997). Carbon–oxygen reductive-elimination from nickel(II) oxametallacycles and factors that control formation of ether, aldehyde, alcohol, or ester products. *J. Am. Chem. Soc.* 119 (34): 8135–8136.

11 Koo, K.M., Hillhouse, G.L., and Rheingold, A.L. (1995). Oxygen-atom transfer from nitrous-oxide to an organonickel(II) phosphine complex - syntheses and reactions of new nickel(II) aryloxides and the crystal-structure

of ($Me_2PCH_2CH_2PMe_2$)Ni(O-*o*-$C_6H_4CMe_2CH_2$). *Organometallics* 14 (1): 456–460.

12 Lin, B.L., Clough, C.R., and Hillhouse, G.L. (2002). Interactions of aziridines with nickel complexes: oxidative-addition and reductive-elimination reactions that break and make C—N bonds. *J. Am. Chem. Soc.* 124 (12): 2890–2891.

13 Kogut, E., Wiencko, H.L., Zhang, L.B. et al. (2005). A terminal Ni(III)-imide with diverse reactivity pathways. *J. Am. Chem. Soc.* 127 (32): 11248–11249.

14 Pandarus, V. and Zargarian, D. (2007). New pincer-type diphosphinito (POCOP) complexes of nickel. *Organometallics* 26 (17): 4321–4334.

15 Pandarus, V. and Zargarian, D. (2007). New pincer-type diphosphinito (POCOP) complexes of Ni^{II} and Ni^{III}. *Chem. Commun.* (9): 978–980.

16 Spasyuk, D.M., Zargarian, D., and van der Est, A. (2009). New POCN-type pincer complexes of nickel(II) and nickel(III). *Organometallics* 28 (22): 6531–6540.

17 Castonguay, A., Beauchamp, A.L., and Zargarian, D. (2008). Preparation and reactivities of PCP-type pincer complexes of nickel. Impact of different ligand skeletons and phosphine substituents. *Organometallics* 27 (21): 5723–5732.

18 Iluc, V.M., Miller, A.J.M., Anderson, J.S. et al. (2011). Synthesis and characterization of three-coordinate Ni(III)-imide complexes. *J. Am. Chem. Soc.* 133 (33): 13055–13063.

19 Jones, G.D., McFarland, C., Anderson, T.J. et al. (2005). Analysis of key steps in the catalytic cross-coupling of alkyl electrophiles under Negishi-like conditions. *Chem. Commun.* (33): 4211–4213.

20 Klein, A., Vicic, D.A., Biewer, C. et al. (2012). Oxidative cleavage of CH_3 and CF_3 radicals from BOXAM nickel complexes. *Organometallics* 31 (15): 5334–5341.

21 Zhang, C.-P., Wang, H., Klein, A. et al. (2013). A five-coordinate nickel(II) fluoroalkyl complex as a precursor to a spectroscopically detectable Ni(III) species. *J. Am. Chem. Soc.* 135 (22): 8141–8144.

22 Lipschutz, M.I., Yang, X., Chatterjee, R. et al. (2013). A structurally rigid bis(amido) ligand framework in low-coordinate Ni(I), Ni(II), and Ni(III) analogues provides access to a Ni(III) methyl complex via oxidative addition. *J. Am. Chem. Soc.* 135: 15298.

23 Lipschutz, M.I. and Tilley, T.D. (2014). Carbon–carbon cross-coupling reactions catalyzed by a two-coordinate nickel(II)-bis(amido) complex via observable Ni^{I}, Ni^{II}, and Ni^{III} intermediates. *Angew. Chem. Int. Ed.* 53 (28): 7290–7294.

24 Zheng, B., Tang, F.Z., Luo, J. et al. (2014). Organometallic nickel(III) complexes relevant to cross-coupling and carbon–heteroatom bond formation reactions. *J. Am. Chem. Soc.* 136 (17): 6499–6504.

25 Tang, F.Z., Rath, N.P., and Mirica, L.M. (2015). Stable bis(trifluoromethyl)nickel(III) complexes. *Chem. Commun.* 51 (15): 3113–3116.

26 Zhou, W., Schultz, J.W., Rath, N.P. et al. (2015). Aromatic methoxylation and hydroxylation by organometallic high-valent nickel complexes. *J. Am. Chem. Soc.* 137 (24): 7604–7607.

27 Yu, S., Dudkina, Y., Wang, H. et al. (2015). Accessing perfluoroalkyl nickel(ii), (iii), and (iv) complexes bearing a readily attached [C_4F_8] ligand. *Dalton Trans.* 44 (45): 19443–19446.
28 Xu, H.W., Diccianni, J.B., Katigbak, J. et al. (2016). Bimetallic C—C bond-forming reductive elimination from nickel. *J. Am. Chem. Soc.* 138 (14): 4779–4786.
29 Diccianni, J.B., Hu, C.H., Diao, T.N., and Bond, N.-N. (2016). Forming reductive elimination via a mixed-valent nickel(II)–nickel(III) intermediate. *Angew. Chem. Int. Ed.* 55 (26): 7534–7538.
30 Zhou, W., Zheng, S.A., Schultz, J.W. et al. (2016). Aromatic cyanoalkylation through double C—H activation mediated by Ni(III). *J. Am. Chem. Soc.* 138 (18): 5777–5780.
31 Schultz, J.W., Fuchigami, K., Zheng, B. et al. (2016). Isolated organometallic nickel(III) and nickel(IV) complexes relevant to carbon—carbon bond formation reactions. *J. Am. Chem. Soc.* 138 (39): 12928–12934.
32 Corona, T., Draksharapu, A., Padamati, S.K. et al. (2016). Rapid hydrogen and oxygen atom transfer by a high-valent nickel–oxygen species. *J. Am. Chem. Soc.* 138 (39): 12987–12996.
33 Bour, J.R., Camasso, N.M., Meucci, E.A. et al. (2016). Carbon–Carbon bond-forming reductive elimination from isolated nickel(III) complexes. *J. Am. Chem. Soc.* 138 (49): 16105–16111.
34 Diccianni, J.B., Hu, C.H., and Diao, T.N. (2017). Binuclear, high-valent nickel complexes: Ni—Ni bonds in aryl-halogen bond formation. *Angew. Chem. Int. Ed.* 56 (13): 3635–3639.
35 Lee, H., Borgel, J., and Ritter, T. (2017). Carbon–fluorine reductive elimination from nickel(III) complexes. *Angew. Chem. Int. Ed.* 56 (24): 6966–6969.
36 Dimitrov, V. and Linden, A. (2003). A pseudotetrahedral, high-oxidation-state organonickel compound: synthesis and structure of bromotris(1-norbornyl)nickel(IV). *Angew. Chem. Int. Ed.* 42 (23): 2631–2633.
37 Carnes, M., Buccella, D., Chen, J.Y.C. et al. (2009). A stable tetraalkyl complex of nickel(IV). *Angew. Chem. Int. Ed.* 48 (2): 290–294.
38 Aihara, Y. and Chatani, N. (2013). Nickel-catalyzed direct alkylation of C—H bonds in benzamides and acrylamides with functionalized alkyl halides via bidentate-chelation assistance. *J. Am. Chem. Soc.* 135 (14): 5308–5311.
39 Rouquet, G. and Chatani, N. (2013). Catalytic functionalization of C(sp^2)—H and C(sp^3)—H bonds by using bidentate directing groups. *Angew. Chem. Int. Ed.* 52 (45): 11726–11743.
40 Camasso, N.M. and Sanford, M.S. (2015). Design, synthesis, and carbon–heteroatom coupling reactions of organometallic nickel(IV) complexes. *Science* 347 (6227): 1218–1220.
41 Camasso, N.M., Canty, A.J., Ariafard, A. et al. (2017). Experimental and computational studies of high-valent nickel and palladium complexes. *Organometallics* 36 (22): 4382–4393.
42 Bour, J.R., Camasso, N.M., and Sanford, M.S. (2015). Oxidation of Ni(II) to Ni(IV) with aryl electrophiles enables Ni-mediated aryl–CF_3 coupling. *J. Am. Chem. Soc.* 137 (25): 8034–8037.

43 Martinez, G.E., Ocampo, C., Park, Y.J. et al. (2016). Accessing pincer bis(carbene) Ni(IV) complexes from Ni(II) via halogen and halogen surrogates. *J. Am. Chem. Soc.* 138 (13): 4290–4293.

44 Chong, E., Kampf, J.W., Ariafard, A. et al. (2017). Oxidatively induced C—H activation at high valent nickel. *J. Am. Chem. Soc.* 139 (17): 6058–6061.

45 Watson, M.B., Rath, N.P., and Mirica, L.M. (2017). Oxidative C—C bond formation reactivity of organometallic Ni(II), Ni(III), and Ni(IV) complexes. *J. Am. Chem. Soc.* 139 (1): 35–38.

46 D'Accriscio, F., Borja, P., Saffon-Merceron, N. et al. (2017). C—H bond trifluoromethylation of arenes enabled by a Robust, high-valent nickel(IV) complex. *Angew. Chem. Int. Ed.* 56 (42): 12898–12902.

47 Robbins, J.L., Edelstein, N., Spencer, B. et al. (1982). Syntheses and electronic-structures of decamethylmetallocenes. *J. Am. Chem. Soc.* 104 (7): 1882–1893.

48 Kolle, U., Khouzami, F., and Lueken, H. (1982). Permethylmetallocenes .3. Decamethylnickelocene - the neutral sandwich complex, the monocation, the dication, and their addition-reactions. *Chem. Ber. Recl.* 115 (3): 1178–1196.

49 Lee, C.M., Chuang, Y.L., Chiang, C.Y. et al. (2006). Mononuclear Ni(III) complexes $[Ni^{III}(L)(P(C_6H_3\text{-}3\text{-}SiMe_3\text{-}2\text{-}S)_3)]^{0/1-}$ (L = thiolate, selenolate, CH_2CN, Cl, PPh_3): relevance to the nickel site of [NiFe] hydrogenases. *Inorg. Chem.* 45 (26): 10895–10904.

50 Chiou, T.W. and Liaw, W.F. (2008). Mononuclear nickel(III) complexes $[Ni^{III}(OR)(P(C_6H_3\text{-}3\text{-}SiMe_3\text{-}2\text{-}S)_3)]^-$ (R = Me, Ph) containing the terminal alkoxide ligand: relevance to the nickel site of oxidized-form [NiFe] hydrogenases. *Inorg. Chem.* 47 (17): 7908–7913.

51 Lee, C.M., Chen, C.H., Liao, F.X. et al. (2010). Mononuclear Ni-III-alkyl complexes (Alkyl = Me and Et): relevance to the acetyl-CoA synthase and methyl-CoM reductase. *J. Am. Chem. Soc.* 132 (27): 9256–9258.

52 Kuwamura, N., Kitano, K., Hirotsu, M. et al. (2011). Redox-controlled, reversible rearrangement of a tris(2-pyridylthio)methyl ligand on nickel to an isomer with an "*N,S*-confused" 2-pyridylthiolate arm. *Chem. Eur. J.* 17 (38): 10708–10715.

53 Pfaff, F.F., Heims, F., Kundu, S. et al. (2012). Spectroscopic capture and reactivity of S=1/2 nickel(III)–oxygen intermediates in the reaction of a Ni^{II}-salt with *m*CPBA. *Chem. Commun.* 48 (31): 3730–3732.

54 Chatterjee, S.K., Roy, S., Barman, S.K. et al. (2012). Shuttling of nickel oxidation states in N_4S_2 coordination geometry versus donor strength of tridentate N_2S donor ligands. *Inorg. Chem.* 51 (14): 7625–7635.

55 Klein, H.F., Bickelhaupt, A., Jung, T. et al. (1994). Syntheses and properties of the 1st octahedral diorganonickel(IV) compounds. *Organometallics* 13 (7): 2557–2559.

56 Klein, H.F., Bickelhaupt, A., and Hammerschmitt, B. (1994). Ligand-induced fragmentation of methylnickel phenolates containing a 2-aldehyde function - structure of (3-*tert*-butyl-5-methyl-2-oxobenzoyl)-tris(trimethylphosphine) nickel. *Organometallics* 13 (8): 2944–2950.

57 Klein, H.F., Bickelhaupt, A., Lemke, M. et al. (1997). Trimethylphosphine complexes of diorganonickel(IV) moieties. *Organometallics* 16 (4): 668–676.

58 Mandal, S. and Gould, E.S. (1995). Electron-transfer .126. Reductions of oxime-bound nickel(IV). *Inorg. Chem.* 34 (15): 3993–3997.
59 Baucom, E.I. and Drago, R.S. (1971). Nickel(II) and nickel(IV) complexes of 2,6-diacetylpyridine dioxime. *J. Am. Chem. Soc.* 93 (24): 6469–6475.
60 Shimada, S., Rao, M.L.N., and Tanaka, N. (1999). Reaction of 1,2-disilylbenzene with bis-[1,2-bis(dimethylphosphino)ethane]nickel(0). Isolation and characterization of the first silylnickel(IV) complex. *Organometallics* 18 (3): 291–293.
61 Chen, W.Z., Shimada, S., Tanaka, M. et al. (2004). Reaction of $[2\text{-}(SiH_3)C_6H_4]_2SiH_2$ with $Ni(Et_2PCH_2CH_2PEt_2)(PEt_3)_2$: characterization of η^2-(Si–H)Ni and Ni^{IV}–H complexes. *J. Am. Chem. Soc.* 126 (26): 8072–8073.
62 Padamati, S.K., Anielone, D., Draksharapu, A. et al. (2017). Transient formation and reactivity of a high-valent nickel(IV) oxido complex. *J. Am. Chem. Soc.* 139 (25): 8718–8724.

Part IV

Carbon Dioxide Fixation

11

Carbon Dioxide Fixation via Nickelacycle

Ryohei Doi and Yoshihiro Sato

Hokkaido University, Faculty of Pharmaceutical Sciences, Nishi 6, Kita 12, Kita-ku, Sapporo 060-0812, Japan

11.1 Introduction: Carbon Dioxide as a C1 Building Block

Efficient utilization of carbon dioxide (CO_2) is an important issue for scientists. CO_2 is produced in large quantity as a waste of petroleum industry along with some biological and geological activities. This gaseous molecule possesses notorious greenhouse effect, which is believed to warm up the earth to cause disruption of the environment. CO_2, on the other hand, is a beneficial chemical used as a convenient coolant (dry ice), fire extinguisher, nontoxic solvent (supercritical fluid), and to carbonate drinks. Therefore, CO_2 is collected as a "product" in an industrial scale. Today, CO_2 is widely available due to its mass production and established manipulation.

From the view point of organic synthesis, CO_2 as a C1 building block holds some ideal properties. For example, CO_2 is practically nontoxic because even humans release CO_2 as a result of expiration. Because of stable CO_2 supply, CO_2 cylinders are readily and inexpensively purchased by the synthetic chemist. Introduction or pressurization of CO_2 into reaction vessels would cause no technical problems for most organic chemists who adopt standard Schlenk techniques; just changing the gas from nitrogen or argon to CO_2 would work well.

However, chemical transformation of CO_2 is hampered by its thermodynamic stability. A traditional method is to use reactive organometallic reagents such as organolithium or Grignard reagents, which suffer from low functional group tolerance due to strong basicity and high nucleophilicity. Kolbe–Schmidt process, which incorporates CO_2 into phenol, requires very high temperature and pressure. Therefore, development of a mild method to incorporate CO_2 is an important issue in organic chemistry, and many contributions to this arena have been published to date [1]. Especially, transition metal complexes have been employed for efficient and mild coupling reactions of CO_2 [2].

In this review, an approach that mediates nickelalactone complex to transform CO_2 into carboxylic acid derivatives is described (Scheme 11.1) [2h, 3]. Nickel is an abundant and cheap transition metal that shows a variety of interesting reactivity by modifying its oxidation state from 0 to IV [4]. One of the characteristic

Nickel Catalysis in Organic Synthesis: Methods and Reactions,
First Edition. Edited by Sensuke Ogoshi.

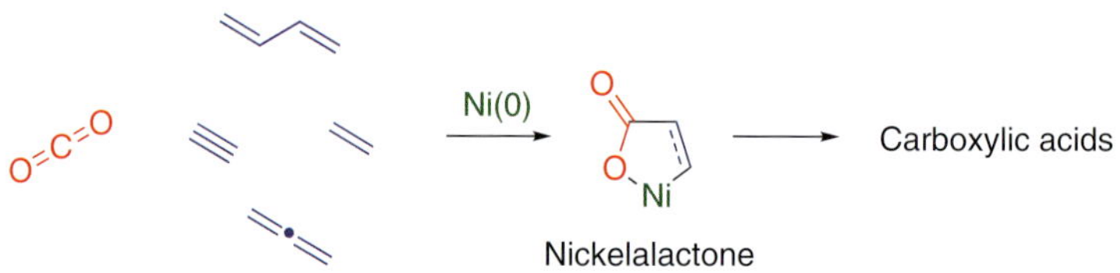

Scheme 11.1 Nickelalactone formation to mediate synthesis of carboxylic acid derivatives.

reactivities of nickel is oxidative cyclization, in which two unsaturated molecules form a nickelacycle complex by making a C—C bond [5]. Interestingly, Ni(0) is known to forge a C—C bond between CO_2 and an unsaturated molecule by oxidative cyclization. We describe the synthesis, characterization, and reactivity of nickelalactone complexes. Furthermore, this review covers the Ni-catalyzed carboxylation of alkyne, allene, diene, and alkene with CO_2 designed on the basis of an oxidative cyclization process [6]. These transformations could offer a brand new and contemporary methodology to utilize CO_2 efficiently in organic synthesis.

11.2 Formation, Structure, and Reactivity of Nickelalactone

11.2.1 Formation and Characterization of Nickelalactone via Oxidative Cyclization with CO_2

11.2.1.1 Reaction with Alkene

In 1982, Hoberg and Schaefer reported the formation of nickelalactone complex **1** by reaction of norbornene, a reactive strained alkene, with CO_2 and $Ni(cod)_2$ in the presence of 2,2′-bipyridine (Bpy) as an ancillary ligand (Scheme 11.2) [7]. The isolated nickel complex exhibited an IR absorption band at 1650 cm^{-1}, which suggests the existence of a carbonyl group. Elemental analysis was conducted to establish the formation of nickelalactone. In addition, they protonated the complex with aqueous HCl to afford *exo*-2-norbornanecarboxylic acid.

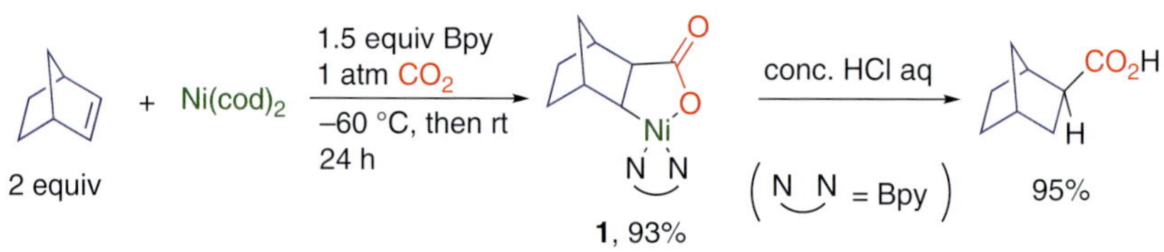

Scheme 11.2 Nickelalactone formation reported by Hoberg.

In an analogous manner, Hoberg and Schaefer succeeded in obtaining nickelalactones with ethylene [8]. The reaction was later recognized as an important elementary step in catalytic acrylic acid production (see Section 11.4). Therefore, oxidative cyclization of ethylene with CO_2 on Ni(0) was found to proceed with a variety of ligands (Figure 11.1) [9–13].

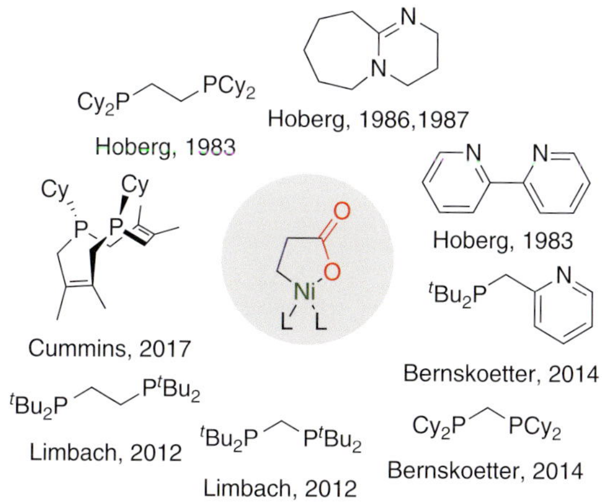

Figure 11.1 Ligands for the synthesis of nickelalactone from ethylene and CO_2.

Computational study of oxidative cyclization with ethylene was reported by Pápai et al. [14]. They revealed that C—C bond formation between ethylene and CO_2 is a single step in which an Ni-ethylene complex reacts with the incoming CO_2 molecule rather than an intramolecular oxidative cyclization to form the nickelacycle (Scheme 11.3). Later, Hofmann and coworkers proposed another mechanism they call "outer sphere" in which C—C bond formation occurs without interaction between CO_2 and Ni [15]. According to their calculation using THF as solvent, the outer sphere mechanism is favored when sterically hindered phosphine ligand was used.

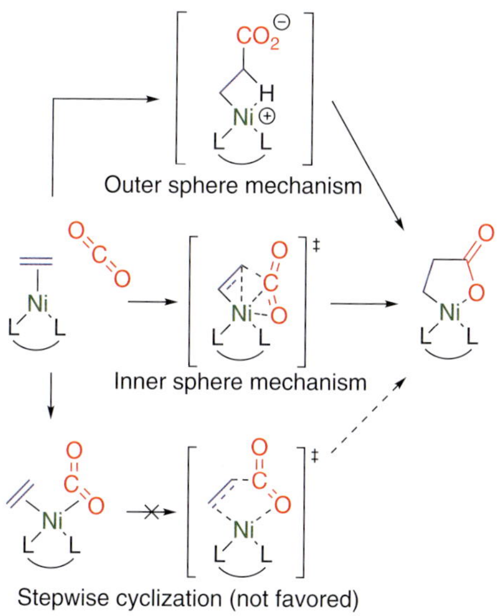

Scheme 11.3 Coupling pathways of ethylene and CO_2 on Ni(0).

Hoberg et al. revealed the structure of nickelalactone by the use of X-ray crystallography (Figure 11.2) [9]. The complex was a 16-electron Ni(II) complex with

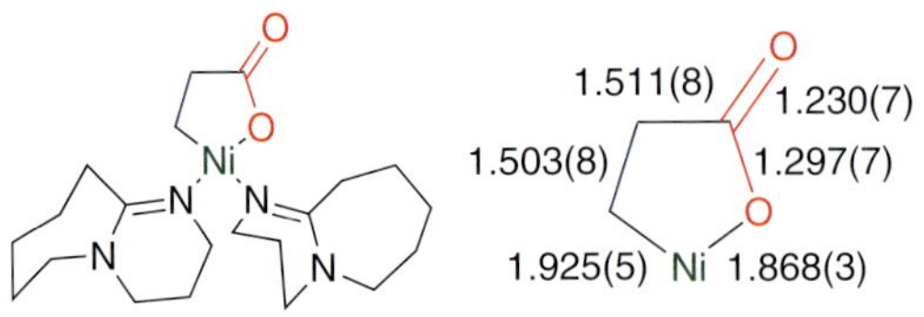

Figure 11.2 The first well-defined nickelalactone.

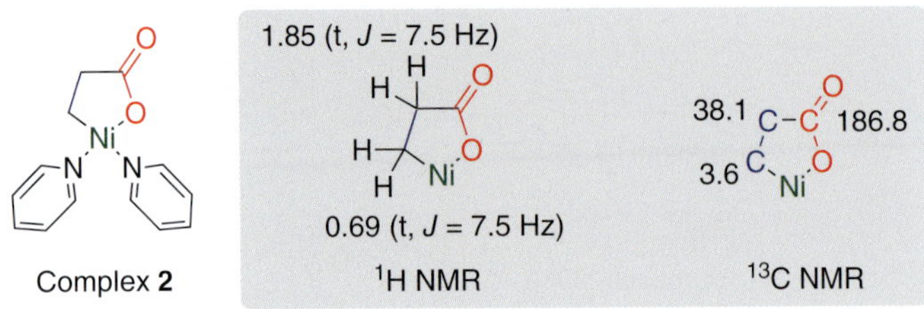

Figure 11.3 Selected resonance of nickelalactone complex **2**.

square-planar coordination. They found a difference in the distance between Ni—N bonds because the *trans* effect of Ni–C was stronger than that of Ni–O. Nickelalactone is mostly planar, and no interaction between C—H bonds and Ni was observed. Later, Hipler and Walther also reported X-ray crystallography of nickelalactones [16, 17].

In addition to IR spectrum that typically exhibits absorption at ~1600 cm^{-1}, some NMR data were provided. For example, signals derived from CH_2 adjacent to Ni in nickelalactone **2** having two pyridine ligands feature a characteristic high field shift in both ^1H and ^{13}C NMR studies [17]. In contrast, signals of CO_2 moiety are found at the low field region, which is typical for ester or carboxylate (Figure 11.3).

In 1986, coupling of styrene with CO_2 was found to proceed in the presence of DBU by Hoberg et al. (Scheme 11.3, DBU = 1,8-diazabicyclo[5.4.0]undec-7-ene) [18]. The nickelalactone itself was not characterized, but hydrolysis afforded the products as a mixture of regioisomers (Scheme 11.4).

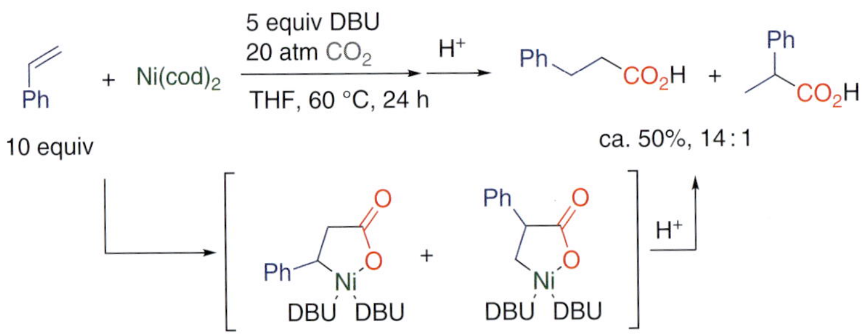

Scheme 11.4 Nickelalactone formation with styrene.

Other monosubstituted aliphatic alkenes were found to proceed with oxidative cyclization with CO_2 [12, 19, 20]. In the presence of Ni(cdt)/DCPE (DCPE = 1,2-bis(dicyclohexylphosphino)ethane), coupling of 1-hexene with

CO_2 gave heptanoic acids in 45% yield after hydrolysis (Scheme 11.5). Interestingly, in contrast to the reaction with styrene, carboxylation at the internal carbon was a major path.

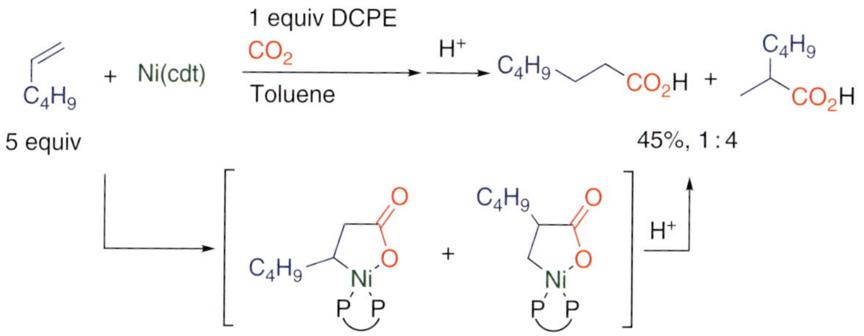

Scheme 11.5 Carboxylation with aliphatic alkene.

Furthermore, Hoberg et al. disclosed the oxidative cyclization of internal alkene [21–23]. For example, cyclooctene, which is an internal, non-strained olefin, was found to form the corresponding nickelalactone with CO_2 [21, 22]. It was found that employment of P,N-ligand **3** induced oxidative cyclization to afford nickelalactone **4** in 88% yield (Scheme 11.6). This ligand was found to be also effective in oxidative cyclization with cyclopentene [23]. In addition, by using DCPE, they also noted that oxidative cyclization occurred with 3-hexene, although they did not describe which isomer (*cis* or *trans*) was used in this reaction [19].

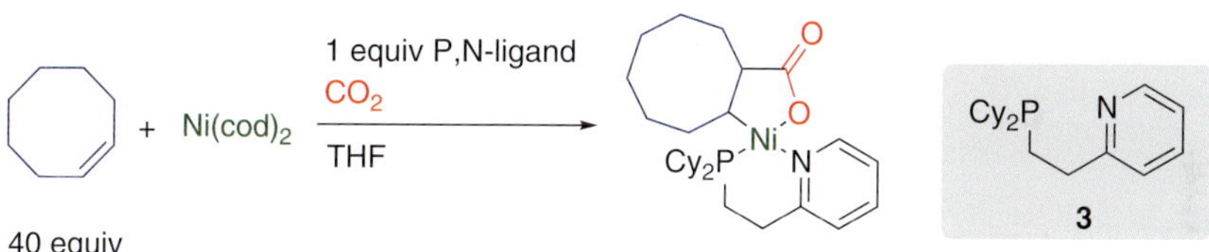

4, 88%

Scheme 11.6 Oxidative cyclization with cyclooctane.

11.2.1.2 Reaction with Allene

Oxidative cyclization of allene was first reported by Hoberg and Oster [24]. They treated Ni(0) complex with allene, 1,1-dimethylallene, and methylallene in the presence of Bpy or DCPE as ligand under CO_2 atmosphere. Regioselective C–C coupling at the C2 position is revealed by analyzing the hydrolysis product of isolated nickelalactone **5** (Scheme 11.7). Substituted allene tends to give an *exo*-cyclization product.

Mori and coworkers reported the oxidative cyclization of allene with CO_2 by using DBU as ligand (Scheme 11.8) [25]. They used DCl instead of HCl to protonate the nickelalactone to confirm the Ni–C connection. Mori and coworkers also reported the regioselective oxidative cyclization of CO_2 with an allene carrying a silyl group [26]. The resulting nickelalactone was treated with $ZnEt_2$ or $PhMe_2SiH$

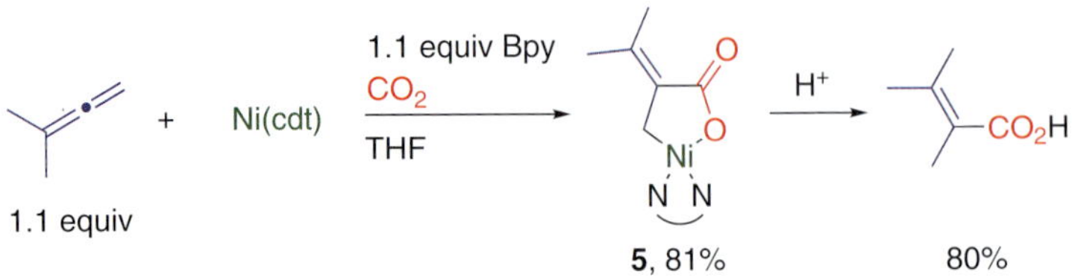

Scheme 11.7 Oxidative cyclization of allene with CO_2.

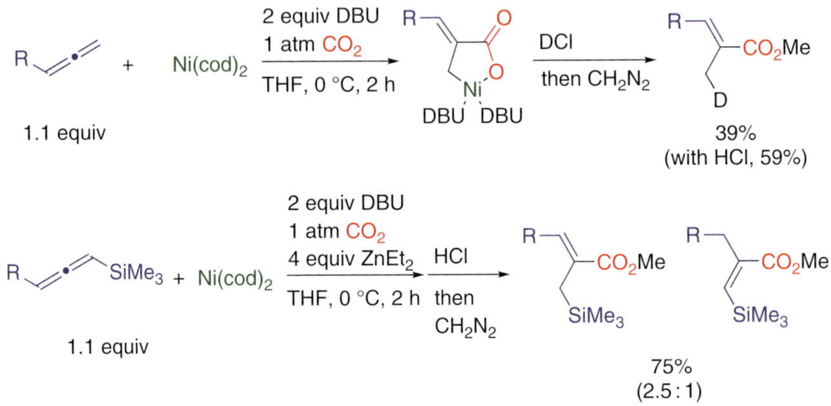

Scheme 11.8 Reaction of allene, nickel, and CO_2. R = CH_2CH_2OBn.

to afford the protonated product that gives slightly better yield than that by simple protonation with acid. In addition, Saito et al. reported the oxidative cyclization of allenamide with CO_2. The reaction proceeded in a regioselective manner due to the electronic bias of allenamide [27].

11.2.1.3 Reaction with Diene

Hoberg et al. reported the first coupling reaction of 1,3-diene with CO_2 in the presence of $Ni(cod)_2$ and DCPE [28, 29]. The nickelalactone **6** was not fully characterized. Instead, they characterized the corresponding methyl ester after treatment with HCl/MeOH (Scheme 11.9). Walther and coworkers obtained X-ray crystallography of nickelalactone generated by oxidative cyclization of diene with CO_2, which successfully confirmed the π-allyl coordination mode [17, 30, 31]. Behr and Kanne reported Ni-mediated coupling of 1,3,7-octatriene with CO_2 to afford the monocarboxylated product along with the double carboxylated product [32]. They also conducted a reaction with 1,3,5-hexatriene to afford several

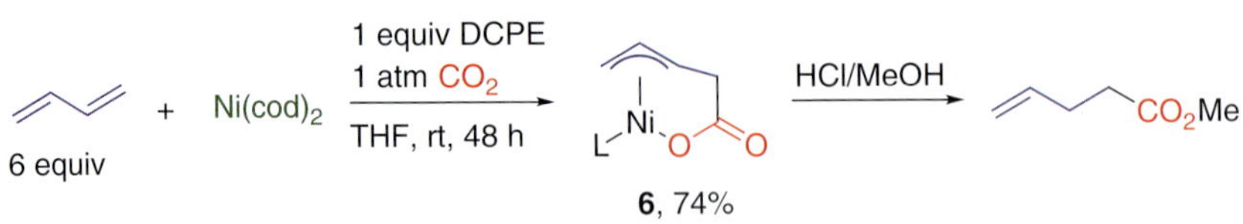

Scheme 11.9 Coupling of Ni(0), diene, and CO_2.

carboxylated products. Although the structure was not well defined, a 2/2 adduct of hexatriene and CO_2 was observed by GCMS.

Mori and coworker found that DBU was a superior ligand for the oxidative cyclization of diene with CO_2 [33]. In the presence of Ni(cod)$_2$, DBU, *trans*-1-phenyl-1,3-butadiene afforded a mixture of regioisomers after protonolysis in good yield (Scheme 11.10).

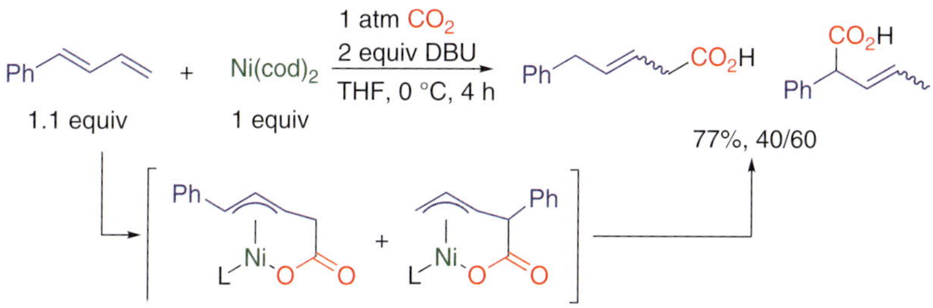

Scheme 11.10 Oxidative cyclization of diene and CO_2 with Ni(0)/DBU.

11.2.1.4 Reaction with Alkyne

Hoberg and coworker reported the oxidative cyclization of nickel, alkyne, and CO_2 to afford the unsaturated nickelalactone **7** (Scheme 11.11) [34]. In this case, tetramethylethylene diamine (TMEDA) was employed as a ligand. The complex **7** was characterized by IR band and analysis of the product after treatment with acid or carbon monoxide. The reaction of **7** with carbon monoxide would be driven by the formation of thermodynamically stable Ni(CO)$_4$. Tsuji and coworkers also reported the synthesis of nickelalactone by the use of 4-octyne with Bpy as an ancillary ligand [35].

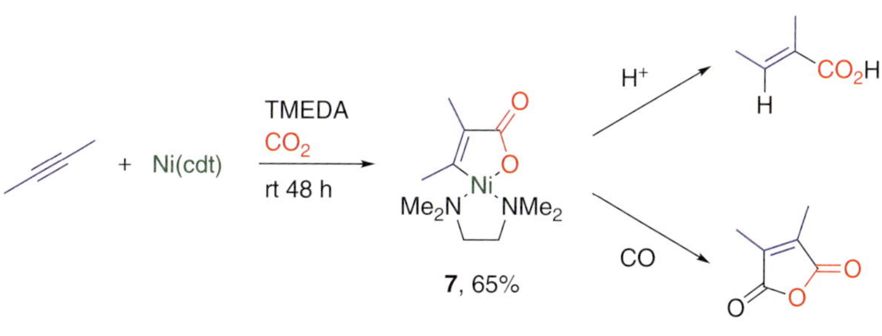

Scheme 11.11 Oxidative cyclization with 2-butyne.

The reaction of symmetric internal alkyne was later reinvestigated by Walther and coworkers to characterize by X-ray crystallography [36, 37]. For instance, Langer succeeded in obtaining X-ray data of unsaturated nickelalactones **8** and **9** having DBU or DCPE as ancillary ligands (Figure 11.4). The nickel atom adopts a square planar coordination. The electron donating nature of DBU makes the nickel center electron rich to shorten the Ni—C1 bond and elongate the C1—C2 bond in complex **8**. The author claims that these bond lengths indicate enhanced

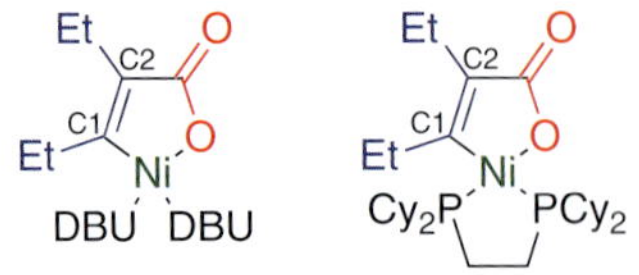

8 9

	8	9
Ni-C1	1.900	1.961
C1-C2	1.352	1.346

Figure 11.4 Structure of complexes **8** and **9**, and selected bond length (Å).

back-bonding of electrons of the nickel into the vacant π*-orbital of the C1—C2 double bond. The complex **9** having DCPE ligand possessed relatively longer Ni—C1 bond and shorter C1—C2 double bond. This can be explained by the partially π-accepting character of phosphine ligand rendering the back-donation of nickel relatively weak.

Yamamoto and coworkers reported carboxylation of terminal alkyne via regioselective nickelacycle formation (Scheme 11.12) [38]. Alkyne was added slowly to a THF solution containing Ni(cod)$_2$ and DBU, under CO_2 atmosphere, at 0 °C. The nickelalactone formed *in situ* was protonated by the addition of acid to obtain α,β-unsaturated carboxylic acid. A variety of alkynes containing aryl, alkyl, and silyl gave the desired product in good yield. In this case, addition of DBU was crucial: reactions with other nitrogen and phosphine ligands were unsuccessful. The reactions of enynes and diynes were also demonstrated to yield the corresponding unsaturated carboxylic acids.

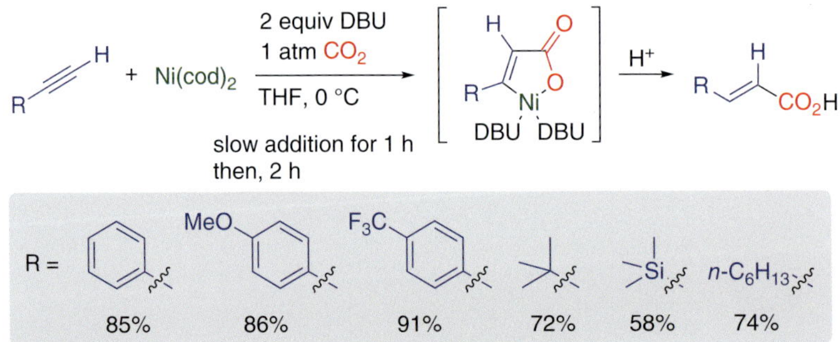

Scheme 11.12 Ni-mediated carboxylation of terminal alkynes.

In 2004, Iwasawa and coworkers adopted bidentate amidine ligands instead of monodentate DBU ligand for the oxidative cyclization of alkyne with CO_2 on Ni(0) to control regioselectivity (Scheme 11.13) [39]. Their trial expanded the scope of carboxylation to some internal alkynes.

Saito et al. reported the regioselective cyclization of ynol or ynamide with CO_2 on Ni(0) (Scheme 11.14) [40, 41]. A polarized C—C triple bond induced by an adjacent heteroatom was found to forge a C—C bond in a regioselective manner at the negatively charged β-position on Ni(0). A putative intermediate nickelacycle derived from ynamide was isolated and characterized by X-ray analysis later on [42].

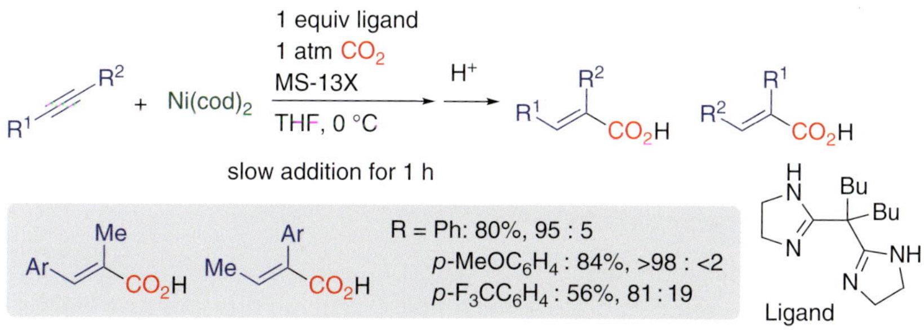

Scheme 11.13 Bidentate amidine ligand for nickelalactone formation.

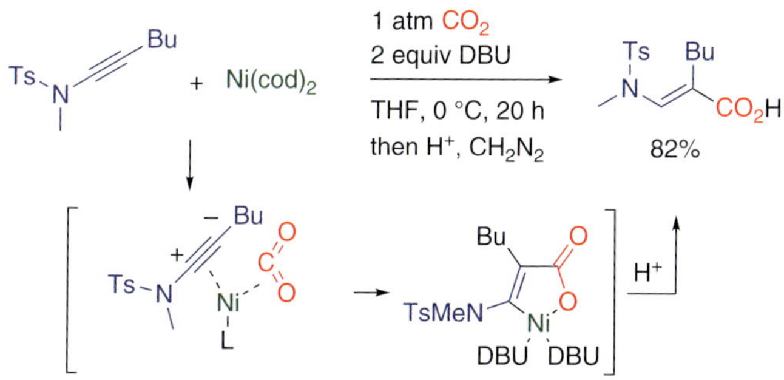

Scheme 11.14 Ni-mediated hydrocarboxylation of ynamide.

Some theoretical calculations to clarify the coupling process of alkyne and CO_2 on Ni(0) have been published [43, 44]. According to the calculation by Buntine and coworkers using acetylene and a simplified model ligand of DBU, concomitant formation of C—C bond and Ni—O bond is proposed (Figure 11.5) [44]. Carboxylation of monosubstituted alkynes was also studied. Except for trimethylsilylacetylene, steric repulsion between the substituent of alkyne and the incoming CO_2 increases the transition-state energy of internal carboxylation. However, 2-substituted nickelalactone formation is thermodynamically

Figure 11.5 Proposed reaction pathway of coupling of alkyne and CO_2.

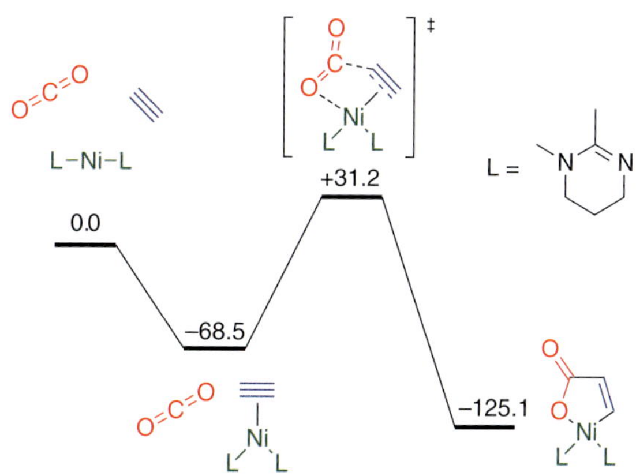

disfavored due to the steric repulsion between the substituent and the ligand. In the case of trimethylsilylacetylene, both steric repulsion with the ligand and CO_2 affect the transition state. Thus, the activation energies of trimethylsilylacetylene showed slight difference. They also revealed that use of DMF instead of THF as solvent lowers the activation barrier due to increased polarity in the transition state by the interaction of weakly polar alkyne and CO_2 with the nickel center.

11.2.1.5 Other Related Reactions

Hoberg et al. discovered that bisallyl Ni complex, generated *in situ* by the treatment of diene with Ni(0), incorporates CO_2 to give a novel nickelalactone complex (Scheme 11.15) [45]. The product was hydrolyzed to determine the structure. In addition, they found that the nickelalactone gives rise to homocoupling at elevated temperature to afford dicarboxylic acid [46].

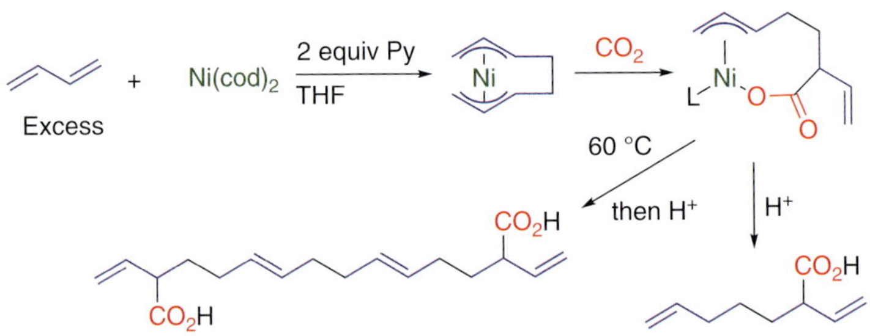

Scheme 11.15 Coupling of bisallyl nickel complex with CO_2.

Mori and coworkers designed a reaction of CO_2 with nickelacycle generated via intramolecular oxidative cyclization of alkene and alkyne (Scheme 11.16) [47, 48]. In the presence of Ni(cod)$_2$ and DBU under CO_2 atmosphere, 1,6-enyne **10** was carboxylated to give the desired product **11** via seven-membered nickelalactone.

Benett et al. reported a series of oxidative cyclization processes of Ni-aryne complexes [49–51]. In 1985, they prepared a Ni-aryne complex by reduction of (*o*-bromophenyl)bromonickel complex. The well-defined aryne complex

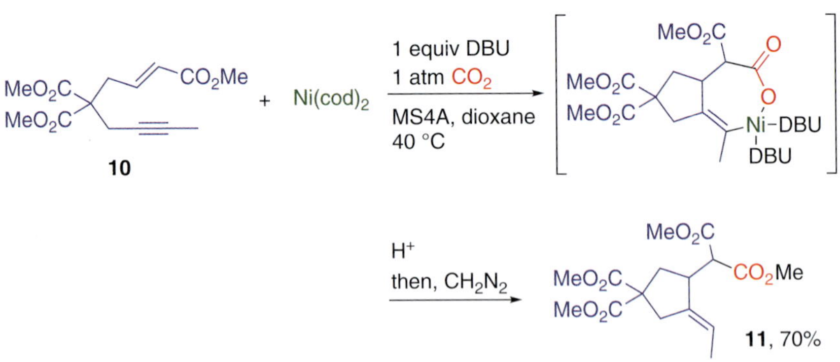

Scheme 11.16 Nickelalactone formation via insertion of CO_2 to five-membered nickelacycle.

was converted into nickelalactone by the addition of CO_2-saturated THF (Scheme 11.17). Later, the corresponding naphthalyne and cyclohexyne complex was isolated and reacted with CO_2 to afford nickelalactones.

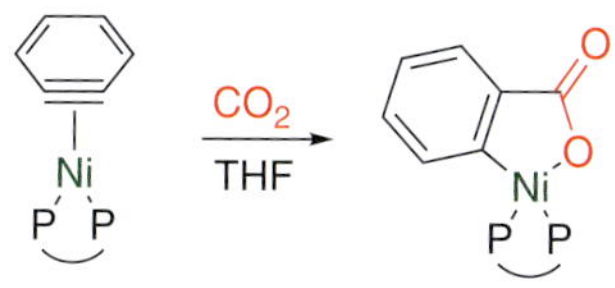

Scheme 11.17 Oxidative cyclization of aryne complex.

11.2.1.6 Generation of Nickelalactone Without CO_2

Some methods that do not use CO_2 to form a nickelalactone have been developed [16, 23, 52–55]. For example, the reaction of Ni(0) with acid anhydride gives nickelalactone via oxidative addition of C—O bond followed by decarbonylation. We do not explain in detail here, but these reactions are useful to prepare nickelalactone complex to examine their reactivity. In addition, Hoberg et al. reported the synthesis of nickelalactone by the reaction of β,γ-unsaturated carboxylic acid with Ni(0) probably via hydronickelation [23].

11.2.2 Reactivity of Nickelalactone

11.2.2.1 Transmetalation with Organometallic Reagent

Mori and coworkers discovered in 2001 that the reaction of nickelalactone generated *in situ* by the reaction of alkyne, CO_2, and Ni(0) with dimethylzinc resulted in the incorporation of the alkyl group (Scheme 11.18) [56]. The reaction might proceed via transmetalation of nickelalactone with organozinc reagent followed by reductive C—C bond forming process. Furthermore, Mori and coworkers utilized this transmetalation method for the synthesis of erythrocarine by using an alkyne carrying nitrogen nucleophile [57].

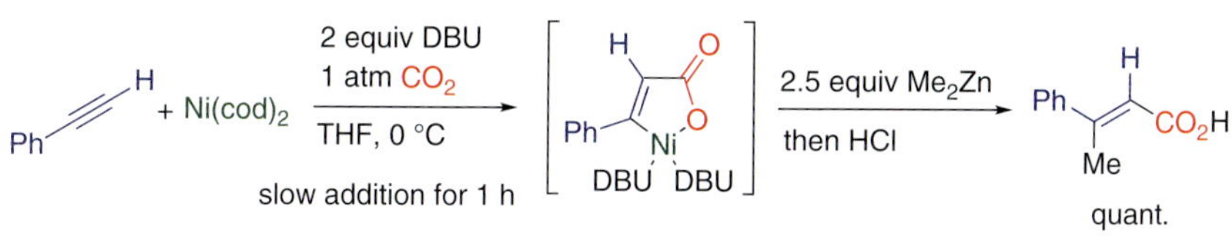

Scheme 11.18 Transmetalation of $ZnMe_2$ with nickelalactone.

Interestingly, Mori and coworker found that addition of $ZnMe_2$ to the nickelalactone generated *in situ* via coupling of substituted diene and CO_2 gives rise to double carboxylation (Scheme 11.19) [33]. It is postulated from the result of double carboxylation of 1,3-cyclohexadiene that the second carboxylation might occur via methyl-π-allylnickel complex, which attacks CO_2 from the backside of the allyl moiety. In contrast to $ZnMe_2$, treatment of nickelalactone with $ZnPh_2$ resulted in the incorporation of both CO_2 and Ph group to afford the monocarboxylated product.

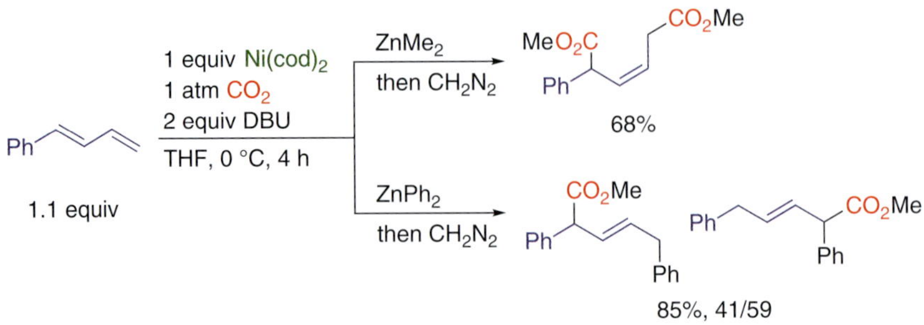

Scheme 11.19 Double carboxylation of diene with Ni(0)/$ZnMe_2$.

Rovis and coworkers achieved a novel nickel-mediated reaction of cyclic anhydride with organozinc reagent to afford a decarbonylative coupling product (Scheme 11.20) [58]. Nickelalactone was formed by the oxidative addition of low-valent nickel to the C—O bond of cyclic anhydride. The putative C—C bond forming step is the reductive elimination of nickel complex generated by transmetalation of organozinc reagent with nickelalactone. In this reaction, the mixed ligand system neocuproine and DPPB was used (DPPB = 1,4-bis(diphenylphosphino)butane). The neocuproine complex is an active species whereas DPPB complex, which might possess an electron-rich nickel center, removes CO from the neocuproine complex to prevent the formation of ketone product.

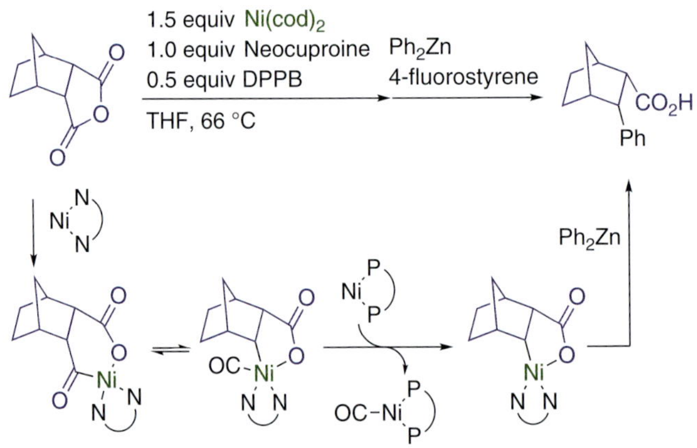

Scheme 11.20 Decarbonylative coupling of cyclic anhydride with organozinc.

In this paper, other interesting isomerizations of nickelalactone are proposed (Scheme 11.21). Unsaturated succinic anhydride derivative **12** was transformed into carboxylic acid **13**, presumably by the insertion of alkene into the initially formed nickelalactone complex. Furthermore, glutaric anhydride gave the β-coupling product **14**. This product might be obtained as a result of isomerization of δ-nickelalactone via β-hydride elimination/hydrometalation sequence.

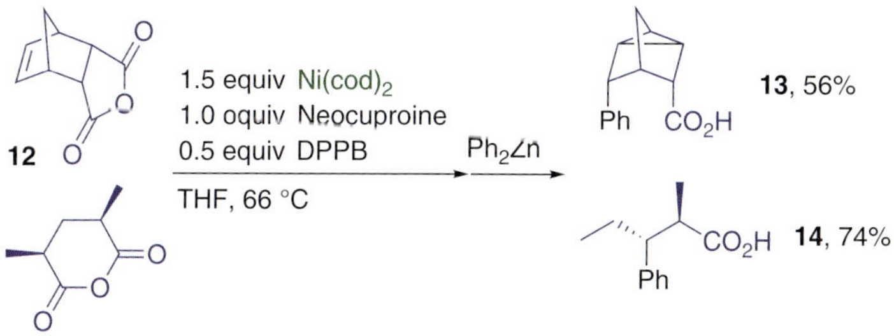

Scheme 11.21 Ni-catalyzed decarbonylative coupling.

11.2.2.2 β-Hydride Elimination

In general, β-hydride elimination of nickelalactone is very slow due to a relatively rigid structure in which efficient interaction for β-hydride elimination between Ni and β-H is unlikely. However, addition of Lewis acid induced β-hydride elimination of nickelalactone [22, 23, 30]. For instance, nickelalactone **4** generated via oxidative cyclization of cyclooctene with CO_2 was converted into the corresponding unsaturated carboxylic acids **15** or **16** by the addition of BF_3•Et_2O or $BeCl_2$, respectively (Scheme 11.22). Thus, addition of BF_3•Et_2O gave α,β-unsaturated carboxylic acid while $BeCl_2$ induced the formation of β,γ-unsaturated carboxylic acid.

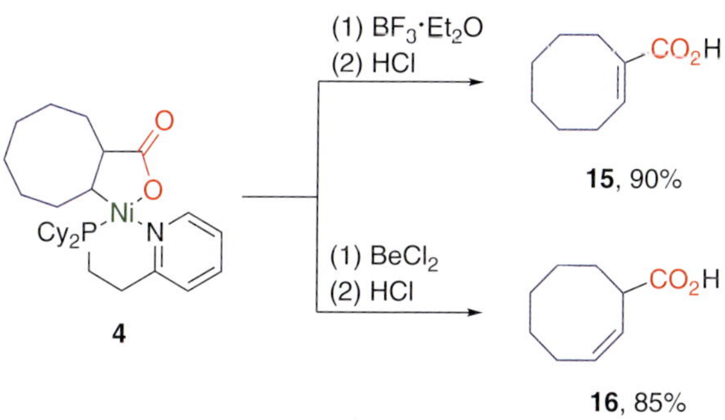

Scheme 11.22 Lewis-acid-induced β-hydride elimination of nickelalactone.

Bernskoetter and coworkers found that addition of tris(pentafluorophenyl) borane caused β-hydride elimination of nickelalactone **17** bearing DPPF (Scheme 11.23) [59]. A Lewis adduct **18** was initially observed by NMR. Then, the complex was completely converted into the four-membered nickelacycle **19** via β-hydride elimination followed by 2,1-insertion. They succeeded in obtaining an acrylate complex by deprotonation of complex **19** with a base.

In the case of saturated nickelalactone, the complex has α-proton, which might be acidic. Thus, addition of a strong base results in deprotonation followed by rearrangement to give the low-valent nickel and enolate. For example, Limbach and coworkers demonstrated that addition of sodium *tert*-butoxide induces quick transformation of nickelalactone into an acrylate complex (Scheme 11.24) [10].

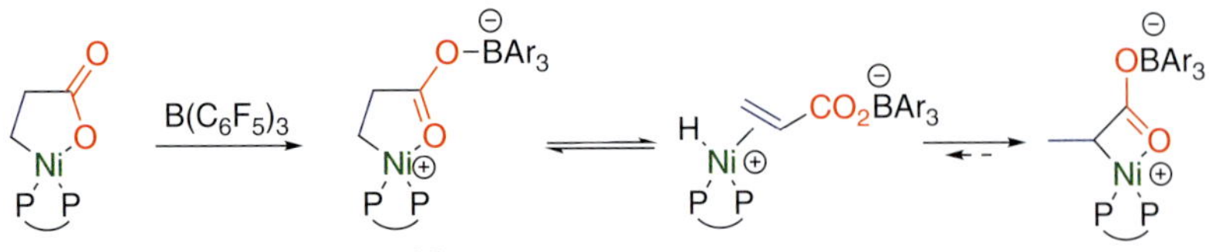

Scheme 11.23 $B(C_6F_5)_3$-induced β-hydride elimination of nickelalactone.

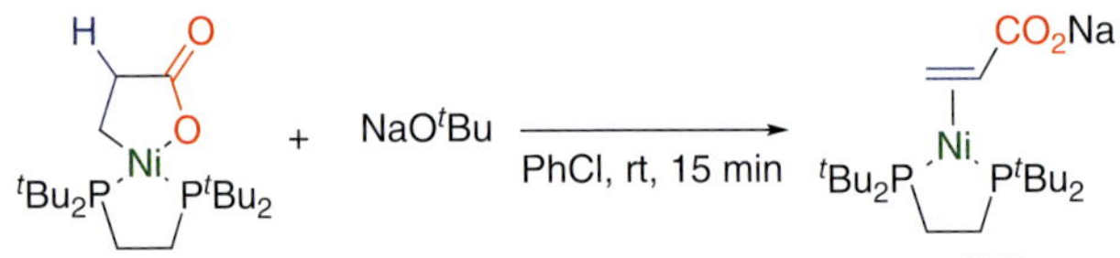

Scheme 11.24 Deprotonation of nickelalactone.

They support a deprotonation pathway based on quantum chemical calculation study.

11.2.2.3 Insertion of Another Unsaturated Molecule

Nickelalactone intermediate is known to cause a variety of insertion processes with unsaturated molecules such as alkyne and alkene. Inoue et al. have found the reaction between two alkynes and CO_2 to produce pyrone in which oxidative cyclization between alkyne and CO_2 occurs first, and then another alkyne insertion occurs (see Section 11.3). Hoberg et al. found that modification of the reaction condition of coupling between ethylene and CO_2 results in the formation of 4-pentenoic acid as a major product (Scheme 11.25) [60]. In this case, insertion of ethylene into nickelalactone was believed to occur to elongate the carbon chain.

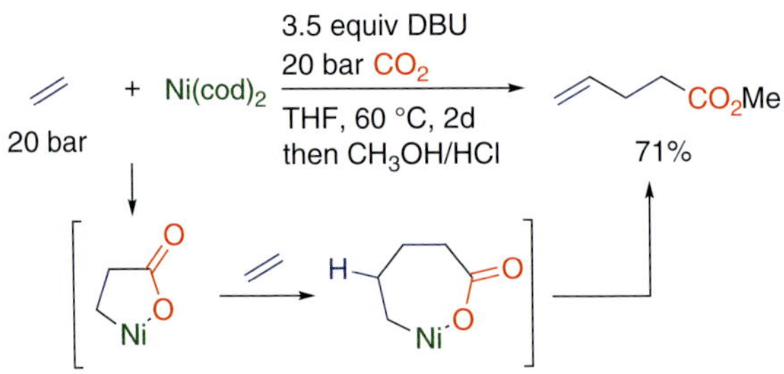

Scheme 11.25 Coupling of ethylene and CO_2.

Insertion of CO is also a well-known process for nickelalactone chemistry to provide the cyclic anhydride. Hoberg and Ballesteros demonstrated the insertion of 4-phenyl-1,2,4-triazoline-3,5-dione [22].

11.2.2.4 Retro-cyclization

Retro-cyclization of nickelalactone has been reported since the early works by Hoberg et al. [19, 30, 61, 62]. They showed evidence of the reversibility of the C–C coupling step by reaction of nickelalactone with alkyne under CO_2 atmosphere (Scheme 11.26) [19]. As a result, carboxylation product of alkyne was confirmed after esterification. In addition, nickelalactone decomposes into an alkene complex along with generation of CO_2 gas by heating. Walther et al. also demonstrated thermal decomposition of nickelalactone into diene and CO_2 via retro-cyclization [30]. Walther and coworkers confirmed that ligand exchange of nickelalactone derived from ethylene and CO_2 results in retro-cyclization to give ethylene and CO_2 [62].

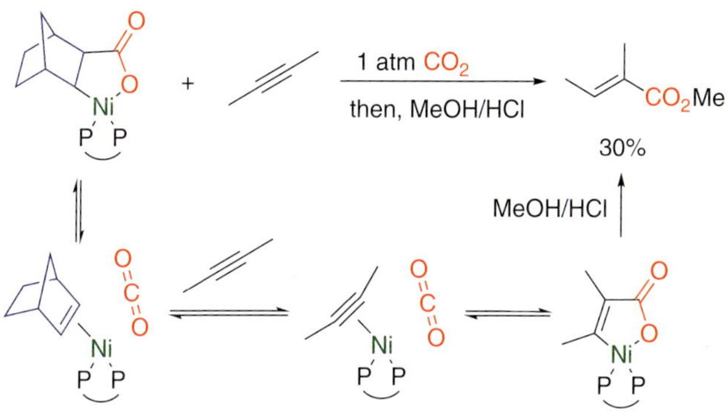

Scheme 11.26 Retro-cyclization of nickelalactone.

11.2.2.5 Nucleophilic Attack

As is already mentioned in Section 11.2.2, a variety of nickelalactones are protonated under acidic condition. In this section, other nucleophilic reactions of nickelalactone are described.

Some alkylations of nickelalactones with alkyl halides have been reported [54, 63–65]. In 1990, Schönecker et al. reported the alkylation of nickelalactone with alkyl halide [64]. Interestingly, they found that addition of MnI_2 drastically improves the yield of alkylation (Scheme 11.27). In the case of ethyl iodide, quantitative formation of pentanoic acid was observed. This transformation means that nickelalactone is a useful homoenolate species. Thus, the same research group used nickelalactone to add an alkyl chain into steroids [65].

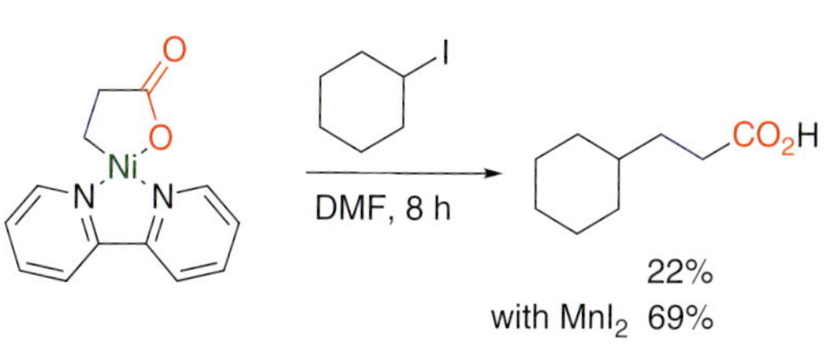

Scheme 11.27 Alkylation of nickelalactone.

Echavarren and coworker demonstrated a radical clock experiment in the alkylation of nickelalactone generated by the reaction of Ni(0) with succinic anhydride [66]. As a result, participation of radical species was indicated, although radical scavengers did not suppress the coupling reaction. Thus, they suggested that the reaction proceeds via electron transfer, followed by collapse of radical pairs to afford a C–C coupling product.

Langer et al. reported the alkylation of unsaturated nickelalactone with α-halocarbonyl compounds without the addition of a promoter (Scheme 11.28) [17, 67].

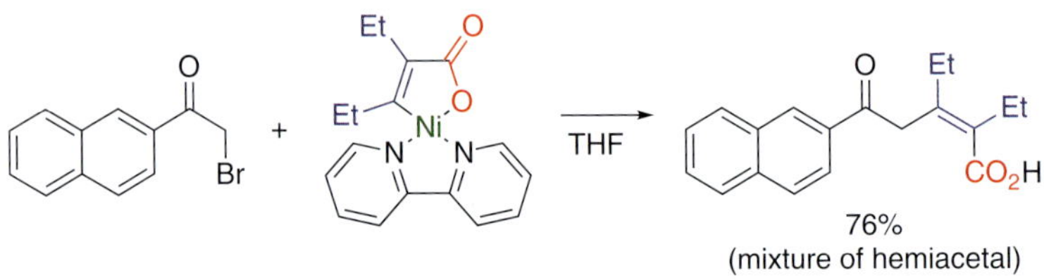

Scheme 11.28 Alkylation of unsaturated nickelalactone.

The reaction of nickelalactone derived from ethylene tends to give an acrylate ester via successive *O*-alkylation and β-hydride elimination. For details, see Section 11.4.2.

It has been known that treatment of $FeCl_3$ to nickelalactone gives chlorinated carboxylic acid [22]. In addition, Echavarren and coworker showed that treatment of nickelalactone with NBS resulted in the formation of brominated carboxylic acid in good yield (Scheme 11.29) [68]. Jordan and coworkers reported the halogenolysis of nickelalactone ligated by dppe to afford halogenated acid anhydrides in good yields [69].

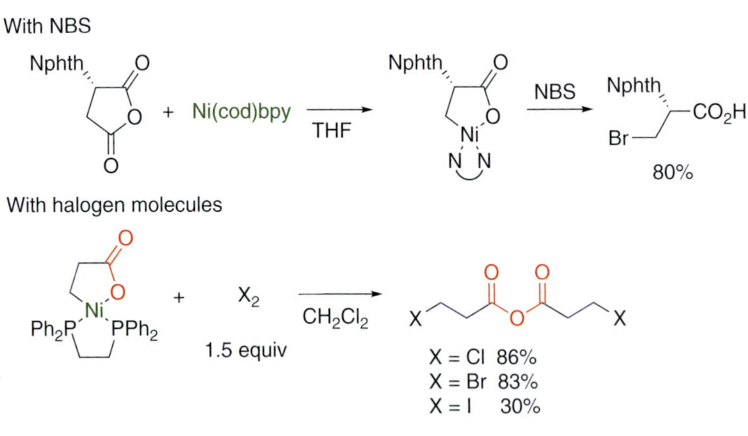

Scheme 11.29 Halogenation of nickelalactone.

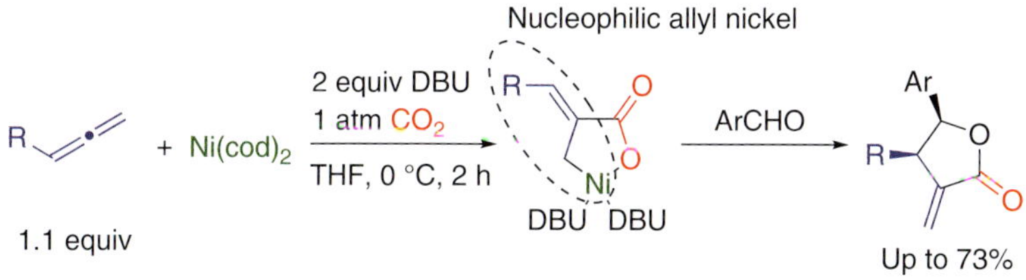

Scheme 11.30 Lactone synthesis via reaction of nickelalactone with aldehyde.

Mori and coworkers reported the oxidative cyclization of allene with CO_2 by using DBU as ligand (Scheme 11.30). The nickelalactone intermediate was found to react with aldehydes to afford *cis*-β,γ-disubstituted α-methylene-γ-lactones [25]. In this case, the intermediate contains a nucleophilic allyl nickel moiety that might be responsible for attacking the aldehyde.

Hoberg and Ballesteros demonstrated that in the presence of pyridine, nickelacycle gives a double carboxylated product [22].

11.2.2.6 Oxidation

Nickelalactone is known to be oxidized to give β-hydroxycarboxylic acid [22]. Echavarren and coworker reported serine synthesis by the treatment of nickelalactone with O_2 or *N*-methylmorpholine *N*-oxide (NMMO) (Scheme 11.31) [66]. In an analogous manner, some nickelalactone has been shown to react with disulfide to afford β-thiocarboxylic acid [17].

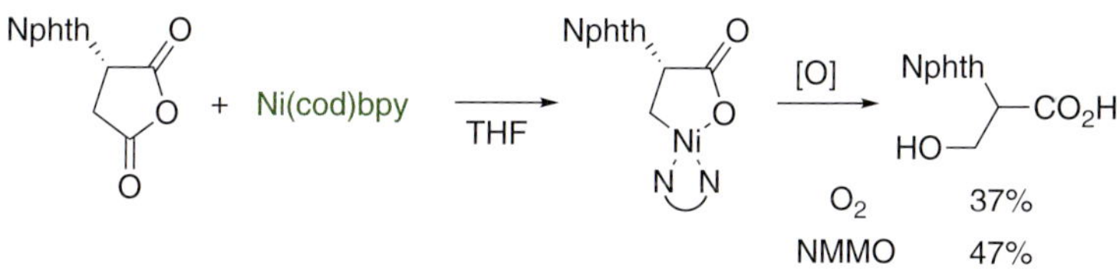

Scheme 11.31 Oxidation of nickelalactone.

11.2.2.7 Ligand Exchange

The supporting ligand of nickelalactone is known to be replaced by others [17, 37, 55, 62, 70–73]. For instance, addition of excess pyridine to a nickelalactone carrying TMEDA ligand affords a bispyridine complex in a quantitative yield (Scheme 11.32) [17]. Walther and coworkers reported the interesting aggregation behavior of monophosphine nickelalactone complex by the use of ligand exchange technique or generation of nickelalactone via decarbonylation [70]. Addition of 1 equiv of PPh_2Et to a bispyridine complex results in the formation of the tetramer of the monophosphine nickelalactone complex.

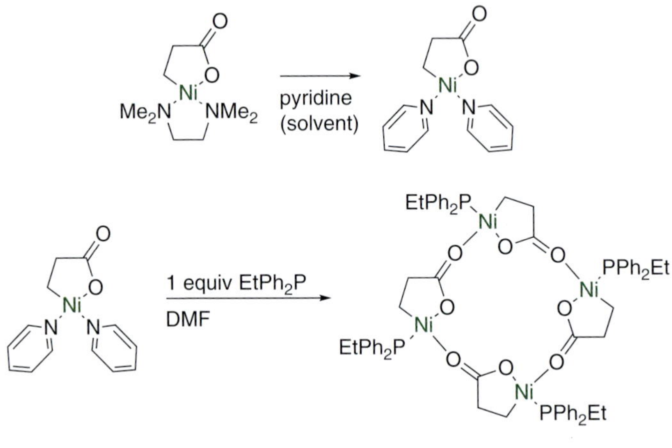

Scheme 11.32 Ligand exchange of nickelalactone.

11.3 Catalytic Transformation via Nickelalactone 1: Reactions of Alkynes

11.3.1 Synthesis of Pyrone

11.3.1.1 Initial Finding

Inoue et al. discovered a novel synthetic method of pyrone ring by a coupling reaction of two alkyne molecules and CO_2 catalyzed by low-valent nickel complexes in the late 1970s [74–76]. Although the reaction suffered from competing trimerization or oligomerization of alkyne, they succeeded in obtaining the desired product in 57% yield by addition of the DPPB ligand, which controls the selectivity (Scheme 11.33). Later the reaction was further optimized by Walther et al. [77].

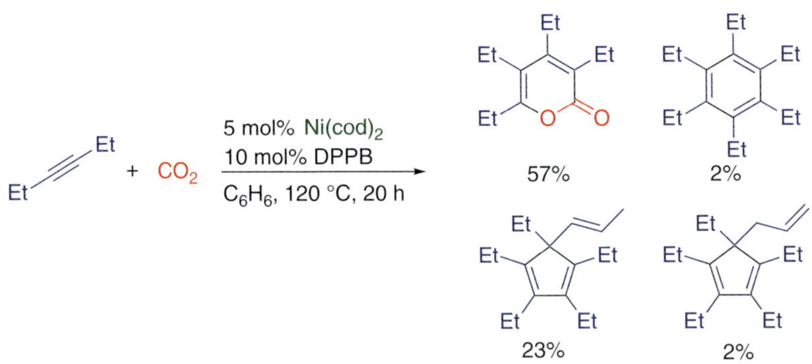

Scheme 11.33 Synthesis of pyrone with CO_2.

11.3.1.2 Reaction of Diynes with CO_2

Tsuda et al. used diyne for high-yield synthesis of pyrone by the use of CO_2 [78–81]. In the presence of a catalytic amount of $Ni(cod)_2$ and monodentate phosphine ligand PCy_3, the diyne was transformed into a bicyclic pyrone in

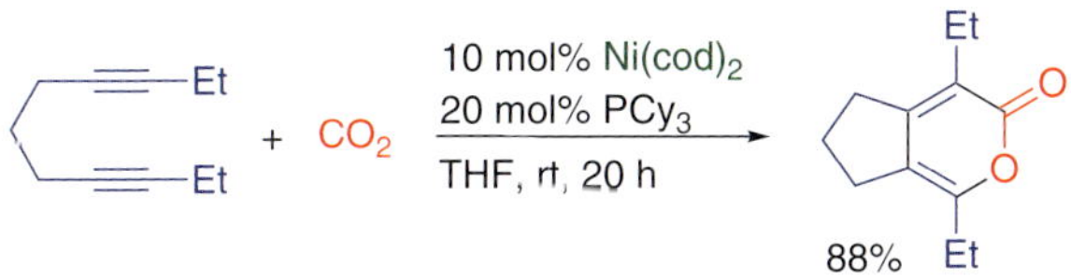

Scheme 11.34 Bicyclic pyrone synthesis under CO_2 pressure.

88% yield under CO_2 pressure (Scheme 11.34). The reaction proceeds at room temperature, and even at 0 °C, although the yield decreases. Later, density functional theory (DFT) calculation was performed by Bi and Liu to clarify the reaction mechanism in detail [82].

The reaction was further investigated by Louie et al. who employed the *N*-heterocyclic ligand [83, 84]. In this case, the reaction proceeded under 1 atm of CO_2 pressure and at relatively low reaction temperature, 60 °C (Scheme 11.35). When an unsymmetrical diyne, having a sterically hindered substituent (e.g. TMS or ^{t}Bu group) at one terminus of alkyne and methyl group at the other terminus, was used as a substrate, the corresponding pyrone was obtained regioselectively. They revealed the structure of the product by X-ray crystallography. The reaction is believed to proceed via nickelalactone formation followed by insertion of alkyne and reductive elimination. Insertion of sterically hindered alkyne would be slower than that of the propyne unit to observe regioselectivity in the catalytic process.

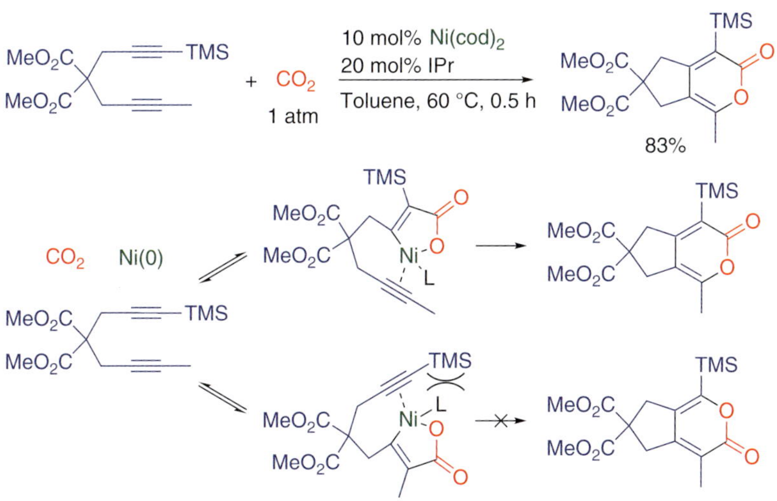

Scheme 11.35 Catalytic pyrone synthesis and its regioselectivity.

11.3.2 Synthesis of α,β-Unsaturated Ester

11.3.2.1 Electrochemical Reactions

Duñach et al. reported a series of reactions concerning the electrochemical carboxylation of alkynes with CO_2 [85–88]. In 1988, they reported the carboxylation of terminal alkynes in the presence of Ni catalyst, in or with the reaction vessel of which attached with the carbon cathode and the sacrificial magnesium anode

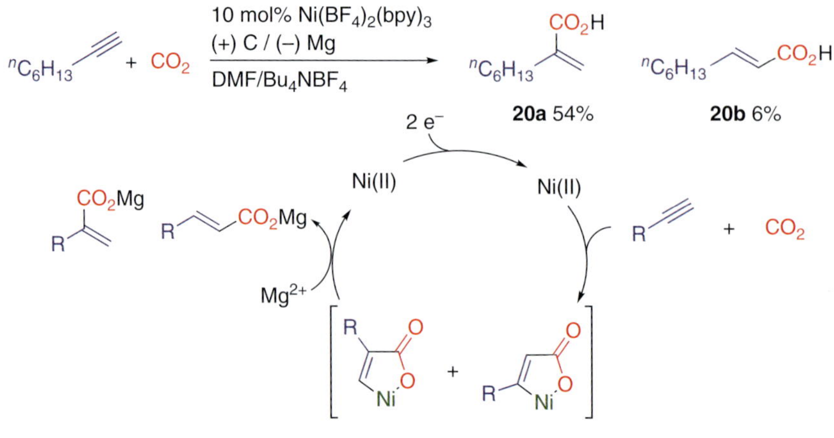

Scheme 11.36 Electrochemical carboxylation of terminal alkyne.

(Scheme 11.36). The reaction proceeded in a regioselective manner, giving **20a** as a major product. The putative reaction mechanism is suggested as follows: Ni(II) is reduced by electrochemical process to give an active Ni(0) complex, which forms nickelalactone by reaction with alkyne and CO_2. The nickelalactone is replaced by Mg salt, followed by proton quench to afford the desired product.

The electrochemical process was successfully used to carboxylate other alkynes, such as internal alkyne [89], diynes [90–92], and 1,3-enyne [93].

11.3.2.2 Reduction with Organometallic Reagents

Mori and coworkers has designed a catalytic carboxylation of alkyne with CO_2 by the use of organozinc reagent as reductant. Their first trial failed, although a stoichiometric reaction was found to proceed to give the desired product in 55% yield [94]. They postulated that coordination of DBU to $ZnMe_2$ consumed the ligand of nickel to prohibit effective catalysis. Thus, in the presence of an excess amount of DBU, the desired carboxylation product was obtained in a catalytic

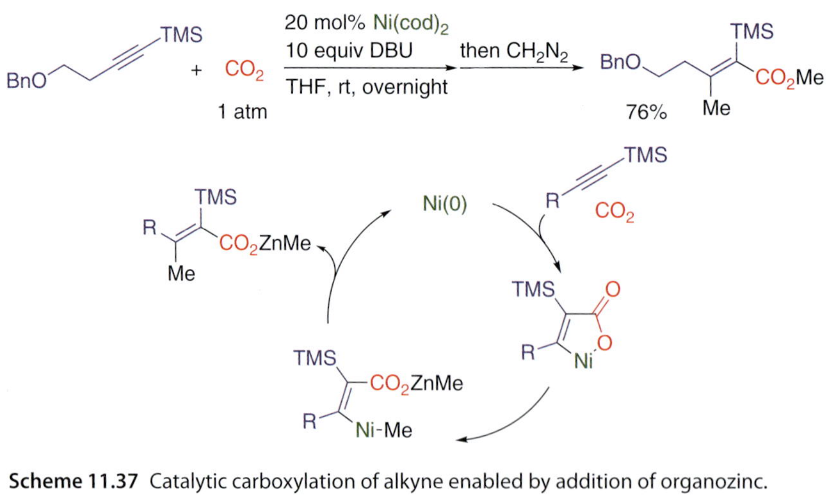

Scheme 11.37 Catalytic carboxylation of alkyne enabled by addition of organozinc.

manner (Scheme 11.37). Later, they successfully applied this methodology for the synthesis of the anticancer drug tamoxifen [95]. In addition, as a key step of the total synthesis of (−)-Corynantheidine, they found that catalytic intramolecular carboxylative coupling of enyne proceed in the presence of $ZnEt_2$ as a reducing agent [96]. In this case, transmetalation of nickelalactone with $ZnEt_2$ gives an ethyl nickel complex, β-hydride elimination of which followed by reductive elimination releases the desired product and regenerates the Ni(0) catalyst.

11.4 Catalytic Transformation via Nickelalactone 2: Reactions of Alkenes and Related Molecules

11.4.1 Transformation of Diene, Allene, and Substituted Alkene

11.4.1.1 Coupling of Diene with CO_2

Hoberg et al. reported the catalytic carboxylation of diene with CO_2 in 1987 [97]. In the presence of $Ni(cod)_2$ and $P(O^iPr)_3$ under CO_2 pressure, 1,3-butadiene was found to give the coupling product **21** in a catalytic manner (Scheme 11.38). The reaction mechanism is proposed as follows: First, the well-known diene–nickel coupling occurs to give a bisallyl nickel complex. Insertion of CO_2, which is previously observed in a stoichiometric reaction, generates the nickelalactone intermediate. Then, insertion of a C—C double bond constructs the five-membered ring. Subsequent β-hydride elimination followed by reductive elimination gives the coupling product. Although the chemical yield is low (in respect to butadiene, the yield was calculated to be 3–4%), this finding is of importance as an early example of catalytic carboxylation via nickelalactone.

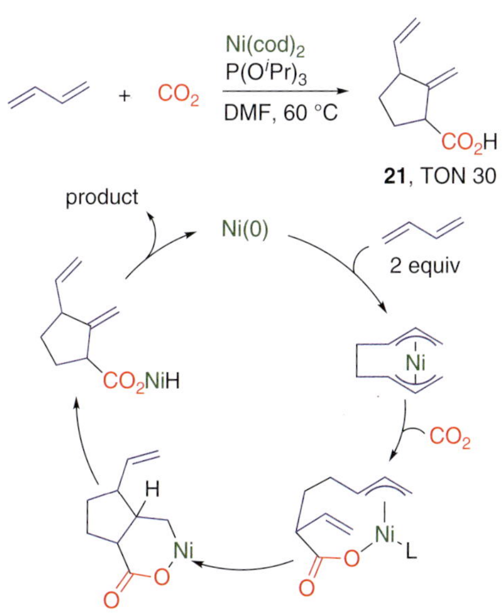

Scheme 11.38 Catalytic coupling of butadiene with CO_2.

11.4.1.2 Electrochemical Process

Duñach and coworkers reported the Ni-catalyzed carboxylation of allene, diene, and alkenes under electrochemical conditions (Scheme 11.39) [98, 99]. The reaction of alkene resulted in the formation of succinic acid derivatives as a major product via double carboxylation. The double carboxylation of 1,3-diene occurs at 1,4-position. The regioselectivity of the reaction of allene depends on the substrate. In this case, the second double bond remained untouched somehow. They also tried the reaction of enynes, which results in selective carboxylation at the alkyne moiety. They believe that the reactions proceed via nickelalactones, which was previously synthesized by Hoberg and others as described in Section 11.2.1.

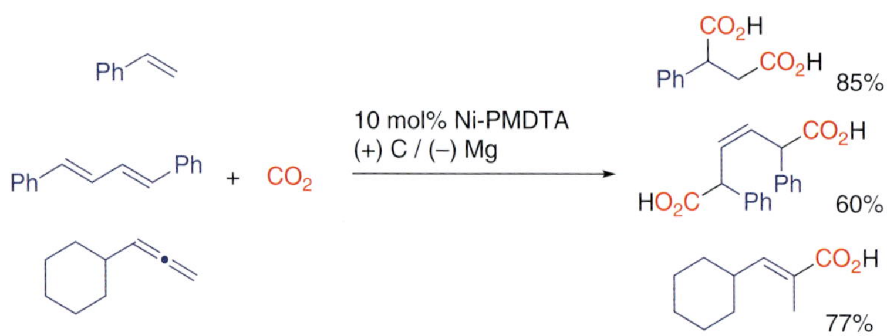

Scheme 11.39 Electrochemical carboxylation of alkene, diene, and allene.

11.4.1.3 Use of Reductant

Mori and coworker reported a novel strategy for catalytic carboxylation in which an organozinc reagent is used to regenerate the nickel complex [100]. The reaction of bisdiene **22** with CO_2 and $ZnMe_2$ in the presence of a catalytic amount of $Ni(cod)_2$ and PPh_3 ligand gave the carboxylated product **23** in excellent yield (Scheme 11.40). A possible mechanism is proposed. Firstly, Ni(0) generated by reduction of $Ni(acac)_2$ with organozinc reagent forms a bis allylnickel complex by reaction with **22**. Carboxylation occurs to give the allyl nickel species. Transmetalation with an organozinc reagent followed by reductive elimination (with $ZnMe_2$, $ZnPh_2$) or β-hydride elimination and reductive elimination affords the desired product.

In 2004, Mori and coworkers succeeded to develop a catalytic enantioselective carboxylation by employing the chiral phosphine ligand (*S*)-MeO-MOP (Scheme 11.41) [101]. The reaction gives the desired product in very high enantiomer excess.

Mori and coworkers used silylallene for catalytic carboxylation by combination with $ZnMe_2$ [102]. The reaction does not give a methylated product but a double carboxylation one. In this case, after the first carboxylation, transmetalation with $ZnMe_2$ induces the carboxylation of the allylnickel species to afford a double carboxylation product.

Although it is difficult to make a distinction between a pathway via nickelalactone and the other ones, Rovis and coworkers reported the Ni-catalyzed hydrocarboxylation of styrene by using $ZnEt_2$ as the reducing agent (Scheme 11.42) [103]. Inspired by Hoberg's nickelalactone formation, they found that styrene

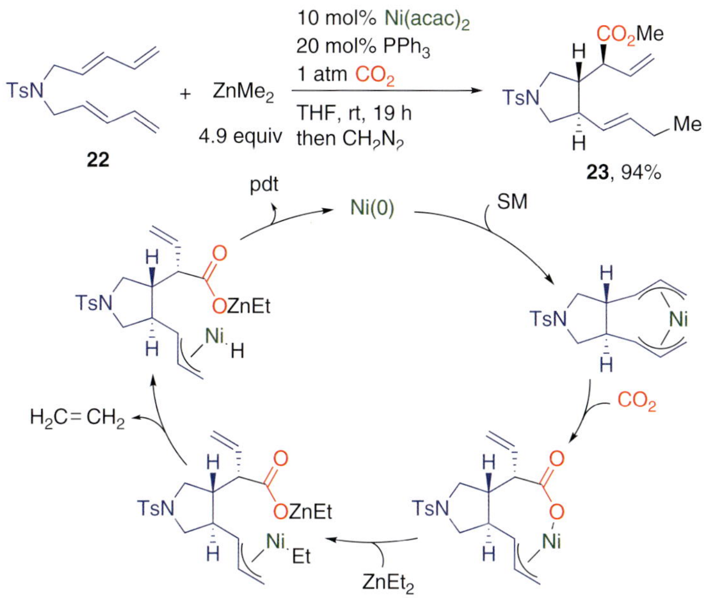

Scheme 11.40 Catalytic carboxylation of **22**.

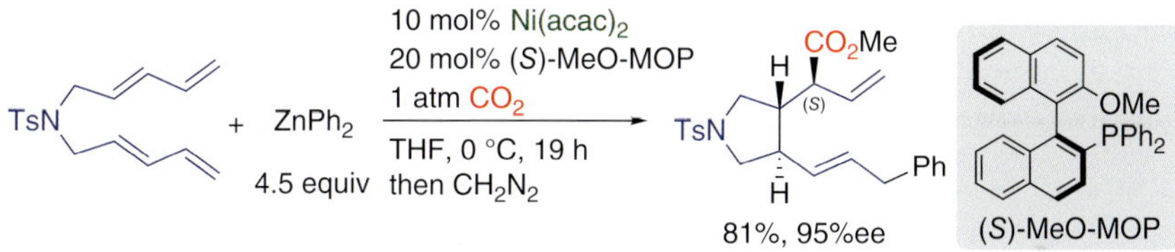

Scheme 11.41 Enantioselective catalytic carboxylation.

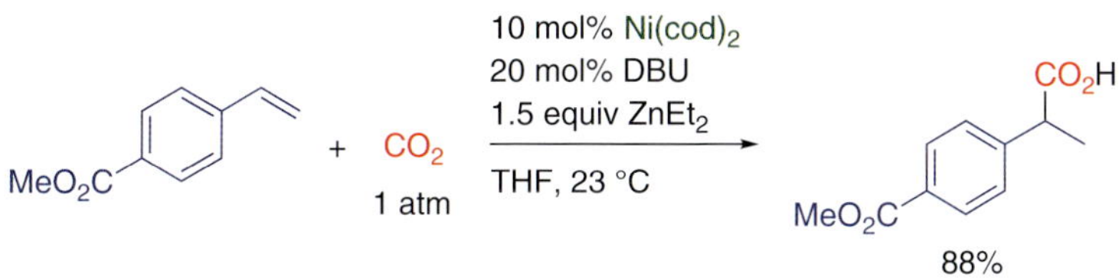

Scheme 11.42 Hydrocarboxylation of styrene.

carrying electron-withdrawing group is hydrocarboxylated in the presence of Ni(0)/DBU catalyst, $ZnEt_2$, and CO_2. However, after further optimization, the reaction was found to proceed in the absence of an ancillary ligand. Thus, Rovis supported the reaction mechanism via Ni-H species.

Later, König and coworkers achieved the regioselective hydrocarboxylation of styrene without use of a metallic reductant [104]. In the presence of 4CzIPN and $NiBr_2$/DPPB catalyst, hydrocarboxylation of styrene occurred to give *anti*-Markovnikov product by using Hantsch ester (4-Me-HEH) as reductant, MS4A as dehydrating agent, and LiOAc as additive under irradiation of blue LED (Scheme 11.43). In contrast, when neocuproine was used as ligand, the reaction gives the Markovnikov product similarly as Rovis' report. They suggest that the reaction might proceed via a nickelalactone complex to afford the *anti*-Markovnikov product, whereas the Markovnikov product would be obtained via hydrometalation of the Ni-H species.

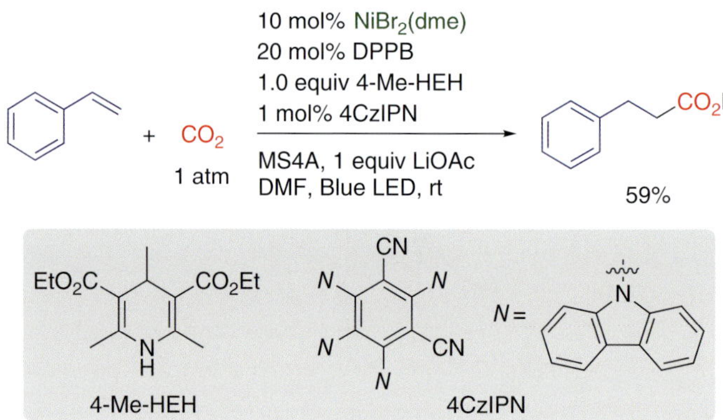

Scheme 11.43 Ni-catalyzed *anti*-Markovnikov selective hydrocarboxylation.

Recently, Martin and coworkers developed a novel catalytic double carboxylation reaction of diene with CO_2 (Scheme 11.44) [105]. They used Mn powder as reductant to regenerate the Ni(II) complex back into the catalytic cycle.

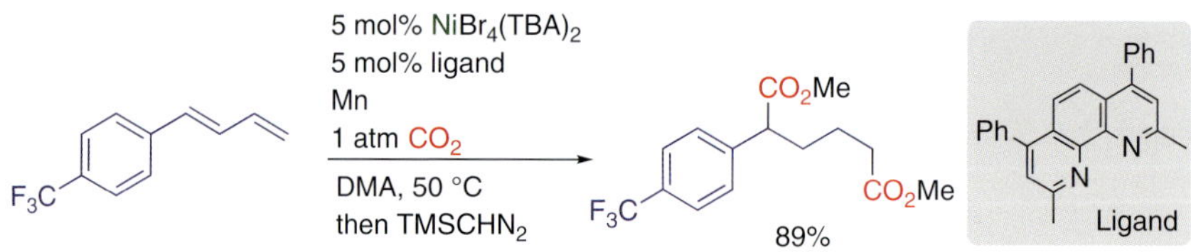

Scheme 11.44 Double carboxylation of diene catalytic in nickel.

11.4.2 Synthesis of Acrylic Acid from Ethylene and CO_2

Acrylic acid is an industrially important ingredient monomer for the manufacture of absorbent polymer. The production of acrylic acid relies on a two-step harsh and high-energy oxidation of propylene. However, industrial production of polypropylene renders the propylene monomer in very high demand. In contrast to propylene, ethylene is one of the bulk chemicals used for the production of a large quantity of polyethylene. Thus, acrylic acid production from ethylene and CO_2, has been known as a "dream reaction" (Scheme 11.45). Although nickel is relatively inexpensive, in respect to acrylic acid synthesis, very high catalytic turnover is required to make this process valuable.

Scheme 11.45 Acrylic acid production from ethylene and CO_2.

11.4.2.1 Before the Dawn

According to the early report of Hoberg et al., as introduced in Section 11.2.1, oxidative cyclization of ethylene and CO_2 on Ni(0) forms a nickelalactone complex. In addition, Hoberg et al. reported the generation of cinnamic acid from nickelalactone prepared with styrene and CO_2 by heating, implying the prospect to develop catalytic acrylic acid production (Scheme 11.46) [18]. In 2006, Walther and coworkers demonstrated the transformation of nickelalactone into an acrylate complex [62]. They added a variety of ligands to a nickelalactone bearing pyridine or a TMEDA complex. When 1,1-bis(diphenylphosphino)methane was added, generation of a binuclear Ni(I) acrylate complex was confirmed (Scheme 11.47).

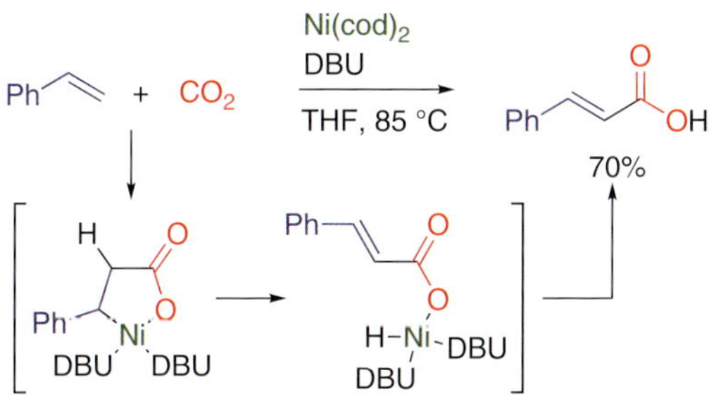

Scheme 11.46 Cinnamic acid was observed on thermal decomposition of nickelalactone.

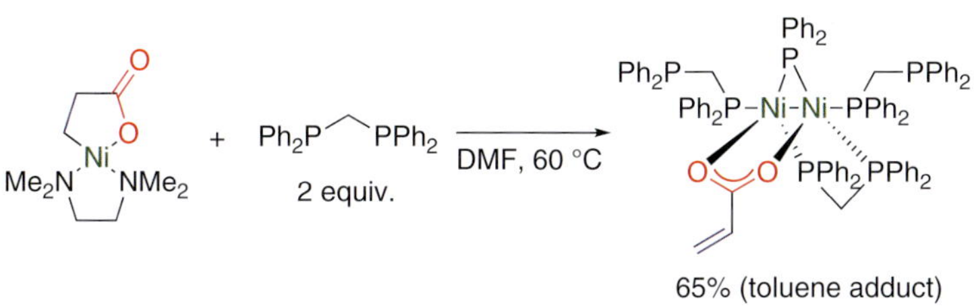

Scheme 11.47 Generation of nickel-acrylate complex from nickelalactone.

Although Hoberg and Walther's reports offered some possibility for the development of catalytic acrylic acid synthesis, a theoretical study by Buntine predicted low feasibility of this transformation [106]. Unfortunately, according to their calculation, the solvation-corrected Gibbs free energy of this coupling reaction is an unfavorable uphill process in +42.7 kJ/mol. In addition, β-hydride elimination of nickelalactone appears to be daunting due to the strain of the five-membered ring.

Another approach by Rieger and coworkers was the addition of methyl iodide to nickelalactone that generates methyl acrylate (Scheme 11.48) [107]. Kühn and coworkers showed that the TMEDA complex gives better yield of methyl acrylate [108]. They monitored the reaction by the use of IR spectroscopy. Based on the observation, they proposed a reaction mechanism via alkyl nickel complex, β-hydrogen elimination of which gives methyl acrylate (Scheme 11.49). The alkyl nickel intermediate, experimentally observed at around $1680\,cm^{-1}$, was calculated to exhibit carbonyl stretching vibration band at $1720 \pm 30\,cm^{-1}$. They concluded that addition of excess methyl iodide is needed due to high reaction barriers in the β-H elimination based on DFT calculation of the reaction profile. This finding was further investigated in detail by Limbach and coworkers [109]. They used 1,2-bis(di-*tert*-butylphosphino)ethane (D*t*BPE) complex for their computational and experimental studies. According to their calculation using D*t*BPE ligand, an S_N2 pathway rather than a concerted mechanism is plausible. In addition, protonation of methyl acrylate complex yielded putative cationic intermediates. Xu and coworkers also published a theoretical study of this process by using other bidentate ligands [110]. Zhou and coworkers used fluorine substituted alkyl iodide to obtain fluorine containing monomers [111].

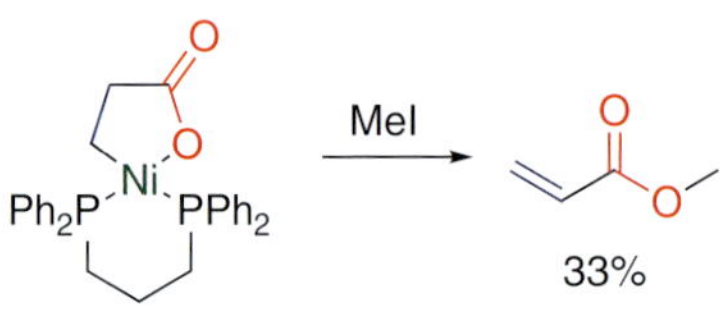

Scheme 11.48 Reaction of nickelalactone with MeI.

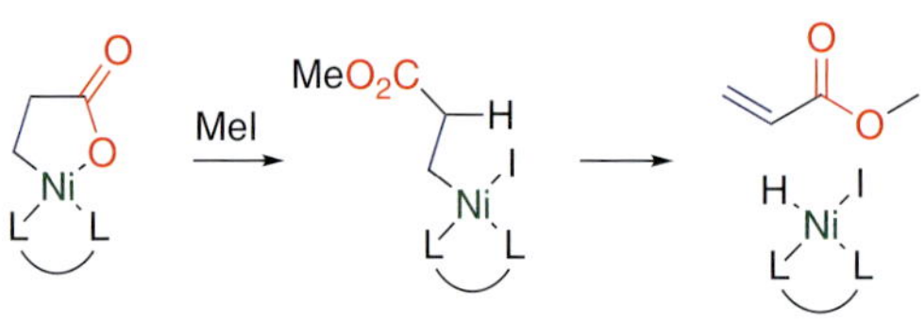

Scheme 11.49 Proposed reaction pathway.

11.4.2.2 Development of Catalytic Reaction

In 2012, Limbach and coworkers achieved the first two-step catalytic acrylate production from ethylene and CO_2 in which the key breakthrough to make this process possible was the addition of a base [10]. They firstly surveyed the effect of a bidentate phosphine ligand for soluble nickelalactone formation. As a result, D*t*BPE was found to be the optimal ligand because of the quick formation of thermally stable nickelalactone. As mentioned in Section 11.2.2, they found that treatment of sodium alkoxide with nickelalactone induces deprotonation leading to a nickel acrylate complex. This finding made the reaction a thermodynamically favored process. In addition, ligand exchange of the acrylate complex under ethylene pressure is faster than that of the acrylic acid complex. They assembled these results to realize the catalytic cycle by dividing the CO_2-rich regime and the CO_2-poor one to avoid irreversible formation of the sodium carbonic acid half ester. A reaction mixture containing ethylene and Ni/D*t*BPE system was pressurized with CO_2 to form nickelalactone (Scheme 11.50). CO_2 gas was released, and then the nickelalactone solution was treated with sodium alkoxide to give sodium acrylate. They achieved greater than 10 catalytic turnovers in respect to nickel by repeating these consecutive cycles.

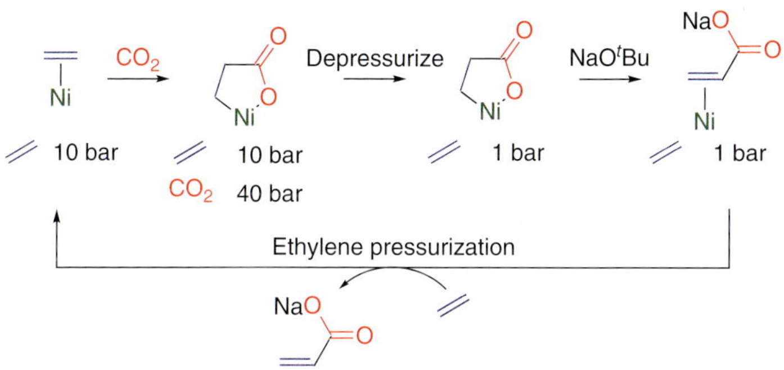

Scheme 11.50 The first catalytic stepwise acrylate production with Ni.

Later, Limbach and coworkers developed a one-pot catalytic process by further examination of the ligand and base tolerating CO_2 [112]. Their strategy was to employ a phenoxide base because of its low nucleophilicity that might work under CO_2 pressure. As a result, in the presence of excess sodium 2-fluorophenoxide (300 equiv relative to Ni), they obtained acrylate with TON of 107 in the presence of Ni(0)/BenzP* catalyst system (Scheme 11.51). Schaub and coworkers further optimized the reaction conditions for the synthesis of sodium acrylate from the view point of process chemistry [113]. Recently, Vavasori et al. used Ni(II) precursor to generate active low-valent species by *in situ* reduction. Ni(II) in combination with diphosphines having large bite angle was superior than Pd-based catalyst [114].

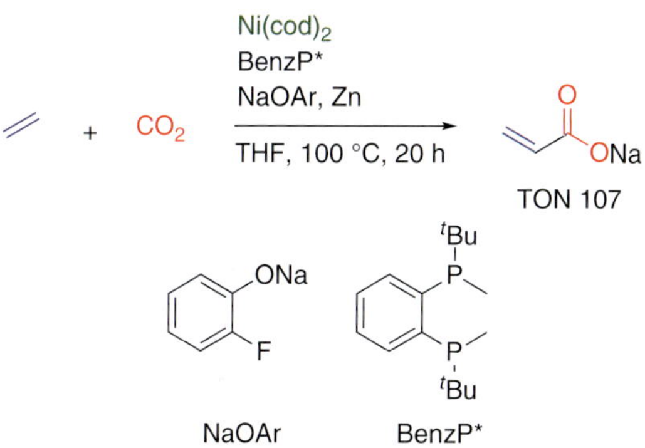

Scheme 11.51 One-pot sodium acrylate production.

Vogt and coworkers also succeeded in developing the one-pot catalytic process in 2014 [115]. They used a combination of amine base and hard Lewis acid to facilitate β-hydride elimination via Ni—O bond cleavage. In the presence of LiI, addition of which was demonstrated to release acrylate from nickelalactone by Rieger and coworkers [107], trimethylamine, zinc, and $Ni(cod)_2$/DCPP catalytic system, ethylene and CO_2 were transformed into lithium acrylate with TON of 21 (Scheme 11.52). Addition of zinc dust was used to reduce the inactive NiI_2 complex produced back to Ni(0) to prolong catalyst lifetime. They suggested a proposed reaction mechanism in which LiI activates the nickelalactone to induce β-hydride elimination. Later, Cummins demonstrated that their originally synthesized macrocyclic bidentate phosphine ligand is compatible or even gives better result than the reaction conditions developed by Limbach or Vogt [13, 112, 113].

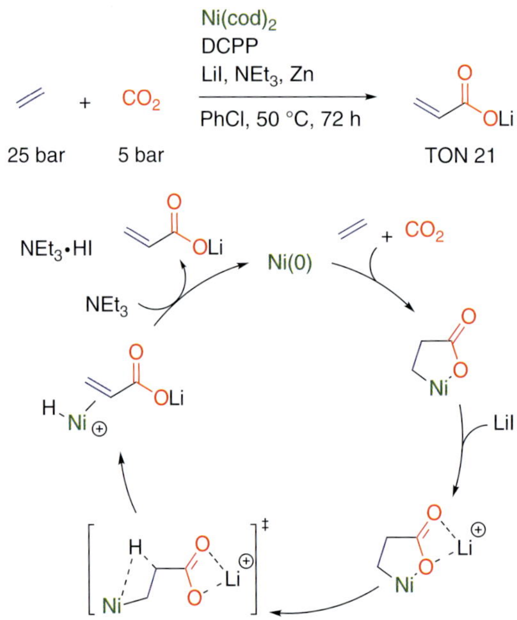

Scheme 11.52 Catalytic acrylic acid production by use of LiI.

11.5 Concluding Remarks

Over the past 40 years, oxidative cyclization of nickel involving CO_2 has opened a new opportunity for the synthesis of carboxylic acid under mild conditions. A variety of unsaturated molecules such as alkene, allene, and alkyne were employed for the preparation of nickelalactone. Some of their structures were characterized by the use of spectroscopic methods including X-ray crystallography. Their reactivities were thoroughly investigated. It was found that nickelalactone undergoes transmetalation with organometallic agents, Lewis acid or base induced β-hydride elimination, insertion, and nucleophilic attack. Upon heating, the complex is known to decompose via retro-cyclization. Ligand exchange realized synthesis of a variety of new nickelalactone complexes. Based upon these findings, several catalytic CO_2 incorporation reactions were developed. Especially carboxylation of ethylene to give acrylate, the so-called "dream reaction" has been achieved, although there is much room left to be improved for industrial applications.

References

1 (a) Sakakura, T., Choi, J.-C., and Yasuda, H. (2007). *Chem. Rev.* 107: 2365. (b) Aresta, M., Dibenedetto, A., and Angelini, A. (2013). *Chem. Rev.* 114: 1709.

2 (a) Braunstein, P., Matt, D., and Nobel, D. (1988). *Chem. Rev.* 88: 747. (b) Yin, X. and Moss, J.R. (1999). *Coord. Chem. Rev.* 181: 27. (c) Aresta, M. and Dibenedetto, A. (2007). *Dalton Trans.* (28): 2975. (d) Riduan, S.N. and Zhang, Y. (2010). *Dalton Trans.* 39: 3347. (e) Huang, K., Sun, C.-L., and Shi, Z.-J. (2011). *Chem. Soc. Rev.* 40: 2435. (f) Cokoja, M., Bruckmeier, C., Rieger, B. et al. (2011). *Angew. Chem. Int. Ed.* 50: 8510. (g) Yu, D., Teong, S.P., and Zhang, Y. (2015). *Coord. Chem. Rev.* 293-294: 279. (h) Luan, Y.-X. and Ye, M. (2018). *Tetrahedron Lett.* 59: 853.

3 Walther, D. (1987). *Coord. Chem. Rev.* 79: 135.

4 Tasker, S.Z., Standley, E.A., and Jamison, T.F. (2014). *Nature* 509: 299.

5 (a) Thakur, A. and Louie, J. (2015). *Acc. Chem. Res.* 48: 2354. (b) Ohashi, M., Hoshimoto, Y., and Ogoshi Dalton, S. (2015). *Trans.* 44: 12060. (c) Ogoshi, S. (2017). *Bull. Chem. Soc. Jpn.* 90: 1401.

6 Mori, M. (2007). *Eur. J. Org. Chem.* 2007: 4981.

7 Hoberg, H. and Schaefer, D. (1982). *J. Organomet. Chem.* 236: C28.

8 Hoberg, H. and Schaefer, D. (1983). *J. Organomet. Chem.* 251: C51.

9 Hoberg, H., Peres, Y., Kruger, C., and Tsay, Y.-H. (1987). *Angew. Chem. Int. Ed. Engl.* 26: 771.

10 Ljkowski, M.L., Lindner, R., Kageyama, T. et al. (2012). *Chem. Eur. J.* 18: 14017.

11 Jin, D., Williard, P.G., Hazari, N., and Bernskoetter, W.H. (2014). *Chem. Eur. J.* 20: 3205.

12 Greenburg, Z.R., Jin, D., Williard, P.G., and Bernskoetter, W.H. (2014). *Dalton Trans.* 43: 15990.

13 Knopf, I., Tofan, D., Beetstra, D. et al. (2017). *Chem. Sci.* 8: 1463.

14 Pápai, I., Schubert, G., Mayer, I. et al. (2004). *Organometallics* 23: 5252.

15 Plessow, P.N., Schäfer, A., Limbach, M., and Hofmann, P. (2014). *Organometallics* 33: 3657.

16 Hipler, B., Döring, M., Dubs, C. et al. (1998). *Z. Anorg. Allg. Chem.* 624: 1329.

17 Langer, J., Fischer, R., Görls, H., and Walther, D. (2004). *J. Organomet. Chem.* 689: 2952.

18 Hoberg, H., Peres, Y., and Milchereit, A. (1986). *J. Organomet. Chem.* 307: C38.
19 Hoberg, H., Schaefer, D., Burkhart, G. et al. (1984). *J. Organomet. Chem.* 266: 203.
20 Hoberg, H., Ballesteros, A., Sigan, A. et al. (1991). *J. Organomet. Chem.* 407: C23.
21 Hoberg, H., Ballesteros, A., and Sigan, A. (1991). *J. Organomet. Chem.* 403: C19.
22 Hoberg, H. and Ballesteros, A. (1991). *J. Organomet. Chem.* 411: C11.
23 Hoberg, H., Ballesteros, A., Sigan, A. et al. (1991). *Synthesis* 1991: 395.
24 Hoberg, H. and Oster, B.W. (1984). *J. Organomet. Chem.* 266: 321.
25 Takimoto, M., Kawamura, M., and Mori, M. (2003). *Org. Lett.* 5: 2599.
26 Takimoto, M., Kawamura, M., and Mori, M. (2004). *Synthesis* 2004: 791.
27 Saito, N., Sugimura, Y., and Sato, Y. (2014). *Synlett* 25: 736.
28 Hoberg, H. and Schaefer, D. (1983). *J. Organomet. Chem.* 255: C15.
29 Hoberg, H., Schaefer, D., and Oster, B.W. (1984). *J. Organomet. Chem.* 266: 313.
30 Walther, D., Dinjus, E., and Görls, H. (1985). *J. Organomet. Chem.* 286: 103.
31 Freund, R.R.A., Görls, H., and Langer, J. (2014). *Dalton Trans.* 43: 13988.
32 Behr, A. and Kanne, U. (1986). *J. Organomet. Chem.* 317: C41.
33 Takimoto, M. and Mori, M. (2001). *J. Am. Chem. Soc.* 123: 2895.
34 Burkhart, G. and Hoberg, H. (1982). *Angew. Chem. Int. Ed. Engl.* 21: 76.
35 Fujihara, T., Horimoto, Y., Mizoe, T. et al. (2014). *Org. Lett.* 16: 4960.
36 Walther, D. and Braunlich, G. (1992). *J. Organomet. Chem.* 436: 109.
37 Langer, J., Görls, H., and Walther, D. (2012). *Polyhedron* 32: 60.
38 Saito, S., Nakagawa, S., Koizumi, T. et al. (1999). *J. Org. Chem.* 64: 3975.
39 Aoki, M., Kaneko, M., Izumi, S. et al. (2004). *Chem. Commun.* (22): 2568.
40 Saito, N., Sun, Z., and Sato, Y. (2015). *Chem. Asian J.* 10: 1170.
41 Saito, N., Abdullah, I., Hayashi, K. et al. (2016). *Org. Biomol. Chem.* 14: 10080.
42 Doi, R., Abdullah, I., Taniguchi, T. et al. (2017). *Chem. Commun.* 53: 7720.
43 Sakaki, S., Mine, K., Taguchi, D., and Arai, T. (1993). *Bull. Chem. Soc. Jpn.* 66: 3289.
44 Graham, D.C., Bruce, M.I., Metha, G.F. et al. (2008). *J. Organomet. Chem.* 693: 2703.
45 Hoberg, H., Peres, Y., Milchereit, A., and Gross, S. (1988). *J. Organomet. Chem.* 345: C17.
46 Hoberg, H. and Bärhausen, D. (1989). *J. Organomet. Chem.* 379: C7.
47 Takimoto, M., Mizuno, T., Sato, Y., and Mori, M. (2005). *Tetrahedron Lett.* 46: 5173.
48 Takimoto, M., Mizuno, T., Mori, M., and Sato, Y. (2006). *Tetrahedron* 62: 7589.
49 Bennett, M.A., Hambley, T.W., Roberts, N.K., and Robertson, G.B. (1985). *Organometallics* 4: 1992.
50 Bennett, M.A., Hockless, D.C.R., and Wenger, E. (1995). *Organometallics* 14: 2091.
51 Bennett, M.A., Johnson, J.A., and Willis, A.C. (1996). *Organometallics* 15: 68.

52 Yamamoto, T., Igarashi, K., Komiya, S., and Yamamoto, A. (1980). *J. Am. Chem. Soc.* 102: 7448.
53 Fischer, R., Nestler, B., and Schültz, H. (1989). *Z. Anorg. Allg. Chem.* 577: 111.
54 Fischer, R., Walther, D., Bräunlich, G. et al. (1992). *J. Organomet. Chem.* 427: 395.
55 Langer, J., Fischer, R., Görls, H., and Walther, D. (2007). *Eur. J. Inorg. Chem.* 2007: 2257.
56 Takimoto, M., Shimizu, K., and Mori, M. (2001). *Org. Lett.* 3: 3345.
57 Shimizu, K., Takimoto, M., and Mori, M. (2003). *Org. Lett.* 5: 2323.
58 O'Brien, E.M., Bercot, E.A., and Rovis, T. (2003). *J. Am. Chem. Soc.* 125: 10498.
59 Jin, D., Schmeier, T.J., Williard, P.G. et al. (2013). *Organometallics* 32: 2152.
60 Hoberg, H., Peres, Y., and Milchereit, A. (1986). *J. Organomet. Chem.* 307: C41.
61 Echavarren, A.M. and Castaño, A.M. (1995). *Tetrahedron* 51: 2369.
62 Fischer, R., Langer, J., Malassa, A. et al. (2006). *Chem. Commun.* (23): 2510.
63 Bräunlich, G., Fischer, R., Nestler, B., and Walther, D. (1989). *Z. Chem.* 29: 417.
64 Schönecker, B., Walther, D., Fischer, R. et al. (1990). *Tetrahedron Lett.* 31: 1257.
65 Fischer, R., Schönecker, B., and Walther, D. (1993). *Synthesis* (12): 1267.
66 Castaño, A.M. and Echavarren, A.M. (1994). *Organometallics* 13: 2262.
67 Langer, J., Gärtner, M., Görls, H., and Walther, D. (2006). *Synthesis* 16: 2697.
68 Castaño, A.M. and Echavarren, A.M. (1990). *Tetrahedron Lett.* 31: 4783.
69 Zarkesh, R.A., Hopkins, M.D., and Jordan, R.F. (2014). *Eur. J. Inorg. Chem.* 2014: 5491.
70 Langer, J., Görls, H., Fischer, R., and Walther, D. (2005). *Organometallics* 24: 272.
71 Langer, J., Walther, D., and Görls, H. (2006). *J. Organomet. Chem.* 691: 4874.
72 Langer, J., Walther, D., Malassa, A. et al. (2010). *Eur. J. Inorg. Chem.* 2010: 275.
73 Langer, J., Görls, H., and Westerhausen, M. (2010). *Inorg. Chem. Commun.* 13: 488.
74 Inoue, Y., Itoh, Y., and Hashimoto, H. (1977). *Chem. Lett.* 6: 855.
75 Inoue, Y., Itoh, Y., and Hashimoto, H. (1978). *Chem. Lett.* 7: 633.
76 Inoue, Y., Itoh, Y., Kazama, H., and Hashimoto, H. (1980). *Bull. Chem. Soc. Jpn.* 53: 3329.
77 Walther, D., Schönberg, H., Dinjus, E., and Sieler, J. (1987). *J. Organomet. Chem.* 334: 337.
78 Tsuda, T., Sumiya, R., and Saegusa, T. (1987). *Synth. Commun.* 17: 147.
79 Tsuda, T., Morikawa, S., Sumiya, R., and Saegusa, T. (1988). *J. Org. Chem.* 53: 3140.
80 Tsuda, T., Morikawa, S., and Saegusa, T. (1989). *J. Chem. Soc., Chem. Commun.* (1): 9.
81 Tsuda, T., Morikawa, S., Hasegawa, N., and Saegusa, T. (1990). *J. Org. Chem.* 55: 2978.

82 Zhao, Y., Liu, Y., Bi, S., and Liu, Y. (2014). *J. Organomet. Chem.* 758: 45.
83 Louie, J., Gibby, J.E., Farnworth, M.V., and Tekavec, T.N. (2002). *J. Am. Chem. Soc.* 124: 15188.
84 Tekavec, T.N., Arif, A.M., and Louie, J. (2004). *Tetrahedron* 60: 7431.
85 Duñach, E. and Périchon, J. (1988). *J. Organomet. Chem.* 352: 239.
86 Labbé, E., Duñach, E., and Périchon, J. (1988). *J. Organomet. Chem.* 353: C51.
87 Duñach, E. and Périchon, J. (1990). *Synlett* 1990: 143.
88 Dérien, S., Duñach, E., and Périchon, J. (1991). *J. Am. Chem. Soc.* 113: 8447.
89 Duñach, E., Dérien, S., and Périchon, J. (1989). *J. Organomet. Chem.* 364: C33.
90 Dérien, S., Duñach, E., and Périchon, J. (1990). *J. Organomet. Chem.* 385: C43.
91 Dérien, S., Clinet, J.-C., Duñach, E., and Périchon, J. (1991). *J. Chem. Soc., Chem. Commun.* (8): 549.
92 Dérien, S., Clinet, J.-C., Duñach, E., and Périchon, J. (1993). *J. Org. Chem.* 58: 2578.
93 Dérien, S., Clinet, J.-C., Duñach, E., and Périchon, J. (1992). *J. Organomet. Chem.* 424: 213.
94 Shimizu, K., Takimoto, M., Sato, Y., and Mori, M. (2005). *Org. Lett.* 7: 195.
95 Shimizu, K., Takimoto, M., Mori, M., and Sato, Y. (2006). *Synlett* 18: 3182.
96 Mizuno, T., Oonishi, Y., Takimoto, M., and Sato, Y. (2011). *Eur. J. Org. Chem.* 2011: 2606.
97 Hoberg, H., Gross, S., and Milchereit, A. (1987). *Angew. Chem. Int. Ed. Engl.* 26: 571.
98 Dérien, S., Clinet, J.-C., Duñach, E., and Périchon, J. (1990). *Synlett* 1990: 361.
99 Dérien, S., Clinet, J.-C., Duñach, E., and Périchon, J. (1992). *Tetrahedron* 48: 5235.
100 Takimoto, M. and Mori, M. (2002). *J. Am. Chem. Soc.* 124: 10008.
101 Takimoto, M., Nakamura, Y., Kimura, K., and Mori, M. (2004). *J. Am. Chem. Soc.* 126: 5956.
102 Takimoto, M., Kawamura, M., Mori, M., and Sato, Y. (2005). *Synlett* 13: 2019.
103 Williams, C.M., Johnson, J.B., and Rovis, T. (2008). *J. Am. Chem. Soc.* 130: 14936.
104 Meng, Q.-Y., Wang, S., Huff, G.S., and König, B. (2018). *J. Am. Chem. Soc.* 140: 3198.
105 Tortajada, A., Ninokata, R., and Martin, R. (2018). *J. Am. Chem. Soc.* 140: 2050.
106 Graham, D.C., Mitchell, C., Bruce, M.I. et al. (2007). *Organometallics* 26: 6784.
107 Bruckmeier, C., Lehenmeier, M.W., Reichardt, R. et al. (2010). *Organometallics* 29: 2200.
108 Lee, S.Y.T., Cokoja, M., Drees, M. et al. (2011). *ChemSusChem* 4: 1275.
109 Plessow, P.N., Weigel, L., Lindner, R. et al. (2013). *Organometallics* 32: 3327.

110 Guo, W., Michel, C., Schwiedernoch, R. et al. (2014). *Organometallics* 33: 6369.

111 Zhang, Z., Guo, F., Kühn, F.K. et al. (2017). *Appl. Organometal. Chem.* 31: e3567.

112 Huguet, N., Jevtovikj, I., Gordillo, A. et al. (2014). *Chem. Eur. J.* 20: 16858.

113 Manzini, S., Huguet, N., Trapp, O., and Schaub, T. (2015). *Eur. J. Org. Chem.* 2015: 7122.

114 Vavasori, A., Calgaro, L., Pietrobon, L., and Ronchin Pure, L. (2018). *Appl. Chem.* 90: 315.

115 Hendriksen, C., Pidko, E.A., Yang, G. et al. (2014). *Chem. Eur. J.* 20: 12037.

12

Relevance of Ni(I) in Catalytic Carboxylation Reactions

Rosie J. Somerville [1,2] *and Ruben Martin* [1,3]

[1] *Institute of Chemical Research of Catalonia (ICIQ), The Barcelona Institute of Science and Technology, Av. Països Catalans 16, 43007 Tarragona, Spain*
[2] *Universitat Rovira i Virgili, Departament de Química Analítica i Química Orgànica, c/Marcel·lí Domingo, 1, 43007 Tarragona, Spain*
[3] *ICREA, Passeig Lluís Companys, 23, 08010 Barcelona, Spain*

12.1 Introduction

Carbon dioxide belongs to a small set of molecules whose profound influence on scientific progress is a testament to Chemistry's status as the central science. To fields that vary from organic synthesis, biology, and climate science to engineering, the chemistry of carbon dioxide poses both challenges and opportunities, and its myriad applications belie its simple molecular structure. The three π-bonded atoms of this "wild spirit" are, for example, essential to Earth's carbon cycle and atmospheric chemistry, a crucial part of industrially relevant molecules, and are assembled into sugars during photosynthesis [1]. From a synthetic chemist's perspective, the conversion of abundant and inexpensive feedstock chemicals such as CO_2 into valuable products is an endeavor that is undoubtedly of value to society. At present, catalytic fixation of CO_2 into chemicals remains predominantly restricted to reductions, which provide a means of forming fuels such as methanol [1b, 2], C—O bond forming reactions that form (poly)carbonates [3], or the formation of ureas via C—N bond formation (Scheme 12.1) [2b, 4]. In contrast, catalytic C—C bond forming reactions from CO_2 are somewhat less developed; thus, the design of catalytic carboxylation reactions would be particularly appealing, as these might provide access to valuable building blocks from simple precursors. This goal is attracting considerable research attention, as CO_2 is a nontoxic, abundant, and renewable reagent and an alternative to the toxic C1 synthons carbon monoxide, phosgene, and cyanide [5].

Since the 1950s, chemists have been working to develop carboxylation reactions that avoid the use of reactive and air-sensitive Grignard reagents or the high temperature and pressure required for the Kolbe–Schmitt reaction [6]. These efforts have gained considerable momentum, and are particularly focused on expanding the scope, tolerance, and operational simplicity of carboxylation in order to facilitate the uptake of mild and general carboxylation reactions

Nickel Catalysis in Organic Synthesis: Methods and Reactions,
First Edition. Edited by Sensuke Ogoshi.

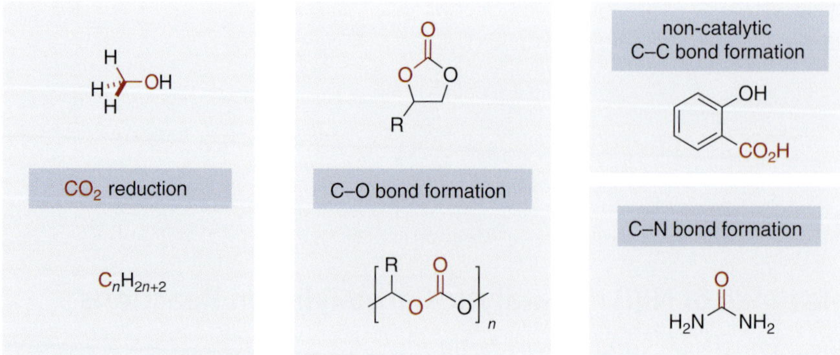

Scheme 12.1 Uses of CO_2.

in synthetic and industrial settings. From a reactivity perspective, CO_2 is not an ideal substrate, as it does not have a dipole moment and is kinetically and thermodynamically inert. However, after the finding by Aresta that CO_2 can coordinate to Ni(0) in an η^2 manner and the first electrochemical syntheses of acids from organic substrates [6, 7], the field of transition-metal-catalyzed carboxylation began to grow. Although noble metals such as Pt, Rh, Ir, and Pd are synonymous with the birth of homogeneous catalysis (particularly with the 2010 Nobel Prize being awarded for the development of Pd-catalyzed cross-couplings), Ni has received the most attention in the field of catalytic carboxylation. Unlike the noble metals, which are by definition rare and expensive, Ni is an Earth-abundant and cheap metal. Economic incentives to employ nickel in catalytic applications are bolstered by its chemical properties. As a first row transition metal, nickel is capable of accessing one-electron transformations and therefore may take on a range of formal oxidation states from Ni(0) to Ni(IV) [8]. One-electron transformations can allow for catalytic cycles to operate with more flexibility than those limited to two-electron transformations as they can, for example, speed up ligand exchange relative to an 18-electron complex [9] and promote radical reactivity. Nickel complexes also support rather weak Ni—C bonds, which is beneficial for catalytic C—C or C—CO_2 bond formation [10]. Although Ni is regularly employed in such reactions, the field of carboxylation is much more diverse, and we direct the reader to the literature that reviews other facets of the field [2, 4, 11]. Here, we focus on a specific area of the carboxylation field – namely, the intriguing role of Ni(I) in nickel-catalyzed carboxylation of organic substrates.

Despite the preparative advances realized, the mechanisms of Ni-catalyzed carboxylation reactions still remain rather speculative, particularly concerning the mode of reactivity of CO_2 with the Ni intermediates. The available literature data suggest that depending on the substrate, nickelacyclic intermediates and/or Ni(I) species are critical intermediates in the mechanism of nickel-catalyzed carboxylation [12]. The formation of Ni(I) may occur during the activation of the substrate by the catalyst, upon reduction (at an electrode or by a metal powder) or upon comproportionation of Ni(II) and Ni(0) species. This being said, if redox non-innocent ligands are employed, it becomes difficult to disentangle the

Ni(I) oxidation state from the charge at the metal and ligand(s). Thus, although placing too much importance on such a "key" oxidation state can be controversial [13], this account will be focused on the role of formally Ni(I) intermediates and their relevance in catalytic carboxylation reactions. In particular, this chapter summarizes the stoichiometric, catalytic, and computational evidence for such intermediates in carboxylation reactions and aims to lay out future directions to aid and encourage those who wish to take part in this open research question. It is organized into three sections. First, the major challenges and key steps involved in mechanisms are broadly sketched out in order to act as a reference for the following sections. Next, we present the evidence for Ni(I) in Ni-catalyzed electrocarboxylation, followed by Ni-catalyzed carboxylation with metal powder reducing agents. We hope to create a reference point for the mechanistic fundamentals that underpin Ni-catalyzed carboxylation and to provide a critical assessment of the evidence for Ni(I) in such reactions.

12.2 Mechanistic Building Blocks

The main challenge we faced when compiling this chapter is that few detailed mechanistic studies have been carried out on Ni-catalyzed carboxylation reactions [11]. Thus, it is not fully clear whether the main steps of any of the well-studied carboxylation catalytic cycles can reasonably be invoked for different methodologies (based on factors such as ligand and substrate). Furthermore, the importance (or not) of assigning a formal oxidation state to the nickel center during the key steps of the catalytic cycle is a magnet for debate. We therefore decided to provide a short at-a-glance summary of the factors that we have identified as being particularly important to the analysis of the proposed carboxylation mechanisms.

12.2.1 Additives

Additives are usually not included in the balanced equation of the catalytic reaction and are employed in stoichiometric or sub-stoichiometric amounts to, in broad terms, disfavor unproductive pathways, remove destructive by-products, or lower energies of favorable mechanistic pathways. Common additives in carboxylation reactions are ammonium salts, LiCl, and $MgCl_2$. Their roles have been studied by both density functional theory (DFT) and brief stoichiometric studies and appear to play their most significant roles in the reduction and/or CO_2 insertion steps.[1]

12.2.2 Coordination of CO_2

Carbon dioxide is able to coordinate to a single metal center in a number of ways (Scheme 12.2a).[21] The first structurally characterized transition metal complex

1 See Refs. [14] (Chen *et al.*), [15] (Börjesson *et al.*), [16] (Fujihara *et al.*), and [17b] (Sakaki) for examples of Ni-catalyzed carboxylation reactions where the role of the additive is investigated.

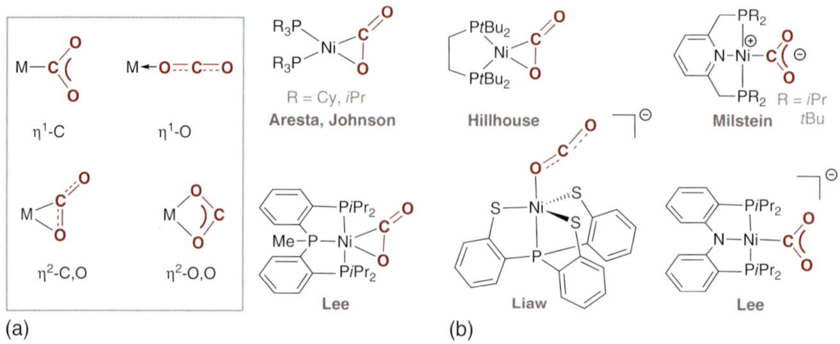

Scheme 12.2 Crystallographically characterized Ni-CO_2 complexes [7, 18–20].

of CO_2, [Ni(PCy_3)$_2$(η^2-CO_2)], was reported by Aresta et al. [7]. Intriguingly, the IR stretches of the coordinated CO_2 suggest that the oxidation state of Ni in [Ni(PCy_3)$_2$(η^2-CO_2)] may be better described as Ni(I), with lengthening of CO bonds due to transfer of an electron from nickel to the coordinated CO_2 [22]. Although the η^2-CO_2 coordination mode dominates for nickel complexes [7, 18], η^1-O [19], and η^1-C [20] CO_2 coordination modes have also been crystallographically characterized (Scheme 12.2b).

One of the features of Ni-catalyzed carboxylation is that monovalent nickel species are usually invoked at the point when CO_2 becomes involved in the catalytic cycle. Thus, studies of CO_2 capture by Ni(I) become necessary. Recently, Johnson et al. employed sophisticated mass spectrometry techniques to first generate a Ni(I) complex with a vacant coordination site, and then expose it to CO_2 (Scheme 12.3) [23]. Analysis of the resulting vibrational predissociation spectrum allowed an η^2-O—C=O geometry to be assigned to the coordinated CO_2 molecule, which was supported with calculations. The bent geometry and vibrational spectrum were consistent with a CO_2 radical anion, and an electron having been transferred from the metal to CO_2 to give the nickel center a Ni(II) formal oxidation state.

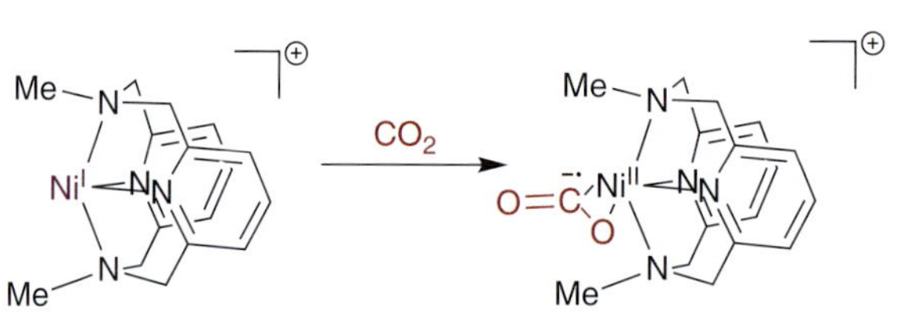

Scheme 12.3 Capture of CO_2 by a Ni(I) complex synthesized in the gas phase.

12.2.3 Insertion/C—C Bond Formation

The formation of carboxylic acids from CO_2 and organic substrates requires the formation of a new C—C bond. This could occur via the following substrate-dependent pathways: (i) direct insertion of CO_2 into a Ni—C bond, (ii) coordination of CO_2 to Ni followed by insertion, (iii) nucleophilic attack of the organometallic ligand on electrophilic CO_2, or, for an allylic substrate (iv) insertion through a six-membered transition state (Scheme 12.4) [10, 11d, 24]. Formation of carbon-based radical anions (either on CO_2 or the substrate) is also a possible C—C bond forming pathway [25]. Reactions are classed as outer sphere insertions when the CO_2 is not involved in the turnover-determining step [26].

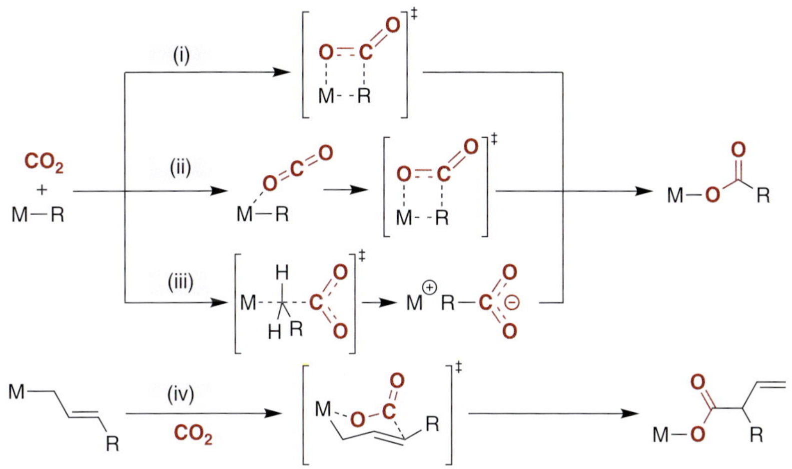

Scheme 12.4 Reaction of organometallic species with CO_2 to form M-carboxylates.

12.2.4 Ligand Effects

Nickel-catalyzed carboxylation reactions usually employ phosphine or α-diimine ligands (Scheme 12.5). Unsurprisingly, ligand choice plays a very important role in determining the yield and selectivity of the reaction. This has, for example, resulted in the development of many new substituted α-diimine ligands that allow highly selective carboxylation reactions (vide infra).

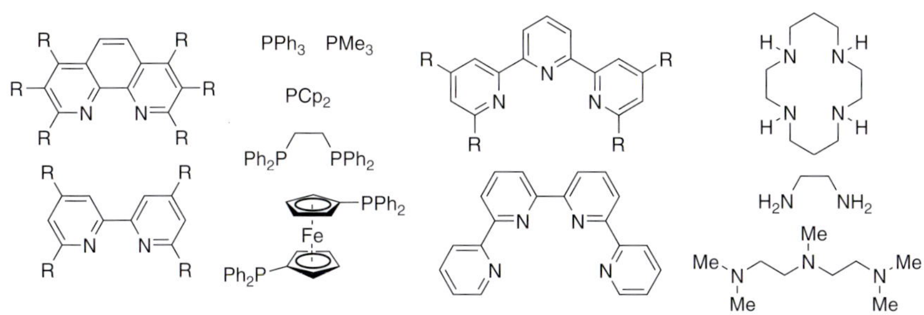

Scheme 12.5 Ligand classes employed in Ni-catalyzed carboxylation reactions.

The ability of certain classes of ligand to act as electron reservoirs is currently an unstudied feature of carboxylation chemistry. In particular, the ligand may play a role in allowing changes to the number of electrons of the complex as a whole, but not to the d electrons held by the metal [27]. As this chapter is focused on Ni(I), the simplest mechanistic picture to invoke such an oxidation state involves oxidation from Ni(0) and reduction from Ni(II). However, if redox-active ligands are present, the oxidation state as indicated by the number of electrons at the metal may differ from Ni(I). For example, spectroscopic data and theoretical evidence suggest that [Ni(Mes)(phen)] (Mes = 1,3,5-trimethylphenyl) is better described as [Ni(II)(Mes)(phen·$^-$)] than [Ni(I)(Mes)(phen)] [28], that [Ni(bipy)(COD)] (COD = 1,5-cyclooctadiene) is better described as [Ni(I)(bipy·$^-$)(COD)] than [Ni(0)(bipy)(COD)] [29], and that "zerovalent" [NiL_2] complexes such as [$Ni(bipy)_2$] are best described as [Ni(bipy·$^-$)(bipy)] [30]. In both cases, the α-diimine ligand accepts electron density from the metal center. The influence of redox non-innocence on the electron transfer steps of the carboxylation reaction can be studied by spectroscopic techniques such as UV/vis and electron paramagnetic resonance (EPR), electrochemical techniques, and computational chemistry [31].

12.2.5 Oxidative Addition

The IUPAC definition of monometallic oxidative addition is "the insertion of a metal complex into a covalent bond involving formally an overall two-electron loss" [32]. This process occurs via different mechanisms depending on the substrate (Scheme 12.6) [17, 33, 34]. A number of factors influence which of these takes precedence, particularly if the metal undergoes facile one-electron oxidation state changes. Regarding nickel, the seminal study by Tsou and Kochi showed that a two-electron transfer from low-valent nickel to aryl halides forms a caged Ni(I)-aryl radical anion, and that depending on a number of factors, different ratios of the Ni(II) oxidative addition complex, or Ni(I)–X and Ar· are formed (Scheme 12.6a) [33b]. Reactions of alkyl halides with low-valent nickel are generally thought to occur via a radical mechanism (Scheme 12.6b) [27b, 34]. In contrast, allyl and benzyl electrophiles tend to undergo two-electron oxidative addition [17].

12.2.6 Oxidation State

As alluded to above, oxidation state is a formalism based on electron counting rules that can in some cases be misleading or ambiguous [13, 35]. Computational tools can be combined with experimental studies to aid in understanding the electron density at the metal and in resolving ambiguity.

12.2.7 Single Electron Transfer (SET)

Radical intermediates have been invoked in many Ni-catalyzed carboxylation reactions. Precise interpretation of "single electron transfer (SET) processes" as invoked in the carboxylation literature should focus on the nature of the oxidative addition reaction (see above), and the feasibility of radical chain processes.

12.2.8 Conclusion

As mentioned above, the majority of proposed carboxylation mechanisms have not undergone detailed mechanistic studies, and numerous questions remain. However, the mechanistic components mentioned above are very important

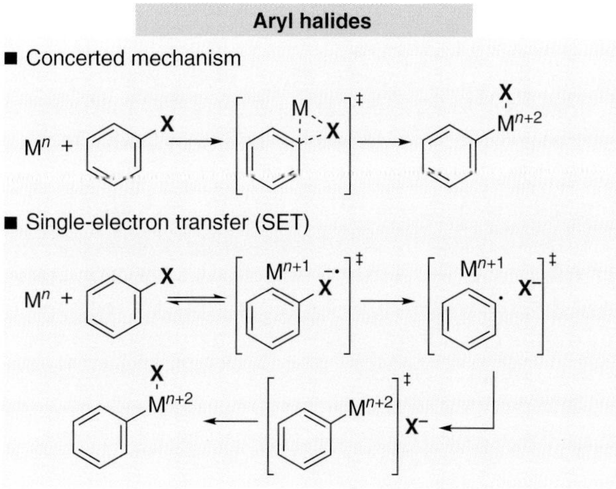

(a)

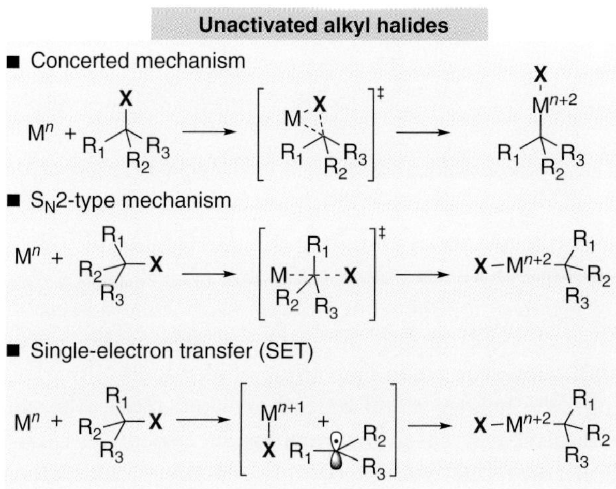

(b)

Scheme 12.6 Oxidative addition.

and will surely form the basis of many efforts to understand the role of Ni(I) in carboxylation. Throughout this chapter – which begins with electrocarboxylation and is followed by methodologies employing metal powders as reductants – these themes will be repeated.

12.3 Electrocarboxylation

12.3.1 Introduction

The origins of nickel-catalyzed fixation of CO_2 into organic molecules lie in two classes of electrochemical syntheses reported more than 30 years ago: the use of metal salts to catalyze the homocoupling of aryl halides, and the formation of carboxylic acids when CO_2 and organic substrates were available at the cathode of an electrochemical cell [6, 36]. The selectivities of the majority of these early carboxylation reactions were low, yields were poor, and the conditions were rarely optimized for particular products. A contributing factor to this is the very negative potential required for the reduction of CO_2 to $CO_2{\cdot}^-$ (approx. −2.2 V in DMF vs. the saturated calomel electrode [SCE]) [37]. In an electrocarboxylation reaction, products are obtained either by one-electron reduction of CO_2 to form $CO_2{\cdot}^-$, which then reacts with the substrate to generate the carboxylate, or two-electron reduction of the substrate followed by attack on CO_2 (Scheme 12.7) [11a, 37, 38]. Depending on the class of substrate and its particular substituents, the $CO_2{\cdot}^-$ radical anion may not react with the substrate. Alternatively, the reduction potential of the organic compound may be very negative, and the reduced product able to abstract protons or hydrogen atoms from donors such as the electrolyte or solvent [37].

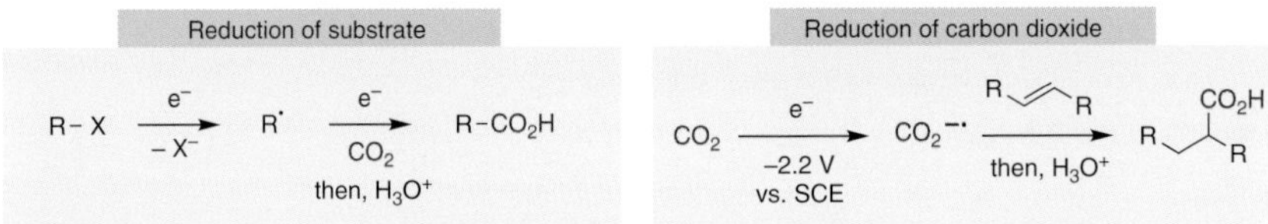

Scheme 12.7 Examples of electrocarboxylation scenarios via reduction of substrate or of CO_2.

The development of sacrificial anodes and, in particular, the use of transition metal catalysts, allowed these challenges to be overcome [37–39]. For example, the catalyst-free carboxylation of iodobenzene (−1.8 V vs. SCE) and bromobenzene (−2.3 V) was limited to yields of only 40%, and chlorobenzene could not be carboxylated [36b, 40]. However, when Fauvarque et al. added $NiBr_2$ and an excess of PPh_3, all three aryl halides could be carboxylated [36a]. The choice of a Ni(II) salt as the catalyst was based on previous success in Ni-catalyzed nucleophilic substitution and reductive dimerization reactions and on the use of Ni(II) salts in the electrochemical synthesis of, for example, biphenyl from aryl halides [41].[2] Overall two-electron reduction from Ni(II) to Ni(0) gave an electrogenerated zerovalent complex able to insert into the C—X bonds of the three aryl halides. Reduction of the subsequent oxidative addition complex $[Ni(II)(Ar)(X)(PPh_3)_2]$ at −1.8 V vs. SCE in the presence of CO_2 (1 atm) gave the desired aryl carboxylate [36b].

These early studies prompted a two-decade period of rapid growth in Ni-catalyzed electrocarboxylation based on the electrochemical generation of reactive nickel complexes. Overall, these have been used to catalyze the carboxylation of a wide variety of substrates, including aryl halides [36, 40b], allenes [42], alkenes [43], epoxides [44], alkynes [45], allylic systems [46], and benzylic halides [47] (Scheme 12.8). The catalysts employed are either a mixture of a Ni(II) salt such as $NiBr_2$, $NiCl_2$ and a phosphine ligand (PPh_3, dppe, dppf) or

2 See Refs. [36, 39a] and references therein.

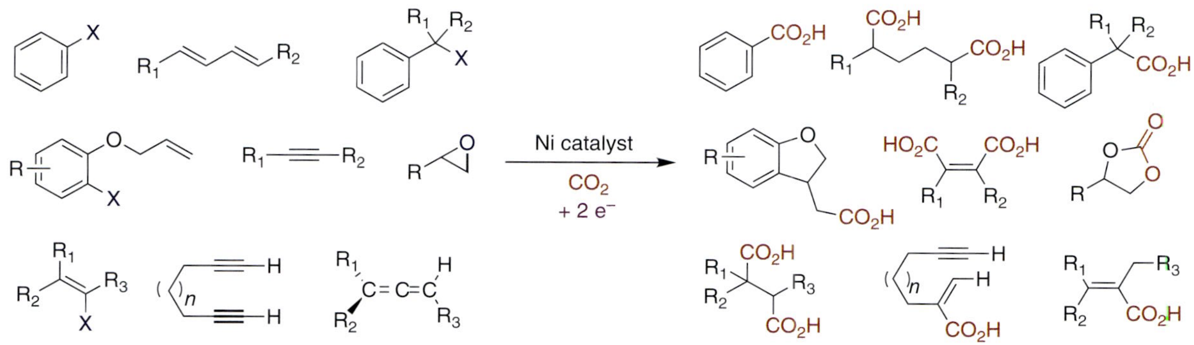

Scheme 12.8 Examples of substrates that undergo electrocarboxylation via electrogenerated nickel complexes.

N-donor ligand (bipy, phen, PMDTA, TMEDA, cyclam), or a preformed Ni(II) complex such as [$NiBr_2$(dppe)] or [$NiBr_2$(bipy)]. Generally, the Ni(II) sources undergo reduction to generate reactive Ni(0) species at potentials near −1.2 V vs. SCE in DMF for bipy and phen ligands, near −1.6 V vs. SCE for cyclam or salen ligands [39a], or in two one-electron steps (−0.8 and −1.35 V vs. SCE) for [$NiCl_2$(dppe)] [48].

Excluding oxanickelacyclopentenone mechanisms, which may be operative during the carboxylation of unsaturated systems [49, 50], mechanistic information about the electrochemical incorporation of CO_2 into organic molecules has been obtained from electrochemical measurements of the reduction of Ni(II) complexes to Ni(I) and/or Ni(0) and of the subsequent overall two-electron carboxylation. The mechanistic information gathered so far in the field of electrocarboxylation has been organized by ligand type in the following sections.

12.3.2 Phosphine Ligands

Phosphine ligands have been employed with varying success in the electrocarboxylation of substrates such as aryl halides [36, 40b], 1-octyne [45b], allenes [42], and diynes [51]. The most detailed mechanistic study of Ni-catalyzed electrocarboxylation focused on the Ni/dppe (dppe = 1,2-bis(diphenylphosphino)ethane) catalyzed electrocarboxylation of PhBr (Scheme 12.9) [41b, 52]. This study was motivated in part by the similarity between the reaction conditions for the electrocatalytic carboxylation of bromobenzene and the homodimerization of PhBr, and the absence of homodimerization in the former [41b, 48a, 53]. Although homodimerization is associated with a radical chain mechanism [48a], it was unclear how CO_2 intercepted the corresponding Ni intermediates. The mechanism proposed by Amatore and Jutand during this study, shown below in Scheme 12.10, laid the foundations for the mechanistic proposals of future Ni-catalyzed carboxylation reactions.

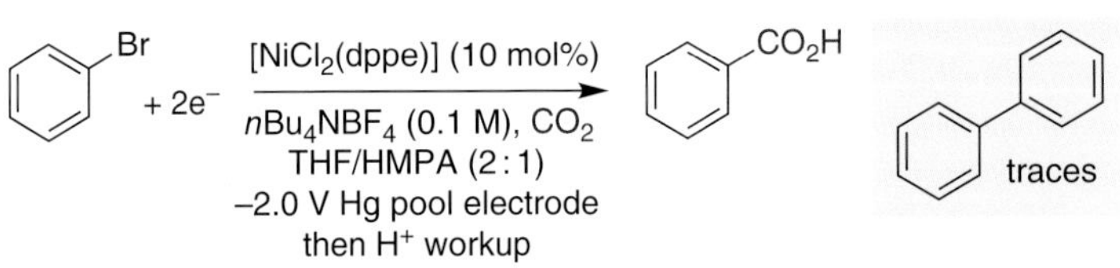

Scheme 12.9 Electrocarboxylation of bromobenzene.

The reaction began by the reduction of the nickel source, [Ni(II)Cl_2(dppe)], in two single-electron steps to form the highly reactive [Ni(0)(dppe)]. The formation of this coordinatively unsaturated species is crucial, as if the reaction is carried out with excess bisphosphine the reduction of the Ni(II) precursor will give the inactive [Ni(0)$(dppe)_2$] as the final product [54]. Oxidative addition of [Ni(0)(dppe)] to the aryl bromide gives [Ni(II)(Ph)(Br)(dppe)] (**1**), a step that is shared by homodimerization. In order to determine how the presence of CO_2 prevented homodimerization, a number of experiments were carried out. First, the behavior of the Ni(II) precatalyst [Ni(II)Cl_2(dppe)] toward PhBr was

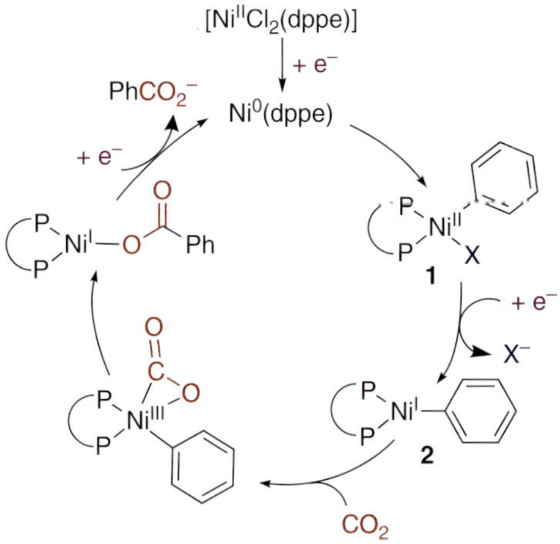

Scheme 12.10 Proposed mechanism for the electrocarboxylation of PhBr.

investigated in the presence of CO_2. The resulting voltammogram contained the same peaks for reduction of the Ni(II) precursor and single-electron reduction of **1** to [Ni(I)(Ph)(dppe)] (**2**) as if CO_2 was absent. However, the peak for single-electron reduction of **1** to **2** was dependent on the concentrations of PhBr and CO_2. Reaction of CO_2 with Ni(II) complex **1** was ruled out as being one of the homodimerization intermediates intercepted by CO_2. The absence of CO_2 insertion products from Ni(II) was in agreement with previous experiments with isolated square planar Ni(II)(X)(Ph) complexes [55].

Next, the relationship between the rate of carboxylation and the rate of reaction of **2** with PhBr was studied to rule out carboxylation from Ni(III). Monovalent complex **2** was instead invoked as the key species due to the voltage at which electrolysis occurs (at the reduction potential of the oxidative addition complex). Amatore and Jutand also determined that **2** reacts much faster with CO_2 than PhBr so that when CO_2 is present, **2** undergoes reaction with CO_2 rather than initiating the synthesis of biphenyl.

The kinetics of CO_2 insertion into the Ni(I)—Ph bond was also investigated. In many carboxylation reports, CO_2 insertion to a Ni—C bond is termed "insertion" without comment on the process itself. Amatore et al. proposed that the reaction of CO_2 with Ni(I) occurs via two steps. First, the reaction of CO_2 with the Ni(I) species forms a Ni(III) species, then insertion into the Ni—Ar bond occurs to form the negatively charged carboxylate ligand and Ni(I). Reduction of the carboxylate complex liberates [Ni(0)(dppe)]. The authors speculated that reaction of **2** with CO_2 is a radical reaction due to the ability of CO_2 to accept radicals. However, there have not been further studies on this "oxidative addition" reaction. Other groups have pointed out that when CO_2 is coordinated to Ni, the coordinated C—O π-bond takes on single bond character, and that the Ni(I) complex is more Ni(II)-like [18, 22, 23]. Therefore, there is still ambiguity on how CO_2 interacts with **2** under electrocatalytic conditions.

12.3.3 Bipyridine and Related α-Diimine Ligands

The Ni(II)/bipyridine system is employed in a large number of electrochemical and non-electrochemical Ni-catalyzed carboxylation reactions. However, the mechanism has not been studied in detail, and the role of Ni(I) remains rather speculative. Following the electrochemical mechanistic study by Amatore and Jutand [48b, 52], Osakada and Yamamoto et al. synthesized the bipyridine-ligated oxidative addition complex [NiBr(Ph)(bipy)] [56]. When this complex was reacted with CO_2 in the absence of reductant, benzoic acid was obtained in a 55% yield after workup, along with a 21% yield of biphenyl (Scheme 12.11). The formation of the acid implied insertion of CO_2 into the Ni(II)—Ph bond, which contrasted with the findings of Amatore and Jutand that have been discussed above [48b, 52]. The sequence of steps for the formation of the benzoate ligand is shown in Scheme 12.11, where cationic species seem to play a critical role for success. The formation of such species by dissociation of Br^- from the Ni(II) oxidative addition complex in coordinating solvents has been confirmed [57]. Although carboxylic acid is indeed produced during the reaction, the concomitant formation of homocoupled species involves Ni species in a range of oxidation states, and it is possible that these may be involved in modulating the oxidation state of the added Ni(II) oxidative addition complex prior to CO_2 insertion.

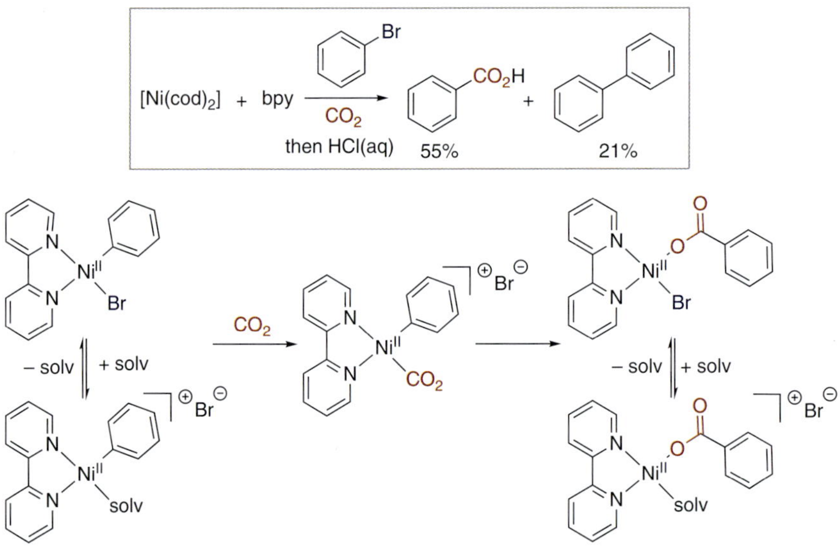

Scheme 12.11 Proposed rationale for the formation of acid from a Ni(II)(bipy) complex in the absence of reductant.

Oxidative addition complexes bearing α-diimine ligands such as phenanthroline and bipyridine display a reduction peak for the formation of a coordinated α-diimine radical anion [58]. Although the π* SOMO may play a role during the carboxylation reaction, this has not yet been studied.[3] Furthermore,

3 In Ref. [29], neutral bipy is invoked during the Ni-catalyzed hydroboration reaction.

various α-diimine-ligated complexes [NiX(Ar)(L)] have been shown to release the halide upon one-electron reduction. For example, upon reduction of [NiBr(Mes)(phen)], formation of [Ni(Mes)(phen)] occurs, where the complex is best described as [Ni(II)(Mes)(phen·⁻)] [58]. Taking into consideration the effect of the solvent on the reduction potentials, coordination of a molecule of solvent in the coordination sphere of [Ni(I)(Ar)(bipy)] has also been proposed [59].

Interestingly, the utilization of benzyl bromides may follow a different mechanism. Amatore showed that the oxidative addition complex [Ni(II)Br(CH_2Ph)(bipy)] underwent two-electron reduction to form the Ni(0)-ate complex $[Ni(0)(CH_2Ph)(bipy)]^-$ [60]. Although these complexes have not been invoked in carboxylation, their reaction with CO_2 cannot be ruled out. The Ni(II)/bipy catalyzed electrocarboxylation of terminal allylic esters and carbonates gave moderate yields and rather high selectivities for the carboxylation of terminal substrates to linear acids [46]. Electrogenerated Ni(0)(bipy) was proposed to react with the allylic substrate to form a Ni(II) π-allyl complex, which was then proposed to undergo one-electron reduction prior to CO_2 insertion.

12.3.4 Salen Ligands

In line with the utility of salen ligands in CO_2 fixation via C—C bond formation [4], a Ni/salen carboxylation of benzyl chlorides via $[Ni(I)(salen)]^-$ was reported by Gennaro, Isse, Maran, and co-workers (Scheme 12.12) [47]. This process is believed to operate via one-electron reduction of [Ni(II)(salen)] to $[Ni(I)(salen)]^-$

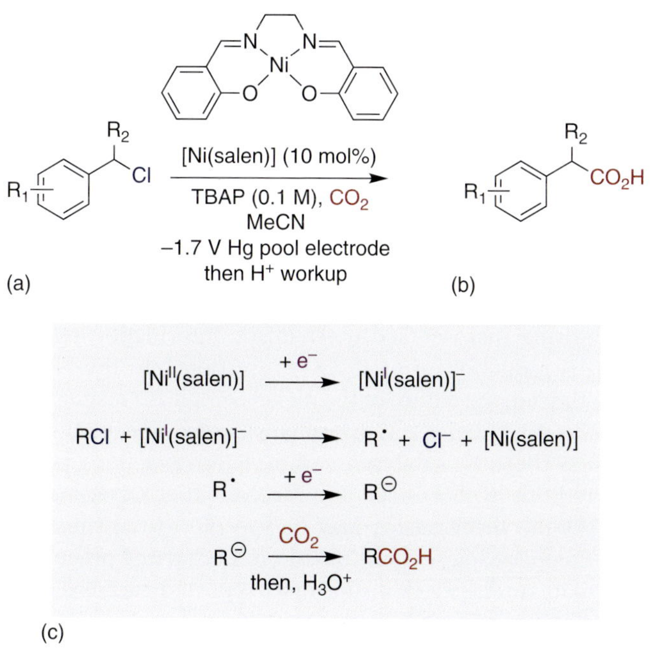

Scheme 12.12 Ni/salen-catalyzed electrocarboxylation via benzyl anion.

(−1.62 V vs. SCE), which provides a reactive species able to reduce the benzyl chloride to a benzyl radical. This radical was further reduced (at −1.7 V vs. SCE) to a benzyl anion, which could subsequently be trapped by CO_2 to form the corresponding carboxylate. In this reaction, the role of the electrochemically generated Ni(I) is better assigned to that of a reducing agent rather than as a key oxidation state for the activation of CO_2.

12.3.5 Conclusion

Electrochemistry has been used to synthesize a large range of carboxylic acids from a wide variety of substrates. It is also a useful technique for studying mechanisms and provides a suite of techniques that can be used to gain information about different steps of the mechanism. Particularly important in this respect is that the number and origin of the redox changes occurring throughout the catalytic cycle can be studied. Although there is still debate over the oxidation state of the nickel intermediate that is responsible for CO_2 insertion into the Ni—C bond, it is likely that electrochemical techniques will be combined with the synthesis and isolation of organometallic intermediates, kinetic studies, and theoretical calculations in order to shed light on the mechanisms of these reactions.

12.4 Non-electrochemical Methods

Recent years have seen a renewed interest in catalytic carboxylation, in which the electrochemical cell has been replaced by a solid reductant or an organometallic reagent (Scheme 12.13) [11j]. This has been touted as a more accessible and general mode of carboxylation; indeed, the substrate scopes and selectivities of these reactions have been optimized and developed to a much greater extent. The newer carboxylation reactions are often also combined with applications of the methodology to, for example, iterative coupling reactions or remote functionalization via chain migration [15, 61].

From a mechanistic standpoint, it is particularly difficult to measure the processes that occur at the surfaces of the metal reductants (typically Mn or Zn), an undoubtedly important feature if these are involved in the reduction of Ni. Thus, it is not particularly surprising that detailed mechanistic information related to the importance of Ni(II) reduction prior to CO_2 insertion into the Ni—C bond is currently lacking. Indeed, the nature of the nickel complexes responsible for CO_2 insertion in Mn/Zn-mediated reactions remains largely speculative. In this section, we will discuss those Ni-catalyzed carboxylation reactions where Ni(I) or SET processes are both hypothesized and receive either experimental or theoretical attention. As the mechanisms may be specific to the substrate class, we have organized the following sections based on the starting precursor. We will not directly speculate on the mechanisms of all reported carboxylation reactions, unless they bear a particular similarity to those mentioned above. Furthermore, unless otherwise noted, the reactions presented herein are carried out under 1 atm CO_2.

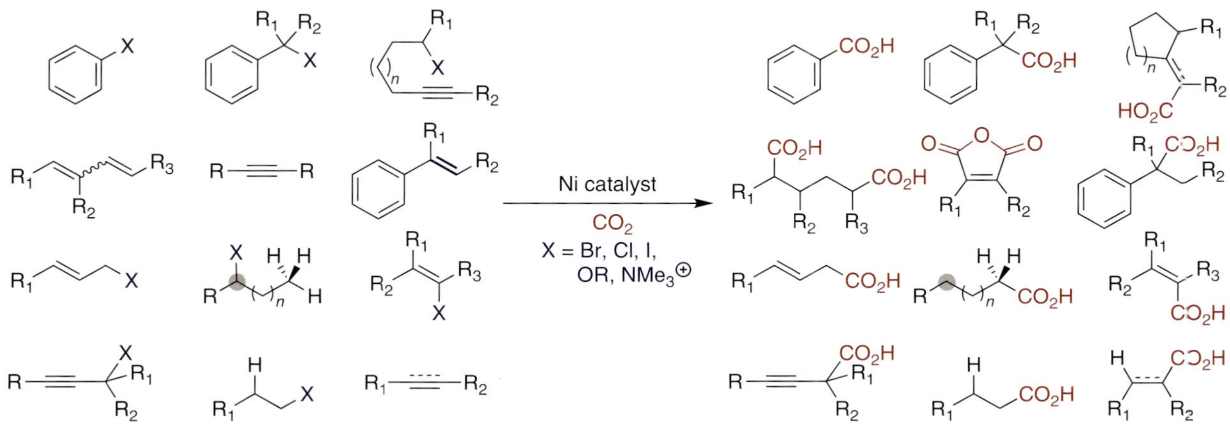

Scheme 12.13 Non-electrochemical Ni-catalyzed carboxylation.

12.4.1 Aryl Halides

The first catalytic reductive carboxylation of aryl halides was reported by Martin et al. with Pd catalysts supported by an electron-rich phosphine ligand (*t*BuXPhos) [62]. Although air-sensitive Et_2Zn was employed as reductant, this methodology can be considered the foundation of modern catalytic carboxylation reactions of aryl halides. Recently, Iwasawa and Martin developed a Pd-catalyzed photoredox technique that used simple tertiary amines as reductants, suggesting the intervention of Pd(I) intermediates prior to CO_2 insertion [63]. Prompted by the work of Martin, the Ni-catalyzed carboxylation of aryl and vinyl chlorides under an atmospheric pressure of CO_2 using solid metal reductants and phosphine ligands was developed by Tsuji et al. (Scheme 12.14) [16]. This report showed that electron-rich aryl and vinyl chlorides could be carboxylated, resulting in a platform with a much-improved substrate scope compared to classical electrochemical reactions. Subsequent work allowed *ortho*-substituted aryl triflates to be carboxylated efficiently [64].

During the optimization, it was found that Zn and Mg both gave drastically lower yields (9% and 0%) compared to manganese. Although Mn (−1.44 V vs. SCE) is a stronger reductant than Zn (−1.02 V vs. SCE), the lack of reactivity of Mg (c. −2.62 V vs. SCE) indicates that the redox potential might not be the only factor coming into play [11j, 65]. The addition of ammonium salts was also found to be crucial, with the nature of the counterion being particularly important. Although the role of the ammonium salt is not completely clear, it seemed to play a role in the Mn-mediated reduction; the authors suggested that the iodide anion from the ammonium salt bridged the Ni(II) center and the Mn during an outer-sphere electron transfer [66]. Confirmation of this hypothesis could be sought by determining the symmetries of the orbitals between which electron transfer is occurring, and the particular affinities of Ni(II) and Mn(0) for iodide [67]. Another aspect that might be important is the ability of the halide ligand to dissociate upon reduction of the Ni(II) oxidative addition complex to Ni(I).

The stoichiometric mechanistic studies detailed in this paper set the basis for the mechanistic conclusions drawn from similar systems and warrant a detailed discussion. After synthesizing the Ni(II) oxidative addition complex $[Ni(Cl)(Ph)(PPh_3)_2]$ (**3**), Tsuji et al. carried out stoichiometric reactions (Scheme 12.15). First, **3** was reacted under the carboxylation conditions to give 47% yield of the desired product (Scheme 12.15a). When Mn or Et_4NI were removed, either traces of product (no Mn) or no product (no Et_4NI) was obtained. The addition of homogeneous reductant ($[Co(\eta^5\text{-}C_5H_5)_2]$, $E = -0.98$ V vs. SCE (MeCN) [68]) recovered the yield, suggesting that SET processes might come into play to generate a transient Ni(I) species that undergoes CO_2 insertion (Scheme 12.12c). Interestingly, replacement of either Mn or $[Co(\eta^5\text{-}C_5H_5)_2]$ by Ni(0) yielded the carboxylation product, suggesting that comproportionation between Ni(0) and **3** may give access to a Ni(I) intermediate that is able to react with CO_2 (Scheme 12.15b).

DFT calculations reported by Sakaki shed light onto the key role of Ni(I) intermediates by comparing the reaction of CO_2 with either $[Ni(II)(Cl)(Ar)(PPh_3)_2]$

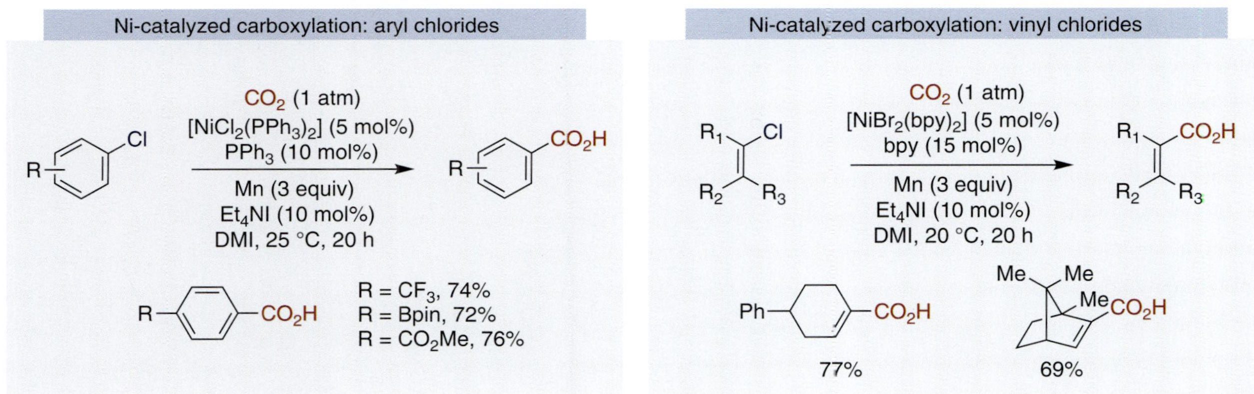

Scheme 12.14 Carboxylation of aryl and vinyl chlorides.

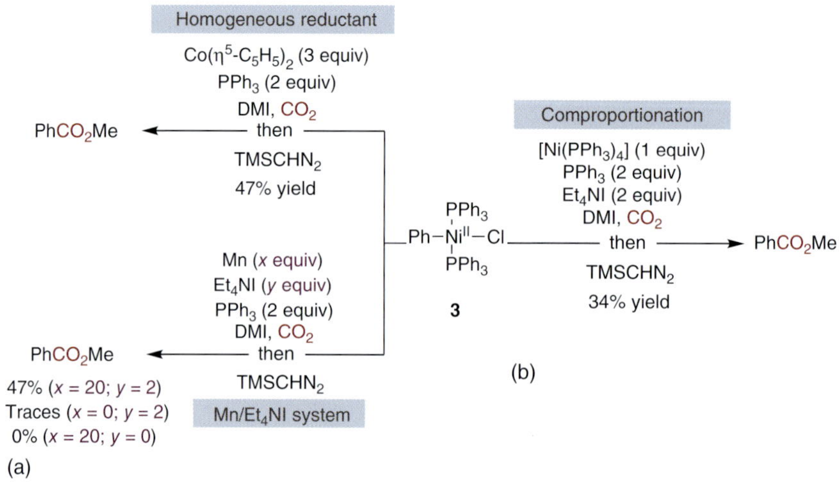

Scheme 12.15 Stoichiometric mechanistic experiments.

(**I**) or [Ni(I)(Ar)(PPh_3)$_2$] (**II**) [69].[4] As anticipated, coordination of CO_2 to **I**, followed by insertion, was an order of magnitude higher in energy than the process of insertion at Ni(I) complex **II**. Thermodynamic arguments for the Ni(I) pathway were also proposed, with the Ni(I)-(O_2CPh-κ^2*O,O′*) complex (**V**) having stronger Ni—O bonds than a related Ni(II)-(O_2CPh-κ^1*O*) species. Scheme 12.16 shows the proposed mechanism: coordination of CO_2 at **II** to form **III**, which then undergoes insertion into the Ni—Ph bond to form **IV**. A κ^1O to κ^2*O,O′* isomerization

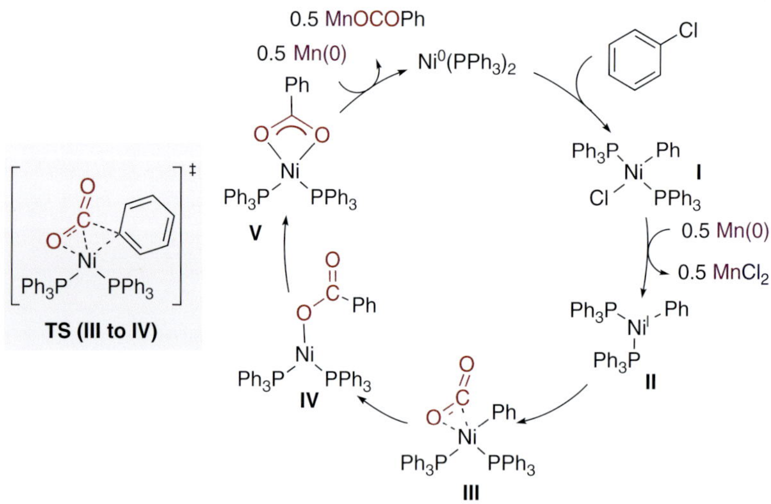

Scheme 12.16 Calculated mechanism for the carboxylation of aryl halides.

4 For a review on theoretical studies on the reactions between CO_2 and transition metals, see Lin *et al.*, Ref [26]. For a review specific to C—C bond forming reactions, see Hopmann *et al.*, Ref. [11i].

to **V** is followed by the final reduction by Mn to yield Ni(0) and a Mn-carboxylate, which is quenched during workup.

In 2014, the available literature data on (electro)carboxylation reactions of aryl electrophiles were limited to those that could be conducted with an organic halide. Aiming at broadening the scope of these reactions, the Martin group reported the carboxylation of aryl and benzyl pivalate esters (Scheme 12.17a) [70]. As for other sp^2 C–O functionalization methodologies, this was limited to the extended π-system of naphthyl substrates [71]. However, the carboxylation of benzylic sp^2 C—O bonds could be extended to regular arenes. Recently, Durandetti et al. reported the carboxylation of aryl and naphthyl tosylates with a Ni(0)/bipyridine catalyst (Scheme 12.17b) [72]. Unlike the previously discussed carboxylation of aryl electrophiles, this methodology employed an α-diimine ligand (bipy), which is reminiscent of previous electrocarboxylation reactions. This sp^2 C—O bond functionalization reaction tolerates ortho substituents and is not limited to extended π systems, an observation that is likely consistent with the ease of C–O cleavage in activated organic sulfonates. Electrochemical measurements were conducted in order to obtain an understanding about the carboxylation mechanism. These demonstrated that at −1.05 V (vs. SCE), a low-valent nickel complex is generated from [NiBr$_2$(bipy)] and that this goes on to react with the sp^2 C—OTs bond to give an oxidative addition complex. As for similar Ar–Ni(II)–X complexes, this intermediate underwent reduction to an Ar–Ni(I) species (−1.20 V vs. SCE). Unfortunately, not enough information was available to determine whether the oxidative addition complex is reduced to Ni(I) prior to reaction with CO_2, or whether reaction of the Ni(II)—Ph bond with CO_2 occurs first. A brief DFT study suggested that coordination of CO_2 to a Ni(I)–Ph complex results in a significantly lower energy complex than if it coordinates to a Ni(II)–Ph complex.

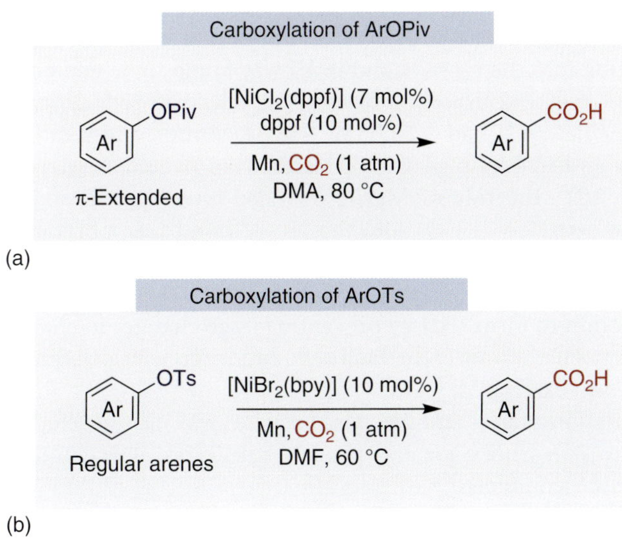

Scheme 12.17 Carboxylation of sp^2 C—O bonds.

12.4.2 Benzyl Electrophiles

Prompted by reports of electrochemical carboxylation, the Martin group developed the Ni-catalyzed carboxylation of a broad range of benzyl halides using Zn powder as the reducing agent and an electron-rich phosphine ligand (PCp_3) along with $NiCl_2$·glyme as the precatalyst (Scheme 12.18a) [73]. Unlike Ni-mediated electrochemical carboxylation, *ortho* substituents on the benzyl halide were well tolerated [47]. Three experiments were carried out in order to obtain mechanistic information about the reaction (Scheme 12.18b). First, a benzyl-Zn reagent **4** was synthesized and subjected to the carboxylation conditions. No product was obtained, suggesting that benzyl–Zn–X species do not intervene in the CO_2 insertion. Next, stoichiometric experiments were carried out with the isolated Ni(II) oxidative addition complex [Ni(II)(Cl)(η^3-C_7H_7)(PCp_3)]. Notably, carboxylation of the benzyl ligand did not occur unless Zn dust was added, which resulted in a 71% yield of benzoic acid. In line with previous carboxylation studies, these results strongly suggested the active participation of Ni(I) intermediates prior to CO_2 insertion.

Although the reaction was inhibited in the presence of radical scavengers (BHT and TEMPO), care must be taken in analyzing these results: it is unclear whether these react with some of the putative reaction intermediates leading to catalyst poisoning rather than radical trapping. The reaction was also carried out with an enantiomerically pure substrate and resulted in a racemic product, suggesting that the oxidative addition reaction occurs via two consecutive single-electron steps, with the first giving Ni(I) and a planar radical that recombines with a Ni(I) species (with or without escape from the solvent cage). In view of these results, a mechanism involving oxidative addition, one-electron reduction to Ni(I), CO_2 insertion, and reduction via formation of a Zn carboxylate was proposed.

Sakaki carried out a DFT study on the mechanism of this reaction (Scheme 12.19) [17b]. Two pathways from the Ni(II) oxidative addition intermediate **VI** were investigated, in which **VI** is common to both pathways. First, direct CO_2 insertion into the Ni—C bond of **VI** was found to be energetically unfavorable ($\Delta G^{\circ\ddagger} = 41.7$ kcal/mol). However, one-electron reduction of **VI** to give Ni(I) intermediate **VII** had a $\Delta G^{\circ\ddagger}$ of −18.5 kcal/mol, and a rather favorable coordination and insertion of CO_2 possessed an overall barrier of only 8.6 kcal/mol from **VII**. The role of $MgCl_2$ was also interrogated, and it was concluded that the insertion of CO_2 into the Ni—C bond was still found to be favored in the Ni(I) case. The positive charge on the Mg and the negative charge of one of the CO_2 oxygen atoms interact to lower the π^* energy of CO_2, thus favoring CO_2 insertion to form **VIII** by preventing large changes in charge during the insertion step. Furthermore, the final reduction step was also found to be favored in the presence of $MgCl_2$.

The catalytic carboxylation of aryl and benzylic pivalates was also developed with a Ni/PMe_3 system [70]. Unlike for aryl pivalate substrates, it was found that the reaction could be conducted with regular arenes. Again, addition of TEMPO resulted in inhibition, and the racemic product was obtained when an enantiomerically enriched benzyl pivalate underwent carboxylation. Although the carboxylation step was not investigated, a mechanism similar to that of

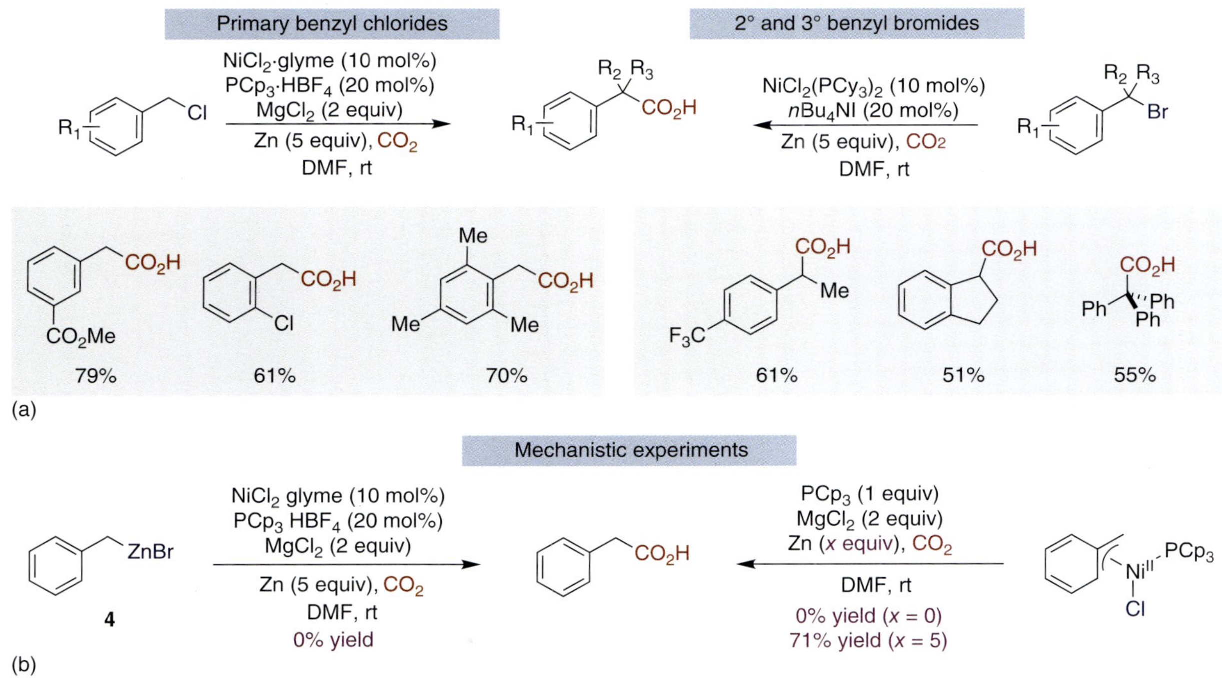

Scheme 12.18 Carboxylation of benzyl halides.

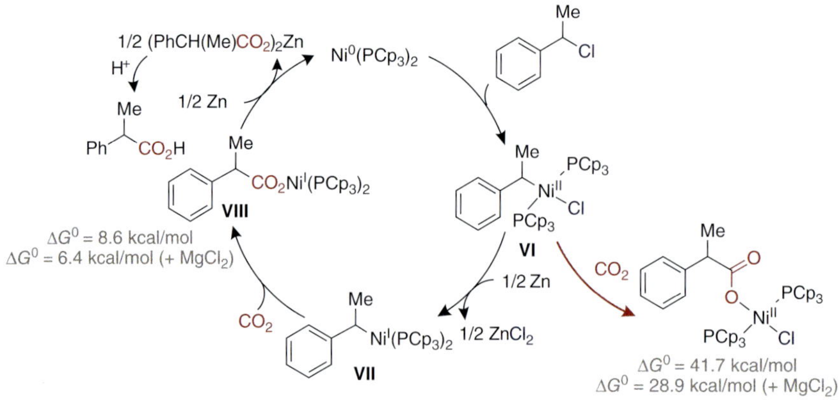

Scheme 12.19 DFT study of benzylic carboxylation.

the carboxylation of benzyl halides was proposed [73]. In both mechanisms, alternative pathways involving comproportionation of the Ni(II) oxidative addition complex and Ni(0) could also be operative.

A Ni/phenanthroline system, rather than a Ni/phosphine system, was employed for the carboxylation of benzylic C—N bonds (Scheme 12.20) [74]. The role of Ni(I) in this reaction was studied by reacting [Ni(**L1**)$_2$](OTf) and [Ni(**L1**)$_2$] with benzylic ammonium salt **5** and CO_2. The lack of reactivity of [Ni(**L1**)$_2$](OTf) in the absence of Mn is deemed to reflect the initiation of the reaction at Ni(0) rather than at Ni(I). As shown in Scheme 12.20b, the neutral [Ni(**L1**)$_2$] complex also furnished the targeted carboxylic acid in the absence of Mn powder. Thus, it remained undetermined whether a Ni(I) benzyl species is required (as supported by evidence from the studies by Sakaki and Martin) [17b, 70, 73], whether Ni(II) was in fact able to react with CO_2, or whether comproportionation between Ni(II) and Ni(0) forms the Ni(I)–benzyl species.

In conclusion, the carboxylation of benzylic electrophiles can be achieved with metal reductants (Mn or Zn) using two different systems: Ni/PR$_3$, or Ni/phen. The general mechanistic picture suggests that SET oxidative addition occurs with substituted phenanthroline ligands and that CO_2 insertion occurs at Ni(I). However, the formation of carboxylic acid in the absence of a stoichiometric reductant suggests that carboxylation may occur from the Ni(II) oxidative addition complex directly, that oxidative addition forms stable benzylic radicals that combine with CO_2, or that the Ni(II) complex undergoes comproportionation with Ni(0) to form Ni(I).

12.4.3 Carboxylation of Unactivated Alkyl Electrophiles

The first example of the carboxylation of an unactivated sp^3 C—X bond was reported by Martin et al. (Scheme 12.21) [75]. In this methodology, unactivated alkyl bromides and sulfonates were carboxylated using a Ni/phenanthroline catalytic system with Mn as the reducing agent. The effect of ligand substitution on the carboxylation yield was rather pronounced (Scheme 12.21a). In particular,

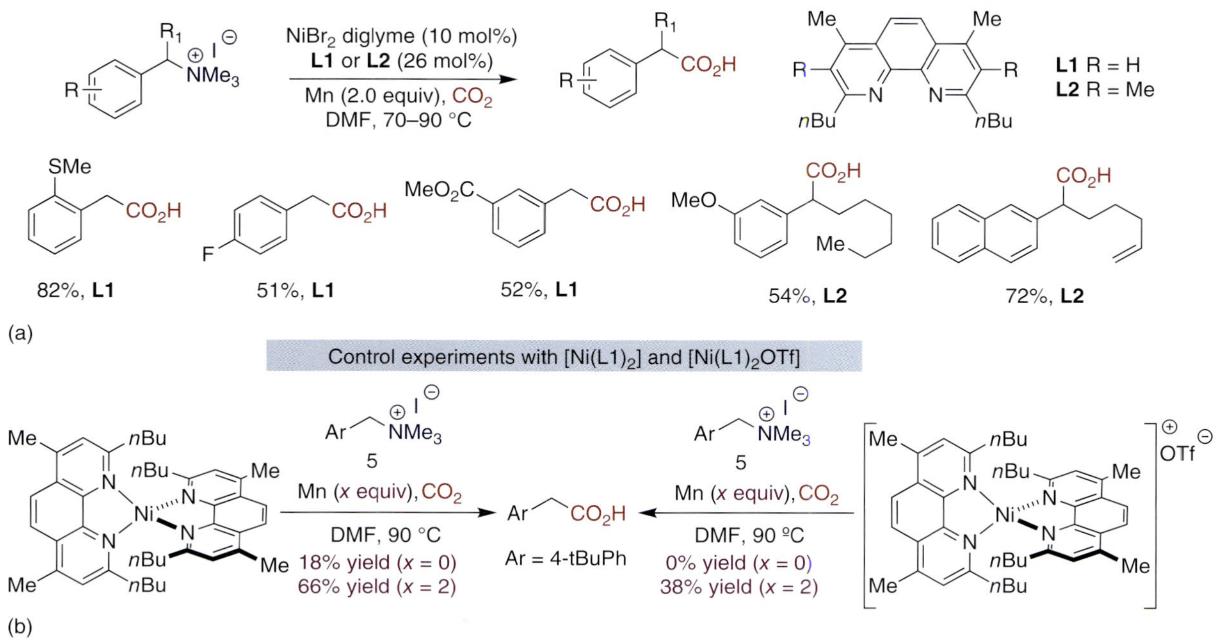

Scheme 12.20 Ni/phen-catalyzed carboxylation of benzylic ammonium salts.

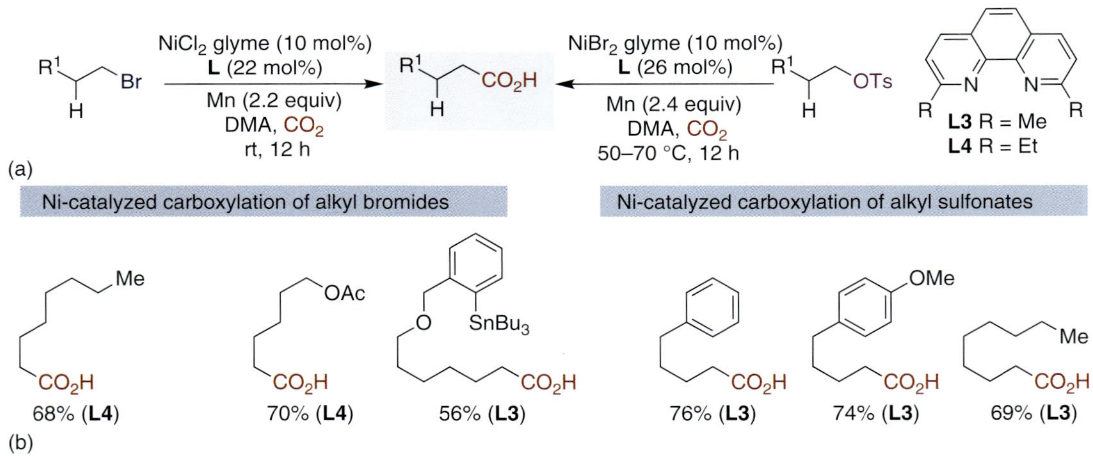

Scheme 12.21 Carboxylation of unactivated alkyl bromides and sulfonates.

it was found that *ortho* substituents on the α-diimine ligand were critical for success, an observation that has been successfully applied in all the subsequent reactions developed by the Martin group (Scheme 12.22a). Although the role of these substituents in improving both the yields and selectivities of carboxylation reactions has not yet been studied, the extra steric bulk imposed by these substituents may promote the loss of the anion upon oxidative addition. Although mechanistic experiments were not carried out to probe the CO_2 insertion step, deuterium labeling experiments suggested that oxidative addition operated via SET rather than a two-electron Ni(0)/Ni(II) pathway (Scheme 12.22b).

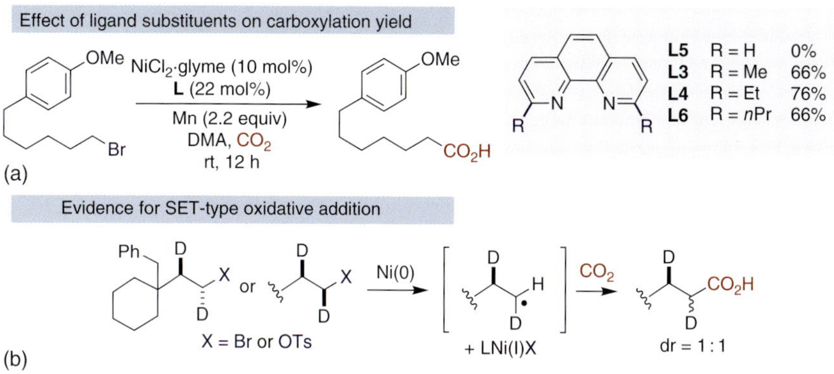

Scheme 12.22 (a) Effect of phenanthroline *ortho* substituents (blue) on the yield of carboxylation. (b) Evidence for SET oxidative addition for alkyl substrates.

The use of *ortho*-substituted α-diimine ligands and a metal powder reducing agent in Ni-catalyzed carboxylation was later extended to strained cyclopropyl systems and to the tandem cyclization/carboxylation of unactivated alkyl halides (Scheme 12.23) [76, 77]. In both cases, reactions with 18-electron species $[Ni(L)_2]$ were conducted with the aim of understanding the reactivity of putative reaction intermediates. As shown in Scheme 12.23b, stoichiometric reactions with these complexes indicated that no reaction occurred in the absence of Mn, suggesting that the Ni(II) oxidative addition complex was unable to react with CO_2. Care must be taken when assigning the oxidation state of nickel in $[Ni(L)_2]$ complexes as Ni(0), as these complexes can also be formulated as $[Ni(I)(L)(L\cdot^-)]$, where the radical anion is delocalized across both ligands [30].

Recently, carboxylation of alkyl bromides was extended to a chain-walking strategy, whereby functionalization occurred at remote unfunctionalized reaction sites (Scheme 12.24) [61]. Similar mechanistic studies were carried out with the corresponding $[Ni(\mathbf{L9})_2]$ complex; as anticipated, Mn was required for the reaction to occur. Unfortunately, it is not clear whether a single electron reduction from Ni(II) to Ni(I) is carried out by Mn or whether comproportionation between Ni(II) and formally Ni(0) species can generate the active Ni(I) species.

Although the low reactivity of the sp^3 C—Cl bond initially suggested that catalytic carboxylation with unactivated alkyl chlorides would be beyond reach, the Martin group recently addressed this challenge by using an electron-rich substituted phenanthroline ligand in combination with ammonium salts as

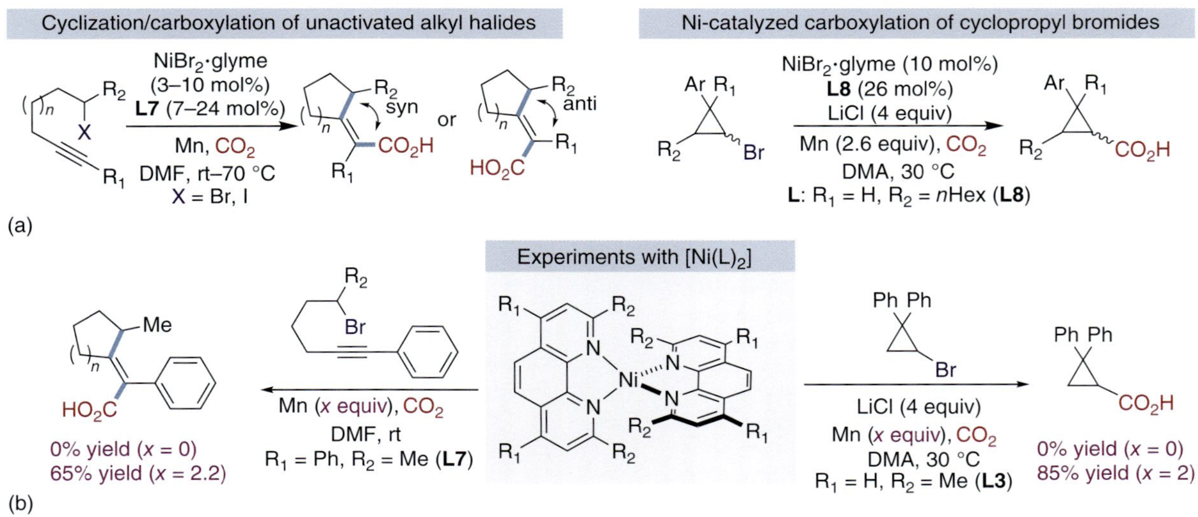

Scheme 12.23 Cyclization–carboxylation of alkyl bromides and carboxylation of cyclopropyl bromides.

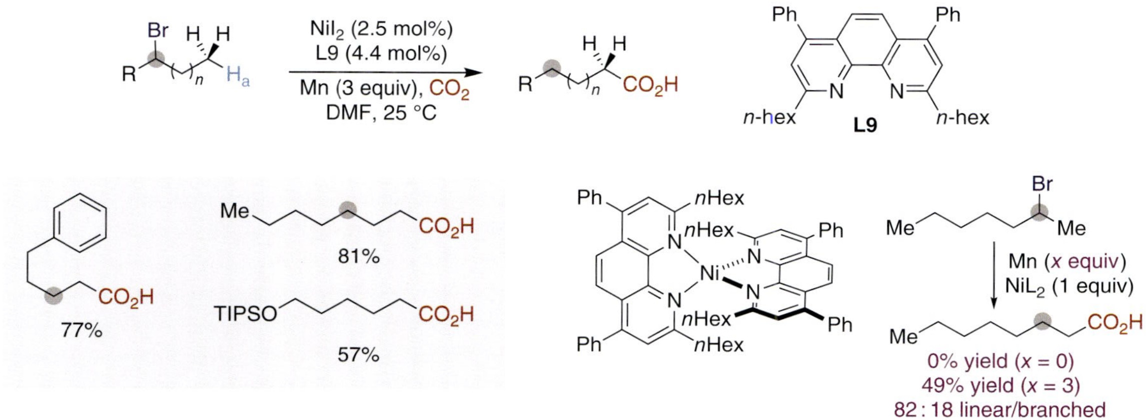

Scheme 12.24 Chain-walking carboxylation of alkyl bromides.

additives (Scheme 12.25) [15]. Unlike the carboxylation of unactivated alkyl bromides, this system allows for the carboxylation of secondary and tertiary alkyl chlorides. As for the carboxylation of alkyl bromides discussed above, the reaction of an α,β-deuterated substrate gave a mixture of isomeric carboxylic acids, suggesting that the oxidative addition likely goes via SET steps. The [Ni(**L1**)$_2$] and [Ni(I)(**L1**)$_2$]OTf complexes were synthesized and reacted with alkyl chloride in the presence and absence of *n*Bu$_4$NBr. Notably, carboxylic acid was obtained in all cases, even in the absence of Mn, demonstrating that the additive and/or the reducing agent might not be required for carboxylation. These results are particularly intriguing when compared with previously discussed carboxylation of alkyl electrophiles, suggesting that these reactions may fall into different categories depending on the substrate and halide leaving group.

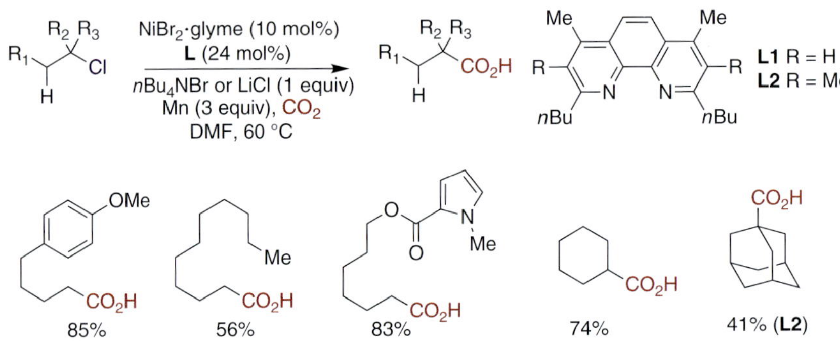

Scheme 12.25 Ni-catalyzed carboxylation of unactivated alkyl chlorides.

12.4.4 Carboxylation of Allyl Electrophiles

The Martin group has also studied the catalytic carboxylation of allylic esters with solid metallic reductants (Scheme 12.26) [78]. These systems are synthetically appealing, as site selectivity might be possible after C–O cleavage followed by CO_2 insertion. This can be tuned and controlled by the judicious choice of ligand and additives. In particular, it was found that linear carboxylation occurred with phenanthroline ligands, whereas quaterpyridine ligands provided branched carboxylation. Although both methods employ different metal reductants and additives, control experiments revealed that site selectivity was controlled by the ligand. These results are particularly interesting when compared to previous Ni(II)/bipy catalyzed electrocarboxylation of terminal allylic esters and carbonates, which gave high linear selectivity but only displayed limited substrate scope [46]. As shown in Scheme 12.26b, stoichiometric reactions of [Ni(L)$_n$] complexes [Ni(**L3**)$_2$] and [Ni(**L10**)] with linear allyl acetate (R = C_5H_{11}) were particularly instructive. While Mn was required for the reaction to occur with [Ni(**L3**)$_2$], the absence of Zn in the case of [Ni(**L10**)] led to significant amounts of branched acid. Although both quaterpyridine and phenanthroline ligands are redox active, the former constrains the coordination geometry and coordination number of Ni to a greater extent, which may cause a different mechanism to operate [79].

A similar Ni-catalyzed system was reported by the Martin group for the carboxylation of allylic alcohols (Scheme 12.27) [80]. In line with the previous

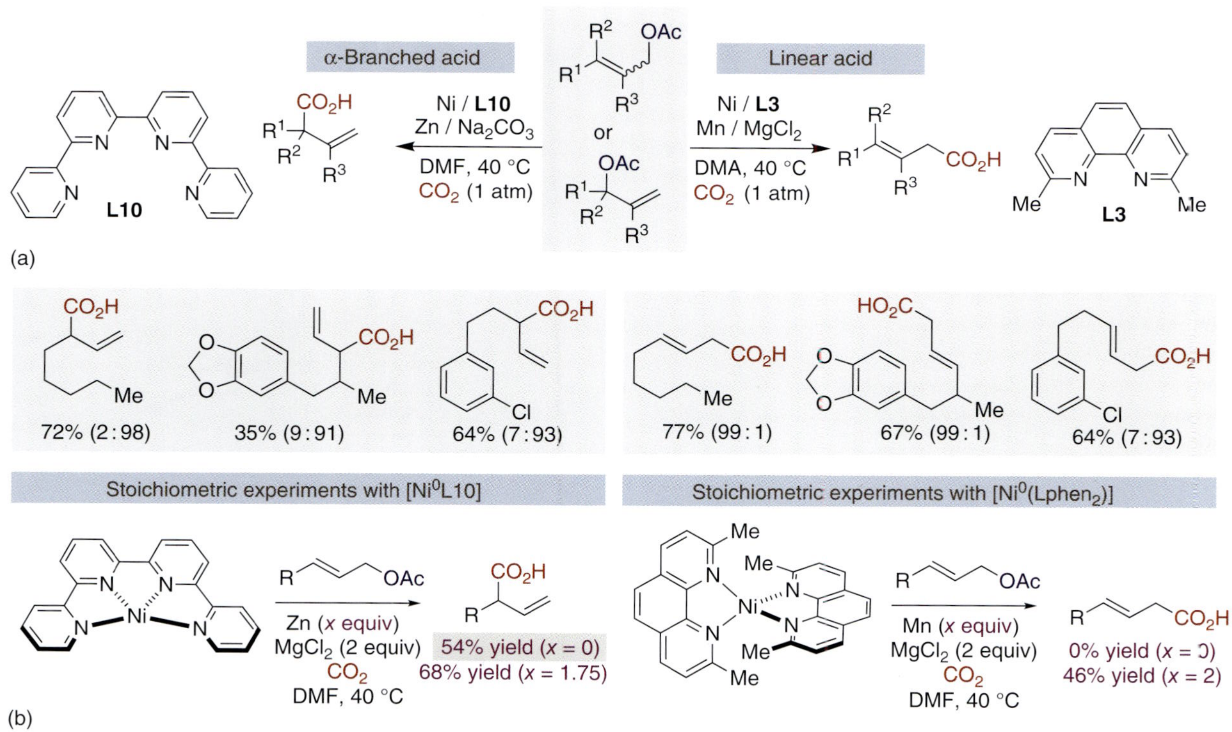

Scheme 12.26 Carboxylation of allyl acetate esters.

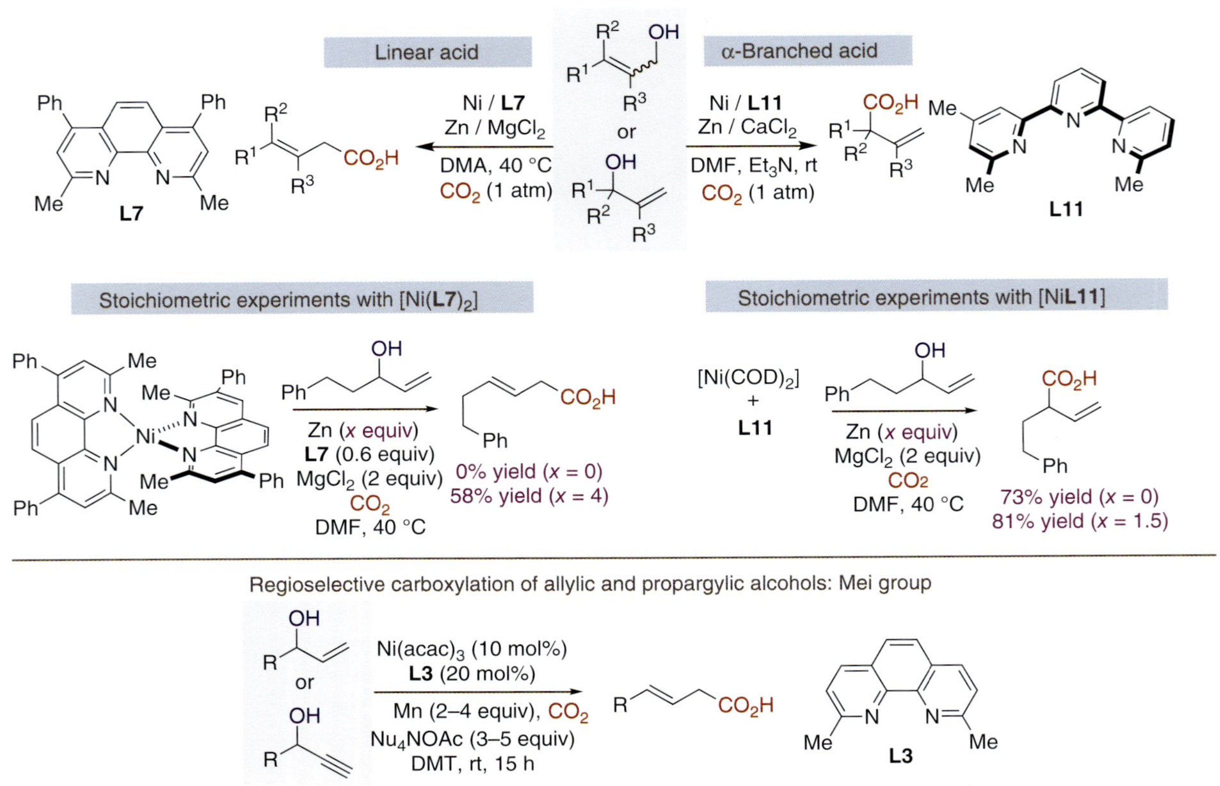

Scheme 12.27 Ni-catalyzed carboxylation of allylic alcohols.

carboxylation of allylic acetates, an *ortho*-substituted phenanthroline ligand gave access to linear products. In this case, however, a substituted terpyridine ligand (**L11**) was employed to obtain the branched carboxylic acids. As for the carboxylation of allylic acetates using a polypyridyl ligand, no reducing agent was required to carboxylate an allylic alcohol in a stoichiometric experiment, suggesting that a complex bearing a polypyridyl entity and an allyl ligand might insert CO_2. This observation is consistent with related Pd pincer complexes in which CO_2 insertion does not occur at the α C—Pd bond, but rather at the γ C—Pd position [77]. As mentioned above, redox non-innocence is often a hallmark of ligands with large electron-accepting π systems such as terpyridine and quaterpyridine and can complicate the assignation of oxidation state [81]. Indeed, the role of related [Ni(CH_3)(terpy)] species has been investigated, showing by both EPR and DFT calculations that this is better visualized as a Ni(II) complex possessing a terpy radical anion rather than a Ni(I) species [81, 82].

Independently, the Mei group reported the Ni-catalyzed carboxylation of allylic and propargylic alcohols to form linear carboxylic acids with phenanthroline-type ligands [14]. It was proposed that a π-allyl Ni(II) complex arises from oxidative addition; upon single-electron transfer, this complex leads to a Ni(I) species responsible for CO_2 insertion at the α C—Ni bond. However, no mechanistic studies were reported, and it is not yet clear how π-allyl Ni(II) complexes form directly from the allylic alcohol via C–OH cleavage. Indeed, Martin did not obtain conversion when [Ni(**L7**)$_2$] was reacted with an allylic alcohol, leaving a reasonable doubt that a direct C–OH cleavage occurs en route to the corresponding Ni(II) oxidative addition complex [80].

12.4.5 Unsaturated Systems

Despite the significant advances that have been realized in the catalytic carboxylation of organic (pseudo)halides with CO_2, the use of unsaturated π-systems offers the advantage of employing simple and readily available starting materials. From a mechanistic standpoint, the carboxylation of unsaturated π-systems typically occurs by an oxidative cyclization with Ni(0) and CO_2, resulting in an Ni(II) oxanickelacyclopentenone rather than the formation of Ni(I) species. Taking into consideration that this account aims to highlight the relevance of Ni(I) intermediates, we refer readers to previous literature on the formation and reactivity of oxanickelacyclopentenone intermediates [83].

A key example of the carboxylation of an unsaturated system that includes both an oxidative cyclization and CO_2 insertion into a Ni(I)—C bond is the double carboxylation of internal alkenes to maleic anhydride derivatives reported by Tsuji (Scheme 12.28) [84]. This methodology improved the substrate scope and selectivity of previously reported electrochemical double carboxylation reactions employing Ni [85]. A DFT study was carried out to study the mechanism of the transformation and to investigate the role of the crucial $MgBr_2$ additive. Indeed, oxidative cyclization of the alkyne and CO_2 with Ni(0)/bipy gave oxanickelacyclopentenone **6**. Intermediate **6** was synthesized by oxidative cyclization and used with success as the Ni source. Calculations suggested that **6** would be reduced by

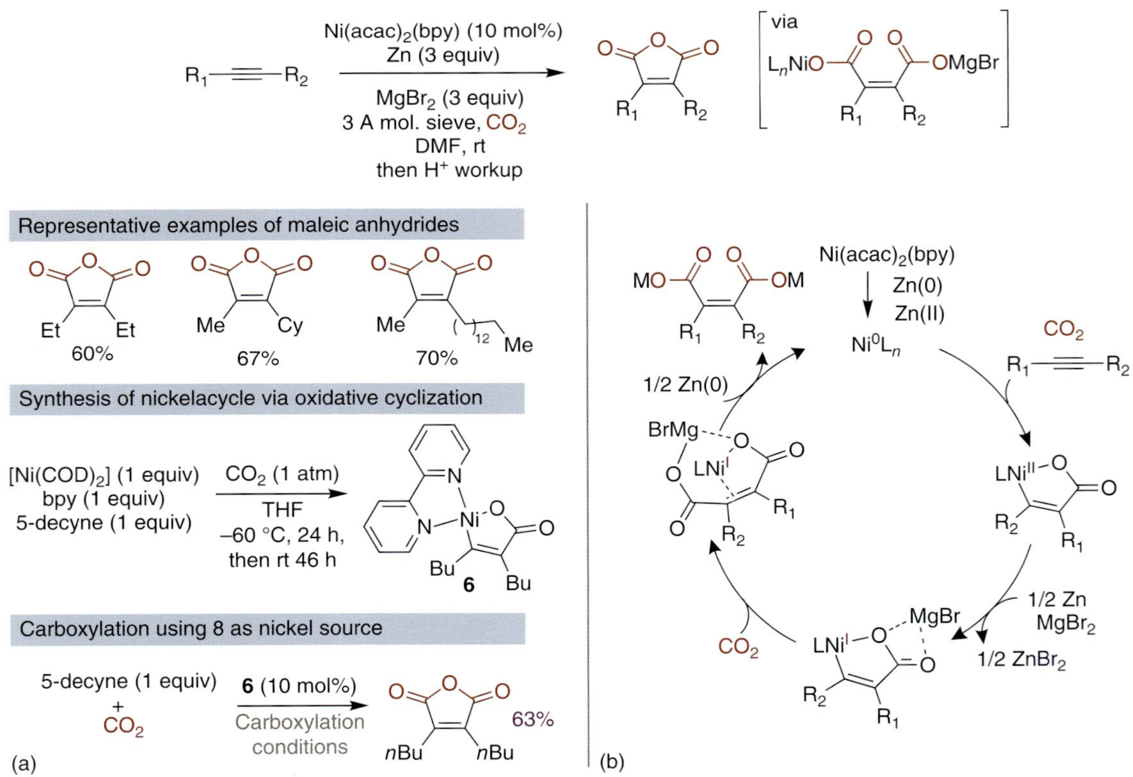

Scheme 12.28 Proposed mechanism for the $MgBr_2$-facilitated Ni-catalyzed carboxylation of internal alkynes.

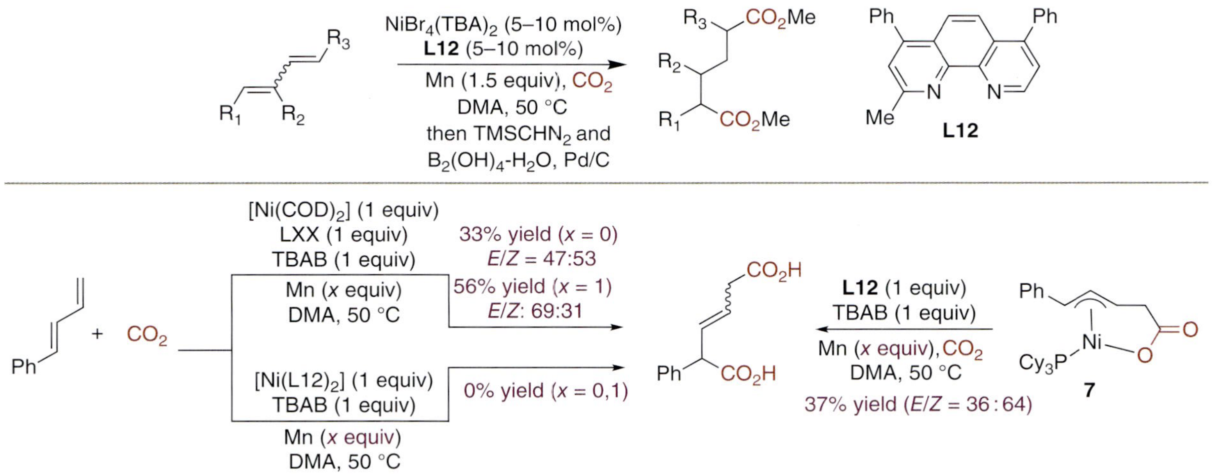

Scheme 12.29 Carboxylation of 1,3-dienes.

Zn/$MgBr_2$ to give a Ni(I) nickelacycle with coordinated MgBr, which would subsequently undergo CO_2 insertion (Scheme 12.28b). Interestingly, this pathway is at least three times lower in energy than a $MgBr_2$-free or Ni(II) pathway. The role of Mg(II) ions in opening oxanickelacyclopentenones has also been studied by Duñach in the carboxylation of alkynes via a cyclization mechanism [50].

The catalytic carboxylation of polyunsaturated systems (diynes, ene-ynes, allenes, and 1,3-dienes) has been carried out electrochemically [86] and with either metal powder reducing agents [87] or organozinc reagents [88]. The intermediacy of nickel(I) species has not been invoked in the majority of these reactions, with a cycloaddition/nickelalactone mechanism usually being operative. Recently, the Martin group described the site-selective carboxylation of 1,3-dienes to form adipic acids (Scheme 12.29) [87]. Mechanistic experiments with [Ni(**L12**)$_2$] or an allyl-Ni(II) carboxylate complex (**7**) gave no carboxylic acid in both the presence and absence of Mn powder. Although these results suggested the intermediacy of Ni(I) species formed upon either reduction or comproportionation, a more detailed mechanistic study is required to unravel the participation of Ni(I) species in the carboxylation of 1,3-dienes.

Unlike the carboxylation reactions of unsaturated systems, which have been proposed to operate via nickelacyclic intermediates, the *hydro*carboxylation of such systems is typically proposed to operate via hydrometalation with Ni–H species. However, the oxidation state of Ni during the reaction has not been confirmed, and although a Ni(0)/Ni(II) mechanism may operate, Ni(I) species may be present, and have not been ruled out [89]. Prompted by Ni and Cu-catalyzed reports by Ma, Tsuji, and Rovis with R_3SiH and Et_2Zn as reductants [90], Martin reported the hydrocarboxylation of alkynes and the hydrocarboxylation of α-olefins using Mn powder as the reductant, suggesting the involvement of Ni(I) species [91]. Driven by the ability of photoredox catalysis to replace metal reductants with simple tertiary amines in reductive quenching scenarios, König reported the Ni/photoredox-catalyzed hydrocarboxylation of styrenes, in which Ni(I) might come into play (Scheme 12.30) [92].

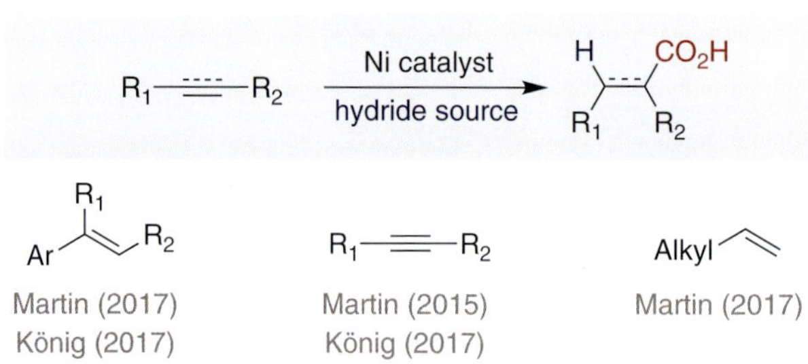

Scheme 12.30 Ni-catalyzed hydrocarboxylation reactions [91, 92].

12.5 Conclusions

Although Ni-catalyzed carboxylation reactions have attracted much research attention in recent years, their mechanisms remain a matter of speculation.

This is due to difficulties in studying these processes, particularly when solid reductants are employed. Additionally, it is worth noting that a wide variety of substrates undergo carboxylation reactions, making it difficult to generalize mechanistic conclusions for these reactions. Although the study of the reactivity of well-defined Ni(II) oxidative addition complexes has provided new vistas in carboxylation reactions, the literature data gathered herein provide evidence that Ni(I) species might be particularly suited for CO_2 insertion. Unfortunately, however, it is unclear exactly how these species are generated, as either SET from metallic reductants or comproportionation events may operate during the reaction. In either case, we anticipate that obtaining reliable access to the putative Ni(I) intermediates might provide an invaluable entry point to studying how the CO_2 insertion takes place. Likewise, a more detailed study on the role exerted by redox non-innocent polypyridyl ligands and their *ortho* substituents is required.

References

1 (a) Almqvist, E. (2003). *History of Industrial Gases*. Springer. (b) Appel, A.M., Bercaw, J.E., Bocarsly, A.B. et al. (2013). Frontiers, opportunities, and challenges in biochemical and chemical catalysis of CO_2 fixation. *Chem. Rev.* 113 (8): 6621–6658. https://doi.org/10.1021/cr300463y.

2 For a selection of reviews, see: (a) Dubois, D.L. (2006). Electrochemical reactions of carbon dioxide. In: *Encyclopedia of Electrochemistry*, 202–225. Wiley. (b) Aresta, M. and Dibenedetto, A. (2007). Utilisation of CO_2 as a chemical feedstock: opportunities and challenges. *Dalton Trans.* 28: 2975–2992. https://doi.org/10.1039/B700658F. (c) Aresta, M., Dibenedetto, A., and Angelini, A. (2014). Catalysis for the valorization of exhaust carbon: from CO_2 to chemicals, materials, and fuels. Technological use of CO_2. *Chem. Rev.* 114 (3): 1709–1742. https://doi.org/10.1021/cr4002758. (d) White, J.L., Baruch, M.F., Pander, J.E. et al. (2015). Light-driven heterogeneous reduction of carbon dioxide: photocatalysts and photoelectrodes. *Chem. Rev.* 115 (23): 12888–12935. https://doi.org/10.1021/acs.chemrev.5b00370. (e) Wang, J., Liu, W., Zhong, D., and Lu, T. (2017). Nickel complexes as molecular catalysts for water splitting and CO_2 reduction. *Coord. Chem. Rev.* https://doi.org/10.1016/j.ccr.2017.12.009.

3 (a) Shaikh, A.-A.G. and Sivaram, S. (1996). Organic carbonates. *Chem. Rev.* 96 (3): 951–976. https://doi.org/10.1021/cr950067i. (b) Ballivet-Tkatchenko, D., Camy, S., and Condoret, J.S. (2005). Carbon dioxide, a solvent and synthon for green chemistry. In: *Environmental Chemistry: Green Chemistry and Pollutants in Ecosystems* (eds. E. Lichtfouse, J. Schwarzbauer and D. Robert), 541–552. Springer-Verlag.

4 (a) Cokoja, M., Bruckmeier, C., Kühn, F.E. et al. (2011). Transformation of carbon dioxide with homogeneous transition-metal catalysts: a molecular solution to a global challenge? *Angew. Chem Int. Ed.* 50 (37): 8510–8537. https://doi.org/10.1002/anie.201102010. (b) Sakakura, T., Choi, J.C., and Yasuda, H. (2007). Transformation of carbon dioxide. *Chem. Rev.* 107 (6):

2365–2387. https://doi.org/10.1021/cr068357u. (c) Artz, J., Müller, T.E., Thenert, K. et al. (2018). Sustainable conversion of carbon dioxide: an integrated review of catalysis and life cycle assessment. *Chem. Rev.* 118 (2): 434–504. https://doi.org/10.1021/acs.chemrev.7b00435.

5 (a) Liu, Q., Wu, L., Jackstell, R., and Beller, M. (2015). Using carbon dioxide as a building block in organic synthesis. *Nat. Commun.* 6: 5933. https://doi.org/10.1038/ncomms6933. (b) Ahamed, M., Verbeek, J., Bormans, G. et al. (2016). Recent progress in metal catalyzed direct carboxylation of aryl halides and pseudo halides employing CO_2: opportunities for ^{11}C radiochemistry. *ChemCatChem* 8 (24): 3692–3700. https://doi.org/10.1002/cctc.201600943. (c) Yuan, G., Qi, C., Wu, W., and Jiang, H. (2017). Recent advances in organic synthesis with CO_2 as C1 synthon. *Curr. Opin. Green Sustainable Chem.* 3: 22–27. https://doi.org/10.1016/j.cogsc.2016.11.006.

6 See Refs. [11], Tyssee et al. and references therein: Tyssee, D.A., Wagenknecht, J.H., Baizer, M.M., and Chruma, J.L. (1972). Some cathodic organic syntheses involving carbon dioxide. *Tetrahedron Lett.* 13 (47): 4809–4812. https://doi.org/10.1016/S0040-4039(01)94435-1.

7 Aresta, M., Nobile, C.F., Albano, V.G. et al. (1975). New nickel-carbon dioxide complex: synthesis, properties, and crystallographic characterization of (carbon dioxide)-bis(tricyclohexylphosphine)nickel. *J. Chem. Soc., Chem. Commun.*: 636–637. https://doi.org/10.1039/C39750000636.

8 For selected reviews on nickel chemistry, see: (a) Tamaru, Y. (ed.) (2005). *Modern Organonickel Chemistry.* Wiley-VCH. (b) Tasker, S., Standley, E., and Jamison, T. (2014). Recent advances in homogeneous nickel catalysis. *Nature* 509 (7500): 299–309. https://doi.org/10.1038/nature13274. (c) Ananikov, V.P. (2015). Nickel: the "spirited horse" of transition metal catalysis. *ACS Catal.* 5 (3): 1964–1971. https://doi.org/10.1021/acscatal.5b00072.

9 Bruns, W., Kaim, W., Ladwig, M. et al. (1993). Using electrochemical information for organometallic coordination chemistry: stoichiometric and catalytic reactions. In: *Molecular Electrochemistry of Inorganic, Bioinorganic and Organometallic Compounds* (eds. A.J.L. Pombeiro and J.A. McCleverty), 255–268. Netherlands: Springer https://doi.org/10.1007/978-94-011-1628-2.

10 See Hazari and Heimann and references therein: Hazari, N. and Heimann, J.E. (2017). Carbon dioxide insertion into group 9 and 10 metal-element σ bonds. *Inorg. Chem.* 56 (22): 13655–13678. https://doi.org/10.1021/acs.inorgchem.7b02315.

11 For selected reviews, see: (a) Tokuda, M. (2006). Efficient fixation of carbon dioxide by electrolysis – facile synthesis of useful carboxylic acids. *J. Nat. Gas Chem.* 15 (4): 275–281. https://doi.org/10.1016/S1003-9953(07)60006-1. (b) Tsuji, Y. and Fujihara, T. (2012). Carbon dioxide as a carbon source in organic transformation: carbon–carbon bond forming reactions by transition-metal catalysts. *Chem. Commun.* 48 (80): 9956–9964. https://doi.org/10.1039/C2CC33848C. (c) Zhang, L. and Hou, Z. (2013). N-Heterocyclic carbene (NHC)–copper-catalysed transformations of carbon dioxide. *Chem. Sci.* 4 (9): 3395–3403. https://doi.org/10.1039/C3SC51070K. (d) Yeung, C.S. and Dong, V.M. (2014). Making C-C bonds from carbon dioxide via transition-metal catalysis. *Top. Catal.* 57 (17–20): 1342–1350. https://

doi.org/10.1007/s11244-014-0301-9. (e) Guo, C.-X., Yu, B., Ma, R., and He, L.-N. (2015). Metal-promoted carboxylation of alkynes/allenes with carbon dioxide. *Curr. Green Chem.* 2 (1): 14–25. https://doi.org/10.2174/2213346102011502091 63930. (f) Börjesson, M., Moragas, T., Gallego, D., and Martin, R. (2016). Metal-catalyzed carboxylation of organic (pseudo)halides with CO_2. *ACS Catal.* 6 (10): 6739–6749. https://doi.org/10.1021/acscatal.6b02124. (g) Wu, X.F. and Zheng, F. (2017). Synthesis of carboxylic acids and esters from CO_2. *Top. Curr. Chem.* 375 (1): 4. https://doi.org/10.1007/s41061-016-0091-6. (h) Martin, R., Tortajada, A., Juliá-Hernández, F. et al. (2018). Transition metal-catalyzed carboxylation reactions with carbon dioxide. *Angew. Chem. Int. Ed.* Accepted Author Manuscript. doi: https://doi.org/10.1002/anie.201803186. (i) Obst, M., Pavlovic, L., and Hopmann, K.H. (2018). Carbon-carbon bonds with CO_2: insights from computational studies. *J. Organomet. Chem.* https://doi.org/10.1016/j.jorganchem.2018.02.020. (j) Richmond, E. and Moran, J. (2018). Recent advances in nickel catalysis enabled by stoichiometric metallic reducing agents. *Synthesis* 50 (3): 499–513. https://doi.org/10.1055/s-0036-1591853. k Chen, Y.G., Xu, X.T., Zhang, K. et al. (2018). Transition-metal-catalyzed carboxylation of organic halides and their surrogates with carbon dioxide. *Synthesis* 50 (1): 35–48. https://doi.org/10.1055/s-0036-1590908.

12 See Ref. [11] and selected reviews concerning Ni(I) complexes: (a) Lin, C.-Y. and Power, P.P. (2017). Complexes of Ni(I): a "rare" oxidation state of growing importance. *Chem. Soc. Rev.* 46: 5347–5399. https://doi.org/10.1039/C7CS00216E. (b) Zimmermann, P. and Limberg, C. (2017). Activation of small molecules at nickel(I) moieties. *J. Am. Chem. Soc.* 139 (12): 4233–4242. https://doi.org/10.1021/jacs.6b12434.

13 For example, see Karen and references therein:Karen, P. (2015). Oxidation state, a long-standing issue! *Angew. Chem. Int. Ed.* 54 (16): 4716–4726. https://doi.org/10.1002/anie.201407561.

14 Chen, Y.G., Shuai, B., Mei, T.S. et al. (2017). Regioselective Ni-catalyzed carboxylation of allylic and propargylic alcohols with carbon dioxide. *Org. Lett.* 19 (11): 2969–2972. https://doi.org/10.1021/acs.orglett.7b01208.

15 Börjesson, M., Moragas, T., and Martin, R. (2016). Ni-catalyzed carboxylation of unactivated alkyl chlorides with CO_2. *J. Am. Chem. Soc.* 138 (24): 7504–7507. https://doi.org/10.1021/jacs.6b04088.

16 Fujihara, T., Terao, J., Tsuji, Y. et al. (2012). Nickel-catalyzed carboxylation of aryl and vinyl chlorides employing carbon dioxide. *J. Am. Chem. Soc.* 134 (22): 9106–9109. https://doi.org/10.1021/ja303514b.

17 (a) Lau, K.S.Y., Wong, P.K., and Stille, J.K. (1976). Oxidative addition of benzyl halides to zero-valent palladium complexes. inversion of configuration at carbon. *J. Am. Chem. Soc.* 98 (19): 5832–5840. https://doi.org/10.1021/ja00435a015. (b) Sayyed, F.B. and Sakaki, S. (2014). The crucial roles of $MgCl_2$ as a non-innocent additive in the Ni-catalyzed carboxylation of benzyl halide with CO_2. *Chem. Commun.* 50 (86): 13026–13029. https://doi.org/10.1039/C4CC04962D. (c) Tollefson, E.J., Hanna, L.E., and Jarvo, E.R. (2015). Stereospecific nickel-catalyzed cross-coupling reactions of benzylic ethers and esters. *Acc. Chem. Res.* 48 (8): 2344–2353. https://doi.org/10.1021/acs.accounts

.5b00223. (d) Jarvo, E.R., Houk, K.N., Hong, X. et al. (2017). Mechanism and origins of ligand-controlled stereoselectivity of Ni-catalyzed Suzuki–Miyaura coupling with benzylic esters: a computational study. *J. Am. Chem. Soc.* 139 (37): 12994–13005. https://doi.org/10.1021/jacs.7b04973.

18 (a) Aresta, M. and Nobile, C.F. (1977). (Carbon dioxide)bis(trialkylphosphine)nickel complexes. *J. Chem. Soc., Dalton Trans.*: 708–711. https://doi.org/10.1039/DT9770000708. (b) Anderson, J.S., Iluc, V.M., and Hillhouse, G.L. (2010). Reactions of CO_2 and CS_2 with 1,2-bis(di-*tert*-butylphosphino)ethane complexes of nickel(0) and nickel(I). *Inorg. Chem.* 49 (21): 10203–10207. https://doi.org/10.1021/ic101652e. (c) Beck, R., Shoshani, M., Johnson, S.A. et al. (2013). Synthesis and chemistry of bis(triisopropylphosphine) nickel(I) and nickel(0) precursors. *Dalton Trans.* 42 (5): 1461–1475. https://doi.org/10.1039/C2DT32008H. (d) Yunho, S., Kim, Y., Kim, J., and Lee, Y. (2014). Formation of a nickel carbon dioxide adduct and its transformation mediated by a Lewis acid. *Chem. Commun.* 50: 11458–11461. https://doi.org/10.1039/c4cc04800h.

19 Chiou, T.-W., Tseng, Y.-M., Lu, T.-T. et al. (2016). [Ni^{III}(OMe)]-mediated reductive activation of CO_2 affording a Ni(κ^1-OCO) complex. *Chem. Sci.* 7 (6): 3640–3644. https://doi.org/10.1039/C5SC04652A.

20 (a) Yoo, C. and Lee, Y. (2017). Carbon dioxide binding at a Ni/Fe center: synthesis and characterization of Ni(η^1-CO_2-κC) and Ni-μ-CO_2-κC:κ^2O,O′-Fe. *Chem. Sci.* 8 (1): 600–605. https://doi.org/10.1039/C6SC03450K. (b) Oren, D., Feller, M., Milstein, D. et al. (2018). Metal–ligand cooperation as key in formation of dearomatized Ni(II)–H pincer complexes and in their reactivity toward CO and CO_2. *Organometallics* 37 (14): 2217–2221. https://doi.org/10.1021/acs.organomet.8b00160.

21 (a) Paparo, A. and Okuda, J. (2017). Carbon dioxide complexes: bonding modes and synthetic methods. *Coord. Chem. Rev.* 334: 136–149. https://doi.org/10.1016/j.ccr.2016.06.005. (b) Schwarz, H. (2017). Metal-mediated activation of carbon dioxide in the gas phase: mechanistic insight derived from a combined experimental/computational approach. *Coord. Chem. Rev.* 334: 112–123. https://doi.org/10.1016/j.ccr.2016.03.009.

22 See Aresta and references therein:Aresta, M. (2017). My journey in the CO_2-chemistry wonderland. *Coord. Chem. Rev.* 334: 150–183. https://doi.org/10.1016/j.ccr.2016.06.004.

23 Menges, F.S., Craig, S.M., Johnson, M.A. et al. (2015). Capture of CO_2 by a cationic nickel(I) complex in the gas phase and characterization of the bound, activated CO_2 molecule by cryogenic ion vibrational predissociation spectroscopy. *Angew. Chem. Int. Ed.* 55 (4): 1282–1285. https://doi.org/10.1002/anie.201507965.

24 Schmeier, T.J., Hazari, N., Incarvito, C.D. et al. (2011). Exploring the reactions of CO_2 with PCP supported nickel complexes. *Chem. Commun.* 47 (6): 1824–1826. https://doi.org/10.1039/C0CC03898A.

25 See Ref. [47] and (a) Yatham, V.R., Shen, Y., and Martin, R. (2017). Catalytic intermolecular dicarbofunctionalization of styrenes with CO_2 and radical precursors. *Angew. Chem. Int. Ed.* 56 (36): 10915–10919. https://doi.org/10.1002/anie.201706263. (b) Seo, H., Liu, A., and Jamison, T.F. (2017). Direct

β-selective hydrocarboxylation of styrenes with CO_2 enabled by continuous flow photoredox catalysis. *J. Am. Chem. Soc.* 139 (40): 13969–13972. https://doi.org/10.1021/jacs.7b05942. (c) Seo, H., Katcher, M.H., and Jamison, T.F. (2017). Photoredox activation of carbon dioxide for amino acid synthesis in continuous flow. *Nat. Chem.* 9 (5): 453–456. https://doi.org/10.1038/nchem.2690.

26 See Refs. [10, 11d], and Fan, T., Chen, X., and Lin, Z. (2012). Theoretical studies of reactions of carbon dioxide mediated and catalysed by transition metal complexes. *Chem. Commun.* 48 (88): 10808–10828. https://doi.org/10.1039/c2cc34542k.

27 (a) Dzik, W.I., van der Vlugt, J.I., Reek, J.N.H., and de Bruin, B. (2011). Ligands that store and release electrons during catalysis. *Angew. Chem. Int. Ed.* 50 (15): 3356–3358. https://doi.org/10.1002/anie.201006778. (b) Hu, X. (1867–1886). Nickel-catalyzed cross coupling of non-activated alkyl halides: a mechanistic perspective. *Chem. Sci.* 2 (10) https://doi.org/10.1039/c1sc00368b. (c) Lyaskovskyy, V. and De Bruin, B. (2012). Redox non-innocent ligands: versatile new tools to control catalytic reactions. *ACS Catal.* 2 (2): 270–279. https://doi.org/10.1021/cs200660v. (d) Kaim, W. (2012). The shrinking world of innocent ligands: conventional and non-conventional redox-active ligands. *Eur. J. Inorg. Chem.* 2012 (3): 343–348. https://doi.org/10.1002/ejic.201101359. (e) Luca, O.R. and Crabtree, R.H. (2013). Redox-active ligands in catalysis. *Chem. Soc. Rev.* 42: 1440–1459. https://doi.org/10.1039/c2cs35228a.

28 Yakhvarov, D.G., Petr, A., Hey-Hawkins, E. et al. (2014). Synthesis, structure and electrochemical properties of the organonickel complex [NiBr(Mes)(phen)] (Mes = 2,4,6-trimethylphenyl, phen = 1,10-phenanthroline). *J. Organomet. Chem.* 750: 59–64. https://doi.org/10.1016/j.jorganchem.2013.11.003.

29 King, A.E., Smythe, N.C., Sutton, A.D. et al. (2016). Ni(bpy)(cod): a convenient entryway into the efficient hydroboration of ketones, aldehydes, and imines. *Eur. J. Inorg. Chem.* 2016 (11): 1635–1640. https://doi.org/10.1002/ejic.201600143.

30 Wang, M., England, J., Wieghardt, K. et al. (2015). Electronic structures of "low-valent" neutral complexes $[NiL_2]^0$ ($S = 0$; L = bpy, phen, tpy) – an experimental and DFT computational study. *Eur. J. Inorg. Chem.* 2015 (9): 1511–1523. https://doi.org/10.1002/ejic.201403144.

31 Budnikova, Y.H., Vicic, D.A., and Klein, A. (2018). Exploring mechanisms in Ni terpyridine catalyzed C–C cross-coupling reactions – a review. *Inorganics* 6 (1): 18. https://doi.org/10.3390/inorganics6010018.

32 IUPAC (1997). *Compendium of Chemical Terminology*, 2e. (the "Gold Book"). Compiled by A. D. McNaught and A. Wilkinson. Oxford: Blackwell Scientific Publications. XML on-line corrected version: http://goldbook.iupac.org (2006-) created by M. Nic, J. Jirat, B. Kosata; updates compiled by A. Jenkins. ISBN 0-9678550-9-8. doi: https://doi.org/10.1351/goldbook.

33 For Ni and aryl halides, see, for example: (a) Bajo, S., Sproules, S., Nelson, D.J. et al. (2017). Oxidative addition of aryl electrophiles to a prototypical nickel(0) complex: mechanism and structure/reactivity relationships. *Organometallics* 36 (8): 1662–1672. https://doi.org/10.1021/acs.organomet

.7b00208. (b) Tsou, T.T. and Kochi, J.K. (1979). Mechanism of oxidative addition. reaction of nickel(0) complexes with aromatic halides. *J. Am. Chem. Soc.* 101 (21): 6319–6332. https://doi.org/10.1021/ja00515a028.
34 For a review, see: Jahn, U. (2012). Radicals in transition metal catalyzed reactions? transition metal catalyzed radical reactions?: a fruitful interplay anyway. *Top. Curr. Chem.* 320: 323–451. https://doi.org/10.1007/128_2011_288.
35 (a) Parkin, G. (2006). Valence, oxidation number, and formal charge: three related but fundamentally different concepts. *J. Chem. Ed.* 83 (5): 791–799. https://doi.org/10.1021/ed083p791. (b) Seddon, E.A. and Seddon, K.R. (2013). *The Chemistry of Ruthenium*, vol. 19. Elsevier. (c) Crabtree, R.H. (2014). *The Organometallic Chemistry of the Transition Metals*, 6e. Wiley.
36 (a) Troupel, M., Rollin, Y., Périchon, J., and Fauvarque, J.F. (1981). *Nouv. J. Chim.* 5 (12): 621. (b) Fauvarque, J.F., Jutand, A., Périchon, J. et al. (1984). Electrosynthese catalytique d'acides benzoiques para-substitues a partir du derive halogene correspondant et de l'anhydride carbonique. *J. Organomet. Chem.* 264 (1–2): 273–281. https://doi.org/10.1016/0022-328X(84)85153-0.
37 (a) Chaussard, J., Périchon, J., Sibille, S. et al. (1990). Use of sacrificial anodes in electrochemical functionalization of organic halides. *Synthesis* 1990 (5): 369–381. https://doi.org/10.1017/CBO9781107415324.004. (b) Matthessen, R., Fransaer, J., De Vos, D.E. et al. (2014). Electrocarboxylation: towards sustainable and efficient synthesis of valuable carboxylic acids. *Beilstein J. Org. Chem.* 10: 2484–2500. https://doi.org/10.3762/bjoc.10.260.
38 Silvestri, G., Gambino, S., and Filardo, G. (1991). Use of sacrificial anodes in synthetic electrochemistry. Processes involving carbon dioxide. *Acta Chem. Scand.* 45: 987–992. https://doi.org/10.3891/acta.chem.scand.45-0987.
39 Further selected reviews: (a) Nédélec, J., Périchon, J., and Troupel, M. (1997). Organic electroreductive coupling reactions using transition metal complexes as catalysts. In: *Electrochemistry VI Electroorganic Synthesis: Bond Formation at Anode and Cathode*. Top. Curr. Chem. vol 185 (ed. E. Steckhan). Springer https://doi.org/10.1007/3-540-61454-0_72. (b) Yan, M., Kawamata, Y., and Baran, P.S. (2017). Synthetic organic electrochemical methods since 2000: on the verge of a renaissance. *Chem. Rev.* 117 (21): 13230–13319. https://doi.org/10.1021/acs.chemrev.7b00397.
40 (a) Barba, F., Guirado, A., and Zapata, A. (1982). Cathodic reduction of aromatic halides. An evidence for an ionic mechanism. *Electrochim. Acta* 27 (9): 1335–1337. https://doi.org/10.1016/0013-4686(82)80157-6. (b) Matsue, T., Kitahara, S., and Osa, T. (1982). *Denki Kagaku* 50 (9): 732.
41 See Amatore et al. and references therein: (a) Amatore, C., Jutand, A., and Mottier, L. (1991). Mechanism of nickel-catalysed electron transfer activation of aromatic halides: Part 1. Biphenyl electrosynthesis from bromobenzene. *J. Electroanal. Chem. Interfacial Electrochem.* 306 (1–2): 125–140. https://doi.org/10.1016/0022-0728(91)85226-F. (b) Amatore, C. and Jutand, A. (1991). Mechanism of nickel-catalyzed electron-transfer activation of aromatic halides. Part 2. electrocarboxylation of bromobenzene. *J. Electroanal. Chem.* 306 (1–2): 141–156. https://doi.org/10.1016/0022-0728(91)85227-G.

42 Dérien, S., Clinet, J.-C., Duñach, E., and Périchon, J. (1990). Coupling of allenes and carbon dioxide catalyzed by electrogenerated nickel complexes. *Synlett* 1990 (6): 361–364. https://doi.org/10.1055/s-1990-21096.

43 Dérien, S., Clinet, J.-C., Duñach, E., and Périchon, J. (1992). Electrochemical incorporation of carbon dioxide into alkenes by nickel complexes. *Tetrahedron* 48 (25): 5235–5248. https://doi.org/10.1016/S0040-4020(01)89021-9.

44 Tascedda, P., Dinjus, E., Duñach, E. et al. (2001). Nickel-catalyzed electrochemical carboxylation of epoxides: mechanistic aspects. *Appl. Organomet. Chem.* 15 (2): 141–144. https://doi.org/10.1002/1099-0739(200102)15:2<141::AID-AOC140>3.0.CO;2-F.

45 (a) Duñach, E. and Périchon, J. (1988). Electrochemical carboxylation of terminal alkynes catalyzed by nickel complexes: unusual regioselectivity. *J. Organomet. Chem.* 352 (1–2): 239–246. https://doi.org/10.1016/0022-328X(88)83038-9. (b) Labbé, E., Duñach, E., and Périchon, J. (1988). Ligand-directed reaction products in the nickel-catalyzed electrochemical carboxylation of terminal alkynes. *J. Organomet. Chem.* 353 (3): 51–56. https://doi.org/10.1016/0022-328X(88)80330-9. (c) Duñach, E., Dérien, S., and Périchon, J. (1989). Nickel-catalyzed reductive electrocarboxylation of disubstituted alkynes. *J. Organomet. Chem.* 364 (3): 33–36. https://doi.org/10.1016/0022-328X(89)87156-6. (d) Li, C., Yuan, G., and Jiang, H. (2010). Electrocarboxylation of alkynes with carbon dioxide in the presence of metal salt catalysts. *Chin. J. Chem.* 28 (9): 1685–1689. https://doi.org/10.1002/cjoc.201090285.

46 Medeiros, M.J., Pintaric, C., Olivero, S., and Duñach, E. (2011). Nickel-catalysed electrochemical carboxylation of allylic acetates and carbonates. *Electrochim. Acta* 56 (11): 4384–4389. https://doi.org/10.1016/j.electacta.2010.12.066.

47 Gennaro, A., Isse, A.A., and Maran, F. (2001). Nickel(I)(salen)-electrocatalyzed reduction of benzyl chlorides in the presence of carbon dioxide. *J. Electroanal. Chem.* 507 (1–2): 124–134. https://doi.org/10.1016/S0022-0728(01)00373-4.

48 (a) Amatore, C. and Jutand, A. (1988). Rates and mechanism of biphenyl synthesis catalyzed by electrogenerated coordinatively unsaturated nickel complexes. *Organometallics* 7 (10): 2203–2214. https://doi.org/10.1021/om00100a019. (b) Amatore, C. and Jutand, A. (1990). Rates and mechanisms of electron transfer/nickel-catalyzed homocoupling and carboxylation reactions. An electrochemical approach. *Acta Chem. Scand.* 44: 755–764.

49 (a) Burkhart, V.G. and Hoberg, H. (1982). Oxanickelacyclopenten-Derivate aus Nickel(0), Kohlendioxid und Alkinen. *Angew. Chem. Suppl.*: 147–152. https://doi.org/10.1002/ange.198201470. (b) Hoberg, H. and Schaefer, D. (1982). Modellkomplexe des nickels für die [2+2+2′]-cycloaddition von alkinen mit kohlendioxid. *J. Organomet. Chem.* 238 (4): 383–387. https://doi.org/10.1016/S0022-328X(00)83800-0. (c) Hoberg, H., Peres, Y., and Milchereit, A. (1986). C-C-verknüpfung von alkenen mit CO_2 an nickel(0); n-pentensäuren aus ethen. *J. Organomet. Chem.* 307: 41–43. https://doi.org/10.1016/0022-328X(86)80488-0.

50 Dérien, S., Duñach, E., and Périchon, J. (1991). From stoichiometry to catalysis: electroreductive coupling of alkynes and carbon dioxide with

nickel-bipyridine complexes. Magnesium ions as the key for catalysis. *J. Am. Chem. Soc.* 113 (22): 8447–8454. https://doi.org/10.1021/ja00022a037.

51 Duñach, E. and Périchon, J. (1990). Electrogenerated nickel(0) catalyzed carbon dioxide incorporation into α,ω-diynes. *J. Organomet. Chem.* 385 (3): 43–46. https://doi.org/10.1016/0022-328X(90)85018-T.

52 Amatore, C. and Jutand, A. (1991). Activation of carbon dioxide by electron transfer and transition metals. Mechanism of nickel-catalyzed electrocarboxylation of aromatic halides. *J. Am. Chem. Soc.* 113 (8): 2819–2825. https://doi.org/10.1021/ja00008a003.

53 (a) Rollin, Y., Meyer, G., Troupel, M., and Fauvarque, J. (1982). Catalyse par des complexes du nickel de l'electrosynthese de 1,1 diarylethane a partir de halogenures aromatiques et d'ethylene. *Tetrahedron Lett.* 23 (35): 3573–3576. https://doi.org/10.1016/S0040-4039(00)87673-X. (b) Schiavon, G., Bontempelli, G., and Corain, B. (1981). Coupling of organic halides electrocatalyzed by the $Ni^{II}/Ni^{I}/Ni^{0}-PPh_3$ system. A mechanistic study based on an electroanalytical approach. *J. Chem. Soc., Dalton Trans.* (5): 1074–1081. https://doi.org/10.1039/DT9810001074.

54 Sock, O., Périchon, J., Jutand, A. et al. (1985). Electrochemistry of nickel complexes with 1,2-bis(diphenylphosphino)ethane. *J. Electroanal. Chem.* 183 (1–2): 237–246. https://doi.org/10.1016/0368-1874(85)85493-9.

55 Behr, A., Keim, W., and Thelen, G. (1983). Stoichiometric insertion of carbon dioxide and ethylene into nickel-carbon bonds. *J. Organomet. Chem.* 249 (2): 38–40. https://doi.org/10.1016/S0022-328X(00)99448-8.

56 Osakada, K., Sato, R., and Yamamoto, T. (1994). Nickel-complex-promoted carboxylation of haloarenes involving insertion of CO_2 into $Ni^{II}-C$ bonds. *Organometallics* 13 (11): 4645–4647. https://doi.org/10.1021/om00023a078.

57 See references in Ref. [56] and, for example: Feth, M.P., Klein, A., and Bertagnolli, H. (2003). Investigation of the ligand exchange behavior of square-planar nickel(II) complexes by X-ray absorption spectroscopy and X-ray diffraction. *Eur. J. Inorg. Chem.* 2003 (5): 839–852. https://doi.org/10.1002/ejic.200390114.

58 (b) Sakhapov, I.F., Rizvanov, I.K., Yakhvarov, D.G. et al. (2015). Electrochemical properties and reactivity of organonickel sigma-complex [NiBr(Mes)(bpy)](Mes = 2,4,6-trimethylphenyl, bpy = 2,2′-bipyridine). *Russ. J. Electrochem.* 51 (11): 1061–1068. https://doi.org/10.1134/S1023193515110142.

59 Klein, A., Schmieder, A., Feth, M.P. et al. (2010). Mono- and binuclear arylnickel complexes of the α-diimine bridging ligand 2,2-bipyrimidine (bpym). *Eur. J. Inorg. Chem.* 2010 (6): 934–941. https://doi.org/10.1002/ejic.200900945.

60 Amatore, C., Jutand, A., Périchon, J., and Rollin, Y. (2000). Mechanism of the nickel-catalyzed electrosynthesis of ketones by heterocoupling of acyl and benzyl halides. *Monatsh. Chem.* 131 (12): 1293–1304. https://doi.org/10.1007/s007060070008.

61 Juliá-Hernández, F., Moragas, T., Cornella, J., and Martin, R. (2017). Remote carboxylation of halogenated aliphatic hydrocarbons with carbon dioxide. *Nature* 545 (7652): 84–88. https://doi.org/10.1038/nature22316.

62 Correa, A. and Martin, R. (2009). Palladium-catalyzed direct carboxylation of aryl bromides with carbon dioxide. *J. Am. Chem. Soc.* 131: 15974–15975. https://doi.org/10.1021/ja905264a.
63 Shimomaki, K., Murata, K., Martin, R., and Iwasawa, N. (2017). Visible-light-driven carboxylation of aryl halides by the combined use of palladium and photoredox catalysts. *J. Am. Chem. Soc.* 139 (28): 9467–9470. https://doi.org/10.1021/jacs.7b04838.
64 Nogi, K., Fujihara, T., Terao, J., and Tsuji, Y. (2015). Cobalt- and nickel-catalyzed carboxylation of alkenyl and sterically hindered aryl triflates utilizing CO_2. *J. Org. Chem.* 80 (22): 11618–11623. https://doi.org/10.1021/acs.joc.5b02307.
65 For standard reduction potentials, see:Vansek, P. (1998). Electrochemical series. In: *CRC Handbook of Chemistry and Physics*, 83e (ed. D.R. Lide), 8-21–8-31. CRC Press.
66 (a) Zembayashi, M., Tamao, K., Yoshida, J.-I., and Kumada, M. (1977). Nickel-phosphine complex-catalyzed homo coupling of aryl halides in the presence of zinc powder. *Tetrahedron Lett.* 18 (47): 4089–4091. https://doi.org/10.1016/S0040-4039(01)83434-1. (b) Colon, I. and Kelsey, D.R. (1986). Coupling of aryl chlorides by nickel and reducing metals. *J. Org. Chem.* 51 (14): 2627–2637. https://doi.org/10.1021/jo00364a002. (c) Iyoda, M., Otsuka, H., Sato, K. et al. (1990). Homocoupling of aryl halides using nickel(II) complex and zinc in the presence of Et_4NI. An efficient method for the synthesis of biaryls and bipyridines. *Bull. Chem. Soc. Jpn.* 63 (1): 80–87. https://doi.org/10.1246/bcsj.63.80.
67 Selected references: (a) Haim, A. (1975). Role of the bridging ligand in inner-sphere electron-transfer reactions. *Acc. Chem. Res.* 8 (8): 264–272. https://doi.org/10.1021/ar50092a002. (b) Weaver, M.J. (1979). On the role of the bridging ligand in electrochemical inner-sphere electron-transfer processes. *Inorg. Chem.* 18 (2): 402–408. https://doi.org/10.1021/ic50192a042. (c) Taube, H. (1984). Electron transfer between metal complex – a retrospective view. *Angew. Chem. Int. Ed. Engl.* 23 (5): 329–339. https://doi.org/10.1002/anie.198403293.
68 Stojanovic, R.S. and Bond, A.M. (1993). Examination of conditions under which the reduction of the cobaltocenium cation can be used as a standard voltammetric reference process in organic and aqueous solvents. *Anal. Chem.* 65 (1): 56–64. https://doi.org/10.1021/ac00049a012.
69 Sayyed, F.B., Tsuji, Y., and Sakaki, S. (2013). The crucial role of a Ni(I) intermediate in Ni-catalyzed carboxylation of aryl chloride with CO_2: a theoretical study. *Chem. Commun.* 49 (91): 10715–10717. https://doi.org/10.1039/c3cc45836a.
70 Correa, A., León, T., and Martin, R. (2014). Ni-catalyzed carboxylation of C(sp^2)— and C(sp^3)—O bonds with CO_2. *J. Am Chem. Soc.* 136 (3): 1062–1069. https://doi.org/10.1021/ja410883p.
71 For examples of C(sp^2)–O functionalization reactions that are limited to regular arene substrates: (a) Tobisu, M., Shimasaki, T., and Chatani, N. (2008). Nickel-catalyzed cross-coupling of aryl methyl ethers with aryl boronic esters. *Angew. Chem. Int. Ed.* 47 (26): 4866–4869. https://doi.org/

10.1002/anie.200801447. (b) Yu, D.G. and Shi, Z.-J. (2011). Mutual activation: Suzuki–Miyaura coupling through direct cleavage of the sp^2 C—O bond of naphtholate. *Angew. Chem. Int. Ed.* 50 (31): 7097–7100. https://doi.org/10.1002/anie.201101461. (c) Guo, L., Liu, X., Baumann, C., and Rueping, M. (2016). Nickel-catalyzed alkoxy–alkyl interconversion with alkylborane reagents through C—O bond activation of aryl and enol ethers. *Angew. Chem. Int. Ed.* 55 (49): 15415–15419. https://doi.org/10.1002/anie.201607646.

72 Rebih, F., Maddaluno, J., Durandetti, M. et al. (2016). Direct carboxylation of aryl tosylates by CO_2 catalyzed by in situ-generated Ni(0). *Chem. Eur. J.* 22 (11): 3758–3763. https://doi.org/10.1002/chem.201503926.

73 León, T., Correa, A., and Martin, R. (2013). Ni-catalyzed direct carboxylation of benzyl halides with CO_2. *J. Am. Chem. Soc.* 135 (4): 1221–1224. https://doi.org/10.1021/ja311045f.

74 Moragas, T., Gaydou, M., and Martin, R. (2016). Nickel-catalyzed carboxylation of benzylic C—N bonds with CO_2. *Angew. Chem. Int. Ed.* 55 (16): 5053–5057. https://doi.org/10.1002/anie.201600697.

75 Liu, Y., Cornella, J., and Martin, R. (2014). Ni-catalyzed carboxylation of unactivated primary alkyl bromides and sulfonates with CO_2. *J. Am. Chem. Soc.* 136 (32): 11212–11215. https://doi.org/10.1021/ja5064586.

76 Moragas, T. and Martin, R. (2016). Nickel-catalyzed reductive carboxylation of cyclopropyl motifs with carbon dioxide. *Synthesis* 48 (17): 2816–2822. https://doi.org/10.1055/s-0035-1560439.

77 Wang, X., Liu, Y., and Martin, R. (2015). Ni-catalyzed divergent cyclization/carboxylation of unactivated primary and secondary alkyl halides with CO_2. *J. Am. Chem. Soc.* 137 (20): 6476–6479. https://doi.org/10.1021/jacs.5b03340.

78 Moragas, T., Cornella, J., and Martin, R. (2014). Ligand-controlled regiodivergent Ni-catalyzed reductive carboxylation of allyl esters with CO_2. *J. Am. Chem. Soc.* 136 (51): 17702–17705. https://doi.org/10.1021/ja509077a.

79 (a) Takaya, J. and Iwasawa, N. (2008). Hydrocarboxylation of allenes with CO_2 catalyzed by silyl pincer-type palladium complex. *J. Am. Chem. Soc.* 130 (46): 15254–15255. https://doi.org/10.1021/ja806677w. (b) Wu, J., Hazari, N., and Incarvito, C.D. (2011). Synthesis, properties, and reactivity with carbon dioxide of $(allyl)_2Ni(L)$ complexes. *Organometallics* 30 (11): 3142–3150. https://doi.org/10.1021/om2002238. (c) Suh, H.-W., Guard, L.M., and Hazari, N. (2014). A mechanistic study of allene carboxylation with CO_2 resulting in the development of a Pd(II) pincer complex for the catalytic hydroboration of CO_2. *Chem. Sci.* 5 (10): 3859–3872. https://doi.org/10.1039/C4SC01110D.

80 van Gemmerem, M., Börjesson, M., Martin, R. et al. (2017). Switchable site-selective catalytic carboxylation of allylic alcohols with CO_2. *Angew. Chem. Int. Ed.* 56 (23): 6558–6562. https://doi.org/10.1002/anie.201702857.

81 See Ref. [31] for a relevant review and: Jones, G.D., Desrochers, P., Vicic, D.A. et al. (2006). Ligand redox effects in the synthesis, electronic structure, and reactivity of an alkyl–alkyl cross-coupling catalyst. *J. Am. Chem. Soc.* 128 (40): 13175–13183. https://doi.org/10.1021/ja063334i.

82 (a) Anderson, T.J., Jones, G.D., and Vicic, D.A. (2004). Evidence for a Ni(I) active species in the catalytic cross-coupling of alkyl electrophiles. *J. Am.*

Chem. Soc. 126 (26): 8100–8101. https://doi.org/10.1021/ja0483903. (b) Hamacher, C., Klein, A., Schüren, A. et al. (2009). Electrochemistry and spectroscopy of organometallic terpyridine nickel complexes. *Inorg. Chem.* 48 (20): 9947–9951. https://doi.org/10.1021/ic900753r.

83 See Ref. [49] and: (a) Fischer, R., Langer, J., Walther, D. et al. (2006). A key step in the formation of acrylic acid from CO_2 and ethylene: the transformation of a nickelalactone into a nickel-acrylate complex. *Chem. Commun.* (23): 2510–2512. https://doi.org/10.1039/B603540J. (b) Bruckmeier, C., Lehenmeier, M.W., Rieger, B. et al. (2010). Formation of methyl acrylate from CO_2 and ethylene via methylation of nickelalactones. *Organometallics* 29 (10): 2199–2202. https://doi.org/10.1021/om100060y.

84 Fujihara, T., Sakaki, S., Tsuji, Y. et al. (2014). Nickel-catalyzed double carboxylation of alkynes employing carbon dioxide. *Org. Lett.* 16 (18): 4960–4963. https://doi.org/10.1021/ol502538r.

85 See Refs. [43, 45c] and (a) Dérien, S., Clinet, J.C., Duñach, E., and Périchon, J. (1992). New C—C bond formation through the nickel-catalysed electrochemical coupling of 1,3-enynes and carbon dioxide. *J. Organomet. Chem.* 424 (2): 213–224. https://doi.org/10.1016/0022-328X(92)83151-7. (b) Yuan, G.Q., Jiang, H.F., and Lin, C. (2008). Efficient electrochemical dicarboxylations of arylacetylenes with carbon dioxide using nickel as the cathode. *Tetrahedron* 64 (25): 5866–5872. https://doi.org/10.1016/j.tet.2008.04.053.

86 See, for examples, Refs. [42, 83a] and (a) Dérien, S., Clinet, J.C., Duñach, E., and Périchon, J. (1993). Activation of carbon dioxide: nickel-catalyzed electrochemical carboxylation of diynes. *J. Org. Chem.* 58 (9): 2578–2588. https://doi.org/10.1021/jo00061a038. (b) Köster, F., Dinjus, E., and Duñach, E. (2001). Electrochemical selective incorporation of CO_2 into terminal alkynes and diynes. *Eur. J. Org. Chem.*: 2507–2511. https://doi.org/10.1002/1099-0690(200107)2001:13%3C2507::AID-EJOC2507%3E3.0.CO;2-P/full.

87 Tortajada, A., Ninokata, R., and Martin, R. (2018). Ni-catalyzed site-selective dicarboxylation of 1,3-dienes with CO_2. *J. Am. Chem. Soc.* 140 (6): 2050–2053. https://doi.org/10.1021/jacs.7b13220.

88 Selected references: (a) Takimoto, M. and Mori, M. (2002). Novel catalytic CO_2 incorporation reaction: nickel-catalyzed regio- and stereoselective ring-closing carboxylation of bis-1,3-dienes. *J. Am. Chem. Soc.* 124 (34): 10008–10009. https://doi.org/10.1021/ja026620c. (b) Takimoto, M., Nakamura, Y., Kimura, K., and Mori, M. (2004). Highly enantioselective catalytic carbon dioxide incorporation reaction: nickel-catalyzed asymmetric carboxylative cyclization of bis-1,3-dienes. *J. Am. Chem. Soc.* 126 (19): 5956–5957. https://doi.org/10.1021/ja049506y. (c) Takimoto, M., Kawamura, M., Mori, M., and Sato, Y. (2005). Nickel-catalyzed regio- and stereoselective double carboxylation of trimethylsilylallene under an atmosphere of carbon dioxide and its application to the synthesis of chaetomellic acid A anhydride. *Synlett* 13: 2019–2022. https://doi.org/10.1055/s-2005-872225. (d) Takimoto, M., Kawamura, M., Mori, M., and Sato, Y. (2011). Nickel-promoted carboxylation/cyclization cascade of allenyl aldehyde under an atmosphere of CO_2. *Synlett* 10: 1423–1426. https://doi.org/10.1055/s-0030-1260565.

89 For an example of Ni(I) involvement in a Ni-catalyzed reaction, see: Nadal, M.L., Bosch, J., Moretó, J.M. et al. (2005). The Ni-mediated cyclocarbonylation of allyl halides and alkynes made catalytic. Evidence supporting the involvement of pseudoradical Ni^{I} species in the mechanism. *J. Am. Chem. Soc.* 127 (30): 10476–10477. https://doi.org/10.1021/ja0525713.
90 (a) Williams, C.M., Johnson, J.B., and Rovis, T. (2008). Nickel-catalyzed reductive carboxylation of styrenes using CO_2. *J. Am. Chem. Soc.* 130 (45): 14936–14937. https://doi.org/10.1021/ja8062925. (b) Li, S., Yuan, W., and Ma, S. (2011). Highly regio- and stereoselective three-component nickel-catalyzed syn-hydrocarboxylation of alkynes with diethyl zinc and carbon dioxide. *Angew. Chem. Int. Ed.* 50 (11): 2578–2582. https://doi.org/10.1002/anie.201007128. (c) Fujihara, T., Terao, J., Tsuji, Y. et al. (2011). Copper-catalyzed hydrocarboxylation of alkynes using carbon dioxide and hydrosilanes. *Angew. Chem. Int. Ed.* 50 (2): 523–527. https://doi.org/10.1002/anie.201006292.
91 (a) Wang, X., Nakajima, M., and Martin, R. (2015). Ni-catalyzed regioselective hydrocarboxylation of alkynes with CO_2 by using simple alcohols as proton sources. *J. Am. Chem. Soc.* 137 (28): 8924–8927. https://doi.org/10.1021/jacs.5b05513. (b) Gaydou, M., Moragas, T., Juliá-Hernández, F., and Martin, R. (2017). Site-selective catalytic carboxylation of unsaturated hydrocarbons with CO_2 and water. *J. Am. Chem. Soc.* 139 (35): 12161–12164. https://doi.org/10.1021/jacs.7b07637.
92 Meng, Q.Y., Wang, S., Huff, G.S., and König, B. (2018). Ligand-controlled regioselective hydrocarboxylation of styrenes with CO_2 by combining visible light and nickel catalysis. *J. Am. Chem. Soc.* 140 (9): 3198–3201. https://doi.org/10.1021/jacs.7b13448.

Index

Nickel Catalysis in Organic Synthesis: Methods and Reactions,
First Edition. Edited by Sensuke Ogoshi.